Springer Series in Materials Science

Volume 187

AF789262

Series Editors

Zhiming M. Wang, Chengdu, People's Republic of China
Chennupati Jagadish, Canberra, ACT, Australia
Robert Hull, Charlottesville, VI, USA
Richard M. Osgood, New York, NY, USA
Jürgen Parisi, Oldenburg, Germany

For further volumes:
http://www.springer.com/series/856

The Springer Series in Materials Science covers the complete spectrum of materials physics, including fundamental principles, physical properties, materials theory and design. Recognizing the increasing importance of materials science in future device technologies, the book titles in this series reflect the state-of-the-art in understanding and controlling the structure and properties of all important classes of materials.

Handong Li · Jiang Wu
Zhiming M. Wang
Editors

Silicon-based Nanomaterials

Springer

Editors
Handong Li
Jiang Wu
Zhiming M. Wang
State Key Laboratory of Electronic
Thin Film and Integrated Devices
University of Electronic Science
and Technology of China
Chengdu
People's Republic of China

ISSN 0933-033X ISSN 2196-2812 (electronic)
ISBN 978-1-4939-5499-5 ISBN 978-1-4614-8169-0 (eBook)
DOI 10.1007/978-1-4614-8169-0
Springer New York Heidelberg Dordrecht London

© Springer Science+Business Media New York 2013
Softcover reprint of the hardcover 1st edition 2013
This work is subject to copyright. All rights are reserved by the Publisher, whether the whole or part of the material is concerned, specifically the rights of translation, reprinting, reuse of illustrations, recitation, broadcasting, reproduction on microfilms or in any other physical way, and transmission or information storage and retrieval, electronic adaptation, computer software, or by similar or dissimilar methodology now known or hereafter developed. Exempted from this legal reservation are brief excerpts in connection with reviews or scholarly analysis or material supplied specifically for the purpose of being entered and executed on a computer system, for exclusive use by the purchaser of the work. Duplication of this publication or parts thereof is permitted only under the provisions of the Copyright Law of the Publisher's location, in its current version, and permission for use must always be obtained from Springer. Permissions for use may be obtained through RightsLink at the Copyright Clearance Center. Violations are liable to prosecution under the respective Copyright Law.
The use of general descriptive names, registered names, trademarks, service marks, etc. in this publication does not imply, even in the absence of a specific statement, that such names are exempt from the relevant protective laws and regulations and therefore free for general use.
While the advice and information in this book are believed to be true and accurate at the date of publication, neither the authors nor the editors nor the publisher can accept any legal responsibility for any errors or omissions that may be made. The publisher makes no warranty, express or implied, with respect to the material contained herein.

Printed on acid-free paper

Springer is part of Springer Science+Business Media (www.springer.com)

Preface

Silicon is one of the "oldest" electronic materials and has served the microelectronics industry for more than half a century. On the other hand, silicon is also a "young" material since both novel exciting aspects in science and new technical applications continue to be revealed. However, subject to the simple crystal and band structure, the functional extension of silicon must rely on artificial modifications to its size and electronic structure. After years of efforts to find proper ways to further expand silicon's application areas, research interests have gradually focused on nanocrystallization, alloying, and integration with other functional materials.

Size and shape control is of central importance in developing silicon-based materials science. Nanosized silicon and porous silicon with characterized pore size dimension on the nanometer scale have been demonstrated to be irreplaceable in photovotaics, photonics, and energy harvesting and storage. However, the full advantage of their unique properties can be exploited or utilized only if controlled preparation of these nanomaterials in terms of orientation, dimension, and size can be achieved.

One characteristic feature of silicon is its highly chemical reactivity to other elements. It can easily react with metal and carbon to form metal silicides and silicon carbide of good crystallinity. Thus a rich functional material family including metals, thermoelectric materials, direct band gap semiconductors, and wideband gap semiconductors with adjustable physical properties can be achieved by alloying methods. More excitingly, nanocrystallization further endows novel properties to these silicon contained compounds.

Besides nanocrystallization and alloying approaches, another straightforward way to expand silicon's versatility is to integrate functional materials on available silicon chips. This can be realized by depositing heterogeneous materials such as nanosized III–V semiconductors on atomically reconstructed surfaces or patterned surfaces of silicon, i.e., a bottom-up growth process. In many cases, the performance of devices based on such heterostructures is completely determined by the epitaxial quality of heterogeneous thin films or nanostructures on silicon surface or existing circuits.

This book aims to provide a comprehensive survey of the research and development status on the above-mentioned silicon-based nanomaterials. It is organized as follows: Chaps. 1–6 are dedicated to elemental-silicon nanostructures. Chapters 1 and 2 introduce the science and applications of porous silicon nanostructures in lithium batteries. Chapter 3 describes novel optical properties of silicon nanowires whose applications in solar energy harvesting and nano-electro-mechanical systems are summarized in Chap. 4. Chapters 5 and 6 deal with the optical properties of silicon nanoparticles embedded in silica matrix and their applications for light emitting devices. Chapters 7–12 concentrate on silicon-contained compounds for nanomaterials. Chapter 7 starts with a theoretical analysis of electronic and optical properties of silicon carbide nanostructures. Chapters 8–10 discuss the fabrication and characterization of various silicon carbide nanostructures. Chapter 11 investigates the transport behaviors of silicon carbide nanostructures. Chapter 12 focuses on synthesis, properties, and applications of metal silicide nanostructures. Chapters 13–15 discuss integration of other functional nanomaterials on silicon. Chapter 13 covers the droplet epitaxy of strain-free III–V quantum dots on silicon. Chapter 14 highlights the monolithic growth of III–V quantum dots on silicon and related devices behaviors. Finally, the authors demonstrate the growth and characterization of gallium nitride quantum wells structures on patterned SiC/Si substrates in Chap. 15.

We would like to thank all of the authors who have contributed chapters to this book. It took nearly one year to put this work together, but we are confident that the effort was worthwhile and that the book will be an important reference for scientists as well as a helpful platform to explore many fundamental questions at the frontier of materials science.

Chengdu, People's Republic of China

Handong Li
Jiang Wu
Zhiming M. Wang

Contents

1 **Porous Silicon as Anode Material for Lithium-Ion Batteries** 1
Madhuri Thakur, Roderick Pernites, Steve L. Sinsabaugh, Michael S. Wong and Sibani L. Biswal

2 **The Development of Si and Ge-Based Nanomaterials for High Performance Lithium Ion Battery Anodes** 25
Xiao-Liang Wang and Wei-Qiang Han

3 **Light Trapping in Coaxial Nanowires of c-Si Cores and a-Si Shells** 45
Jeong Il Oh, Wenfu Liu, Weiqiang Xie and Wenzhong Shen

4 **Applications of Ordered Si Nanowire Array to Solar Energy Harvesting and NEMS** 67
Yuerui Lu and Amit Lal

5 **Synchrotron-Excited Photoluminescence Spectroscopy of Silicon- and Carbon-Containing Quantum Dots in Low Dimensional SiO_2 Matrices** 89
Anatoly F. Zatsepin and Evgeny A. Buntov

6 **Silicon Nanoparticles-Based Light Emitting Capacitors** 119
A. Morales Sánchez, J. Barreto, C. Domínguez Horna, M. Aceves Mijares, J. A. Luna López and L. Licea Jiménez

7 **Electronic and Optical Properties of Silicon Carbide Nanostructures** 139
Hung-Chung Hsueh, Guang-Yu Guo and Steven G. Louie

8 **Plasma Enabled Fabrication of Silicon Carbide Nanostructures** 161
Jinghua Fang, Igor Levchenko, Morteza Aramesh, Amanda E. Rider, Steven Prawer and Kostya (Ken) Ostrikov

9 Catalyst-Free Chemical Vapor Deposition for Synthesis of SiC Nanowires with Controlled Morphology 179
Jyoti Prakash, Sunil Kumar Ghosh and Dakshinamoorthy Sathiyamoorthy

10 Adhesion and Indentation Fracture Behavior of Silicon Carbonitride Nanocomposite Coatings Deposited by Magnetron Sputtering 215
S. K. Mishra and A. S. Bhattacharyya

11 Impact of Defects and Doping on Electron Transport in SiCNTs 243
Sudhanshu Choudhary and S. Qureshi

12 Synthesis, Properties, and Applications of One-Dimensional Transition Metal Silicide Nanostructures. 265
Guangwei She, Hailong Liu, Lixuan Mu and Wensheng Shi

13 Integration of Strain Free III–V Quantum Dots on Silicon. 327
Stefano Sanguinetti, Sergio Bietti and Giovanni Isella

14 III–V Quantum-Dot Materials and Devices Monolithically Grown on Si Substrates 357
Huiyun Liu

15 Cubic GaN on Nanopatterned 3C-SiC/Si (001) Substrates 381
Ricarda Maria Kemper, Donat Josef As and Jörg K. N. Lindner

Index 407

Contributors

Donat Josef As Department of Physics, University of Paderborn, Warburger Str. 100, 33098 Paderborn, Germany

Morteza Aramesh University of Melbourne, Melbourne VIC 3010, Australia

A. S. Bhattacharyya Central University of Jharkhand, Ranchi 835205, India

Evgeny A. Buntov Institute of Physics and Technology, Ural Federal University, Mira street 19, Ekaterinburg, Russia 620002

J Barreto Centro de Investigación en Materiales Avanzados, Parque de Investigación e Innovación Tecnológica, Alianza Norte 202, 66600 Apodaca, Nuevo León C.P., México

S. Bietti Polo di Como, Politecnico di Milano, Via Anzani 42, 22100 Como, Italy

S. L. Biswal Department of Chemical and Biomolecular Engineering, Rice University, Houston, TX 77005, USA

Sudhanshu Choudhary Department of Electrical Engineering, I. I. T. Kanpur, Kanpur, Kanpur 208016, India

Jinghua Fang CSIRO Materials Science and Engineering—Lindfield, Bradfield Road, West Lindfield NSW 2070, Australia

Guang-Yu Guo Department of Physics, National Taiwan University, Taipei 10617, Taiwan

Sunil Kumar Ghosh Bio Organic Division, Bhabha Atomic Research Centre, Trombay, Mumbai 400085, India

C. Domínguez Horna Centro de Investigación en Materiales Avanzados, Parque de Investigación e Innovación Tecnológica, Alianza Norte 202, 66600 Apodaca, Nuevo León C.P., México

Hung-Chung Hsueh Department of Physics, Tamkang University, New Taipei City, Tamsui 25137, Taiwan

Wei-Qiang Han Ningbo Institute of Material Technology and Engineering, Chinese Academy of Sciences, 315202 Ningbo, People's Republic of China

G. Isella Politecnico di Milano, Polo di Como, Via Anzani 42, 22100 Como, Italy

L. Licea Jiménez Centro de Investigación en Materiales Avanzados, Parque de Investigación e Innovación Tecnológica, Alianza Norte 202, 66600 Apodaca, Nuevo León C.P., México

R. M. Kemper Department of Physics, University of Paderborn, Warburger Str. 100, 33098 Paderborn, Germany

Amit Lal SonicMEMS Laboratory, School of Electrical and Computer Engineering, Cornell University, Ithaca, NY 08541, USA

Handong Li State Key Laboratory of Electronic Thin Film and Integrated Devices, University of Electronic Science and Technology of China, 610054 Chengdu, People's Republic of China

Hailong Liu Key Laboratory of Photochemical Conversion and Optoelectronic Materials, Technical Institute of Physics and Chemistry, Chinese Academy of Sciences, 100190 Beijing, People's Republic of China

Huiyun Liu Department of Electronic and Electrical Engineering, University College London, Torrington Place, London WC1E 7JE, UK

Igor Levchenko CSIRO Materials Science and Engineering—Lindfield, Bradfield Road, West Lindfield NSW 2070, Australia

J A Luna López Centro de Investigación en Materiales Avanzados, Parque de Investigación e Innovación Tecnológica, Alianza Norte 202, 66600 Apodaca, Nuevo León C.P., México

Jörg K. N. Lindner Department of Physics, University of Paderborn, Warburger Str. 100, 33098 Paderborn, Germany

Siyuan Luo State Key Laboratory of Electronic Thin Film and Integrated Devices, University of Electronic Science and Technology of China, 610054 Chengdu, People's Republic of China

Steven G. Louie Department of Physics, University of California at Berkeley, Berkeley, California 94720, USA; Materials Sciences Division, Lawrence Berkeley National Laboratory, Berkeley, CA 94720, USA

Wenfu Liu Laboratory of Condensed Matter Spectroscopy and Opto-Electronic Physics, Key Laboratory of Artificial Structures and Quantum Control (Ministry of Education), Department of Physics, Institute of Solar Energy, Shanghai Jiao Tong University, 800 Dong Chuan Road, 200240 Shanghai, China

Yuerui Lu SonicMEMS Laboratory, School of Electrical and Computer Engineering, Cornell University, Ithaca, NY 08540, USA

Lixuan Mu Key Laboratory of Photochemical Conversion and Optoelectronic Materials, Technical Institute of Physics and Chemistry, Chinese Academy of Sciences, Beijing 100190, People's Republic of China

M Aceves Mijares Centro de Investigación en Materiales Avanzados, Parque de Investigación e Innovación Tecnológica, Alianza Norte 202, 66600 Apodaca, Nuevo León C.P., México

S. K. Mishra CSIR-National Metallurgical Laboratory, Jamshedpur, Jharkhand 831007, India

Jeong Il Oh Laboratory of Condensed Matter Spectroscopy and Opto-Electronic Physics, Key Laboratory of Artificial Structures and Quantum Control (Ministry of Education), Department of Physics, Institute of Solar Energy, Shanghai Jiao Tong University, 800 Dong Chuan Road, 200240 Shanghai, China

Kostya (Ken) Ostrikov CSIRO Materials Science and Engineering—Lindfield, Bradfield Road, West Lindfield NSW 2070, Australia

Jyoti Prakash Powder Metallurgy Division, Bhabha Atomic Research Centre, Trombay, Mumbai 400085, India

R. Pernites Department of Chemical and Biomolecular Engineering, Rice University, Houston, TX 77005, USA

Steven Prawer University of Melbourne, Melbourne, VIC 3010, Australia

S. Qureshi Department of Electrical Engineering, I. I. T. Kanpur, Kanpur 208016, India

Amanda E. Rider CSIRO Materials Science and Engineering—Lindfield, Bradfield Road, West Lindfield NSW 2070, Australia

A Morales Sánchez Centro de Investigación en Materiales Avanzados, Parque de Investigación e Innovación Tecnológica, Alianza Norte 202, 66600 Apodaca, Nuevo León C.P., México

Dakshinamoorthy Sathiyamoorthy Powder Metallurgy Division, Bhabha Atomic Research Centre, Trombay, Mumbai 400085, India

Guangwei She Key Laboratory of Photochemical Conversion and Optoelectronic Materials, Technical Institute of Physics and Chemistry, Chinese Academy of Sciences, Beijing 100190, People's Republic of China

S. Sanguinetti Dipartimento di Scienza dei Materiali, Universita' degli Studi di Milano-Bicocca, via R. Cozzi 53, 20125 Milan, Italy

S. L. Sinsabaugh Lockeed Martin MS2, Akron, OH, USA

Wensheng Shi Key Laboratory of Photochemical Conversion and Optoelectronic Materials, Technical Institute of Physics and Chemistry, Chinese Academy of Sciences, 100190 Beijing, People's Republic of China

Wenzhong Shen Laboratory of Condensed Matter Spectroscopy and Opto-Electronic Physics, Key Laboratory of Artificial Structures and Quantum Control (Ministry of Education), Department of Physics, Institute of Solar Energy, Shanghai Jiao Tong University, 800 Dong Chuan Road, 200240 Shanghai, China

M. Thakur Department of Chemical and Biomolecular Engineering, Rice University, Houston, TX 77005, USA

M. S. Wong Department of Chemical and Biomolecular Engineering, Rice University, Houston, TX 77005, USA

Zhiming M. Wang State Key Laboratory of Electronic Thin Film and Integrated Devices, University of Electronic Science and Technology of China, 610054 Chengdu, People's Republic of China

Weiqiang Xie Laboratory of Condensed Matter Spectroscopy and Opto-Electronic Physics, Key Laboratory of Artificial Structures and Quantum Control (Ministry of Education), Department of Physics, Institute of Solar Energy, Shanghai Jiao Tong University, 800 Dong Chuan Road, 200240 Shanghai, China

Xiao-Liang Wang Ningbo Institute of Material Technology and Engineering, Chinese Academy of Sciences, 315202 Ningbo, People's Republic of China

Anatoly F. Zatsepin Institute of Physics and Technology, Ural Federal University, Mira street 19, Ekaterinburg, Russia 620002

Chapter 1
Porous Silicon as Anode Material for Lithium-Ion Batteries

Madhuri Thakur, Roderick Pernites, Steve L. Sinsabaugh, Michael S. Wong and Sibani L. Biswal

Abstract Lithium-ion batteries are ubiquitous in our modern society, powering everything from cell phones, laptops, and power tools.They are also powering emerging applications such as electric vehicles and used for on-grid power stabilization. Lithium-ion batteries are a significant and growing part of this market due to their high specific energy. The worldwide market for lithium-ion batteries is projected to reach more than USD 9 billion by 2015. While lithium-ion batteries are often selected for their high specific energy, the market is demanding yet higher performance, usually in terms of energy stored per unit mass of battery. Many groups have recently turned their attention toward developing a silicon-based anode material to increase lithium-ion battery density. Silicon continues to draw great interest as an anode for lithium-ion batteries due to its large specific capacity as compared to the conventional graphite. Despite this exciting property, its practical use has been limited due to a large volume change associated with the insertion and extraction of lithium, which oftentimes leads to cracking and pulverization of the anode, limiting its cycle life. To overcome this problem, significant research has been focused toward developing various silicon nanostructures to accommodate the severe volume expansion and contraction. The structuring of the silicon often involves costly processing

M. Thakur · R. Pernites · M. S. Wong · S. L. Biswal (✉)
Department of Chemical and Biomolecular Engineering, Rice University,
Houston, TX 77005, USA
e-mail: biswal@rice.edu

R. Pernites
e-mail: rbp3@rice.edu

M. S. Wong
e-mail: mswong@rice.edu

S. L. Sinsabaugh
Lockeed Martin, MS2, Akron, OH, USA
e-mail: steven.sinsabaugh@lmco.com

M. Thakur
e-mail: mt5@rice.edu

H. Li et al. (eds.), *Silicon-based Nanomaterials*,
Springer Series in Materials Science 187, DOI: 10.1007/978-1-4614-8169-0_1,
© Springer Science+Business Media New York 2013

steps, limiting its application in price sensitive commercial lithium-ion batteries. To achieve commercial viability, work is being pursued on silicon battery anode structures and processes with a special emphasis on the cost and environment. In this review book chapter, we will summarize recent development of a cost-effective electrochemically etched porous silicon as an anode material for lithium-ion batteries. Briefly, the new approach involves creating hierarchical micron-and nanometer-sized pores on the surface of micron-sized silicon particulates, which are combined with an excellent conductor binder.

1.1 Introduction

Since introduction of the first commercial lithium-ion battery by Sony in 1990s, significant progress has been made in improving the performance of the battery with improvements in the electrode materials, particularly graphitic anodes and various lithiated metal oxide cathodes. With these conventional electrode materials achieving maximum practical performance, new and improved materials are needed to achieve the next major step in lithium-ion battery performance.

Silicon is one of the most promising anode materials for next generation lithium-ion batteries because of its ability to form multiple alloys with lithium, with a maximum uptake of 4.4 lithium atoms per silicon atom [1]. As such, silicon can provide the highest achievable specific capacity of about 3,579 mAh g^{-1} (Fig. 1.1) [2] at room temperature. This is an order of magnitude greater than the current lithium-ion battery graphite anodes can achieve at 372 mAh g^{-1} [3]. Furthermore, silicon has been reported to show low voltage of <0.3 V versus Li/Li^{+} [4], high coulombic efficiency (>98 %) [5], and allows fast lithium-ion diffusion with known diffusion

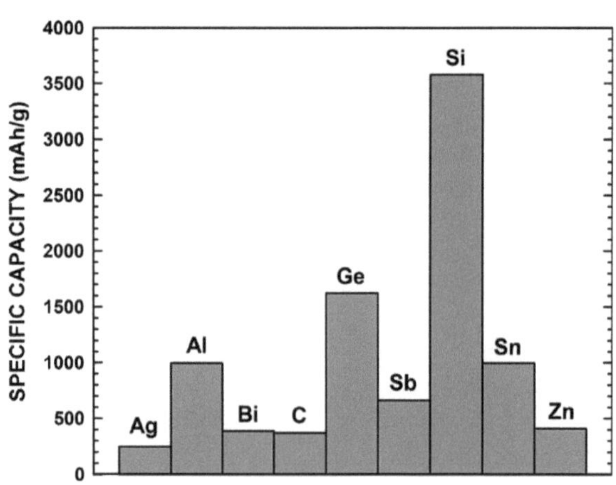

Fig. 1.1 The specific capacity of a number of electrochemically elements [2]

coefficient equivalent to about $7.2 \times 10^{-5}\,cm^2\,s^{-1}$ at 415 °C [6]. Also, silicon is an abundant commonly available element and thus is relatively cheap and readily available in the market. Therefore, it is not a surprise that significant research has been devoted to developing silicon-based battery anodes [7–9].

Although it has exciting properties for use as a lithium-ion battery anode, practical application of silicon is limited due to large volume change of 280 % during charge and discharge, if the silicon is fully lithiated to $Li_{15}Si_4$. Full lithiation of a solid silicon electrode results in severe fracturing and crumbling of the electrode due to stresses created in the crystal lattice. This crumbling results in portions of the silicon electrode losing electrical continuity with the current collector, and typically very rapid cell failure. The maximum theoretical volumetric capacity for fully lithiated silicon is about 2,190 mAh cm^{-3} [10]. To overcome the challenge of early silicon anode failure during cyclic lithiation and delithiation, contemporary strategies involve the creation of pores in the silicon structure [11] and/or formation of micro/nano structures of silicon [12–14]. Both strategies are intended to accommodate the cyclic stresses. Another approach is to considerably decrease the size of the silicon into nanoscale particulates, which is believed to tolerate the physical strains during Li alloying/dealloying and also minimize fracturing and particle pulverization during volume expansion [15, 16]. These methods are also combined with various binding materials including carbon powders [17] and different types of polymers [18].

At present, porous silicon and related composites are being developed as high performance anode materials to create new advanced lithium-ion batteries that can deliver large specific capacity and high power density for a longer lifetime. This research is driven by the strong demand for improved batteries in everything from consumer electronics to electric vehicles to grid storage for intermittent renewables like wind and solar. This new material will significantly increase the energy stored in a battery, providing longer discharges and more watt-hours stored per liter and per kilogram. Key performance parameters in this development are capacity, cycle life, scalability to low-cost large-scale material production, and compatibility with existing battery cell manufacturing processes.

In the remainder of this chapter, we summarize the recent developments of electrochemically etched porous silicon as an anode material. Our group's development process has evolved through several major steps, starting with (i) gold-coated porous silicon film that is still attached to bulk silicon, then (ii) complete macroporous silicon lift-off film with no bulk silicon underneath, and more recently to (iii) macroporous silicon particulates (Fig. 1.2). Common to all of these configurations is a hierarchical porosity that is both micron and nanometer scale. The porous silicon particulates provide the greatest promise for commercialization due to its ease of handling and having demonstrated discharge capacity of at least 1,200 mAh g^{-1} for long cycle life (> 150 cycles), discussed later in this chapter. The fabrication method of these porous silicon particulates is amenable to low-cost large-scale manufacture as an electrode precursor suitable for insertion into a typical lithium-ion battery roll-to-roll production line. In this book chapter, the three generation of porous silicon anode materials and their unique method of fabrication will be reviewed.

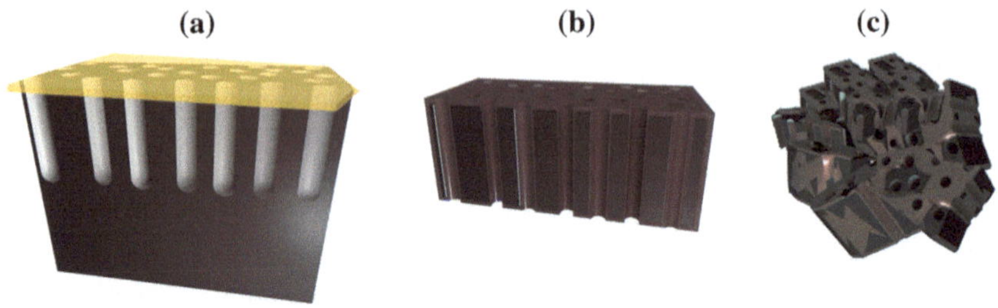

Fig. 1.2 Three generations of high performance porous silicon anode materials for rechargeable lithium-ion battery applications: **a** gold-coated porous silicon film attached to bulk silicon under layer, **b** complete macroporous silicon lift-off film, and **c** macroporous silicon particulates

1.2 Overview of Porous Silicon

1.2.1 Brief History

Porous silicon was inadvertently discovered in 1950s by the Uhlirs in the Bell Labs USA when they were trying to develop an electrochemical method to machine silicon wafers for microelectronic circuit. The silicon wafer did not dissolve uniformly as expected but instead they found out the appearance of fine holes on the surface of the wafer that are propagating primarily in the ⟨100⟩ direction [19]. For many years, their findings remained largely unnoticed due to lack of scientific significance. Although the first discovery occurred about 40 years ago, a renewal of interest in porous silicon began only in 1990s following the breakthrough detection of efficient emission of light from the infrared region to the ultraviolet region induced by porous silicon [20, 21]. Briefly, Lehmann and Gosele [21] of Duke University recognized the quantum confinement effects in the absorption spectrum of porous silicon while Leigh Canham [20] of the Defense Research Agency in England introduced the efficient bright red–orange overlap photoluminescence from the material. Since their pioneering works, research in the structural, optical, and electrical properties of porous silicon [22] and its potential applications has surged worldwide.

1.2.2 Properties

Porous silicon materials are less dense than the crystalline material from which they originate because of the voids they contain. Depending on etching parameters, porous silicon has been reported to contain complex, anisotropic architecture of high surface area with wide range of porous layer thicknesses [23]. Its aspect ratio (pore depth to width) can be controlled during electrochemical etching by varying HF concentration, current density, and etch time [1]. If the variables are set such that 80 % or more of the silicon is etched away, the adjacent vertical pores overlap, resulting in a fragile layer consisting of extremely fine structure composed of very thin pillars with nanometer scale cross-sections [24]. The original silicon crystalline structure remains intact, with the same good electrical transport properties and thermal stability comparable to the original bulk silicon wafer.

Like the other porous solids, the porosity of silicon can be classified into three regimes based on pore diameter as recommended by the International of Pure and Applied Chemistry (IUPAC). Pores less than 2 nm in diameter are called micropores. Pores with diameters between 2 and 50 nm are classified as mesopores. Finally, pores with diameters greater than 50 nm are called macropores. In the context of nanotechnology these days, though, materials with pore sizes smaller than 100 nm are often referred as being nanoporous [25]. Lehmann and Rönnebeck [26] have described the differences in the physics of macropore formation in *p*-type silicon compared to *n*-type silicon. They also studied the dependence of the morphology on substrate doping, electrolyte composition, and applied current density. In their findings, porosity is weakly dependent on current density while the pore size increases with decreasing doping level. There is also preference in pore growth toward the (100) orientation though this preference is stronger in *n*-type silicon than in *p*-type. And macropore growth velocity in silicon increases significantly with the applied current density and is similar to or slightly greater than the growth velocity of micropores in silicon under similar etching conditions.

1.2.3 Methods of Preparing

Porous silicon can be synthesized directly from commercially available single-crystal silicon wafers via two standard approaches—chemical and electrochemical etching procedures. These are both wet etching processes due to the removal of materials from the starting silicon with highly corrosive hydrogen fluoride (HF). A third type of etching is via photochemistry. A discussion of the photochemical etching procedure, which is a dry etching due to the removal of materials via plasma phase, is described elsewhere [27].

Electrochemical [1, 11, 28] and chemical etching [29–31] are commonly applied methods to make porous silicon for various applications because they are more cost-effective and practical for larger or industrial scale. Unlike photochemical etching, wet etching method requires mask patterns to generate the pores on the surface of the silicon wafer. Depending on the applied etching condition, generated pore sizes can range from a few nanometers to a few micrometers.

Electrochemical etching requires an external current source, an etching cell, and at least two electrodes. The silicon wafer serves as the working electrode, where a constant current is applied, and a platinum wire serves as the counter electrode. Both electrodes are submerged into an HF acid electrolyte dissolved in an organic solvent like dimethylformamide (DMF) or ethanol. During etching, two reactions are simultaneously taking place: Oxidation of silicon at the anode and reduction at the cathode.

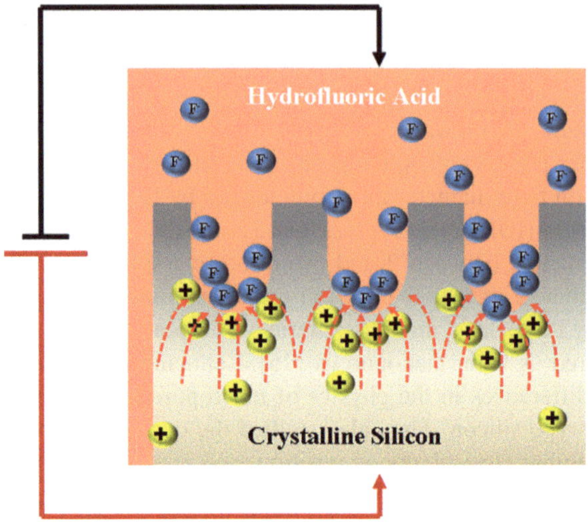

Fig. 1.3 Schematic of electrochemical etching of porous silicon. Etching occurs at the tip of the pores where the valence band holes (+) are focused by the electric field

Electrochemical etching, described schematically in Fig. 1.3, is performed under a lateral electric field, which greatly improves the etching rate [32]. This type of etching is generally the preferred approach for reproducibly obtaining wide ranges of porosity and thickness, but careful design of the electrochemical cell is critical to achieve them [33]. The electrochemical etching of silicon has been known for decades and usually forms microporous structures [34]. Although the macroporous structures in *n*-type wafer was first discovered in 1972 using aqueous HF, [35] these structures were only observed for *p*-type silicon in the early 1990s [36] using a nonaqueous mixture of pure HF and an organic solvent. This early procedure was then modified using organic electrolytes with some water [37] or even in pure aqueous solution of HF [38, 39], which also yield macropores in *p*-type silicon wafer.

Chemical etching does not require an external bias to initiate the etching of silicon. An example of chemical etching is using HF/HNO_3, which is the most widely used isotropic etchant for silicon [40]. The etching of silicon using this mixture is actually described as a two-step chemical process [41] including (i) the formal oxidation of silicon to SiO_2 by HNO_3 and (ii) subsequent stripping of formed SiO_2 by HF. The HF/HNO_3 etching has been used in MEMS fabrication [42] to smoothen surfaces and remove defects in silicon.

1.2.4 Applications

Initially, the main interest in porous silicon is due to its potential application in the emerging field of optoelectronics, particularly with silicon's ubiquitous role in microelectronics. This suggests the potential for integrating silicon-based optical components on a single chip with microelectronics [15]. However, porous silicon is needed to overcome challenges such as lack of stability and low electroluminescence (EL) quantum efficiencies [43]. Over the years, researchers have discoverd other potential applications for porous silicon beyond optoelectronics [44, 45], such as solar cells [46–48], drug carriers [49], chemical and biological sensors [50, 51], bioimaging [52, 53], cancer therapy [54], bioactive surfaces [55, 56], and anode materials for rechargeable lithium-ion batteries [1, 11, 29, 30]. These other applications are due to the many exciting properties of porous silicon earlier mentioned and its ease of synthesis. For this chapter, the discussion will focus on the application of the electrochemically etched porous silicon in rechargeable lithium-ion batteries.

1.3 Background of Electrochemically Etched Porous Silicon in Lithium-Ion Batteries

As discussed earlier, silicon can be used as an anode material in lithium-ion batteries because it is highly reactive with lithium at room temperature. This property extends to porous silicon [57]. In 2005, Shin et al. reported for the first time the application of electrochemically etched porous silicon (PS) as an anode material for lithium-ion batteries [1, 57]. The PS was fabricated by electrochemical etching of *p*-type silicon wafer in 49 % hydrofluoric acid (HF) and organic solvent in a standard electrochemical cell made from a high-density polyethylene [58]. They fabricated different porous structure by changing the etching condition as shown by the surface morphology in Fig. 1.4 [1].

They tested this porous silicon as an anode in lithium-ion batteries. The results showed that lithiation/delithiation reactions readily take place in porous silicon with minimal structural change. The cycle life of this material was 40 cycles with a discharge capacity of 60 μA h cm^{-2}, when charge/discharge was between 0.1 and 2 V.

Later on in 2005, Kang et al. from the same group reported that by changing the organic solvent and etching parameter they were able to fabricate porous silicon with a hybrid micro-nano-porous structure. The surface morphology of the porous silicon with hybrid micro-nano-porous structure is shown in Fig. 1.5 [11].

They tested these porous structures as an anode for lithium-ion battery and discussed the effect of the area/structure of the tunnel wall on electrochemical activity. The advantage of the hybrid structure was that the micropores with a one-dimensional tunnel structure facilitate the transport of lithium ions through the structure while the nanoporous structure of the wall surface (i.e., nanosized particles) successfully relieves the lithiation/delithiation-induced stress, which is the critical shortcoming in the application of silicon to the battery electrodes. However, it should be noted that in both reports the performance of the porous silicon electrode in lithium-ion batteries has been limited to only 50 or even fewer charge/discharge cycles with a very limited capacity.

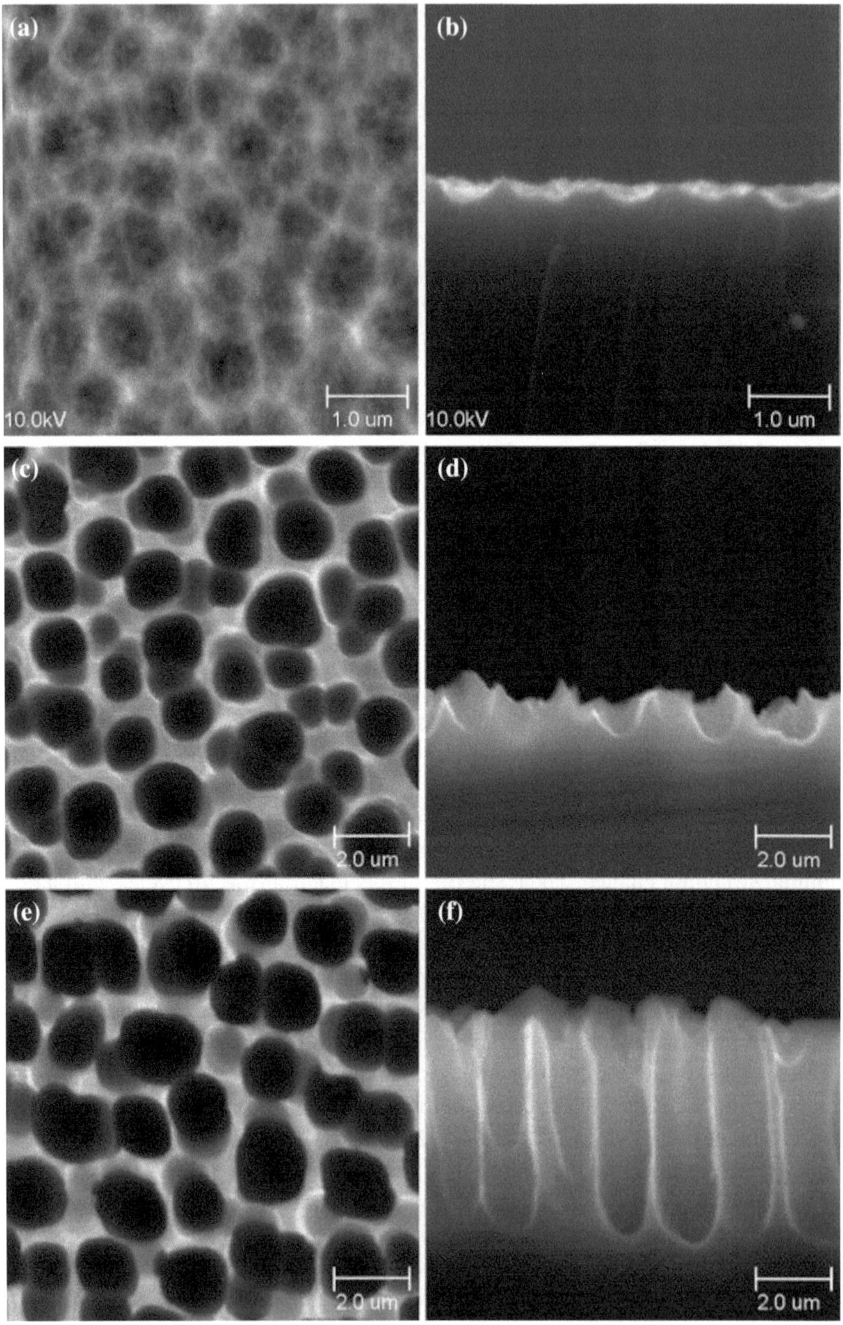

Fig. 1.4 *Top* (*left*) and *cross-sectional* view (*right*) of porous silicon (PS) electrodes with different channel depths: **a**, **b** was etched for 50 min; **c**, **d** and **e**, **f** are two separate regions of the same wafer etched for 40 min under the general conditions indicated in the experimental section [1]

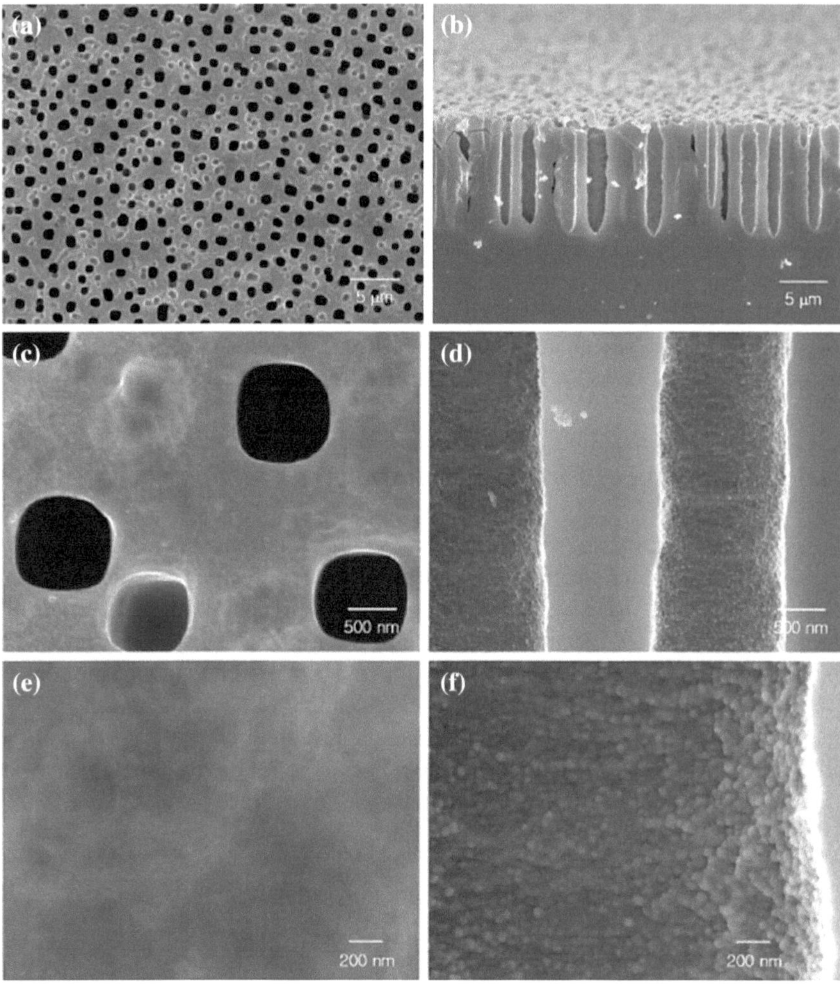

Fig. 1.5 *Top* (**a**, **c**, **e**) and *cross-sectional* (**b**, **d**, **f**) view of the porous silicon at different magnifications [11]

In this chapter we review the latest development of electrochemically etched macroporous silicon (macroporous is the structure where the pore size is >50 nm) to improve the performance of the lithium-ion batteries.

1.4 First Generation: Gold-Coated Porous Silicon

Recently, it has been shown that thin metal coatings such as silver can increase the capacity and cycle life of silicon anode materials [59]. The increase in capacity and cycle life was attributed to a stable solid electrolyte interface (SEI) layer that did not increase in thickness as the material was cycled. However, many of these studies have not combined these thin films with silicon structuring. Additionally, in porous silicon geometries [1, 11], there has been little description on the role of the underlying support structure. In 2012, an electrochemically etched gold-coated porous silicon film was reported as an anode material for lithium-ion batteries [28]. Combining a porous silicon architecture attached to a bulk silicon substrate with a thin gold coating provides the benefits of structuring with a protective metal coating.

1.4.1 Preparation of First Generation Porous Silicon

Porous silicon electrodes were prepared using commercially available silicon wafers (Fig. 1.6a) and standard electrochemical etching procedures. For instance, a cylindrical Teflon cell was used with a hydrofluoric acid (HF) etch solution. More details about the design of the electrochemical etching cell can be found elsewhere [19]. The backside of the wafer was coated with thin aluminum film to reduce the contact resistance with the aluminum plate of the cell. Scanning electron microscopy (SEM) images of a representative porous silicon sample reveal pores on the surface of the etched wafer with an average diameter of ~1 μm (Fig. 1.6b) and average depth of ~5.5 μm (Fig. 1.6c). The porous silicon mass was determined to be 0.78 mg.

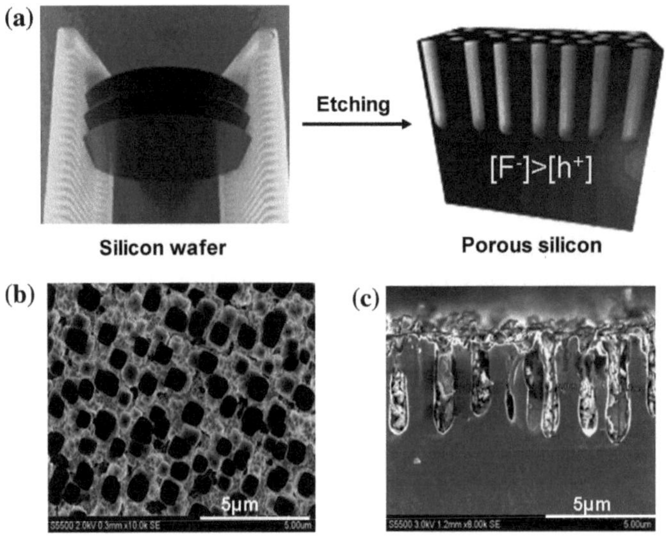

Fig. 1.6 **a** Schematic of the fabrication of porous silicon film. **b** SEM image for the *top view* of the porous silicon and **c** SEM image of the *cross-sectional* view [28]

1.4.2 Gold-Coated Porous Silicon in Lithium-Ion Batteries

The cycling of gold-coated porous silicon had been examined using galvanostatic electrochemical techniques. To calculate the specific capacity (mAh g^{-1}), the active mass of the porous silicon was calculated from the porosity and thickness of the porous silicon layer [28, 60]. Charge/discharge tests were performed at constant current using an electrochemical cell (HS Test cell, Hohsen Corp. Japan) with a multichannel BT2000 battery tester (Arbin Instruments BT2000, USA) equipped with MITSPRO version 4.21. The gold-coated porous silicon film was used as a working electrode with lithium foil as the counter electrode. Between the electrodes is a fiber glass filter paper (Fisher Scientific) wetted with electrolyte, which act as a separator. The electrolyte used is a 1.0 M $LiPF_6$ salt in a 1:1 v/v ethylene carbonate: diethyl carbonate (EC/DC) (Ferro Corporation) solution. All cells were assembled in an Argon-filled glove box.

To see the effect of the gold coating on the performance of the cell, we compared the cycle performance of porous silicon with and without a gold coating [28]. The discharge capacity and coulombic efficiency versus cycle number of the porous silicon with and without a gold coating is shown in Fig. 1.7.

The anode was cycled between 0.1 and 2 V. The lower cutoff voltage was set at 0.1 V to avoid the deep charging of the porous silicon. Shin et al. have reported that for the deep charging there was a very wide potential plateau below 0.1 V versus Li/Li^+ after the second inflection point on the voltage profile [1, 57]. Therefore, the lower cutoff potential was limited to that point where the second inflection point on the voltage profile begins to appear (typically around 0.1 V versus Li/Li^+). Figure 1.7 shows that the material without gold coating gives discharge capacity of only 500 mAh g^{-1} for the first cycle and then attenuates to 76 mAh g^{-1} after the 10th cycle at a constant current of 50 $\mu A\ cm^{-2}$. These results are consistent or comparable with the results observed by Shin and coworkers [1, 11], where they also charge/discharge their pretreated porous silicon anode between 0.1 and 2 V.

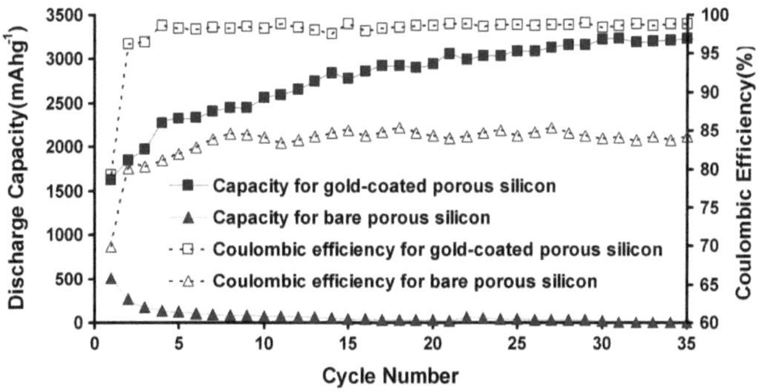

Fig. 1.7 Comparison of discharge capacity (*filled symbols*) and coulombic efficiency (*open symbols*) versus cycle number for gold-coated porous silicon (*blue squares*) and uncoated porous silicon (*red triangles*) [28]

The specific capacity of the sample with gold coating was significantly greater than the uncoated samples. A porous silicon material with a gold coating demonstrated a capacity of 1,619 mAh g^{-1} in the first cycle and 2,452 mAh g^{-1} after the 10th cycle at a higher current of 100 $\mu A\ cm^{-2}$. The porous silicon without the gold coating was unable to cycle at this higher constant current.

This enhanced electrochemical performance of the anode was due to both the internal pores and the gold coating, which creates a more stable interface between the electrolyte and the porous silicon electrode. These results are supported by Takamura et al. [61], where they showed that thin evaporated metals such as gold and silver form a Li alloy and allow Li movement through the metal while providing a stable SEI interface between the carbon and the electrolyte. Figure 1.8 compares the surface morphology of the porous silicon with and without gold coating after 10 cycles. The SEM image shows that the non-coated porous silicon surface (Fig. 1.8a) was severely cracked as compared to the gold-coated porous silicon (Fig. 1.8b). As the volume of silicon expands during the lithiation, the non-coated porous silicon cracks to relieve the internal stress, whereas there was no visual cracking or fragmentation observed on the gold-coated porous silicon electrode after 10 cycles (Fig. 1.8b). The cracks in the gold-coated porous silicon were observed only after 100 cycles. The gold coating hinders the structural degradation in the porous silicon as it undergoes volume expansion during lithiation. However, over many cycles, there is a gradual deterioration in the gold film that is caused by Li intercalation and deintercalation, leading to loss in performance of the electrode.

To show the increase in cycle performance, the gold-coated sample was charged and discharged at different C-rates. The sample was cycled at an initial constant current of 100 $\mu A\ cm^{-2}$($\sim$C/20) for 55 cycles, followed by a current of 150 $\mu A\ cm^{-2}$($\sim$C/15) for cycles 55–65, and 200 $\mu A\ cm^{-2}$($\sim$C/10) after the 65th cycle (data shown in Fig. 1.9). The cell containing the gold-coated porous silicon

Fig. 1.8 SEM surface morphology of **a** non-coated porous silicon and **b** gold-coated porous silicon electrode after 10th cycle of charge/discharge [28]

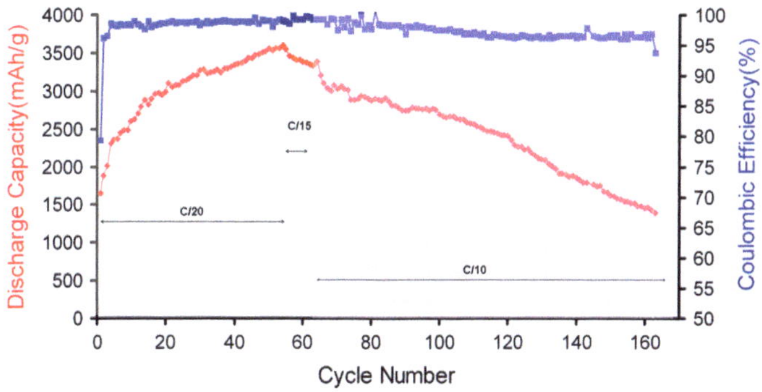

Fig. 1.9 Discharge capacity and efficiency versus cycle number of the porous silicon anode during galvanostatic charge/discharge tested at $100\,\mu A\,cm^{-2}$(~C/20), $150\,\mu A\,cm^{-2}$(~C/15), and $200\,\mu A\,cm^{-2}$(~C/10 rates) [28]

anode ran for 100 cycles with specific capacities of over 2,500 mAh g^{-1} (peaking at 3,500 mAh g^{-1}) with columbic efficiencies over 97 %. It was observed that the specific capacity decreased immediately upon increasing the C-rate and then fades as the number of cycle increases due to the degradation of the material. The capacity of the material was calculated based on the mass of the active porous silicon layer. The bulk silicon may be playing a role in alleviating stress generated during the lithiation/delithiation processes but did not contribute significantly to the charge capacity. The detailed discussion about the role of the bulk silicon on the capacity of the materials was discussed in Ref. [28].

1.5 Second Generation: Freestanding Macroporous Silicon Film (MPSF)

The major disadvantage of the first generation porous silicon is that the bulk silicon substrate does not contribute to the specific capacity [11, 28], but leads to increase in the overall weight of the anode as well as its cost. It was also unclear how practical this process would be for insertion into conventional battery manufacture. The bulk silicon can be removed by backside chemical etching processes [62]. A more efficient solution is to utilize a multi-step lift-off process, in which the same wafer can be reused multiple times to create several porous silicon films [63].

1.5.1 Preparation of Second Generation Freestanding MPSF

Freestanding macroporous silicon film (MPSF) was fabricated by electrochemical etching of a silicon wafer in hydrofluoric acid (HF) solution via a multi-step lift-off

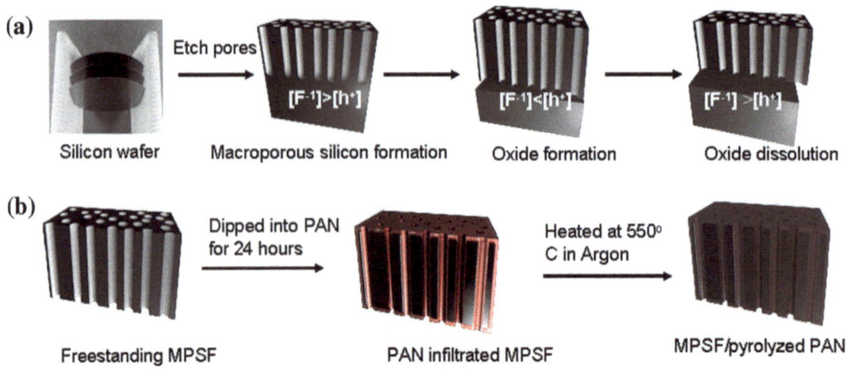

Fig. 1.10 **a** Schematic of the fabrication of the freestanding MPSF and **b** for the infiltration of the PAN into the pores and subsequent pyrolyzation [63]

process using the same setup that was earlier described to create the first generation porous silicon. The concept for the lift-off procedure is depicted in Fig. 1.10a.

For electrochemically etched porous silicon, the formation of the pores takes place when the number of fluoride ions is greater than the number of holes ($[F^-] > [h^+]$). For small increments of the current density in the steps, the depth of the pores increases and the availability of fluoride ions at the pore tip decreased ($[F^-] < [h^+]$). The final boost in the current density leads to increased etch rate, expanding the pore diameters at the tips. As the pores expand at the pore tips, they eventually connect and release the porous layer as a freestanding macroporous silicon film (MPSF). After the liftoff, the MPSF is washed with methanol to remove residual HF. Multi-step lift-off electrochemical etching procedure was more reliable than the one-and two-step lift-off processes [64, 65] because it can easily separate the MPSF from the underlying bulk substrate in its entirety. The process can then be repeated on the same bulk silicon substrate to generate multiple MPSFs. Therefore, this approach is more cost-effective with a little material waste.

After a MPSF was generated and cleaned, a polymer called polyacrylonitrile (PAN) is infiltrated onto the pores, followed by pyrolysis at 550 °C. When PAN was pyrolyzed, it has been reported to form a highly conjugated cyclic chemical structure with increased electrical conductivity compared to native unpyrolyzed PAN [66, 67]. Figure 1.10b illustrates the infiltration of the PAN into MPSF and pyrolysis of MPSF/PAN composite. The SEM images of the bare freestanding MPSF are shown in Fig. 1.11a including the top view (Fig. 1.11b), backside view (Fig. 1.11c), side view (Fig. 1.11d) as well as the backside and side view of MPSF/PPAN composite (Fig. 1.11e–f). From the SEM analysis, the MPSF/pyrolyzed PAN (PPAN) composite was determined to have an average pore diameter close to ~1.5 μm and a thickness of ~50 μm.

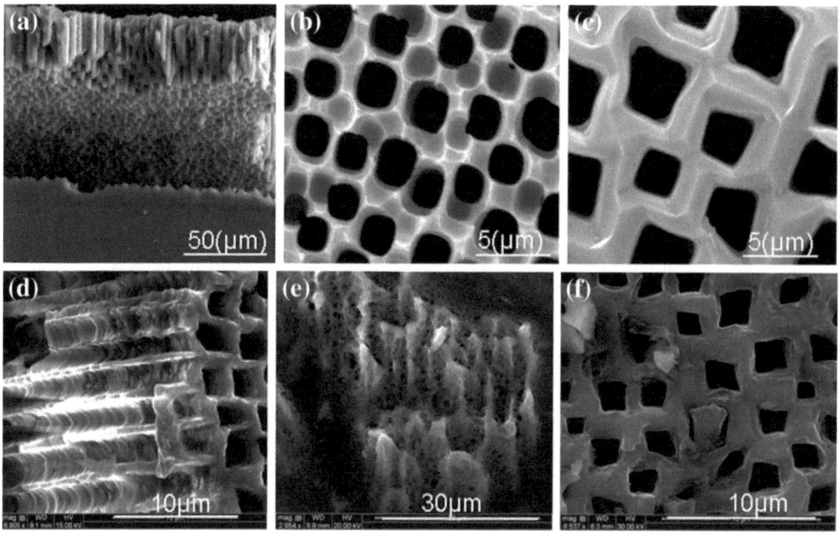

Fig. 1.11 SEM images of the surface morphology of **a** bare freestanding macroporous silicon film (MPSF) with its **b** *top view* and **c** *backside view*. **d** *Side view* of MPSF with PAN before pyrolysis. **e** and **f** are *side view* and *backside view* of MPSF/PPAN composite [63]

1.5.2 Freestanding MPSF in Lithium-Ion Battery

The performance of the freestanding MPSF/PPAN composite has been examined using galvanostatic electrochemical techniques. The freestanding MPSF/PPAN composite was used as a working electrode and lithium foil as a counter electrode. The two electrodes were separated by a Celgard 2325 separator. The cell was charged and discharged between 0.07 and 1.5 V at 200 $\mu A\ cm^{-2}$. The cycling performance of bare freestanding MPSF and MPSF with pyrolyzed PAN(PPAN) is shown in Fig. 1.12a [63]. From the results, the capacity as well as the cycle life for MPSF/PPAN composite was significantly better than bare MPSF. The increase in the battery cycle life and capacity of the MPSF/PPAN composite is due to PPAN, which improves the stability of the solid electrolyte interface (SEI) [68] and offers structural integrity as well as high electric conductivity [69].

To assess the contribution of pyrolyzed PAN to the total capacity of the MPSF/pyrolyzed PAN composite, pyrolyzed PAN was tested without MPSF as an anode material. The cycle performance of pyrolyzed PAN without MPSF is shown in Fig. 1.12b. The pyrolyzed PAN has a very low capacity of only 55 $mAh\ g^{-1}$ after the first cycle (130/35 $mAh\ g^{-1}$), which indicates that the total capacity contributed by the pyrolyzed PAN was relatively negligible. The discharge capacity for first cycle of MPSF/pyrolyzed PAN is 850 $mAh\ g^{-1}$, whereas for the bare MPSF is 757 $mAh\ g^{-1}$. The bare MPSF rapidly drops in capacity after the second cycle, falling to only 200 $mAh\ g^{-1}$ by the 10th cycle and completely failing after 15 cycles. For the

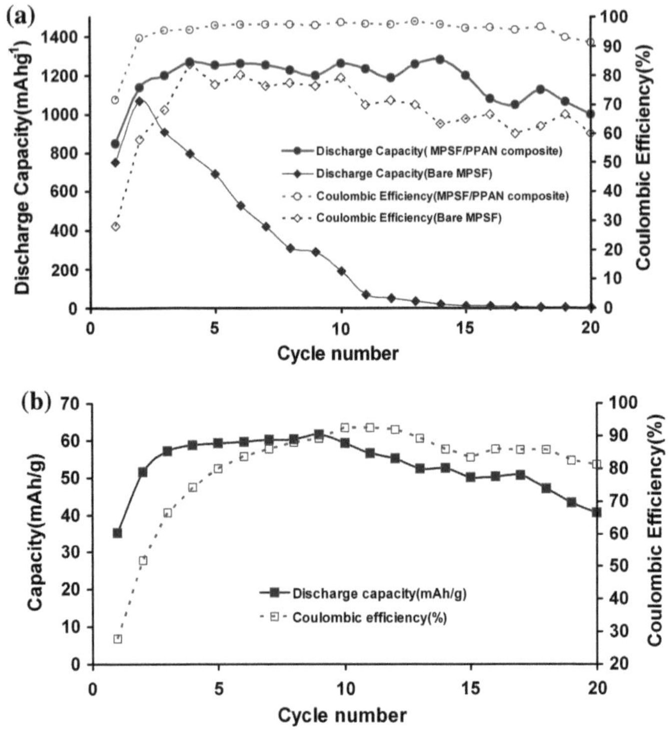

Fig. 1.12 **a** Comparison of discharge capacity and coulombic efficiency versus cycle number for the freestanding MPSF with and without pyrolyzed PAN during galvanostatic charge/discharge tested between 0.07 and 1.5 V at 200 μA cm^{-2}. **b** Capacity and coulombic efficiency versus cycle number for only PAN pyrolyzed at 550 °C under an argon atmosphere [63]

MPSF/PPAN composite, the discharge capacity increases for the first four cycles to 1,260 mAh g^{-1} and remains constant until the 20th cycle. This initial increase in capacity is typical for porous silicon films [28].

The average coulombic efficiency for MPSF with pyrolyzed PAN was above 90 % after the first cycle while the bare MPSF reaches a maximum of only 83 % after 4 cycles, then dropping to 70 %. These results indicate that the pyrolyzed PAN can improve the cycleability and coulombic efficiency of the electrode, but still there was a loss in capacity due to irreversibility with each cycle that may be related to the electrolyte decomposition and the increase in thickness of the SEI layer. To study the loss of the capacity in the MPSF/PPAN composite, an Energy Dispersive X-Ray (EDAX) measurement was carried out to see the infiltration of the PAN into the pores. Data is shown in Fig. 1.13. Results shown in Fig. 1.13 indicate that the 50 μm thick pores of the MPSF were not completely infiltrated with the pyrolyzed PAN. As such, it caused the loss of the capacity as well as the cycle life of the MPSF/PPAN composite after few cycles.

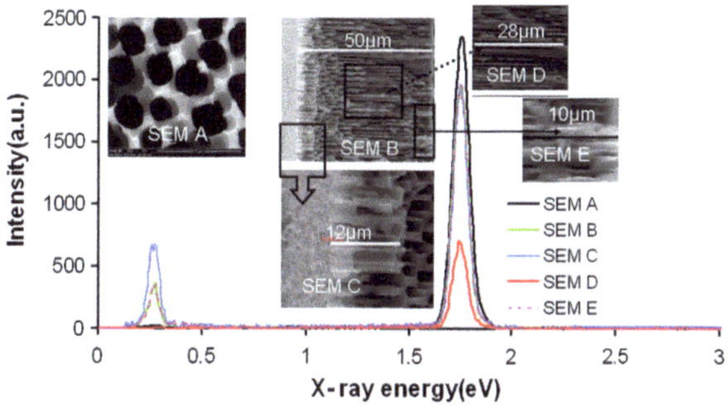

Fig. 1.13 EDAX spectrum of MPSF without PAN (SEM A) that shows a characteristic crystalline peak. The spectrum for the 50 μm thick MPSF infiltrated with pyrolyzed PAN (SEM B-50 μm), *top* 12 μm (SEM C-12 μm), and *bottom* 10 μm (SEM E-10 μm) show both carbon and silicon peaks, indicative of PAN coating. EDAX spectrum of the MPSF with pyrolyzed PAN between 12 and 40 μm (SEM D-28 μm) shows only the silicon peak without evidence of PAN infiltration [63]

1.6 Third Generation: Macroporous Silicon Micro Particulates (MPSP)

Though freestanding MPSF offered large surface area to volume ratios with controllable pore diameters [2], these films are difficult to incorporate into the current battery processing techniques, which typically use material slurries and roll-to-roll processing. To design a more practical material for commercial use, we changed our porous silicon architecture from a film structure to a particulate structure, which can be readily combined with PAN or any other binder to form a slurry that can be processed with standard coating technologies. An inexpensive wet etch processing technique can be used to generate grams of macroporous silicon particulates (MPSPs) [70]. When combined with PAN and then pyrolyzed, the material containing the MPSP/PPAN composite anode can be cycled at a fixed capacity of 1,200 mAh g^{-1} for over 150 cycles.

1.6.1 Preparation of Third Generation MPSP

As described in the second generation, freestanding macroporous silicon films were synthesized via electrochemical etching using a multi-step lift-off process. The thickness of the freestanding macroporous silicon films can be modified by controlling the etching parameters such as applied current, wafer resistivity, concentration of electrolyte and doping of the wafer. To create the third generation porous silicon (called MPSP), the earlier freestanding macroporous silicon film (MPSF) were ultrasonicated to form particulates. Figure 1.14a illustrates the preparation of the MPSPs, the sonication of freestanding macroporous silicon films to form MPSPs, and the

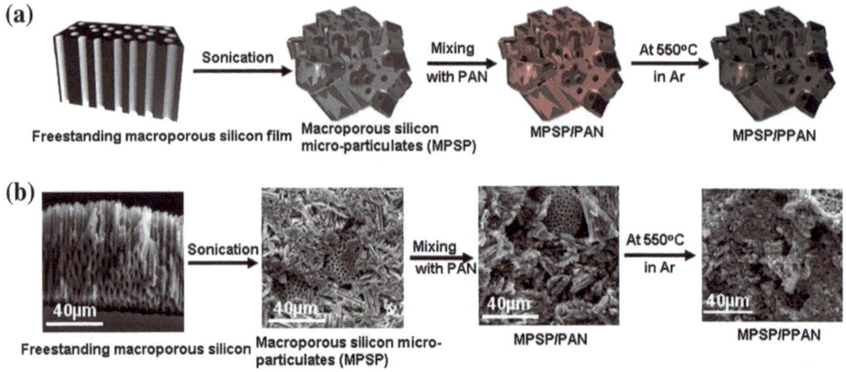

Fig. 1.14 **a** Overview of the process to generate MPSP/PPAN composite from a freestanding macroporous silicon film. **b** Corresponding scanning electron microscope (SEM) images for freestanding macroporous silicon film (MPSF), MPSP, MPSP with PAN and MPSP/PPAN composite [70]

combination of MPSPs with PAN and subsequent pyrolysis. Figure 1.14b shows SEM images for the freestanding macroporous silicon film, the macroporous silicon microparticulates, particulates with infiltrated PAN, and the particulates after the pyrolysis of PAN.

1.6.2 Macroporous Silicon Particulates in Lithium-Ion Batteries

Similar to our previous generations, the charge/discharge cycle performance of the MPSP/PPAN composite was examined using galvanostatic electrochemical techniques. Briefly, the electrochemical cell was constructed using the MPSP/PPAN composite as the anode, lithium foil as a counter electrode, and a Celgard 2325 as a separator that is wet with same electrolyte. Figure 1.15a compares the electrochemical performance of a MPSP/PPAN composite anode versus an anode fabricated from untreated crushed silicon particulates from silicon wafer (control sample). Note that both samples had a nominal particulate size of the order of 10–50 μm and similar silicon to PAN mass ratios. For the first two cycles, the porous (MPSP) and non-porous silicon microparticulates (NPSP) with PPAN were cycled at 200 $\mu A\,cm^{-2}$ between 0.07 and 1.0 V. From the third cycle onward, the cell was cycled between 0 and 1 V at a constant charge capacity of 1,200 $mAh\,g^{-1}$. The performance of the MPSP/PPAN composite electrode shows good cycling stability with an average discharge capacity retention of 99 % even after 150 cycles, as shown by the blue lines in Fig. 1.15a. In contrast, the crushed silicon electrode lost 75 % of its discharge capacity even in five cycles as shown by the brown lines in Fig. 1.15a. Remarkably, this new anode material with MPSP/PPAN composite has a high capacity and longer cycle life as compared to the crushed silicon particulates. To further contrast, the earlier report of Cui et al. [71] has shown that the silicon microparticles (8 and 15 μm diameters) with carbonized PAN and a uniform coating of amorphous

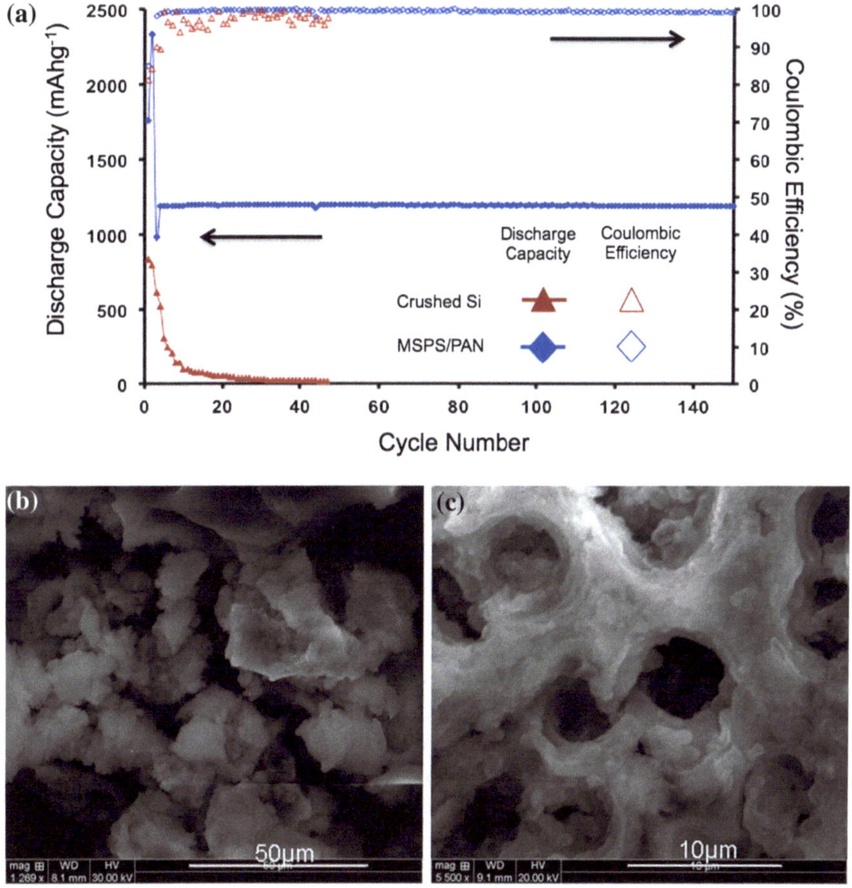

Fig. 1.15 **a** Comparison of the discharge capacity and coulombic efficiency of composite of MPSP versus crushed silicon particulates mixed with PPAN versus cycle number **b** SEM images of the crushed silicon/PPAN **c** and the MPSP/PPAN after 50 cycles

silicon by chemical vapor deposition (CVD) have poor cycleability. The significant increase in the performance of the macroporous material (MPSP) in comparison to the non-porous materials is due to the greater porosity of the MPSP particulates as well as the larger surface area, which can better accommodate the large volume expansion of the silicon upon alloying with lithium. The increase in the surface area of the MPSP as compared to the non-porous silicon particulate was measured by Brunauer-Emmett-Teller (BET) method. The BET surface area of macroporous silicon particulates is $46.85\ m^2\ (STP)g^{-1}\ (\pm 0.12)$, whereas the crushed silicon microparticulates have a surface area of only $0.71\ m^2\ (STP)g^{-1}\ (\pm 0.08\ m^2\ g^{-1})$. The surface area of the macroporous silicon was significantly larger then the non-porous silicon for the similar dimensions of the two particulates, indicating a highly porous silicon structure. SEM analysis of the crushed and macroporous silicon microparticulates after

50 cycles shows that the crushed silicon has large observable cracks while the macroporous silicon maintains a porous structure (Fig. 1.15b, c).

1.7 Conclusion

This chapter describes the application of the electrochemically etched porous silicon in lithium-ion batteries with an emphasis on the three generations of porous silicon architectures: Gold-coated porous silicon, freestanding macroporous silicon (MPSF), and macroporous silicon particulates (MPSP). The gold-coated porous silicon with bulk substrate has been shown to significantly enhance the capacity and cycle life as an anode in lithium-ion batteries over conventional porous silicon. But this material is not economical for commercialization due to the metal coating and attached bulk silicon underneath, which does not contribute to its battery performance. Thus, the first generation etching approach was modified to create the lift-off second generation freestanding porous silicon film (MPSF). The MPSF with pyrolyzed polyacrylonitrile (PPAN) composite was tested as an anode material, where PPAN provides both structural integrity and increased electrical conductivity, leading to better battery performance. While showing better performance, the MPSF/PPAN composite films were fragile to handle and not conducive to conventional battery manufacture processes. To overcome these difficulties the freestanding MPSF can be ultrasonicated to create macroporous silicon particulates (MPSP). The MPSP was mixed with PAN and pyrolyzed to form the composite anode. The MPSP/PPAN composite anode has high capacity and large cycle life, which is very promising for commercial rechargeable lithium-ion batteries. Fabrication of MPSP/PPAN anodes should be low-cost and scalable, and can be inserted into the conventional battery manufacturing processes. For these reasons, the MPSP/PPAN anodes have the potential to provide the next major step in increased lithium-ion battery performance.

Acknowledgments This work is supported by LANCER, the Lockheed Martin Advanced Nanotechnology Center of Excellence at Rice University.

References

1. Shin, H.C., Corno, J.A., Gole, J.L., Liu, M.L.: Porous silicon negative electrodes for rechargeable lithium batteries. J. Power Sources **139**, 314–320 (2005)
2. Obrovac, M.N., Christensen, L., Le, D.B., Dahnb, J.R.: Alloy design for lithium-ion battery anodes. J. Electrochem. Soc. **154**, A849–A855 (2007)
3. Winter, M., Besenhard, J.O., Spahr, M.E., Novak, P.: Insertion electrode materials for rechargeable lithium batteries. Adv. Mater. **10**, 725–763 (1998)
4. Anani, A., Huggins, R.A.: Multinary alloy electrodes for solid-state batteries. 1. A phase-diagram approach for the selection and storage properties determination of candidate electrode materials. J. Power Sources **38**, 351–362 (1992)
5. Chen, X., Gerasopoulos, K., Guo, J., et al.: Virus-enabled silicon anode for lithium-ion batteries. ACS Nano **4**, 5366–5372 (2010)

6. Wen, C.J., Huggins, R.A.: Chemical diffusion in intermediate phases in the lithium-silicon system. J. Solid State Chem. **37**, 271–278 (1981)
7. Yang, Y., McDowell, M.T., Jackson, A., et al.: New nanostructured Li_2S/silicon rechargeable battery with high specific energy. Nano Lett. **10**, 1486–1491 (2010)
8. Rong, J., Masarapu, C., Ni, J., Zhang, Z., Wei, B.: Tandem structure of porous silicon film on single-walled carbon nanotube macrofilms for lithium-ion battery applications. ACS Nano **4**, 4683–4690 (2010)
9. Liu, N., Hu, L., McDowell, M.T., Jackson, A., Cui, Y.: Prelithiated silicon nanowires as an anode for lithium ion batteries. ACS Nano **5**, 6487–6493 (2011)
10. Obrovac, M.N., Krause, L.J.: Reversible cycling of crystalline silicon powder. J. Electrochem. Soc. **154**, A103–A108 (2007)
11. Kang, D.K., Corno, J.A., Gole, J.L., Shin, H.C.: Microstructured nanopore-walled porous silicon as an anode material for rechargeable lithium batteries. J. Electrochem. Soc. **155**, A276–A281 (2008)
12. Chan, C.K., Peng, H., Liu, G., et al.: High-performance lithium battery anodes using silicon nanowires. Nature Nanotechnol. **3**, 31–35 (2008)
13. Qiu, M.C., Yang, L.W., Qi, X., Li, J., Zhong, J.X.: Fabrication of ordered NiO coated Si nanowire array films as electrodes for a high performance lithium ion battery. ACS Appl. Mater. Interfaces **2**, 3614–3618 (2010)
14. Zhong, X., Qu, Y., Lin, Y.-C., Liao, L., Duan, X.: Unveiling the formation pathway of single crystalline porous silicon nanowires. ACS Appl. Mater. Interfaces **3**, 261–270 (2011)
15. Hatchard, T.D., Dahn, J.R.: In situ XRD and electrochemical study of the reaction of lithium with amorphous silicon. J. Electrochem. Soc. **151**, A838–A842 (2004)
16. Baranchugov, V., Markevich, E., Pollak, E., Salitra, G., Aurbach, D.: Amorphous silicon thin films as a high capacity anodes for Li-ion batteries in ionic liquid electrolytes. Electrochem. Commun. **9**, 796–800 (2007)
17. Du, C., Gao, C., Yin, G., Chen, M., Wang, L.: Facile fabrication of a nanoporous silicon electrode with superior stability for lithium ion batteries. Energy Environ. Sci. **4**, 1037–1042 (2011)
18. Magasinski, A., Zdyrko, B., Kovalenko, I., et al.: Toward efficient binders for Li-ion battery Si-based anodes: polyacrylic acid. ACS Appl. Mater. Interfaces **2**, 3004–3010 (2010)
19. Sailor, M.J.: Porous Silicon in Practice. Wiley-VCH, Weinheim (2012)
20. Canham, L.T.: Silicon quantum wire array fabrication by electrochemical and chemical dissolution of wafers. Appl. Phys. Lett. **57**, 1046–1048 (1990)
21. Lehmann, V., Gosele, U.: Porous silicon formation—a quantum wire effect. Appl. Phys. Lett. **58**, 856–858 (1991)
22. Cullis, A.G., Canham, L.T., Calcott, P.D.J.: The structural and luminescence properties of porous silicon. J. Appl. Phys. **82**, 909–965 (1997)
23. Stewart, M.P., Buriak, J.M.: Chemical and biological applications of porous silicon technology. Adv. Mater. **12**, 859–869 (2000)
24. Menna, P., Di Francia, G., Laferrara, V.: Porous silicon in solar-cells - a review and a description of its application as an ar coating. Sol. Energy Mater. Sol. Cells **37**, 13–24 (1995)
25. Wong, M.S., Knowles, W.V.: Surfactant-templated mesostructured materials:synthesis and compositional control,chap5. In: Lu, G.Q., Zhao, X.S.(eds.) Nanoporous Materials - Science and Engineering, 125–164. Imperial College Press, London (2004)
26. Lehmann, V., Rönnebeck, S.: The physics of macropore formation in low-doped p-type silicon. J. Electrochem. Soc. **146**, 2968–2975 (1999)
27. Houle, F.A.: Photochemical etching of silicon–the influence of photogenerated charge-carriers. Phys. Rev. B **39**, 10120–10132 (1989)
28. Thakur, M., Isaacson, M., Sinsabaugh, S.L., Wong, M.S., Biswal, S.L.: Gold-coated porous silicon films as anodes for lithium ion batteries. J. Power Sources **205**, 426–432 (2012)
29. Kim, H., Cho, J.: Superior lithium electroactive mesoporous Si@carbon core-shell nanowires for lithium battery anode material. Nano Lett. **8**, 3688–3691 (2008)

30. Kim, H., Han, B., Choo, J., Cho, J.: Three-dimensional porous silicon particles for use in high-performance lithium secondary batteries. Angew. Chem. Int. Ed. **47**, 10151–10154 (2008)
31. Bang, B.M., Lee, J.-I., Kim, H., et al.: High-performance macroporous bulk silicon anodes synthesized by template-free chemical etching. Adv. Eng. Mater. **2**, 878–883 (2012)
32. Shen, P., Uesawa, N., Inasawa, S., Yamaguchi, Y.: Characterization of flowerlike silicon particles obtained from chemical etching: visible fluorescence and superhydrophobicity. Langmuir **26**, 13522–13527 (2010)
33. Halimaoui, A.: Porous Silicon Science and Technology. Springer, Berlin (1995)
34. Turner, D.R.: Electropolishing silicon in hydrofluoric acid solutions. J. Electrochem. Soc. **105**, 402–408 (1958)
35. Theunissen, M.J.J.: Etch channel formation during anodic dissolution of n-type silicon in aqueous hydrofluoric acid. J. Electrochem. Soc. **119**, 351–360 (1972)
36. Propst, E.K., Kohl, P.A.: The electrochemical oxidation of silicon and formation of porous silicon in acetonitrile. J. Electrochem. Soc. **141**, 1006–1013 (1994)
37. Ponomarev, E.A., Levy-Clement, C.: Macropore formation on p-type Si in fluoride containing organic electrolytes. Electrochem. Solid-State Lett. **1**, 42–45 (1998)
38. Wehrspohn, R.B., Chazalviel, J.N., Ozanam, F., Solomon, I.: Electrochemistry and photoluminescence of porous amorphous silicon. Thin Solid Films **297**, 5–8 (1997)
39. Wehrspohn, R.B., Chazalviel, J.N., Ozanam, F.: Macropore formation in highly resistive p-type crystalline silicon. J. Electrochem. Soc. **145**, 2958–2961 (1998)
40. Steinert, M., Acker, J., Krause, M., Oswald, S., Wetzig, K.: Reactive species generated during wet chemical etching of silicon in HF/HNO_3 mixtures. J. Phys. Chem. B **110**, 11377–11382 (2006)
41. Steinert, M., Acker, J., Oswald, S., Wetzig, K.: Study on the mechanism of silicon etching in HNO_3-rich HF/HNO_3 mixtures. J. Phys. Chem. C **111**, 2133–2140 (2007)
42. Lee, J.-H., Seo, Y., Lim, T.-S., Bishop, P.L., Papautsky, I.: MEMS needle-type sensor array for in situ measurements of dissolved oxygen and redox potential. Environ. Sci. Technol. **41**, 7857–7863 (2007)
43. Bsiesy, A.: In: Canham, L.T. (ed.) Properties of Porous Silicon, pp. 283–289. INSPEC, London (1997)
44. Iyer, S.S., Xie, Y.H.: Light-emission from silicon. Science **260**, 40–46 (1993)
45. Takasuka, E., Kamei, K.: Microstructure of porous silicon and its correlation with photoluminescence. Appl. Phys. Lett. **65**, 484–486 (1994)
46. Peng, K.Q., Xu, Y., Wu, Y., et al.: Aligned single-crystalline Si nanowire arrays for photovoltaic applications. Small **1**, 1062–1067 (2005)
47. Garnett, E.C., Yang, P.: Silicon nanowire radial p-n junction solar cells. J. Am. Chem. Soc. **130**, 9224–9225 (2008)
48. Fang, H., Li, X., Song, S., Xu, Y., Zhu, J.: Fabrication of slantingly-aligned silicon nanowire arrays for solar cell applications. Nanotechnology **19**(25), 255703. doi:10.1088/0957-4484/19/25/255703
49. Wu, E.C., Park, J.-H., Park, J., et al.: Oxidation-triggered release of fluorescent molecules or drugs from mesoporous Si microparticles. ACS Nano **2**, 2401–2409 (2008)
50. Harper, J., Sailor, M.J.: Detection of nitric oxide and nitrogen dioxide with photoluminescent porous silicon. Anal. Chem. **68**, 3713–3717 (1996)
51. Patolsky, F., Zheng, G.F., Lieber, C.M.: Nanowire-based biosensors. Anal. Chem. **78**, 4260–4269 (2006)
52. Tasciotti, E., Liu, X., Bhavane, R., et al.: Mesoporous silicon particles as a multistage delivery system for imaging and therapeutic applications. Nat. Nanotechnol. **3**, 151–157 (2008)
53. Park, J.-H., Gu, L., von Maltzahn, G., et al.: Biodegradable luminescent porous silicon nanoparticles for in vivo applications. Nature Mater. **8**, 331–336 (2009)
54. Lee, C., Kim, H., Hong, C., et al.: Porous silicon as an agent for cancer thermotherapy based on near-infrared light irradiation. J. Mater. Chem. **18**, 4790–4795 (2008)
55. De Stefano, L., Rea, I., Giardina, P., Armenante, A., Rendina, I.: Protein-modified porous silicon nanostructures. Adv. Mater. **20**, 1529–1533 (2008)

56. Canham, L.T.: Bioactive silicon structure fabrication through nanoetching techniques. Adv. Mater. **7**, 1033–1037 (1995)
57. Shin, H.C., Shi, Z., Gole, J.L., Liu, M.L.: Porous silicon based electrodes for lithium batteries. In: Wachman, E., Swider-Lyons, K., Carolan, M.F., Garzon, F.H., Liu, M., Stetter, J.R. (eds.) Solid State Ionic Devices III, PV 2002-26, 2003, pp. 518–525. The Electrochemical Society, Pennington, NJ (2003)
58. Gole, J.L., Seals, L.T., Lillehei, P.T.: Patterned metallization of porous silicon from electroless solution for direct electrical contact. J. Electrochem. Soc. **147**, 3785–3789 (2000)
59. Yu, Y., Gu, L., Zhu, C.B., et al.: Reversible storage of lithium in silver-coated three-dimensional macroporous silicon. Adv. Mater. **22**, 2247–2250 (2010)
60. Halimaoui, A.: Porous silicon formation by anodisation. In: Canhan, L. (ed.) Properties of Porous Silicon, p. 18. The Institution of Electrical Engineering, London (1997)
61. Takamura, T., Sumiya, K., Suzuki, J., Yamada, C., Sekine, K.: Enhancement of Li doping/undoping reaction rate of carbonaceous materials by coating with an evaporated metal film. J. Power Sources **81**, 368–372 (1999)
62. Astrova, E.V., Fedulova, G.V., Smirnova, I.A., et al.: Porous silicon based negative electrodes for lithium ion batteries. Tech. Phys. Let **37**, 731–734 (2011)
63. Thakur, M., Pernites, R.B., Nitta, N., et al.: Freestanding macroporous silicon and pyrolyzed polyacrylonitrile composite as an anode for lithium ion batteries. Chem. Mater. **24**, 2998–3003 (2012)
64. Solanki, C.S., Bilyalov, R.R., Bender, H., Poortmans, J.: New approach for the formation and separation of a thin porous silicon layer. Phys. Status Solidi A **182**, 97–101 (2000)
65. Solanki, C.S., Bilyalov, R.R., Poortmans, J., et al.: Self-standing porous silicon films by one-step anodizing. J. Electrochem. Soc. **151**, C307–C314 (2004)
66. Xue, J.S., Dahn, J.R.: Dramatic effect of oxidation on lithium insertion in carbons made from epoxy-resins. J. Electrochem. Soc. **142**, 3668–3677 (1995)
67. Kim, C., Yang, K.S., Kim, Y.J., Endo, M.: Heat treatment temperature effects on structural and electrochemical properties of PVDC-based disordered carbons. J. Mater. Sci. **38**, 2987–2991 (2003)
68. Yen, Y.-C., Chao, S.-C., Wu, H.-C., Wu, N.-L.: Study on solid-electrolyte-interphase of Si and C-coated Si electrodes in lithium cells. J. Electrochem. Soc. **156**, A95–A102 (2009)
69. Pospisil, J., Samoc, M., Zieba, J.: Third-order nonlinear optical properties of a ladder polymer obtained by pyrolysis of polyacrylonitrile. Eur. Polym. J. **34**, 899–904 (1998)
70. Thakur, M., Sinsabaugh, S.L., Isaacson, M., Wong, M.S., Biswal, S.L.: Inexpensive method for producing macroporous silicon particulates (MPSPs) with pyrolyzed polyacrylonitrile for lithium ion batteries. Sci. Rep. **2**(795), (2012)(Accepted). doi:10.1038/srep00795
71. Cui, L.-F., Hu, L., Wu, H., Choi, J.W., Cui, Y.: Inorganic glue enabling high performance of silicon particles as lithium ion battery anode. J. Electrochem. Soc. **158**, A592–A596 (2011)

Chapter 2
The Development of Si and Ge-Based Nanomaterials for High Performance Lithium Ion Battery Anodes

Xiao-Liang Wang and Wei-Qiang Han

Abstract Silicon and germanium are among the most promising anode materials for high performance lithium ion batteries, due to their unprecedented high capacities. In recent few years, increasingly enormous efforts have been dedicated to these two important anodes, leading to significant improvement in their cycling life, practical capacity, rate capability, and coulombic efficiency. Nanostructuring is playing a crucial role in enabling the improvement and will lead to their widespread use in various battery markets. Nanoscale particles can better tolerate the wild volume change upon cycling and maintain their integrity than micron-sized particles. They can also shorten the diffusion distance of lithium ions and electrons and thus have high capacity. Further, one-dimensional nanowires exhibit superior stress behavior and electron transport. Porous and hierarchical nanostructures can provide extra space to accommodate the volume change. Wisely manipulating these handles have produced impressively better-performing systems. Porous single-crystal silicon nanowires have shown more stable capacity than solid nanowires. Hierarchical porous amorphous GeO_x is another system with very long cycle life and high capacity.

2.1 Introduction

The demand for electric vehicles has offered a great opportunity for lithium ion battery R&D. One important goal is to develop high energy-density systems, which can support sufficient mileage per charge. In a lithium-ion battery, energy is stored

X.-L. Wang · W.-Q. Han (✉)
Ningbo Institute of Material Technology and Engineering (NIMTE), Chinese Academy of Sciences, 315201 Ningbo, People's Republic of China
e-mail: hanweiqiang@nimte.ac.cn; whan@bnl.gov

Present address:
X.-L. Wang
Seeo Inc., 3906 Trust Way, Hayward, CA 94545, USA

H. Li et al. (eds.), *Silicon-based Nanomaterials*,
Springer Series in Materials Science 187, DOI: 10.1007/978-1-4614-8169-0_2,

© Springer Science+Business Media New York 2013

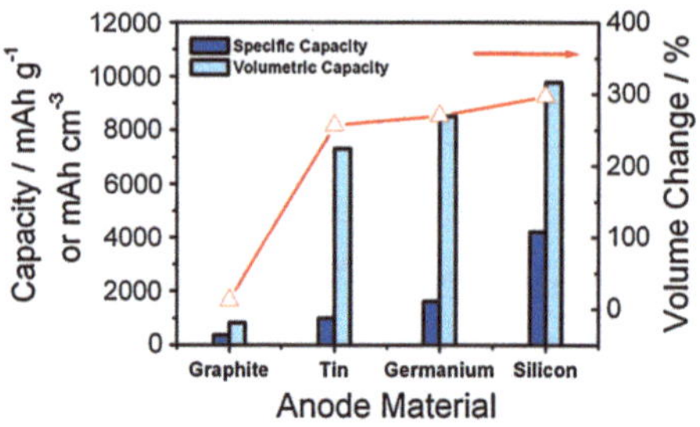

Fig. 2.1 Theoretical capacities and the volume change of graphite, tin, germanium and silicon

in two electrodes—the anode and the cathode. As we use the battery (discharging), the anode gets oxidized and generates lithium ions and electrons. Lithium ions travel through the electrolyte and electrons travel in the external circuit, with the help of the potential difference between the anode and the cathode. At the same time, lithium ions and electrons participate in the reduction of the cathode.

One important approach to achieving high energy density is to use high capacity anode materials. In this regard, silicon, germanium and tin-based systems are very promising. As shown in Fig. 2.1, these materials have much higher capacities than the state-of-the-art graphite does. The specific capacities of silicon, germanium, and tin are 4,198 mAh g^{-1}, 1,624 mAh g^{-1}, and 993 mAh g^{-1}, while the capacity of graphite is only 373 mAh g^{-1}. Their volumetric capacities also are way higher than graphite's capacity; 9,781 mAh cm^{-3} for silicon, 8,516 mAh cm^{-3} for germanium and 7,313 mAh cm^{-3} for tin, versus 833 mAh cm^{-3} for graphite.

The principal challenge in using these anodes is their poor cycle life. The high capacity causes huge volume change. Figure 2.1 also compares the volume increase after fully lithiation. In the case of graphite, there is only 12 % in volume increase, while the values for silicon, germanium and tin are 297 %, 270 %, and 257 %, respectively. As a result of the huge volume change, particles tend to break, which (1) creates new surfaces for side reactions involving lithium ions, (2) produces isolated anode fragments, and (3) disconnects charge transport pathways. Then, the capacity fades too fast. As an example, 1–5 μm silicon particles showed a reversible capacity of ~1,700 mAh g^{-1} in the first cycle, but then the capacities were ~1,000 mAh g^{-1}in the second cycle, ~800 mAh g^{-1} in the third cycle and ~600 mAh g^{-1} in the fourth cycle, respectively [1].

As a summary, the following fundamental performance issues need to be addressed before these high-capacity anodes can be commercialized:

Poor cycle life. We have discussed this issue in the last paragraph.

Low practical capacity. Although the theoretical capacities are high, it usually is quite difficult for these materials to reach such high capacities. One reason is that the charge transportation is not efficient enough for the whole anode particles to get reacted. For example, undoped silicon has low electric conductivity and low lithium diffusion coefficient.

Low coulombic efficiency. The surface chemistry of these high-capacity anodes may be quite different from that of graphite. This affects side reactions and the formation of solid electrolyte interface (SEI). It is commonly seen that the initial coulombic efficiency is far below 90 % and the efficiency upon cycling has difficulty reaching 100 %.

Poor rate capability. This also results from the poor charge transportation in these anodes. The accessible capacity decreases significantly as C-rate increases.

2.2 The Development of Nanostructured High-Capacity Anodes

Different nanostructuring strategies have been developed to address these issues and many have been proven to be quite effective. To cope with the volume expansion, there are tolerance enhancements such as decreasing particle size and making amorphous materials, accommodation such as porous nanomaterials and buffering such as nanocomposites, surface coating, and alloys. Low dimensionality and carbon coating also facilitate the charge transportation. Coulombic efficiency can be increased by carbon coating and composites.

Among the first reports on nanosilicon anode systems, Hong Li et al. studied the cell performance of silicon powder with an average size of 78 nm [1]. Although the cycle life still was inferior, the capacity fading was slower than that of micron-sized silicon. The authors also found that the cycling stability became better as they decreased the ratio of silicon in the electrode and limited the delithiation voltage.

Other than the particles being nanosized, silicon/carbon nanocomposites form another important category. One early work was done by Chun-Sheng Wang et al. [2] by ball-milling micron-sized silicon with graphite, a composite consists of nanosilicon particles encapsulated in amorphous carbon was formed. Such composites showed higher capacities and lower irreversible capacities as compared with micron-sized silicon.

As nanosilicon enhances the cycling stability, it is important to know where the bounds of nano-size effects are. In other words, what would be the optimal size in the regard of cell performance. Hyejung Kim et al. synthesized silicon nanoparticles with different sizes of 5, 10, and 20 nm, utilizing reverse micelles at high pressure and temperature [3]. The 10 nm system displayed better performance; the reversible capacity of the first cycle was ~3,400 mAh g^{-1} and the capacity on the 40th cycle was ~2,800 mAh g^{-1}. As compared, the 20 nm system had a similar initial capacity but lost capacity faster, the value on the 40th cycle being ~2,400 mAh g^{-1}. The 5 nm system delivered a low initial capacity of ~2,600 mAh g^{-1}, and the capacity decreased to ~2,000 mAh g^{-1} after 40 cycles. It is worth noting that the cycling stability was further improved after carbon-coating via a chemical vapor deposition (CVD) process. C-coated 10 nm silicon particles had an initial capacity of ~3,500 mAh g^{-1} and showed negligible capacity loss for 40 cycles. In addition, carbon-coating further improved the coulombic efficiency.

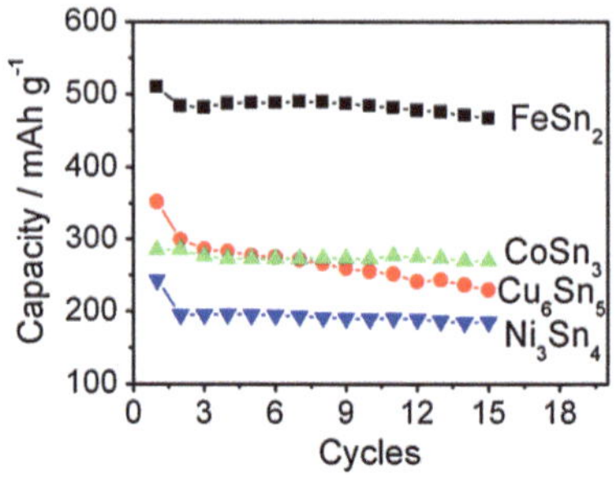

Fig. 2.2 Delithiation capacities of coin cells with intermetallic nanospheres as the working electrodes, and lithium metal as both the reference- and counter-electrodes. The cycling rate was C/20 based on the theoretical capacity of each system. The voltage range was 0.05–1.5 V [12] (Reprinted with permission from Wang et al. [12]. Copyright 2010, American Chemical Society)

Our group also were interested in this question and synthesized Sn/SnO_x core-shell nanospheres with controlled sizes of 30, 45, 79, and 100–500 nm [4], by a modified polyol wet-chemistry method. The 45 nm anode behaved better in the sense of improved cycling stability and high capacity among samples studied. The 79 nm anode, although still in the nano-regime, showed rapid capacity decay, similar to the behavior of micron-sized particles. The 30 nm anode, on the other hand, delivered less capacity than the 45 nm anode did.

One major reason behind the enhanced stability of nanoparticles lies in the fact that small particles tend to tolerate the volume change better than large particles do. However, the Sn/SnO_x anode experienced abrupt failure after ~15 cycles due to the fracture of the nanospheres. Nanowires are believed to be suitable for the accommodation of the volume change without fracture. In an in-situ transmission electron microscopy (TEM) study of the electrochemical lithiation of a single SnO_2 nanowire by Jian-Yu Huang et al., the as-prepared straight nanowire turned into a twisted one after lithiation, because of the plastic deformation and microstructural changes [5]. Large stress and mechanical plasticity occurred at the reaction front, as a result of the volume expansion.

Candace K. Chan et al. reported a high-performance silicon nanowire system grown on stainless steel substrate via a vapor-liquid-solid (VLS) process [6]. The system exhibited high and stable capacity of ~3,000 mAh g^{-1} for 10 cycles. The performance was much better than the silicon thin film consisting of 12-nm crystals. Electron microscopy observation suggested that the nanowires remained intact after cycling, although significant volume expansion was evident.

Another way to overcome the capacity fading of nanocrystal thin film is to make amorphous thin films. Jason Graetz et al. achieved stable capacities of 1,700 mAh g^{-1} for 60 cycles with an amorphous germanium thin film [7]. On the contrary, the nanocrystal thin film had a reversible capacity of ~1,400 mAh g^{-1} with 60 % capacity retention after 50 cycles. The rate capability of the amorphous thin film also was elevated.

While thin films may be used in microbatteries, powder form with high packing density would be easily adapted by the battery industry. Hierarchical structures with a micron-sized granule appearance but also with nanoscale features like nanopores and nanocomposites would be an efficient solution. A Si/C nanocomposite granule anode was developed by A. Magasinski et al. [8] the synthesis involved the pre-annealing of carbon-black nanoparticles into branched short chains, the CVD assembly of silicon on the surface of carbon black and the self assembly into micron-sized spherical granules with nanopores as a result of the carbon CVD coating process. The anode possessed capacities of ~2,000 mAh g^{-1} at C/20, ~1,600 mAh g^{-1} at 1C and ~800 mAh g^{-1} at 8C. The cycling was stable for 100 cycles at 1C. The rate performance was superior to that of the graphite anode, which discharged ~40 mAh g^{-1} at 8C.

Quan Fan et al. characterized a commercialized composite system composed of amorphous Sn–Co alloy, graphitic carbon, and titanium [9]. The characteristic particle size was 5 nm. This anode showed better cell performance than the crystalline Sn–Co alloy did. The capacity stabilized at 400 mAh g^{-1} while the crystalline Sn–Co anode exhibited low and fast-decaying capacities.

Forming alloys has long been considered as an important approach to obtaining better-performing high-capacity anodes. In one of the pioneering studies, Jurgen O. Besenhard et al. found that the electroplated sub-micron Sn–Ag and Sn–Sb alloys cycled much more stably than sub-micron Sn did [10]. As the charge input was limited to 1.7 Li/M, Sn started to acceleratedly lose capacity after ~20 cycles, while the alloy systems could be cycled over 150 times. The cycle life of the Sn–Sb system is longer than that of the Sn–Ag system.

In another pioneering work, Ou Mao et al. prepared a series of Sn–Fe–C composites [11]. Two major components were active $FeSn_2$ and inactive $SnFe_3C$. The capacities of these composites were higher than that of $SnFe_3C$ and more stable than that of $FeSn_2$. The presence of the inactive phase helped stabilize the anodes.

In the Sn alloy research, the cell performance of the same composition always depends on the nanostructure and the synthesis method. To identify intrinsically better-performing alloys would be of great significance for the development of the Sn anode. Employing the modified polyol synthesis, we successfully converted Sn nanospheres into $FeSn_2$, $CoSn_3$, Cu_6Sn_5 and Ni_3Sn_4 nanospheres with similar sizes of ~40 nm [12]. Moreover, they had the same nanostructure—single-crystal Sn–M/amorphous Sn–M–O core-shell structure. Under this well-controlled condition, as depicted in Fig. 2.2 the $FeSn_2$ system had the highest stable capacity of ~500 mAh g^{-1} for 15 cycles. The capacity of the $CoSn_3$ anode also was stable, but was only ~280 mAh g^{-1}. The Cu_6Sn_5 system showed fading capacities from ~350 mAh g^{-1} to ~220 mAh g^{-1} over 150 cycles. The Ni_3Sn_4 exhibited the lowest stable capacity of ~200 mAh g^{-1}.

Later, by using the same synthesis method but lowering the molar ratio of the $FeCl_3$ and $SnCl_2$ precursors to 0.1, we fabricated nanospheres of the $FeSn_5$ phase, which had not been established in the Sn–Fe phase diagram and represented a new binary structure type [13]. The Rietveld refinement of Fe occupancy implied vacancies at Fe sites and a Sn/Fe molar ratio of 6.8. Therefore, the material had a chemical formula

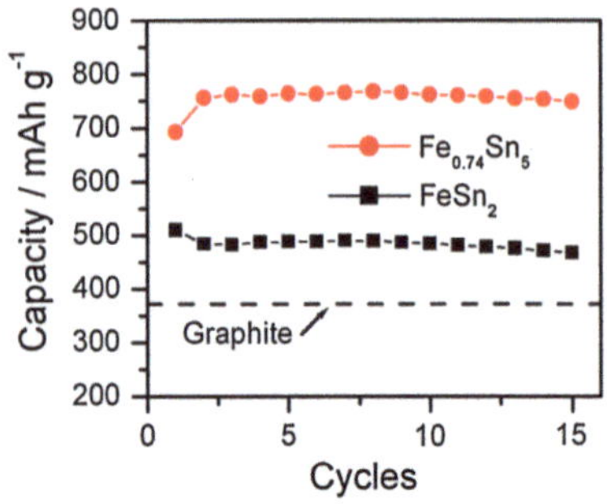

Fig. 2.3 Delithiation capacities of $Fe_{0.74}Sn_5$- and $FeSn_2$-nanospheres as anodes [13] (Reprinted with permission from Wang et al. [13]. Copyright 2011, American Chemical Society)

of $Fe_{0.74}Sn_5$. This fact made the anode to have the highest theoretical capacity of 929 mAh g^{-1} among inactive M-Sn intermetallic anodes. The system delivered stable capacities of ~750 mAh g^{-1} for 15 cycles, ~50 % higher than the values of the $FeSn_2$ system (Fig. 2.3). The initial coulombic efficiency also was higher for $Fe_{0.74}Sn_5$, i.e., 72 % versus 50 % for $FeSn_2$.

Further, interestingly, we also synthesized the new $CoSn_5$ phase [14]. Its crystal structure was the same as $FeSn_5$. The $Co_{0.83}Sn_5$ nanospheres had a theoretical capacity of 917 mAh g^{-1}, which was only lower than that of $Fe_{0.74}Sn_5$ in the category. In practice, the reversible capacity was ~500 mAh g^{-1}, which was much higher than the capacity of the $CoSn_3$ nanospheres. Moreover, the capacity was quite stable over 100 cycles, while in the case of the $Fe_{0.74}Sn_5$ nanospheres, the capacity started to drop from cycle 18.

In future development of nanostructured high-capacity anodes, the following areas are worth paying great attention to:

Packing density. Nanoparticles usually are loose and have low packing density. This would affect the volumetric energy density of batteries as well as add complexity to the battery fabrication process.

Low coulombic efficiency. This partly stems from the high surface area of nanostructures, which would cause irreversible consumption of lithium ions from side reactions and SEI formation. It could also depend on the stability of the SEI layer. The mechanism of these processes has not been well established.

Cost. The commercialization of high-capacity anodes requires inexpensive preparation processes. It should be kept in mind that the overall cost includes the material cost and the process cost. This does not mean that one should stay away from complicated synthesis. Model systems would provide knowledge otherwise hard to gain and serve as a guidance to the mass production of high-performance anode materials.

2.3 Porous Single-Crystal Silicon Nanowires with Enhanced Cell Performance

One-dimensional nano-anodes possess efficient 1D electron transport and facile strain-relaxation, thus making them promising for high performance Li-ion

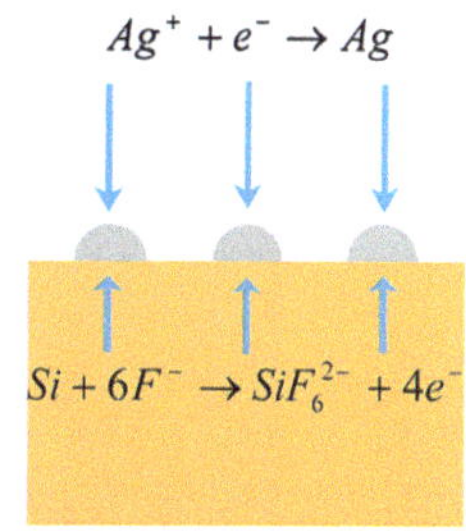

Scheme 2.1 Schematic of the electroless $HF/AgNO_3$ etching process. *Light brown* silicon; *grey* silver

batteries [6]. The use of nanoporous structures is another extensively studied strategy because they can help accommodate the changes in volume and mechanic strain.

We synthesized porous single-crystal silicon nanowires via an electroless $HF/AgNO_3$ etching process (Scheme 2.1) [15] for which we used single-crystal silicon wafers [p-type, 0–100 Ω cm, (100)] as the silicon source. The etching aqueous solution contained HF and $AgNO_3$. After immersing the wafers in the solution, the Ag^+ ions easily were reduced, coating Ag nanoparticles on wafers' surface and oxidizing the surrounding Si. As the reaction progressed, Ag nanoparticles were etched all the way down into the wafer leaving porous nanowires behind them. Thereafter, the Ag was washed away by nitric acid and the nanowire powder was collected by the doctor-blade technique.

The Si nanowires thus grown were ~10 μm long (Fig. 2.4a). Their XRD patterns verified that the nanowires were pure Si phase (Fig. 2.4b). They were highly porous; the pore size and the wall thickness were ~5 nm and ~10 nm respectively, as shown in Fig. 2.4c. Moreover, selected area electron diffraction (SAED) reflected that nanowires are single-crystal with a ⟨100⟩ axial direction. Therefore, generating Si nanowires by electroless etching preserved both the single crystal and the orientation of the parent wafer.

Si nanowires with carbon black (NW + CB) as the conductive additive showed an initial delithiation capacity of 1,066 mAh g^{-1}, as shown in Fig. 2.5a. The initial Coulombic efficiency was 49 %. The capacity on the second cycle was 1,256 mAh g^{-1}; after 20 cycles, it was 815 mAh g^{-1}, i.e., 76 % of the initial value. Capacity retention was much better than that of solid nanowires. In the latter case, the capacity dropped by ~50 % after 10 cycles [16].

This nanoporous structure also enhanced the anode's electrochemical activity due to the highly accessible surfaces and short diffusion-distance. As shown in Fig. 2.6a, the NW + CB system had a bigger lithiation peak than did the NP + CB system; the major lithiation process also started at a higher voltage of ~0.22 V. This peak was related to the crystalline Si-amorphous alloy transformation. Upon delithiation, both systems exhibited two peaks. We assigned the peaks at ~0.35 V and at ~0.47 V, respectively, to the low-voltage delithiation (2.1), and the high-voltage delithiation (2.2) of amorphous Li-Si alloys [17].

$$\mathrm{Li_{x'+x''}Si} \rightarrow \mathrm{Li_{x'}Si} + \mathrm{x''Li^+} + \mathrm{x''}e^- \qquad (2.1)$$

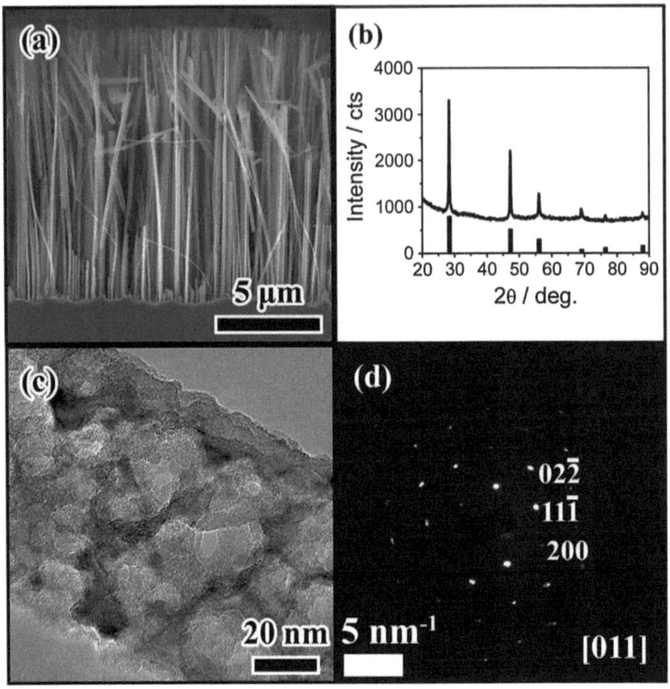

Fig. 2.4 Porous single-crystal Si nanowires. **a** SEM image of the wafer; **b** XRD patterns; **c** TEM image; and **d** SAED pattern along the [011] direction [15] (Reprinted with permission from Wang et al. [15]. Copyright 2010, American Chemical Society)

$$Li_{x'}Si \rightarrow Si + x'Li^{+} + x'e^{-} \tag{2.2}$$

In NW + CB, the delithiation peaks were stronger than those of NP + CB and the 0.34 V peak shifted to a lower voltage and became more intense than the 0.47 V peak, indicating that more Li-enriched $Li_{x'+x''}$ Si had formed and more Si participated in Li storage.

Furthermore, graphene nanosheets, massless charge carriers with high mobility were employed as effective 2D conductive additive to assure the high performance of Si nanowires.

Graphene nanosheets were prepared by the liquid-phase exfoliation of graphite. In this synthesis, we sonicated graphite in a N,N-dimethylformamide (DMF) solution of poly(m-phenylenevinylene-co-2,5-dictoxy-pphenylenevinylene) to break it up into graphene nanosheets that were not oxidized and had few defects as compared with nanosheets obtained by other methods.

Figure 2.7 shows TEM morphology of these nanosheets that typically were micron-sized. The high-resolution TEM (HRTEM) image, and the fast-Fourier-transform electron-diffraction patterns of a five-layered nanosheet (Fig. 2.7b) unveiled

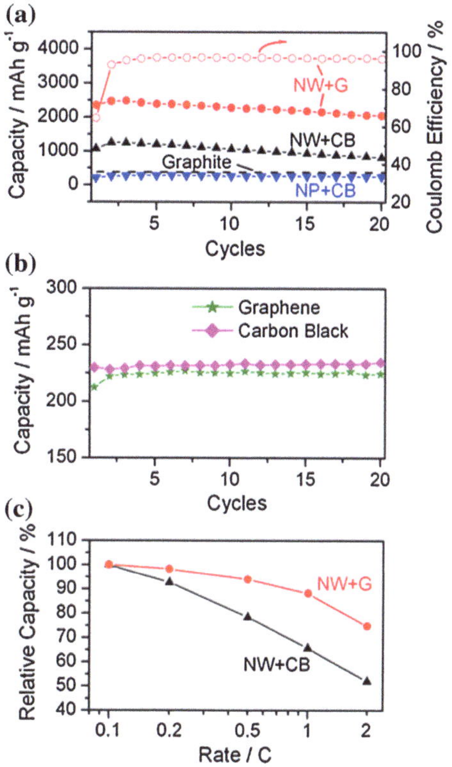

Fig. 2.5 Anode performance. **a** Delithiation capacities and coulombic efficiency of cells with noted Si anodes for 20 cycles. *NW* Si nanowires; *G* graphene; *CB* carbon black; *NP* Si nanoparticles. We removed the contribution to capacity from graphene and carbon black (**b**). **c** Rate capacities from 0.1C to 2C, with the rate for discharge fixed to 0.1C [15] (Reprinted with permission from Wang et al. [15]. Copyright 2010, American Chemical Society)

a single crystal, viewed along the [001] direction. The fringe gap in the sheet was ~0.34 nm, in agreement with the distance between two (002) planes of graphite.

The NW + G (Si nanowire/graphene) anode had an initial delithiation capacity of 2,347 mAh g^{-1} (Fig. 2.5a). The capacity reached 2,470 mAh g^{-1} on the third cycle and retained 87 % of this value after 20 cycles. The coulombic efficiency was 64 % on the first cycle and quickly increased upon cycling, reaching close to 100 %. Graphene also enhanced the rate performance of Si nanowires. In Fig. 2.5c, the 2C capacity of the NW + G anode was ~75 % of the value at 0.1C. In comparison, the NW + CB anode only retained ~52 % of its initial capacity.

The electrochemical activity of Si nanowires with graphene was better than that of those with carbon black. As shown in Fig. 2.6a, the peaks of the NW + G anode were stronger than those of the NW + CB anode. The major lithiation process in both began at a similar voltage of ~0.22 V. The additional delithiation peak at ~0.13 V in the former compared with the latter could be from graphene (Fig. 2.6b).

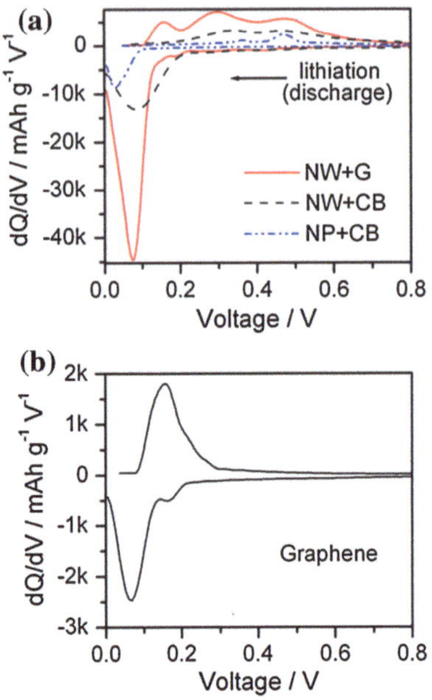

Fig. 2.6 Differential capacity dQ/dV *curves*. **a** First charge/discharge cycle of cells with the noted Si anodes. *NW* Si nanowires; *G* graphene; *CB* carbon black; *NP* Si nanoparticles. **b** First cycle of the graphene cell [15] (Reprinted with permission from Wang and Han [15]. Copyright 2010, American Chemical Society)

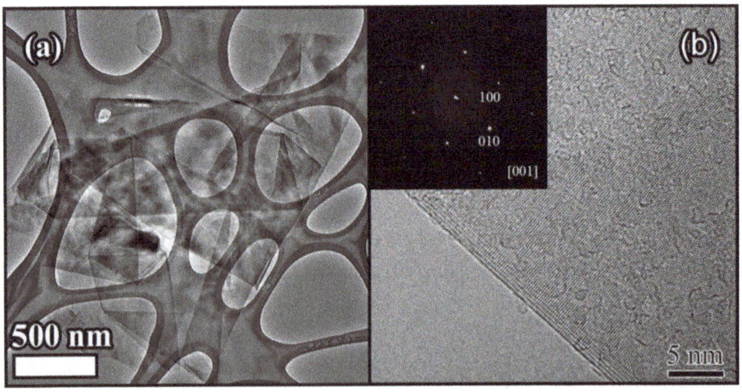

Fig. 2.7 TEM images of graphene nanosheets. **a** Low magnification; **b** high magnification viewed along the [001] direction, where the *inset* shows the fast-Fourier-transform electron-diffraction patterns [15] (Reprinted with permission from Wang and Han [15]. Copyright 2010, American Chemical Society)

2.4 Amorphous Hierarchical Porous GeO_x with High Capacity and Very Long Cycle Life

Three-dimensional nanostructures also largely enhance the performance of these high-capacity anodes. We synthesized amorphous hierarchical porous-GeO_x anode powders that had a stable high capacity of ~1,250 mAh g^{-1} for 600 cycles [18].

GeO_x powders were generated via a simple wet-chemistry method at room temperature. A GeO_2 suspension in water reacted with NH_4OH to form germanate (dihydrogen) ions, that then were reduced by $NaBH_4$. The product was filtrated, washed with distilled water, and dried under vacuum.

Figure 2.8a shows micron-sized agglomerates containing nanopores. Under higher magnification (Fig. 2.8b), another kind of nano-agglomerate, ~50 nm in size, is apparent. Furthermore, these nano-agglomerates consisted of ~3.7 ± 1.0 nm primary nanoparticles (Fig. 2.8c). This GeO_x powder was amorphous, as confirmed by XRD in Fig. 2.8d. There were two broad peaks near 13° and 24°, coinciding respectively with the Ge(111) and Ge(220) peaks of Ge. Therefore, the atomic arrangement was similar to that in Ge than in GeO_2. The profile from synchrotron extended X-ray absorption fine structure (EXAFS) at the Ge K-edge (Fig. 2.8e) revealed that Ge atoms similar to those in crystalline Ge formed the first coordination shell, and O atoms similar to those in GeO_2 formed the second coordination shell. The STEM-EDS measurement obtained a Ge/O molar ratio of 6/4. This GeO_x powder had a Brunauer-Emmett-Teller (BET) surface area of 187 m^2g^{-1}.

After heating the GeO_x powder in H_2 at 300 °C for 30 min, the micron-sized amorphous agglomerates, consisting of ~20 nm nanoparticles, were retained, but there were no longer any 3.7 nm primary particles. The surface area became 58.3 m^2 g^{-1}. Exposing the powder to 700 °C for 30 min led to regular micron-sized crystalline particles.

The initial C/20 (80 mA g^{-1}) delithiation capacity of the starting GeO_x powder was 1,728 mAh g^{-1} (Fig. 2.9a). The C/5- and C/2-capacities were 575 and 1,268 mAh g^{-1}, respectively. After cycling at C/2 (for both charge and discharge) for 600 full cycles, the capacity was very well maintained; capacity on the 600th C/2 cycle was an impressive 96.7 % of that of the first C/2 cycle.

Amorphous agglomerates with larger particles (the 300 °C sample) had lower yet stable capacities; 688 mAh g^{-1} at C/20, 661 mAh g^{-1} at C/5, and 607 mAh g^{-1} at C/2. The 700 °C sample was not porous or amorphous, and its capacity faded fast stabilizing only at a lower stage of ~300 mAh g^{-1}. Figure 2.9b illustrates the charge/discharge curves of the initial GeO_x sample and the 700 °C sample.

The high-capacity GeO_x with its very long cycle life reflected four beneficial characteristics: small primary particles, porous structure, amorphous state, and the incorporation of oxygen.

Among them, the small size might play the crucial role. First, it might tolerate the change in volume change, so preventing the anode from volume-change-induced pulverization since the size of the primary particles was only ~3.7 nm. Micron-sized Ge could be broken into 5–15 nm particles after cycling [19]. Second, the absolute

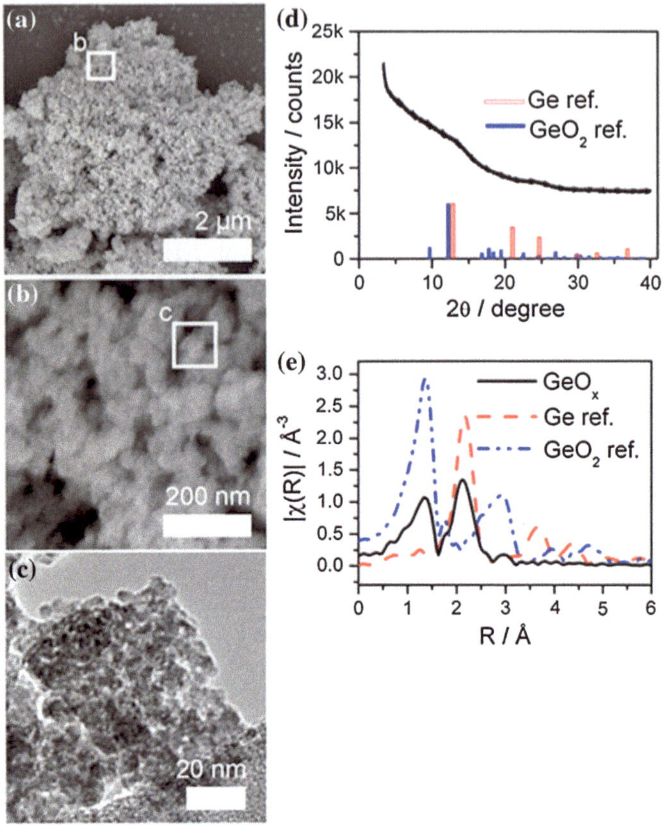

Fig. 2.8 Nano- and atom-structure of the initial Ge ultrafine powder. **a**–**c** Hierarchical nanoporous structure, with **a** Low-magnification SEM image, **b** Enlarged SEM image corresponding to the area enclosed by a square in (**a**) and **c** TEM image corresponding to the area in the square in **b** **d** Synchrotron XRD profile ($\lambda = 0.72958$ Å). **e** Synchrotron EXAFS profile. Powder Diffraction File (PDF) peaks of Ge (#00-004-0545) and GeO_2 (#01-085-0473), and synchrotron EXAFSs of Ge and GeO_2 references, respectively, also are shown in (**d**) and (**e**) [18] (Reprinted with permission from Wang et al. [18]. Copyright 2011, American Chemical Society)

volume change of each primary particle was small. The electrical contact thus could be maintained, together with that of the particle's integrity. Third, smallness may enhance electrochemical activity by shortening charge-transportation distances and offering having more surfaces for charge transfer. In summary, the small primary particles were the basis for the hierarchical porous structure's stability and integrity upon high-capacity cycling.

The amorphous state was another important factor. Ultrafine crystallites (<10 nm) still showed fading capacities [20] that might be related to the harmful anisotropic expansion/extraction, and the stresses from phase transitions occurring in crystalline anodes.

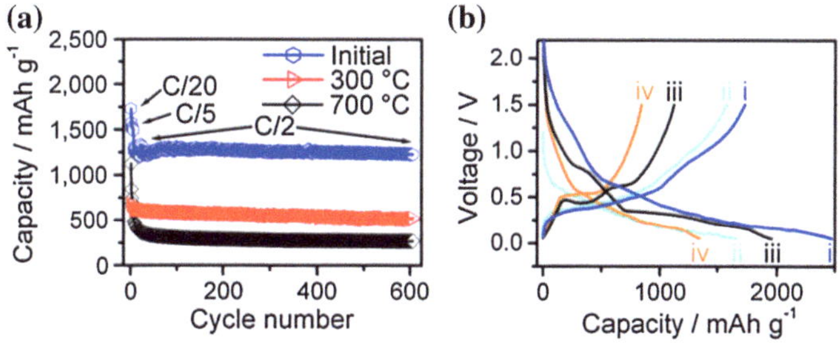

Fig. 2.9 Anode performance in half cells. **a** Delithiation capacities. **b** Charge-discharge curves. (i) The first cycle of the initial GeO_x (C/20). (ii) The second cycle of the initial GeO_x (C/5). (iii) The first cycle of the 700 °C sample (C/20). (iv) The second cycle of the 700 °C sample (C/5) [18] (Reprinted with permission from Wang et al. [18]. Copyright 2011, American Chemical Society)

Porosity also played an important role. Even small amorphous anode particles need room for volume changes upon cycling. The hierarchical porous structure accommodated such alterations and preserved the intactness of the micron-sized GeO_x agglomerates, thus preventing changes in the microscopic charge-transport pathway and the particles' locations whilst maintaining the electrical contact. Porosity also helped avoid the build-up of contractive stresses and the formation of cracks resulting from the lack of room for expansion during lithiation.

The oxygen in the initial GeO_x diluted volume expansion, thus assuring that the open space in the agglomerates was sufficient for accommodation. The powder had a Barrett-Joyner-Halenda (BJH) pore volume of $0.34\,cm^3g^{-1}$. Assuming the density of crystalline Ge, the pores of a 1 g sample could encompass an extra 1.81 g. Accounting for the change in the volume of oxygen during cycling, the porosity could be more than needed. This role of oxygen was similar to that of second metals in alloy anodes, and of oxygen in conversion electrodes and alloy-metal oxides.

We also evaluated the initial GeO_x in full-cell configuration, a very challenging key step toward real-world applications. Before fabricating full cells, we applied an air-stable lithium powder from FMC Lithium to the surface of the anode web to compensate for the initial irreversible capacity-loss. We registered an initial coulombic efficiency of 99.5 % with an open-circuit voltage (OCV) of 0.74 V. $Li(NiCoMn)_{1/3}O_2$ (NCM) was used as the active cathode in full cells.

As shown in Fig. 2.10, a full cell discharged 164 mAh (g of NCM)$^{-1}$ at C/20 (based on $Li(NiCoMn)_{1/3}O_2$). The voltages were lower and more slanted than those of the NCM half-cell. The full cell also showed stable cycling; the average capacity loss over 200 cycles at C/2 was only 0.028 % per cycle. Therefore, the initial GeO_x anode in a full cell had excellent reversibility and stability.

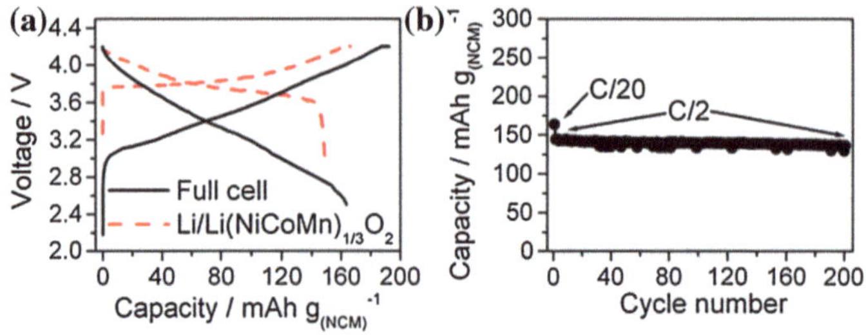

Fig. 2.10 Cell performance of the initial GeO_x anode. **a** Initial profiles of the Li-compensated GeO_x/NCM full cell in comparison with those of the Li metal/NCM half-cell. **b** Reversible battery discharge capacity of NCM in the full cell [18] (Reprinted with permission from Wang et al. [18]. Copyright 2011, American Chemical Society)

2.5 Carbon-Coated Magnéli-Phase Ti_nO_{2n-1} Nanobelts with High Capacity

Although are not with high capacity, TiO_2 anodes have a redox potential of ~1.8 V, preventing the formation of Li dendrites. This example would mainly showcase how carbon-coating improves cell performance.

Different TiO_2 polymorphs have been studied as Li-ion battery anodes, including anatase, rutile, brookite, and bronze (TiO_2-B). However, TiO_2 phases are wide-band-gap nano-conducting materials whose cell performance is unfavorable.

The Magnéli-phase Ti_nO_{2n-1} ($n = 4$–10) is characteristic of two-dimensional chains of octahedral TiO_2 with every *n*th layer missing oxygen atoms. It possesses a high electronic-conductivity comparable to that of graphite [21]. We used $H_2Ti_3O_7$ nanobelts as templates to synthesize graphitic carbon-coated Magnéli-phase Ti_9O_{17} nanobelts [22].

We generated the $H_2Ti_3O_7$ nanobelts by treating anatase TiO_2 powder with NaOH in an autoclave at elevated temperatures of 150–180 °C. They were 30–200 nm wide, 15–40 nm thick, and up to 10 μm long. Then, the nanobelts were heated at 950 °C under a mixture gas of ethane (50 SCCM, SCCM denotes cubic centimeter per minute at STP) and hydrogen (100 SCCM). Their color turned black, indicating the formation of C-coated Ti_9O_{17} nanobelts. The carbon layer played a key role in restraining the sintering growth of the core during the phase transformation and in retaining the nanobelt morphology. Bare thick fibers of Ti_9O_{17}, several microns in size, were obtained using the same experimental conditions except for using only hydrogen as the reaction gas.

Figure 2.11a displays the XRD spectrum of these fibers. Its peaks denoted that the sample was triclinic Ti_9O_{17}. The SEM image (Fig. 2.11b) and the low-magnification TEM image (Fig. 2.11c) revealed that the samples were straight nanobelts with widths similar to that of the $H_2Ti_3O_7$ nanobelts. The HRTEM image (Fig. 2.11d) and the fast Fourier transform electron-diffraction pattern (Fig. 2.11e) exemplify part of a single-crystal nanobelt. The labeled distance corresponded to the (01-2) plane of

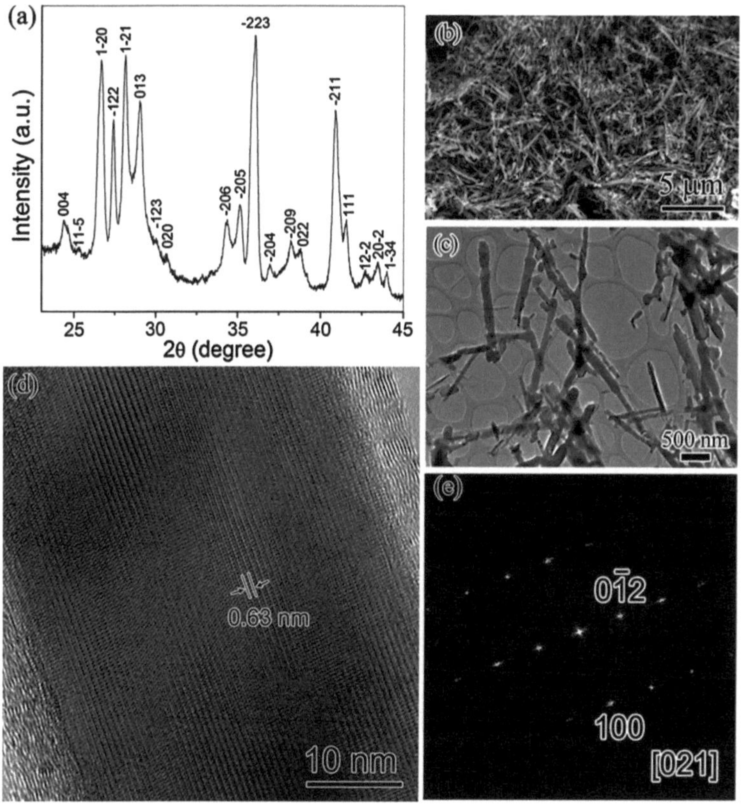

Fig. 2.11 Structure of graphitic carbon-coated Ti_9O_{17} nanobelts. **a** XRD spectrum; **b** SEM image; **c** a low-magnification TEM image; **d** a HRTEM image of a part of a nanobelt; and, **e** fast-Fourier transform electron-diffraction patterns of part of the Ti_9O_{17} nanobelt in **d** [22] (Reprinted with permission from Han and Wang [22]. Copyright 2010, American Institute of Physics)

Ti_9O_{17}. The carbon layer was 4.5 nm thick, and the carbon coating was ~16 wt%, determined by thermogravimetric analysis (TGA).

Bare Ti_9O_{17} fibers could realize only about 70 mAh g^{-1}, as shown in Fig. 2.12a. In contrast, carbon-coated nanobelts exhibited a much higher capacity that could stabilize at ~130 mAh g^{-1} after 50 cycles. Cyclic voltammetry (CV) measurements (Fig. 2.12b) showed a pair of major redox peaks at ~0.95 V versus Li. In Fig. 2.12c, the peak current scaled with the first power of the scan rate, indicative of a Faradaic pseudo-capacitive charging-behavior. This finding was a reasonable one since the triclinic phase lacked open structures in its lattices. The Faradiac reaction mainly occurred at the surfaces. The nanobelt morphology ensured that at least one of its dimensions was at the nanoscale, and thus enabled more efficient Li storage as compared with the bare fibers. The carbon coating also helped facilitate charge transportation and enhanced cell performance.

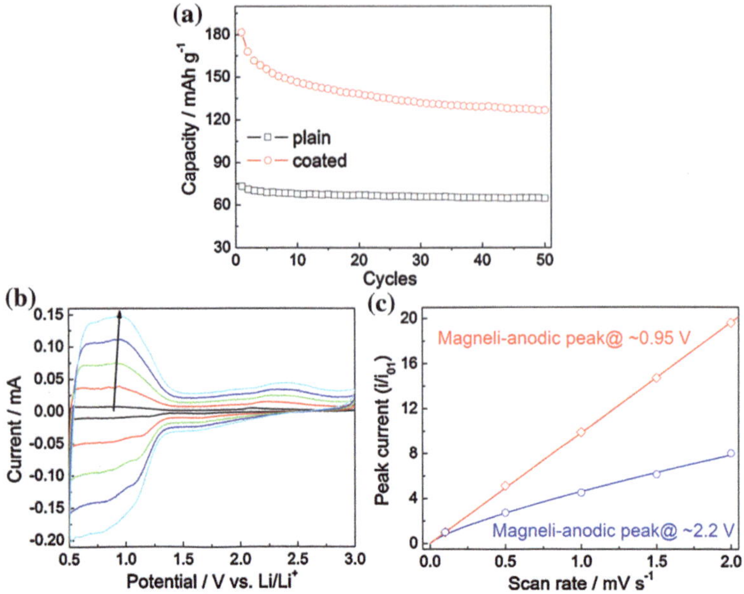

Fig. 2.12 Li-ion battery anode performance of the carbon-coated Ti_9O_{17} nanobelts. **a** Delithiation capacities of coin cells with the carbon-coated Ti_9O_{17} nanobelts as the working electrodes, and lithium metal as both the reference- and counter-electrodes. The cycling rate was C/10 based on theoretical capacity (167 mAh g^{-1}); **b** cyclic voltammograms (CV) of the carbon-coated Ti_9O_{17} nanobelts at scan rates of (along the arrows) 0.1-, 0.5-, 1-, 1.5-, and 2-mV s^{-1}; **c** normalized anodic peak-current with respect to the value at 0.1 mV s^{-1}, in which the lines are the fitted results [22] (Reprinted with permission from Han and Wang [22]. Copyright 2010, American Institute of Physics)

Hybrid electrochemical cells (HECs) with carbon-coated Ti_9O_{17} nanobelts as the anode-active material and activated carbon as the cathode-active material also were tested. HECs integrate the high energy-density of Li-ion battery anodes and the high power-density of electric-double-layer cathodes [23, 24]. As is shown in Fig. 2.13a, the distorted CV shape reflected Li redox reactions in the anode. The cell could have a specific cell-capacitance of 26 F g^{-1} at the scan rate of 5 mVs^{-1}. The Ragone plot (inset of Fig. 2.13a) implied that a power density of ~1,900 W kg^{-1} could be reached with an energy density of >10 Wh kg^{-1}.

The HECs also showed good cyclability. Figure 2.13b denotes that the energy- and power-densities were retained well after 450 cycles at 100 mV s^{-1}. The CV shapes were almost identical.

Their performance was better than that of carbon nanotube-carbon nanotube supercapacitors and comparable to that of the TiO_2-B nanowire-carbon nanotube system [23].

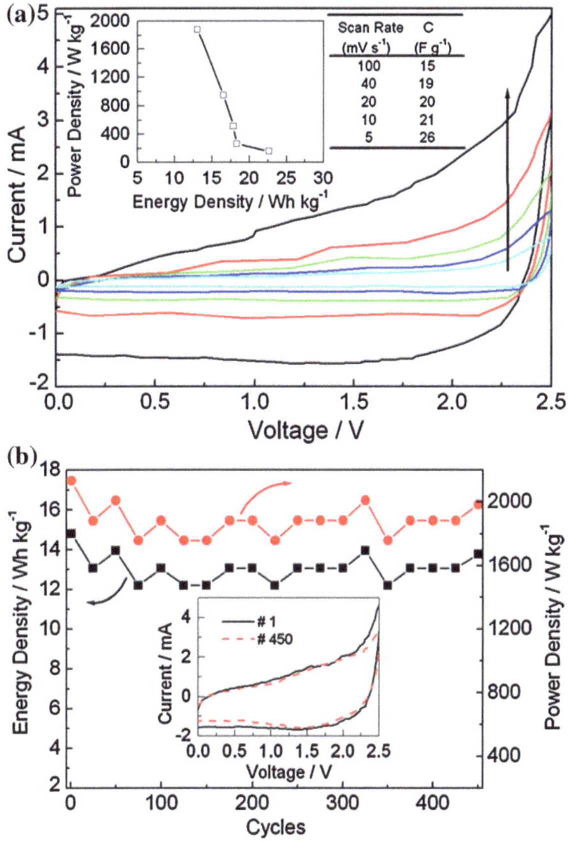

Fig. 2.13 Hybrid electrochemical cell (HEC) anode performance of the carbon-coated Ti_9O_{17} nanobelts. **a** Cyclic voltammograms of the Ti_9O_{17}/activated carbon-cell at scan rates of (along the *arrow*) 5-, 10-, 20-, 40-, and 100-mV s^{-1}, where the *inset* figure shows the Ragone plot of power density versus energy density, and the *inset* table shows the capacitance; **b** cycling performance of the cell at 100 mV s^{-1}, wherein the *inset* shows the CV profiles at the 1st and 450th cycles [22] (Reprinted with permission from Han and Wang [22]. Copyright 2010, American Institute of Physics)

2.6 Conclusions

We demonstrated the enhanced cell performance of porous single-crystal Si nanowires obtained by an electroless etching process. The accommodating porous structure and serviceable 1D structure were key factors. Graphene nanosheets, as the conductive additive, further increased the capacity of Si.

Moreover, a 3D GeO_x anode system realized a very long cycle life of 600 full cycles with high capacity of ~1,250 mAh g^{-1}. The excellent performance stemmed from several key characteristics: Small primary particles and the amorphous state the enhanced tolerance, its porous structure that was accommodating, and the incorporation of oxygen that served in buffering.

We also discussed graphitic carbon-coated Magnéli-Phase Ti_nO_{2n-1} nanobelts wherein the carbon layer was responsible for restraining the sintering growth of the nanobelts, boosting higher capacity and buffering the changes in volume.

Acknowledgments We thank our colleagues Drs. J. J. Chen, J. Graetz, M. Feygenson, M. C. Aronson, H. Y. Chen, C. H. Lin, W. Ku, J. M. Bai, T. A. Tyson, X. Q. Yu, X. J. Wang, and X. Q. Yang for their helpful advice. We also are grateful to Dr. H. Li (Institute of Physics, Chinese Academy of Sciences) and Drs. J. Hong, F. Wang, L. H. Zhang, L. J. Wu. and C. Ma (Brookhaven National Laboratory) for their technical support and valuable discussions.

References

1. Li, H., Huang, X.J., Chen, L.Q., Wu, Z.G., Liang, Y.: A high capacity nano-Si composite anode material for lithium rechargeable batteries. Electrochem. Solid State Lett. **2**, 547–549 (1999)
2. Wang, C.S., Wu, G.T., Zhang, X.B., Qi, Z.F., Li, W.Z.: Lithium insertion in carbon-silicon composite materials produced by mechanical milling. J. Electrochem. Soc. **145**, 2751–2758 (1998)
3. Kim, H., Seo, M., Park, M.H., Cho, J.: A critical size of silicon nano-anodes for lithium rechargeable batteries. Angew. Chem. Int. Ed. **49**, 2146–2149 (2010)
4. Wang, X.L., Feygenson, M., Aronson, M.C., Han, W.Q.: Sn/SnO_x core-shell nanospheres: synthesis, anode performance in Li ion batteries, and superconductivity. J. Phys. Chem. C **114**, 14697–14703 (2010)
5. Huang, J.Y., Zhong, L., Wang, C.M., Sullivan, J.P., Xu, W., Zhang, L.Q., Mao, S.X., Hudak, N.S., Liu, X.H., Subramanian, A., Fan, H.Y., Qi, L.A., Kushima, A., Li, J.: In situ observation of the electrochemical lithiation of a single SnO_2 Nanowire electrode. Science **330**, 1515–1520 (2010)
6. Chan, C.K., Peng, H.L., Liu, G., McIlwrath, K., Zhang, X.F., Huggins, R.A., Cui, Y.: High-performance lithium battery anodes using silicon nanowires. Nat. Nanotechnol. **3**, 31–35 (2008)
7. Graetz, J., Ahn, C.C., Yazami, R., Fultz, B.: Nanocrystalline and thin film germanium electrodes with high lithium capacity and high rate capabilities. J. Electrochem. Soc. **151**, A698–A702 (2004)
8. Magasinski, A., Dixon, P., Hertzberg, B., Kvit, A., Ayala, J., Yushin, G.: High-performance lithium-ion anodes using a hierarchical bottom-up approach. Nat. Mater. **9**, 353–358 (2010)
9. Fan, Q., Chupas, P.J., Whittingham, M.S.: Characterization of amorphous and crystalline tin-cobalt anodes. Electrochem. Solid State Lett. **10**, A274–A278 (2007)
10. Besenhard, J.O., Yang, J., Winter, M.: Will advanced lithium-alloy anodes have a chance in lithium-ion batteries? J. Power Sources **68**, 87–90 (1997)
11. Mao, O., Dunlap, R.A., Dahn, J.R.: Mechanically alloyed Sn-Fe(-C) powders as anode materials for Li-ion batteries-I. The Sn_2Fe-C system. J. Electrochem. Soc. **146**, 405–413 (1999)
12. Wang, X.L., Han, W.Q., Chen, J.J., Graetz, J.: Single-crystal intermetallic M-Sn (M = Fe, Cu, Co, Ni) nanospheres as negative electrodes for lithium-ion batteries. ACS Appl. Mater. Interfaces **2**, 1548–1551 (2010)
13. Wang, X.L., Feygenson, M., Chen, H.Y., Lin, C.H., Ku, W., Bai, J.M., Aronson, M.C., Tyson, T.A., Han, W.Q.: Nanospheres of a new intermetallic FeSn(5) phase: synthesis, magnetic properties and anode performance in Li-ion batteries. J. Am. Chem. Soc. **133**, 11213–11219 (2011)
14. Wang, X.L., Chen, H.Y., Bai, J.M., Han, W.Q.: CoSn5 phase: crystal structure resolving and stable high capacity as anodes for Li ion batteries. J. Phys. Chem. Lett. **3**, 1488–1492 (2012)

15. Wang, X.L., Han, W.Q.: Graphene enhances Li storage capacity of porous single-crystalline silicon nanowires. ACS Appl. Mater. Interfaces **2**, 3709–3713 (2010)
16. Huang, R., Fan, X., Shen, W.C., Zhu, J.: Carbon-coated silicon nanowire array films for high-performance lithium-ion battery anodes. Appl. Phys. Lett. **95**, 133119 (2009)
17. Obrovac, M.N., Krause, L.J.: Reversible cycling of crystalline silicon powder. J. Electrochem. Soc. **154**, A103–A108 (2007)
18. Wang, X.L., Han, W.Q., Chen, H.Y., Bai, J.M., Tyson, T.A., Yu, X.Q., Wang, X.J., Yang, X.Q.: Amorphous hierarchical porous GeO_x as high-capacity anodes for Li ion batteries with very long cycling life. J. Am. Chem. Soc. **133**, 20692–20695 (2011)
19. Yoon, S., Park, C.M., Sohn, H.J.: Electrochemical characterizations of germanium and carbon-coated germanium composite anode for lithium-ion batteries. Electrochem. Solid State Lett. **11**, A42–A45 (2008)
20. Kim, H., Cho, J.: Superior lithium electroactive mesoporous Si@carbon core-shell nanowires for lithium battery anode material. Nano Lett. **8**, 3688–3691 (2008)
21. Han, W.Q., Zhang, Y.: Magnéli phases Ti_nO_{2n-1} nanowires: formation, optical, and transport properties. Appl. Phys. Lett. **92**, 203117 (2008)
22. Han, W.Q., Wang, X.L.: Carbon-coated magnéli-phase Ti(n)O(2n−1) nanobelts as anodes for Li-ion batteries and hybrid electrochemical cells. Appl. Phys. Lett. **97**, 243104 (2010)
23. Wang, Q., Wen, Z.H., Li, J.H.: A hybrid supercapacitor fabricated with a carbon nanotube cathode and a TiO_2-B nanowire anode. Adv. Funct. Mater. **16**, 2141–2146 (2006)
24. Wang, D.W., Fang, H.T., Li, F., Chen, Z.G., Zhong, Q.S., Lu, G.Q., Cheng, H.M.: Aligned titania nanotubes as an intercalation anode material for hybrid electrochemical energy storage. Adv. Funct. Mater. **18**, 3787–3793 (2008)

Chapter 3
Light Trapping in Coaxial Nanowires of c-Si Cores and a-Si Shells

Jeong Il Oh, Wenfu Liu, Weiqiang Xie and Wenzhong Shen

Abstract Light absorption is investigated in coaxial nanowires (NWs) of crystalline silicon (c-Si) cores and amorphous silicon (a-Si) shells, including both cases of single coaxial NWs and coaxial NW arrays, for an incident light spectrum of 1.0–4.0 eV covering the major solar band for photovoltaic cells. Based on the Lorenz-Mie light scattering theory for the single coaxial NWs and the rigorous coupled-wave analysis method for the coaxial NW arrays, it is found that the incident light is effectively trapped in the coaxial NWs through absorption resonances so that the light absorption of the coaxial NWs can be significantly enhanced compared to that of c-Si NWs. In the coaxial NWs, the absorption resonances occur due to their subwavelength dimensions, as in the c-Si NWs, whereas the absorption enhancement originates from the a-Si shells. By tuning their structural parameters, the light absorption in the coaxial NWs can be readily optimized for photovoltaic applications. At the optimal absorption conditions, the photocurrent in the coaxial NWs can be enhanced up to 560 % (single case) and 14 % (array case) compared to that in the c-Si NWs. The underlying physics of the light absorption in the coaxial NWs is discussed in terms of the excitation of leaky-mode resonances. The practical use of the coaxial NWs for photovoltaic cells is also addressed.

3.1 Introduction

Light trapping is a very important technology to increase the light absorption of and thereby to enhance the power conversion efficiency of photovoltaic (PV) cells [1–35]. Surface texturing combined with front-surface antireflection coating and rear-surface reflector is conventionally used to elongate the optical path length of the incident light

J. I. Oh · W. F. Liu · W. Q. Xie · W. Z. Shen (✉)
Laboratory of Condensed Matter Spectroscopy and Opto-Electronic Physics,
Key Laboratory of Artificial Structures and Quantum Control (Ministry of Education),
Department of Physics and Institute of Solar Energy, Shanghai Jiao Tong University,
800 Dong Chuan Road, Shanghai 200240 , China
e-mail: wzshen@sjtu.edu.cn

H. Li et al. (eds.), *Silicon-based Nanomaterials*,
Springer Series in Materials Science 187, DOI: 10.1007/978-1-4614-8169-0_3,

© Springer Science+Business Media New York 2013

in commercial PV cells [2], whereas light interaction with subwavelength-scaled matters is becoming a new approach to effective light trapping for next generation PV cells, including plasmonic PV cells [3–8], thin-film PV cells with submicron-sized light scatterers [9], and NW-based PV cells [10–35]. Among these, the latter PV cells, especially based on c-Si NWs [15–35], have recently drawn considerable attention because of their great potential to reduce the cost of the power generation mainly owing to the new light-trapping scheme.

Sunlight can be effectively trapped in c-Si NWs mainly through absorption resonances, leading to significant absorption enhancement compared to the bulk counterparts. The absorption resonances originate from the interaction between the incident sunlight and subwavelength-scaled c-Si NWs. [32–34] The light absorption in c-Si NWs can exceed the conventional Lambertian limit of light trapping [35–38] so that the light-trapping scheme in c-Si NWs is particularly important in designing next generation PV cells based on Si nanomaterials. The light absorption in c-Si NWs can be further enhanced by coating highly absorptive a-Si shells. [29] These hybrid NWs of coaxial structure, comprising c-Si cores and a-Si shells, would be very attractive since they could take advantage of superior light absorption property of a-Si [39] while retaining long charge carrier diffusion lengths of c-Si ($>200\,\mu m$) [40]. More specifically, in PV cells made of such coaxial NWs of c-Si cores and a-Si shells, the Si cores could serve as an efficient charge collector [41] to compensate for short diffusion lengths of a-Si ($\sim 100\,nm$) [40], and the a-Si shells could be used not only as an excellent surface passivator [42] but also as an excellent energy absorber to overcome rather poor light absorption of c-Si [39].

In this chapter, the light absorption and its underlying mechanism in the coaxial NWs of c-Si cores and a-Si shells are investigated for both cases of single coaxial NWs and coaxial NW arrays, based on the Lorenz-Mie light scattering theory [43] and the rigorous coupled-wave analysis method [44–46], respectively. Three important findings of this investigation are: (1) The excitation of LMRs governs the light absorption in the coaxial NWs; (2) the significant absorption enhancement in the coaxial NWs, as compared to that in the c-Si NWs, stems from the a-Si shells; and (3) the dimensions of the coaxial NWs can be optimally tuned for PV applications. This chapter will first describe the calculation methods of the light absorption in the coaxial NWs in the next section, extensively deal with the light trapping mechanism in the coaxial NWs and its potential uses in PV applications in the following two sections, and close with the summary and conclusions at the last section.

3.2 Light Absorption Calculation Methods

3.2.1 Lorentz-Mie Scattering Theory

The light absorption efficiency of single NWs can be calculated as follows: Within the framework of the Lorentz-Mie light scattering theory, where single NWs are treated as infinitely long cylinders, normally illuminated by a plane wave of the incident propagation vector k_0 [43]. For unpolarized light, the light absorption efficiency Q_{abs}, defined as the ratio of the absorption cross section to the geometrical cross section of NWs, can be expressed as

$$Q_{abs} = (Q_{abs}^{TE} + Q_{abs}^{TM})/2, \tag{3.1}$$

where Q_{abs}^{TE} and Q_{abs}^{TM} are the light absorption efficiencies for transverse-electric (TE, electric field perpendicular to the NW axis as in Fig. 3.1a) and transverse-magnetic (TM, electric field parallel to the NW axis as in Fig. 3.1a) illuminations, respectively. These are decomposed into

$$Q_{abs}^{TE} = Q_{ext}^{TE} - Q_{sca}^{TE}, \quad Q_{abs}^{TM} = Q_{ext}^{TM} - Q_{sca}^{TM}, \tag{3.2}$$

where $Q_{ext}^{TE}(Q_{sca}^{TE})$ and $Q_{ext}^{TM}(Q_{sca}^{TM})$ are the extinction (scattering) efficiencies for TE and TM illuminations, respectively. These efficiencies are given by

$$Q_{ext}^{TE} = \frac{2}{k_0 r}\mathrm{Re}\left\{\sum_{n=-\infty}^{\infty} a_n\right\}, \quad Q_{ext}^{TM} = \frac{2}{k_0 r}\mathrm{Re}\left\{\sum_{n=-\infty}^{\infty} b_n\right\}, \tag{3.3a}$$

$$Q_{sca}^{TE} = \frac{2}{k_0 r}\left\{\sum_{n=-\infty}^{\infty} |a_n|^2\right\}, \quad Q_{sca}^{TM} = \frac{2}{k_0 r}\left\{\sum_{n=-\infty}^{\infty} |b_n|^2\right\}, \tag{3.3b}$$

where r is the radial dimension of the absorption part of a NW. In the case of a coaxial NW of Si core and SiO_2 shell (a case of Fig. 3.1b) under a major solar illumination (1.0–4.0 eV), for instance, r is r_2 but not r_1, since the bandgap of SiO_2 is over 4.0 eV so that the SiO_2 shell acts as a nonabsorbing medium in such a coaxial NW [13]. a_n and b_n can be readily obtained by solving Maxwell's equations with the appropriate boundary conditions at such interfaces as air/NW (Fig. 3.1a), air/shell and shell/core (Fig. 3.1b), and air/outer shell, outer shell/inner shell, and inner shell/core (Fig. 3.1c). The coefficients a_n and b_n can be written as

$$a_n = a_{n0}/\Delta_n^{TE}, \quad b_n = b_{n0}/\Delta_n^{TM}. \tag{3.4}$$

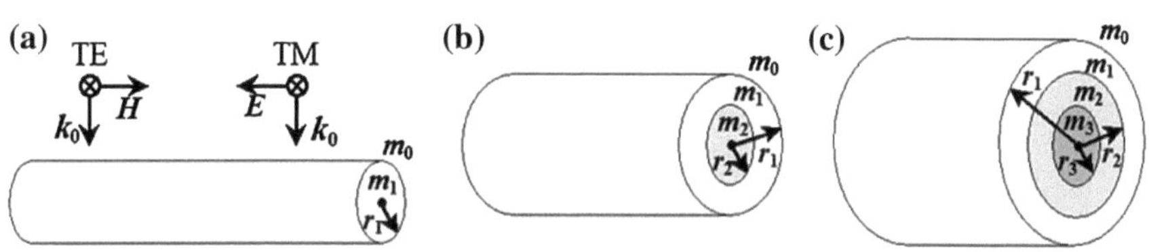

Fig. 3.1 Schematic diagrams of cylindrical NWs. Incident light polarizations with respect to the NW axis (TE and TM), radii (r), and complex refractive indices (m) are indicated. **a** Single material NW of r_1 and m_1. **b** Coaxial NW of core (r_2 and m_2) and shell ($r_1 - r_2$ and m_1). **c** Coaxial NW of core (r_3 and m_3), inner shell ($r_2 - r_3$ and m_2), and outer shell ($r_1 - r_2$ and m_1)

For single material NWs as in Fig. 3.1a,

$$a_{n0} = \begin{vmatrix} J_n'(k_0 r_1) & J_n'(k_1 r_1) \\ m_0 J_n(k_0 r_1) & m_1 J_n(k_1 r_1) \end{vmatrix}, \tag{3.5a}$$

$$\Delta_n^{\mathrm{TE}} = \begin{vmatrix} H_n'(k_0 r_1) & J_n'(k_1 r_1) \\ m_0 H_n(k_0 r_1) & m_1 J_n(k_1 r_1) \end{vmatrix}, \tag{3.5b}$$

$$b_{n0} = \begin{vmatrix} J_n(k_0 r_1) & J_n(k_1 r_1) \\ m_0 J_n'(k_0 r_1) & m_1 J_n'(k_1 r_1) \end{vmatrix}, \tag{3.5c}$$

$$\Delta_n^{\mathrm{TM}} = \begin{vmatrix} H_n(k_0 r_1) & J_n(k_1 r_1) \\ m_0 H_n'(k_0 r_1) & m_1 J_n'(k_1 r_1) \end{vmatrix}. \tag{3.5d}$$

For coaxial NWs of core/shell structure as in Fig. 3.1b,

$$a_{n0} = \begin{vmatrix} 0 & J_n'(k_1 r_2) & H_n'(k_2 r_2) & -J_n'(k_2 r_2) \\ 0 & m_1 J_n(k_1 r_2) & m_1 H_n(k_2 r_2) & -m_2 J_n(k_2 r_2) \\ J_n'(k_0 r_1) & J_n'(k_1 r_1) & H_n'(k_1 r_1) & 0 \\ m_0 J_n(k_0 r_1) & m_1 J_n(k_1 r_1) & m_1 H_n(k_1 r_1) & 0 \end{vmatrix}, \tag{3.6a}$$

$$\Delta_n^{\mathrm{TE}} = \begin{vmatrix} 0 & J_n'(k_1 r_2) & H_n'(k_1 r_2) & -J_n'(k_2 r_2) \\ 0 & m_1 J_n(k_1 r_2) & m_1 H_n(k_1 r_2) & -m_2 J_n(k_2 r_2) \\ H_n'(k_0 r_1) & J_n'(k_1 r_1) & H_n'(k_1 r_1) & 0 \\ m_0 H_n(k_0 r_1) & m_1 J_n(k_1 r_1) & m_1 H_n(k_1 r_1) & 0 \end{vmatrix}, \tag{3.6b}$$

$$b_{n0} = \begin{vmatrix} 0 & J_n(k_1 r_2) & H_n(k_2 r_2) & -J_n(k_2 r_2) \\ 0 & m_1 J_n'(k_1 r_2) & m_1 H_n'(k_2 r_2) & -m_2 J_n'(k_2 r_2) \\ J_n(k_0 r_1) & J_n(k_1 r_1) & H_n(k_1 r_1) & 0 \\ m_0 J_n'(k_0 r_1) & m_1 J_n'(k_1 r_1) & m_1 H_n'(k_1 r_1) & 0 \end{vmatrix}, \tag{3.6c}$$

$$\Delta_n^{\mathrm{TM}} = \begin{vmatrix} 0 & J_n(k_1 r_2) & H_n(k_1 r_2) & -J_n(k_2 r_2) \\ 0 & m_1 J_n'(k_1 r_2) & m_1 H_n'(k_1 r_2) & -m_2 J_n'(k_2 r_2) \\ H_n(k_0 r_1) & J_n(k_1 r_1) & H_n(k_1 r_1) & 0 \\ m_0 H_n'(k_0 r_1) & m_1 J_n'(k_1 r_1) & m_1 H_n'(k_1 r_1) & 0 \end{vmatrix}. \tag{3.6d}$$

For coaxial NWs of double shell structure as in Fig. 3.1c,

$$a_{n0} = \begin{vmatrix} 0 & 0 & 0 & -J_n'(k_2r_3) & -H_n'(k_2r_3) & J_n'(k_3r_3) \\ 0 & 0 & 0 & -m_2J_n(k_2r_3) & -m_2H_n(k_2r_3) & m_3J_n(k_3r_3) \\ 0 & J_n'(k_1r_2) & H_n'(k_1r_2) & -J_n'(k_2r_2) & -H_n'(k_2r_2) & 0 \\ 0 & m_1J_n(k_1r_2) & m_1H_n(k_1r_2) & -m_2J_n(k_2r_2) & -m_2H_n(k_2r_2) & 0 \\ J_n'(k_0r_1) & J_n'(k_1r_1) & H_n'(k_1r_1) & 0 & 0 & 0 \\ m_0J_n(k_0r_1) & m_1J_n(k_1r_1) & m_1H_n(k_1r_1) & 0 & 0 & 0 \end{vmatrix}, \tag{3.7a}$$

$$\Delta_n^{\mathrm{TE}} = \begin{vmatrix} 0 & 0 & 0 & -J_n'(k_2r_3) & -H_n'(k_2r_3) & J_n'(k_3r_3) \\ 0 & 0 & 0 & -m_2J_n(k_2r_3) & -m_2H_n(k_2r_3) & m_3J_n(k_3r_3) \\ 0 & J_n'(k_1r_2) & H_n'(k_1r_2) & -J_n'(k_2r_2) & -H_n'(k_2r_2) & 0 \\ 0 & m_1J_n(k_1r_2) & m_1H_n(k_1r_2) & -m_2J_n(k_2r_2) & -m_2H_n(k_2r_2) & 0 \\ H_n'(k_0r_1) & J_n'(k_1r_1) & H_n'(k_1r_1) & 0 & 0 & 0 \\ m_0H_n(k_0r_1) & m_1J_n(k_1r_1) & m_1H_n(k_1r_1) & 0 & 0 & 0 \end{vmatrix}, \tag{3.7b}$$

$$b_{n0} = \begin{vmatrix} 0 & 0 & 0 & -J_n(k_2r_3) & -H_n(k_2r_3) & J_n(k_3r_3) \\ 0 & 0 & 0 & -m_2J_n'(k_2r_3) & -m_2H_n'(k_2r_3) & m_3J_n'(k_3r_3) \\ 0 & J_n(k_1r_2) & H_n(k_1r_2) & -J_n(k_2r_2) & -H_n(k_2r_2) & 0 \\ 0 & m_1J_n'(k_1r_2) & m_1H_n'(k_1r_2) & -m_2J_n'(k_2r_2) & -m_2H_n'(k_2r_2) & 0 \\ J_n(k_0r_1) & J_n(k_1r_1) & H_n(k_1r_1) & 0 & 0 & 0 \\ m_0J_n'(k_0r_1) & m_1J_n'(k_1r_1) & m_1H_n'(k_1r_1) & 0 & 0 & 0 \end{vmatrix}, \tag{3.7c}$$

$$\Delta_n^{\mathrm{TM}} = \begin{vmatrix} 0 & 0 & 0 & -J_n(k_2r_3) & -H_n(k_2r_3) & J_n(k_3r_3) \\ 0 & 0 & 0 & -m_2J_n'(k_2r_3) & -m_2H_n'(k_2r_3) & m_3J_n'(k_3r_3) \\ 0 & J_n(k_1r_2) & H_n(k_1r_2) & -J_n(k_2r_2) & -H_n(k_2r_2) & 0 \\ 0 & m_1J_n'(k_1r_2) & m_1H_n'(k_1r_2) & -m_2J_n'(k_2r_2) & -m_2H_n'(k_2r_2) & 0 \\ H_n(k_0r_1) & J_n(k_1r_1) & H_n(k_1r_1) & 0 & 0 & 0 \\ m_0H_n'(k_0r_1) & m_1J_n'(k_1r_1) & m_1H_n'(k_1r_1) & 0 & 0 & 0 \end{vmatrix}, \tag{3.7d}$$

where J_n is the nth-order Bessel function of the first kind, and H_n is the nth-order Hankel function of the first kind. The prime superscript stands for differentiation with respect to the propagation vector $k_i = m_i k_0$ ($i = 0$, 1, 2, and 3) for the air environment ($m_0 = 1$). It is worth noting that the optical properties of single NWs retain qualitatively same for a wide range of the angle of incident light so that the angle of incident does not play an important role in designing NW-based optical devices [32, 47].

3.2.2 Rigorous Coupled-Wave Analysis

The light absorption of coaxial NW arrays can be calculated by the rigorous coupled-wave analysis (RCWA) method [44–46]. It has been initially developed for calculating the diffraction orders of one-dimensional (1D) gratings by solving Maxwell's

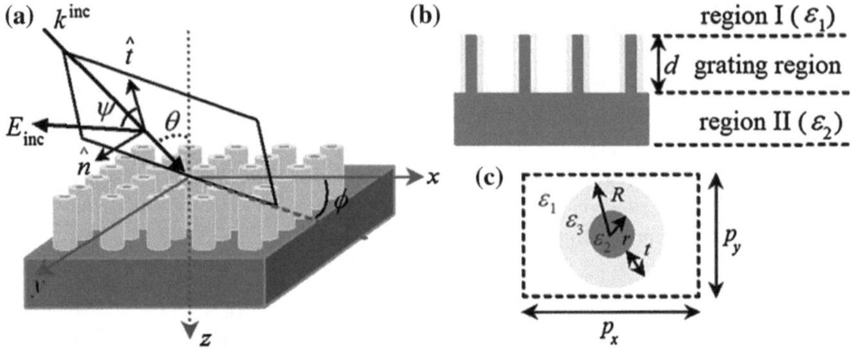

Fig. 3.2 Schematic illustrations of a 2D grating, comprising a rectangular array of coaxial pillars with periodicities p_x and p_y. **a** Incident light (k^{inc}) and electric field (E_{inc}) are shown with respect to the grating. **b** *Side view* of the grating, showing three regions of the grating. **c** *Top view* of the unit cell of the grating

equations and later extended to two-dimensional (2D) periodic structures with sub-wavelength scale such as surface-relief gratings [46] and periodic nanostructures [48]. As a frequency-domain simulation, it can readily incorporate material dispersion, thereby offering reflectance (*R*), transmittance (*T*), and absorptance (*A*) for investigating the optical properties of subject materials. The RCWA approach to calculating the light absorption of coaxial NW arrays can be programmed as described in the following procedures.

Figure 3.2 illustrates a 2D NW pillar grating, consisting of three regions: The incident region or reflected region with the relative permittivity ε_1 (Region I), the transmitted region with ε_2 (Region II), and the grating region of a coaxial NW array with three relative permittivities whose geometric parameters are shown in Figs. 3.2b and 3.2c. A linearly polarized plane wave (k^{inc}) at an arbitrary angle of incidence (θ, ϕ) is considered. The angle between the electric field (E_{inc}) and the plane of incidence is denoted by ψ so that $\psi = 0°$ and $90°$ correspond to TE and TM polarizations, respectively. The periodic relative permittivity $\varepsilon(x, y)$ in the grating region $(0 < z < d)$ can be expanded in a Fourier series as

$$\varepsilon(x, y) = \varepsilon(x + p_x, y + p_y) = \sum_{m,n} \varepsilon_{mn} \exp\left(j\frac{2\pi mx}{p_x} + j\frac{2\pi ny}{p_y}\right), \tag{3.8}$$

where ε_{mn} is the (*m*,*n*)th Fourier coefficient, which is determined only by the geometry of the grating as shown in Fig. 3.2c. The electric-field vectors in the regions I and II are given by

$$E_{\mathrm{I}} = E_{\mathrm{inc}} + \sum_{m,n} R_{mn} \exp[-j(k_{x,m}x + k_{y,n}y - k^{\mathrm{I}}_{z,mn}z)], \tag{3.9a}$$

$$E_{\mathrm{II}} = \sum_{m,n} T_{mn} \exp\{-j[k_{x,m}x + k_{y,n}y + k_{z,mn}^{\mathrm{II}}(z-d)]\}, \tag{3.9b}$$

where E_{inc} is the normalized incident electric-field vector (i.e., $|E_{\mathrm{inc}}|^2 = 1$) and $r_{mn}(T_{mn})$ is the normalized amplitude of electric-field vector in the (m,n)th reflected (transmitted) diffraction order. In Equations 3.9a and 3.9b, the components of wave vector are determined from the Floquet condition as

$$k_{x,m} = k_x^{\mathrm{inc}} - 2\pi m/p_x, \tag{3.10a}$$

$$k_{y,n} = k_y^{\mathrm{inc}} - 2\pi n/p_y, \tag{3.10b}$$

$$k_{z,mn}^{i} = \begin{cases} \sqrt{k_0^2\varepsilon_i - (k_{x,m}^2 + k_{y,n}^2)} & (k_{x,m}^2 + k_{y,n}^2 < k_0^2\varepsilon_i) \\ -j\sqrt{(k_{x,m}^2 + k_{y,n}^2) - k_0^2\varepsilon_i} & (k_{x,m}^2 + k_{y,n}^2 > k_0^2\varepsilon_i) \end{cases} \quad i = \mathrm{I} \text{ and } \mathrm{II}, \tag{3.10c}$$

where k^{inc} is the wave vector of the incident light and $k_0 = 2\pi/\lambda_0$ with λ_0 the wavelength of the incident light in free space. The electric- and magnetic-field vectors (E_g and H_g) in the grating region can be expressed in terms of the space-harmonic fields as

$$E_g = \sum_{m,n} S_{mn} \exp[-j(k_{x,m}x + k_{y,n}y)], \tag{3.11a}$$

$$H_g = -j\left(\frac{\varepsilon_0}{\mu_0}\right)^{1/2} \sum_{m,n} U_{mn} \exp[-j(k_{x,m}x + k_{y,n}y)], \tag{3.11b}$$

where S_{mn} and U_{mn} are the normalized amplitudes of the (m,n)th space-harmonic fields, ε_0 is the permittivity of free space, and μ_0 is the permeability of free space. E_g and H_g satisfy Maxwell's curl equations in the source-free space as

$$\nabla \times E_g = -j\omega\mu_0 H_g, \quad \nabla \times H_g = j\omega\varepsilon_0\varepsilon E_g. \tag{3.12}$$

Inserting Equations 3.11a and 3.11b into 3.12, one can obtain a set of coupled-wave equations, which can be readily solved by applying the boundary conditions for electric- and magnetic-field components. The reflected and transmitted efficiencies for the (m,n)th diffraction order are defined as

$$\mathrm{DE}_{mn}^{R} = |R_{mn}|^2 \mathrm{Re}\left(\frac{k_{z,mn}^{\mathrm{I}}}{k_z^{\mathrm{inc}}}\right), \quad \mathrm{DE}_{mn}^{T} = |T_{mn}|^2 \mathrm{Re}\left(\frac{k_{z,mn}^{\mathrm{II}}}{k_z^{\mathrm{inc}}}\right) \tag{3.13}$$

The total reflectance (R) and transmittance (T) are obtained by summing all diffraction orders as

$$R = \sum_{m,n} \mathrm{DE}_{mn}^{R}, \quad T = \sum_{m,n} \mathrm{DE}_{mn}^{T} \tag{3.14}$$

For lossy dielectric materials, the total absorptance (A) is simply calculated by $A = 1 - R - T$. The accuracy of the RCWA simulation depends on the number of the space-harmonic fields retained in Equations 3.11a and 3.11b. It is worth noting that although it is a powerful simulation tool for periodic subwavelength structures, the RCWA method is also well applied to capturing the main optical properties of aperiodic structures by modeling the structures appropriately [49, 50].

3.3 Light Trapping in Single Coaxial Nanowires

The light absorption in single coaxial NWs of c-Si cores and a-Si shells can be well described within the framework of the Lorenz-Mie light scattering theory [43]. The term ‘single’ will be omitted hereafter if there is no confusion in this section. Such a hybrid system, which is a case of Fig. 3.1b, is simply characterized by the core radius r of c-Si, the shell thickness t of a-Si, and the total radius $R = r + t$. The cross section of the coaxial NWs is illustrated in Fig. 3.3a, together with the incident light perpendicular to the NW axis (thick arrows). The light absorption efficiency Q_{abs}, which is a measure of the capability to absorb light, can be obtained from equation 3.6

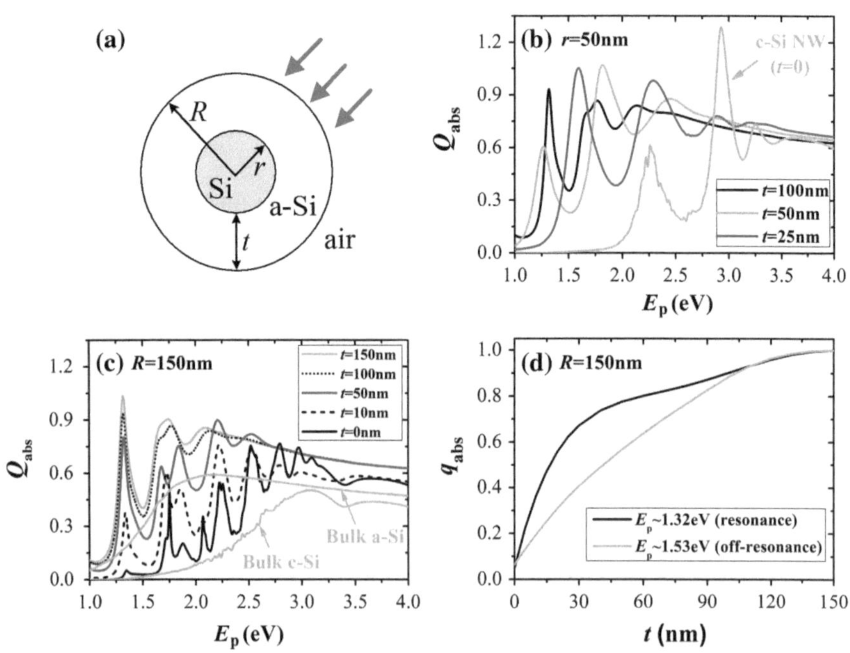

Fig. 3.3 **a** A schematic diagram of the cross section of coaxial NWs. *Thick arrows* indicate the incident light. **b** Q_{abs} versus E_{p} for various t at $r = 50$ nm. **c** Q_{abs} versus E_{p} for various t at $R = 150$ nm. Q_{abs} for bulk Si and bulk a-Si is shown for comparison. **d** Q_{abs} versus t at $E_{\mathrm{p}} \sim 1.32$ and 1.53 eV. (Reproduced with permission from Liu et al. [53]. © 2011 IOP Publishing Ltd.)

for the coaxial NWs. For Q_{ab} calculations, the complex refractive indices of c-Si (m_2) and a-Si (m_1) were taken from [39].

The absorption efficiency Q_{abs} is presented versus photon energy E_p in Fig. 3.3b, in coaxial NWs of a fixed core radius ($r = 50$ nm) for various shell thicknesses ($t = 0$, 25, 50, and 100 nm) in air ($m_0 = 1$). Regardless of the shell thicknesses, there occur distinct absorption resonances even for c-Si NWs ($t = 0$), indicating that they do not simply come from the hybrid nature of coaxial NWs. These absorption resonances are in fact an interesting property of NWs, so-called leaky-mode resonances (LMRs) that can arise when light is incident onto subwavelength structures of high refractive indices, as will be discussed in detail later. In the figure, it has to be emphasized that the coaxial NWs ($t \neq 0$) have significantly enhanced light absorption for the weak absorption region (E_p <3.0 eV) of the c-Si NWs ($t = 0$), likely due to the a-Si shells.

The absorption efficiency Q_{abs} is also presented versus photon energy E_p in Fig. 3.3c, in coaxial NWs of a fixed total radius ($R = 150$ nm) for various shell thicknesses ($t = 0$, 10, 50, 100, and 150 nm) in air, together with those in bulk c-Si and a-Si of 300 nm (=2R) thickness for comparison. Note that $t = 0$ and 150 nm represent c-Si and a-Si NWs of the radius 150 nm, respectively. The light absorption in the c-Si and a-Si NWs, as compared to in their bulk counterparts, is significantly enhanced for the entire solar band investigated ($E_p = 1.0 - 4.0$ eV) and, more importantly, this absorption enhancement is more significant for the weak absorption solar bands of the bulk materials ($E_p < 3.0$ eV for c-Si and <1.8 eV for a-Si) mainly owing to the occurrence of absorption resonances in these NWs. As the a-Si shell thickness increases, on the other hand, the overall light absorption in the coaxial NWs is quickly enhanced and becomes rather saturated for $t > 100$ nm. Such saturating behavior can be clearly seen from Fig. 3.3d, which displays a scaled absorption efficiency $Q_{abs} = Q_{abs}(t)/Q_{abs}(t = 150\,\text{nm})$ at two representative photon energies, i.e., at $E_p \sim 1.53$ eV (off-resonance) and ~1.32 eV (resonance). Note that $Q_{abs}(t = 150\,\text{nm})$ is the absorption efficiency of the a-Si NWs of the radius 150 nm. The absorption efficiency of the coaxial NWs reaches ~90 % of that of the a-Si NWs (90.1 % at $E_p \sim 1.32$ eV and 88.3 % at ~1.53 eV) at $t = 100$ nm (=2r). It is important to address that the light absorption of coaxial NWs can further increase up to additional ~10 % for $t > 100$ nm, however, the performance of PV cells made of such coaxial NWs may be degraded due to the poor charge collection of thick a-Si shells: charge carrier diffusion lengths of a-Si ~100 nm. Thus, the saturating behavior of the absorption efficiency may help in designing coaxial NWs-based PV cells of enhanced light absorption (owing to the a-Si shells) with still high charge collection (owing to the c-Si cores).

The absorption resonance behavior of coaxial NWs presented above can be understood in terms of LMRs that may occur in subwavelength structures of high refractive indices such as semiconductor NWs [13, 32] and nanospheres [51]. For high-refractive index NWs, the incident light can be trapped in the NWs by multiple internal reflections, between which constructive interference arises and leads to LMRs when the wavelength λ of the incident light matches up with one of LMRs supported by the NWs. LMRs can be expressed as TM_{ml} or TE_{ml}, where m and l are the azimuthal mode number and the radial order of LMRs, respectively. As in

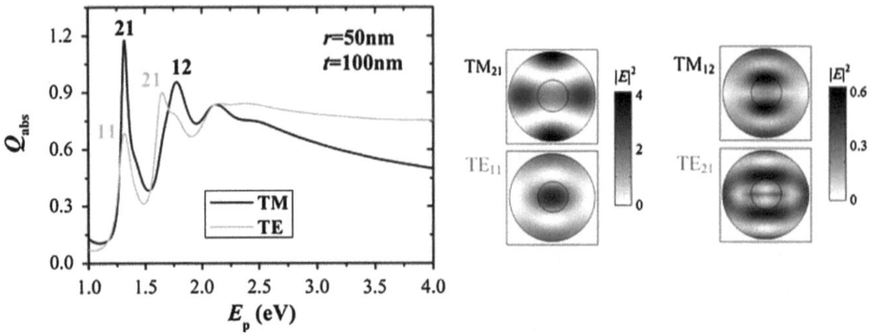

Fig. 3.4 (*Left*) Q_{abs} versus E_p at $r = 50$ nm and $t = 100$ nm for TM- and TE-polarized illuminations. (*Right*) Distributions of the electric field intensity $|E|^2$ (normalized to the incident light) inside the coaxial NWs are shown for the LMR peaks indicated. (Reproduced with permission from Liu et al. [53]. © 2011 IOP Publishing Ltd.)

Fig. 3.4, showing the absorption efficiency Q_{abs} for coaxial NWs of $r = 50$ nm and $t = 100$ nm under TM-/TE-polarized illumination, each absorption resonance corresponds to such a specific LMR as TM_{12}/TE_{21} (nondegenerate) near $E_p \sim 1.70$ eV ($\lambda \sim 730$ nm) and TM_{21}/TE_{11} (exact degenerate) at $E_p \sim 1.32$ eV ($\lambda \sim 940$ nm). Fig. 3.4 also shows the cross-sectional distributions of the electric field intensity $|E|^2$ inside the coaxial NWs at the resonance peaks. The amplitudes of $|E|^2$ are clearly stronger in the a-Si shells than in the c-Si cores. Thus, Fig. 3.4 provides compelling evidence that the incident light can be effectively trapped in the coaxial NWs through LMRs and also that the light absorption enhancement in the coaxial NWs arises owing to the a-Si shells.

The light absorption in coaxial NWs can be better seen from two-dimensional (2D) Q_{abs} maps, as plotted in Fig. 3.5, versus R and E_p at a fixed $g = 2$ for TM-/TE-polarized illumination, where g is defined as the ratio of the a-Si shell thickness to the c-Si core radius, i.e., $g = t/r$. Each absorption resonance corresponds to a LMR, as labeled in the figure. These 2D Q_{abs} maps provide extensive tunability of the light absorption in coaxial NWs. As the total radius R increases, there occur more LMRs and the position of each mode is shift to lower energy, indicating a redshift. At a given R, the coaxial NWs show relatively high absorption at a high photon energy region ($E_p > E_C$), e.g., $E_C \sim 1.32$eV at $R = 150$ nm, as indicated with an arrow in Fig. 3.5a. Such high absorption has also been reported in similar yet arrayed coaxial NWs [29]. Importantly, it is evident from the figure that the light absorption can be particularly high if the total radius R is tuned such that the light absorption spectrum contains TM_{01} (the strongest LMR) or TM_{11} (the second strongest LMR) mode, respectively.

The tunability of the light absorption gives a useful hint to design high-performance PV cells based on coaxial NWs. In order to investigate a potential benefit of such PV cells, one can calculate the photocurrent density or short-circuit current density J_{SC} in coaxial NWs as

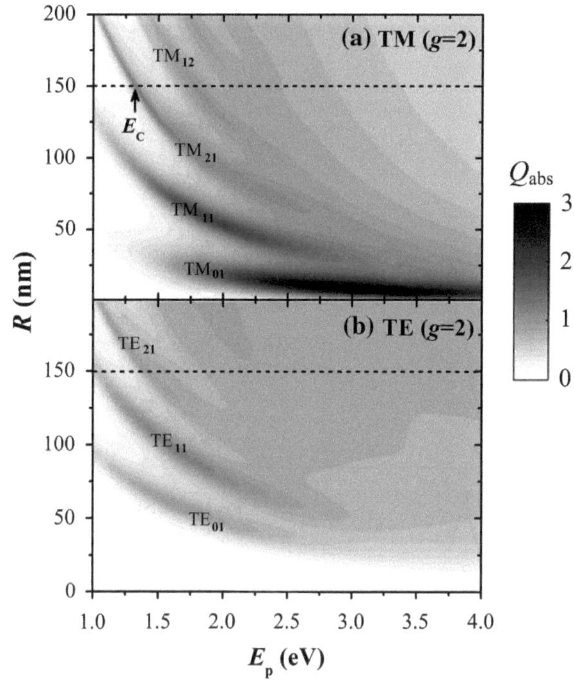

Fig. 3.5 Q_{abs} versus R and E_{p} at $g = 2$ for **a** TM- and **b** TE-polarized illuminations. The *dashed lines* represent the results shown in Fig. 3.4. LMRs are indicated. See text for E_c. (Reproduced with permission from Liu et al. [53]. © 2011 IOP Publishing Ltd.)

$$J_{SC}(R, g) = q \int Q_{\mathrm{abs}}(E_{\mathrm{P}}, R, g)\, F(E_{\mathrm{P}})\, dE_{\mathrm{P}}, \tag{3.15}$$

where q is the elementary charge, $F(E_{\mathrm{p}})$ is the reference AM 1.5G spectra [52], and 100 % charge collection efficiency is considered to obtain the ultimate photocurrent density. A 2D J_{SC} versus R and g is presented in Fig. 3.6a. At a given R, J_{SC} is fast enhanced with g increased and its enhancement then becomes slow for $g > 2$ (i.e., $t > 2r$), e.g., $J_{\mathrm{SC}}(g = 2)$ reaches 93.6 % of $J_{\mathrm{SC}}(g \to \infty)$ at $R = 150$ nm, where $J_{\mathrm{SC}}(g \to \infty)$ implies $J_{SC}$of a-Si NWs. In Fig. 3.6b, the R-dependent J_{SC} in the coaxial NWs is presented at a fixed $g = 2$, together with those in Si NWs, bulk Si, and bulk a-Si, having the same volume of materials, for comparison. The coaxial NWs can clearly produce much larger photocurrent than their counterparts. For example, the coaxial NWs yield $J_{\mathrm{SC}} = 28.2\,\mathrm{mA/cm^2}$ at $R = 150$ nm (or $r = 50$ nm and $t = 100$ nm), which appears significant as compared to 8.5 mA/cm^{2}in the Si NWs. Such photocurrent enhancement of the coaxial NWs with respect to the counterparts can be readily seen from Fig. 3.6c, where the photocurrent enhancement factor (PEF) is plotted versus R. The PEF is defined as $(I_{SC,cNW}/V_{cNW} - I_{SC,CP}/V_{CP})/(I_{SC,CP}/V_{CP})$, where $I_{SC,cNW}(V_{cNW})$ and

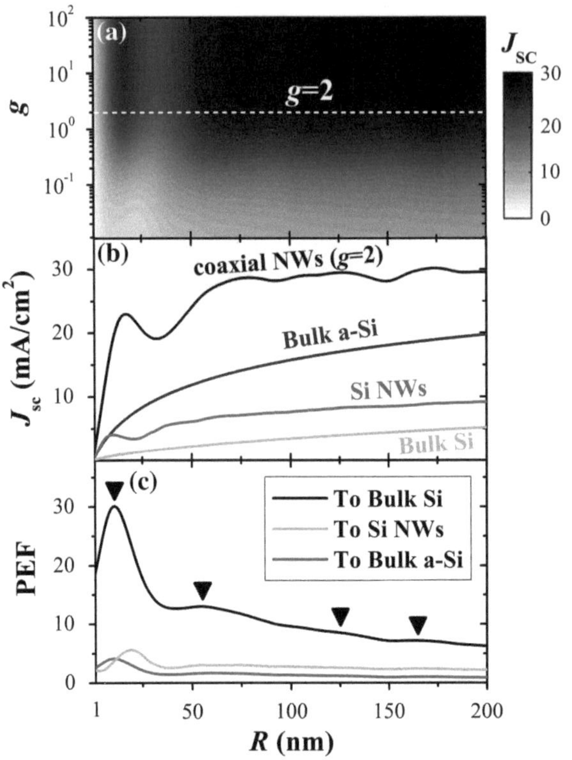

Fig. 3.6 **a** J_{sc} versus g and R. The *dashed line* indicates the case of $g = 2$. **b** J_{sc} versus R for coaxial NWs ($g = 2$), together with for Si NWs, bulk Si, and bulk a-Si for comparison. **c** PEF versus R with respect to Si NWs, bulk Si, and bulk a-Si. See text for *triangles*. (Reproduced with permission from Liu et al. [53]. © 2011 IOP Publishing Ltd.)

$I_{SC,CP}(V_{CP})$ are photocurrents (volumes) for the coaxial NWs ($g = 2$) and the counterparts (Si NWs, bulk Si, and bulk a-Si), respectively. The PEF curves with respect to the bulk counterparts exhibit a series of local maxima at certain R's as indicated with the triangles in Fig. 3.6c. These local maxima result primarily from the occurrence of such LMRs as TM_{01}, TM_{11}/TE_{01}, TM_{21}/TE_{11}, and TM_{12}/TE_{21} from the left to the right. At these R's (11, 55, 125, and 165 nm), the PEF against the bulk Si (bulk a-Si) are 30.1 (4.1), 13.0 (1.7), 8.5 (1.2), and 7.2 (1.0), respectively. Also, it ranges between 2.1 and 5.6 with respect to the Si NWs for the entire R investigated.

A practical use of coaxial NW PV cells may be limited since the electrical power generated is extremely low due to their nanoscale dimensions. They may be however still useful to power nanoelectronic devices that may require ultralow power. There has been a pioneer work recently, demonstrating that coaxial NW PV cells can serve as power sources to drive functional nanoelectronic sensors and logic gates [15]. The coaxial NW PV cells were made of p-type c-Si cores, intrinsic nanocrystalline Si

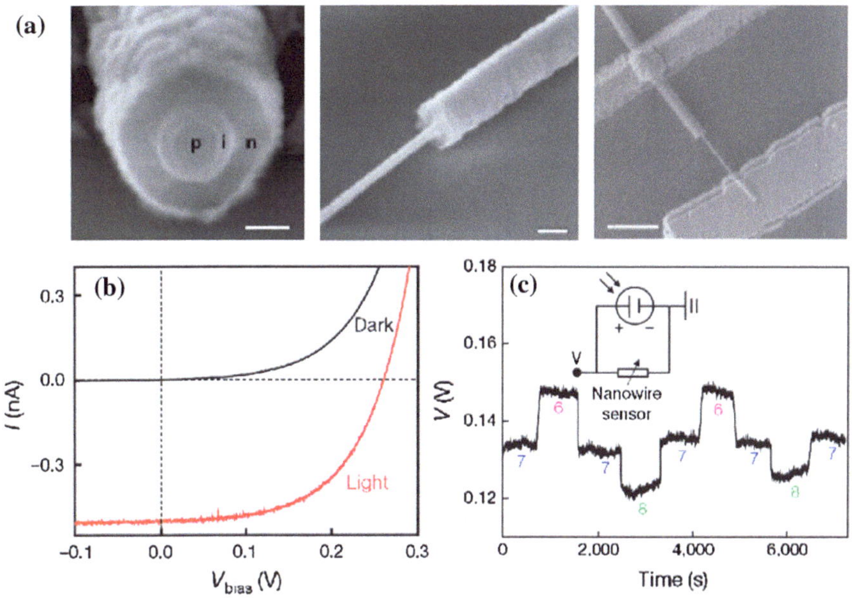

Fig. 3.7 **a** SEM images of a coaxial NW PV cell: (*left*) *p*-type c-Si core, intrinsic nc-Si shell, and *n*-type nc-Si shell. (middle) selective etching to expose the *p*-core. (*right*) metal contacts deposited on the *p*-core and *n*-shell. *Scale bars* are 100 nm (*left*), 200 nm (*middle*), and 1.5 μm (*right*). **b** Dark and light I–V curves. **c** Real-time detection of the voltage drop across an aminopropyltriethoxysilane-modified silicon nanowire at different pH values. The silicon nanowire pH sensor is powered by a coaxial NW PV cell operating under 8-sun illumination (V_{oc} = 50.34 V, I_{sc} = 58.75 nA). *Inset* shows circuit schematics. (Reprinted with permission from Tian et al. [15]. © 2007 Nature Publishing Group.)

(nc-Si) shells, and n-type nc-Si shells, as shown in Fig. 3.7a. *I*-*V* curves are also shown in Fig. 3.7b, providing that an open-circuit voltage V_{OC} of 0.260 V, a short-circuit current I_{SC} of 0.503 nA, a fill factor of 55.0 %, and the maximum power output of 72 pW at 1-sun. The upper bound of a short-circuit current density estimated was about 23.9 mA/cm^2, similar to our calculated result (28.2 mA/cm^2 at R = 150 nm) presented above. Importantly, the functionality of the coaxial NW PV cells has been demonstrated to power a nanoelectronic sensor, as can be seen from Fig. 3.7c.

Although the coaxial NW PV cells may not be used for large-scale power generation, the light trapping mechanism in coaxial NWs may help in designing large-scale PV cells. A feasible way is to make use of planar semi-coaxial NW arrays, as illustrated in Fig. 3.8. Such planar NW arrays can be fabricated on c-Si wafers by means of current lithography technique (for c-Si cores) combined with chemical vapor deposition (for a-Si shells). The underlying mechanism of light absorption in planar NW arrays may not be expected to be much different from that in the single coaxial NWs (i.e., LMRs) discussed above, except that coupled LMRs might be considered if neighboring NWs are too close [33]. The optimal dimensions of the planar NW

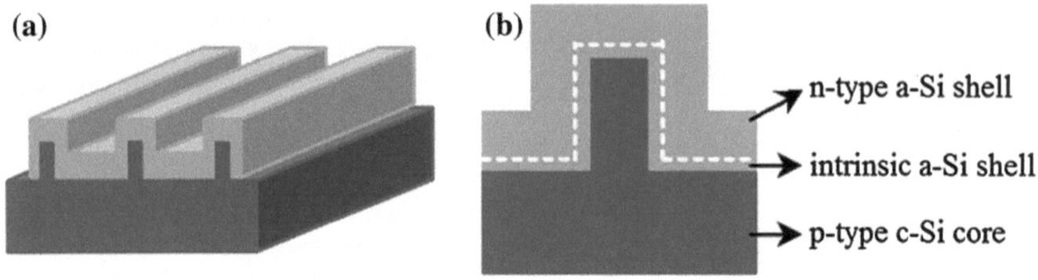

Fig. 3.8 Schematic illustrations of wafer-scale PV cells made of planar semi-coaxial NW arrays. **a** Overall look. **b** *Side view*, showing *p*-type c-Si core, intrinsic a-Si shell, and *n*-type a-Si shell

arrays for PV applications, however, need to be further investigated, which is beyond the scope of this chapter.

3.4 Light Trapping in Coaxial Nanowire Arrays

The light absorption in coaxial NW arrays of c-Si cores and a-Si:H shells can be numerically calculated by the RCWA method [44–46]. Here, vertically aligned square arrays in air are considered, which are a case of the 2D grating depicted in Fig. 3.2, so that the regions I and II can be treated as air, the pitch $p = p_x = p_y$, and the length of the NWs $L = d$. Such coaxial NW arrays are simply characterized by the pitch *p*, the length *L*, the core radius *r* of c-Si, the total radius *R* (or the a-Si:H shell thickness $t = R - r$), and the filling ratio $f = \pi R^2/p^2$. For RCWA simulations, TE- and TM-polarized lights were considered to be incident at an angle of θ in the $x - z$ plane ($\phi = 0$ in Fig. 3.2a) and the complex refractive indices of c-Si ($m_2 = \varepsilon_2^{1/2}$ in Fig. 3.2c) and a-Si:H ($m_3 = \varepsilon_3^{1/2}$ in Fig. 3.2c) were taken from [39].

Absorptance and reflectance are presented versus photon energy E_p in Fig. 3.9, in coaxial NW arrays of $p = 200$ nm, $r = 50$ nm, and $L = 1.0\,\mu$m for various shell thicknesses ($t = 0 - 50$ nm, corresponding to the filling ratio $f = 0.196 - 0.785$) at $\theta = 0°$ (i.e., parallel to the *z* axis or the NW axis). As shown in the top panel of Fig. 3.9a, the light absorption in c-Si NW arrays ($t = 0$) is clearly enhanced for high photon energy (E_p >2.5 eV) compared to in c-Si thin films, likely resulting from lower reflection in the former than in the latter (see the top panel of Fig. 3.9b). This absorption enhancement may be associated with intrinsically lower reflection in c-Si NW arrays due to their open structure. The NW array structure has a large open area, naturally leading to low reflection that perhaps results in high absorption. This absorption enhancement may also be associated with interwire light scattering [10, 27], which occurs when dimensions of NW arrays (diameter and pitch) are comparable to the wavelength of incident light, elongating the optical path length. Although it is enhanced for the high photon energy, the light absorption in c-SiNW arrays remarkably drops for the photon energy range of 1.5–2.5 eV that is the most important solar band for PV applications. This absorption drop has been known to

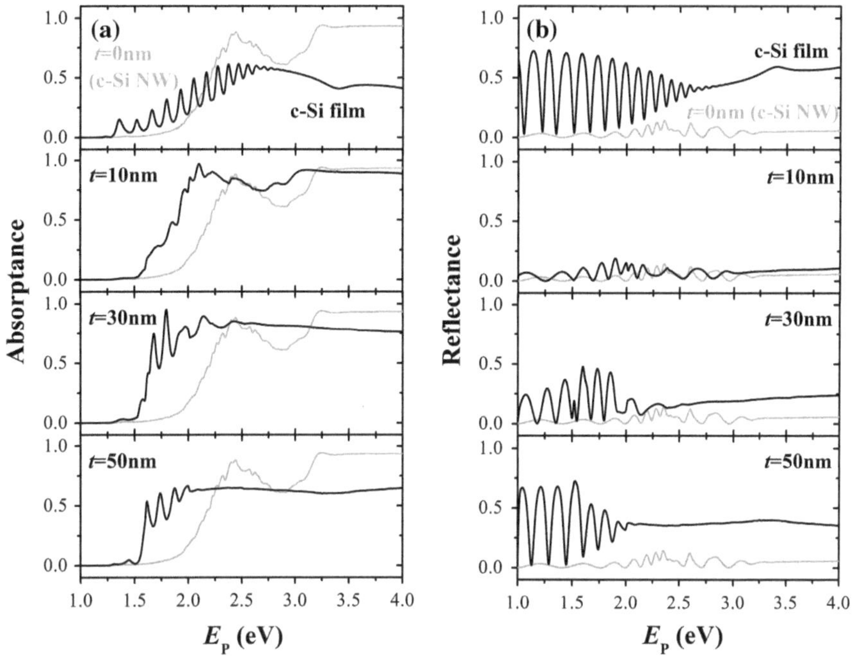

Fig. 3.9 **a** Absorptance and **b** reflectance versus E_p for various t at $p = 200\,\text{nm}$, $r = 50\,\text{nm}$, $L = 1.0\,\mu\text{m}$, and $\theta = 0°$. (Reproduced with permission from Xie et al. [54]. © 2011 American Institute of Physics.)

come from the indirect bandgap nature of c-Si that leads to a poor capability to absorb low energy photons ($E_p < 2.5\,\text{eV}$) [19]. Although this low energy absorption suppression has been known to be improved to some extent by tuning the geometry of NW arrays (length, diameter, or/and pitch), its improvement is rather limited [19–21].

It is evident in Fig. 3.9a that the light absorption in c-SiNW arrays for the low energy (1.5–2.5 eV) is significantly reinforced by coating a small amount of a-Si:H ($t = 10\,\text{nm}$). Such absorption reinforcement in coaxial NW arrays may be expected from the fact that the imaginary part of complex refractive index, which is proportional to the absorption coefficient, is ~10 times bigger in a-Si:H than in c-Si for the low energy: e.g., it is 0.217 (0.022) for a-Si:H (c-Si) at $E_p = 2.0\,\text{eV}$ [39]. As the shell thickness increases to 30 nm, the light absorption is further enhanced mainly owing to distinct absorption resonances. For a further increase of the shell thickness to 50 nm, however, the overall absorption becomes smaller than that for $t = 30\,\text{nm}$, since a larger material filling ratio for $t = 50\,\text{nm}$ leads to a smaller open area which then results in larger reflection as can be seen in Fig. 3.9b. It has to be emphasized here that the distinct absorption resonances, which are in fact LMRs discussed below, are clearly a major source of the absorption enhancement in coaxial NW arrays for the low photon energy. In Fig. 3.10a, the E_p-dependent absorptance is presented for coaxial NW arrays ($R = 80\,\text{nm}$, $r = 50\,\text{nm}$, and $L = 1.0\,\mu\text{m}$), c-Si NW arrays

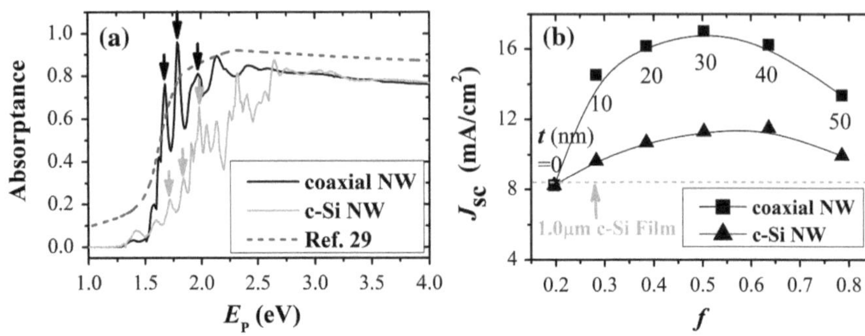

Fig. 3.10 **a** Absorptance versus E_p for coaxial NW arrays ($R = 80\,\text{nm}$, $t = 30\,\text{nm}$) and c-Si NW arrays ($r = 80\,\text{nm}$) at $p = 200\,\text{nm}$, $L = 1.0\,\mu\text{m}$, and $\theta = 0°$, together with for disordered coaxial NW arrays ($R = 74$–$90\,\text{nm}$, $r = 15$–$30\,\text{nm}$, $L = 1.282\,\mu\text{m}$ from [29]). **b** J_{SC} versus f for coaxial NW arrays ($r = 50\,\text{nm}$) and c-Si NW arrays at $p = 200\,\text{nm}$, $L = 1.0\,\mu\text{m}$, and $\theta = 0°$, together with for c-Si thin film of $1.0\,\mu\text{m}$ thick. (Reproduced with permission from Xie et al. [54]. © 2011 American Institute of Physics.)

($r = 80\,\text{nm}$ and $L = 1.0\,\mu\text{m}$), and disordered coaxial NW arrays ($R = 74$–$90\,\text{nm}$, $r = 15$–$30\,\text{nm}$, and $L = 1.282\,\mu\text{m}$ from [29]). The amplitudes of resonances indicted with arrows are evidently bigger in the coaxial NW arrays than the c-Si NW arrays, similar to a behavior of LMRs occurring in the single coaxial NW system discussed in the previous section. The number of resonance peaks is less in the coaxial NW arrays than in the c-Si NW arrays, also similar to another behavior of LMRs occurring in the single coaxial NW system. Interestingly, absorption resonances won't occur in the disordered coaxial NW arrays [29], implying that absorption resonances appear to be a signature of ordered NW arrays.

In addition to a large open area and the interwire light scattering as mentioned above, absorption resonances evidently play an important role in the absorption enhancement in coaxial NW arrays. As discussed in the previous section, LMRs occur and serve as a primary source to trap light in the single NW system. Furthermore, LMRs have recently been shown to be a major source of absorption enhancement in the NW array system [34]. Note that, in this chapter, the incident light is considered to be perpendicular to the NW axis in the case of the single coaxial NW system, whereas it is parallel to the NW axis in the case of the coaxial NW array system. Although the configuration of incident light versus the NW axis results in different kinds of excitation of LMRs, i.e., TM and TE modes (perpendicular) and hybrid TM-dominant HE modes (parallel) [34], a primary role of LMRs in the light absorption is essentially same for both cases. Note also that LMRs, which result from the interaction between light and subwavelength matters, are physically different from Fabry-Pérot resonances, which come from interference between multiple reflections of light from two reflecting surfaces. The latter is typically observed in c-Si thin films (see regular interference patterns in the top panels of Fig. 3.9).

As in single coaxial NWs, one can calculate the ultimate photocurrent density or short-circuit current density J_{SC} to evaluate the absorption enhancement in coaxial NW arrays for photovoltaic applications as

$$J_{SC}(f, p) = q \int A\,(E_P, f, p)\, F\,(E_P)\, dE_P, \tag{3.16}$$

which is the same equation as equation 3.15, expressed with absorptance A instead of absorption efficiency Q_{abs}. The f-dependent (or t-dependent) J_{SC} is presented in Fig. 3.10b, in coaxial NW arrays of $r = 50$ nm, $p = 200$ nm, and $L = 1.0\,\mu$m. J_{SC} in the coaxial NW arrays has the maximum of 17.0 mA/cm^2 at $f = 0.5$ (or $t = 30$ nm), at which J_{SC} is 11.3 mA/cm^2 (8.5 mA/cm^2) in c-SiNW arrays of the same dimensions (c-Si thin film of 1.0 μm thick). At this optimal filling ratio $f = 0.5$, the light absorption in the coaxial NW arrays is highest as can be seen in Fig. 3.9a. As a result, the PEF of the coaxial NW arrays is about 0.5 (3.0) with respect to the c-SiNW arrays (c-Si thin film).

To further investigate the optimal geometry of coaxial NW arrays for PV applications, absorptance and reflectance are shown versus E_p in Fig. 3.11, in coaxial NW arrays of $f = 0.5$, $r/p = 1/4$, and $L = 1.0\,\mu$m for various pitches ($p = 100$–900 nm, corresponding to $t = 14.9$–134.1 nm) at $\theta = 0°$. The top views of coaxial NW arrays with different pitches are schematically shown in Fig. 3.12a. For the smallest pitch ($p = 100$ nm), the absorption spectrum exhibits a clifflike behavior: a plateau of high absorption ($A = 80 \sim 87\,\%$), a cliff edge ($E_{CE} \sim 2.0$ eV), and a sharp cliff hill.

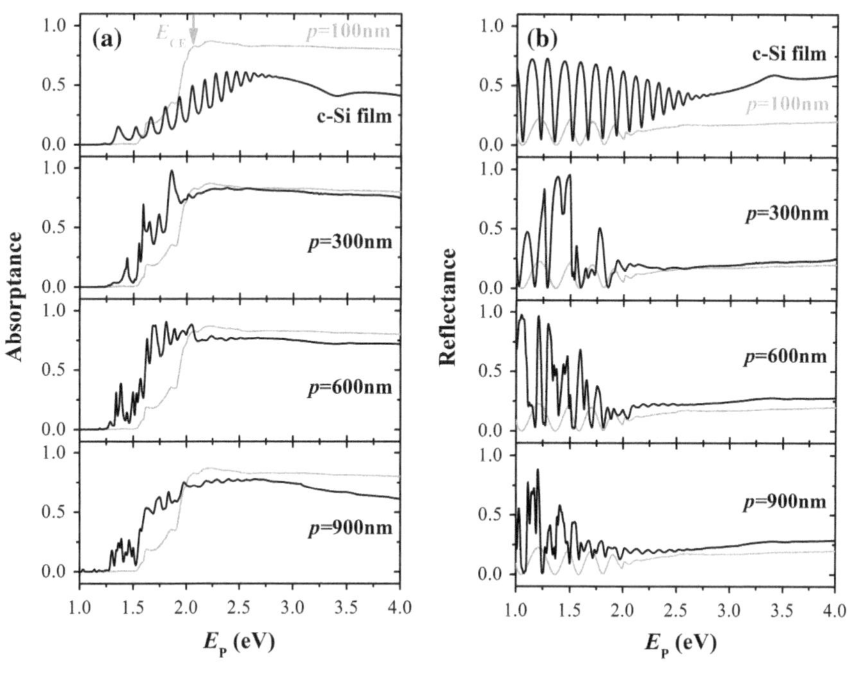

Fig. 3.11 **a** Absorptance and **b** reflectance versus E_p for various p at $f = 1/2$, $r/p = 1/4$, $L = 1.0\,\mu$m, and $\theta = 0°$. See texts for E_{CE}. (Reproduced with permission from Xie et al. [54]. © 2011 American Institute of Physics.)

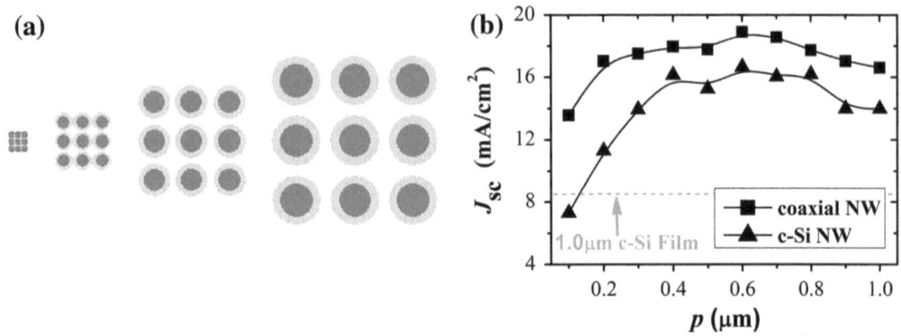

Fig. 3.12 **a** Schematic top views of coaxial NW arrays with $p = 100, 300, 600, 900$ nm from *left* to *right* at $f = 1/2$ and $r/p = 1/4$. **b** J_{SC} versus p for coaxial NW arrays ($r/p = 1/4$) and c-Si NW arrays $[r/p = (1/2\pi)^{1/2}]$ at $f = 1/2$, $L = 1.0\,\mu$m, and $\theta = 0°$, together with for c-Si thin film of $1.0\,\mu$m thick. (Fig. 3.12b is reproduced with permission from Xie et al. [54]. © 2011 American Institute of Physics.)

For $p = 300$ and 600 nm, the absorption resonances strongly occur along the cliff hill and the cliff edge clearly becomes shifted to the low energy (e.g., $E_{CE} \sim 1.5$ eV for $p = 600$ nm), leading to absorption enhancement. For $p = 900$ nm, however, the overall absorption is evidently smaller than that for $p = 600$ nm. Very interestingly, the p-dependent reflectance behaves just like the absorptance, i.e., it increases as $p = 100 \rightarrow 600$ nm and decreases as $p = 600 \rightarrow 900$ nm. This implies that the incident light becomes more effectively trapped into coaxial NW arrays through LMRs as $p = 100 \rightarrow 600$ nm so that it is less transmitted. Also, the absorption drop for $p = 600 \rightarrow 900$ nm has to result from transmission increase, since the overall reflection for $p = 900$ nm is slightly smaller than that for $p = 600$ nm. This implies that such subwavelength light-trapping effect diminishes as the pitch becomes larger ($p = 600 \rightarrow 900$ nm), in good agreement with [38]. The optimal pitch of coaxial NW arrays for photovoltaic applications is clearly seen to be ~600 nm from Fig. 3.12b, which shows the p-dependent J_{SC} in coaxial NW arrays. At this optimal pitch $p = 600$ nm, J_{SC} in coaxial NW arrays reaches the maximum of 18.9 mA/cm^2, whereas J_{SC} is 16.6 mA/cm^2 (8.5 mA/cm^2) in c-SiNW arrays (c-Si thin film of $1.0\,\mu$m thick). Thus, the PEF of the coaxial NW arrays is about 0.14 (3.45) with respect to the c-SiNW arrays (c-Si thin film).

To examine the effect of the NW length L of coaxial NW arrays on the light absorption or the photocurrent density, J_{SC} is plotted versus L in Fig. 3.13a at the optimal filling ratio ($f = 0.5$) and pitch ($p = 600$ nm). The L-dependent J_{SC} in coaxial NW arrays sharply increases as L varies from 0.1 to $5.0\,\mu$m and shows a gradually growing behavior for a further increase of L. To also examine the effect of the incident angle θ of light on the light absorption, J_{SC} in the 1.0μm-length coaxial NW arrays is plotted versus θ at the optimal configuration ($f = 0.5$ and $p = 600$ nm) for the TE and TM polarizations in Fig. 3.13b, together with for the unpolarized light that is simply an average value of those for TE and TM. It is evident from the figure that the photocurrent density is maintained at a high constant (~20 mA/cm^2) over a

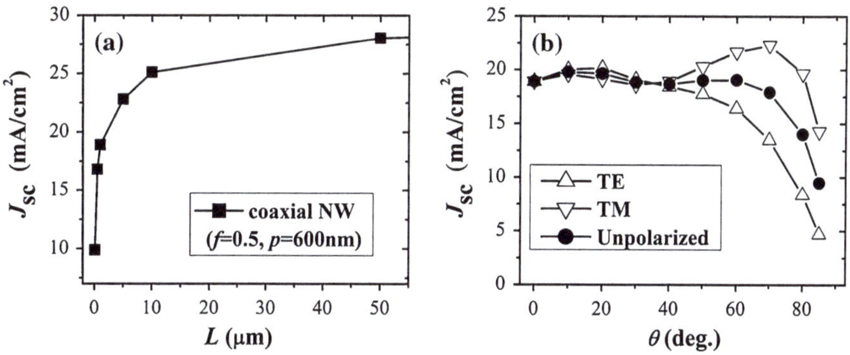

Fig. 3.13 **a** J_{SC} versus L at $f = 0.5$, $p = 600\,\text{nm}$, and at $\theta = 0°$. **b** J_{SC} versus θ for TE-, TM-, and unpolarized-illuminations at $f = 0.5$, $p = 600\,\text{nm}$, and $L = 1.0\,\mu\text{m}$. (Reproduced with permission from Xie et al. [54]. © 2011 American Institute of Physics.)

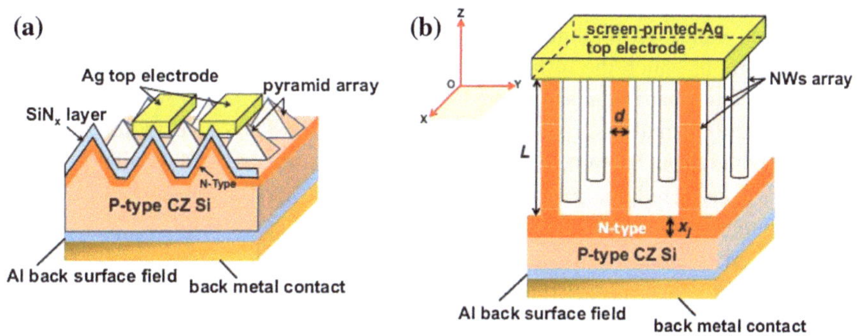

Fig. 3.14 The three-dimensional schematic view of (**a**) the conventional pyramid-array-textured c-Si solar cells and (**b**) the Si NW-array-textured solar cells, in which L represents the length of the NWs, d is the diameter of the single NW, and X_j is the junction depth of the p-n junction after phosphorus diffusion. (Reprinted with permission from Chen et al. [30]. © 2010 American Institute of Physics.)

large range of the incident angle of light ($\theta = 0-70°$), indicating an omnidirectional light absorption in coaxial NW arrays. Note that the absorption difference between the cases of the incident light in and out-of the x-z plane is known to be negligible [19].

Coaxial NW arrays may be practically used as a light absorbing layer in wafer-scale PV cells. There has been a recent work, fabricating c-Si NW-textured PV cells [30]. There, c-Si NW arrays replaced conventional pyramid-textured layers, as shown in Fig. 3.14. Such NW-array-textured wafer-scale ($125 \times 125\,\text{mm}^2$) PV cells (Fig. 3.14b) had a 16.5 % power conversion efficiency, which was 35.4 % enhancement as compared to the pyramid-textured control PV cells (Fig. 3.14a). That enhancement turned out to come from significant absorption enhancement of c-Si NW arrays. Nonetheless, more efforts need to be made to develop high-performance

PV cells based on NW arrays. The light absorption properties of coaxial NW arrays discussed in this section could help in designing high-performance NW-based PV cells.

3.5 Summary and Conclusions

The light-trapping scheme has been investigated in coaxial NWs of c-Si cores and a-Si shells for both cases of single coaxial NWs and coaxial NW arrays. Based on numerical calculations, it has been shown that the incident light can be effectively trapped in the coaxial NWs through the excitation of LMRs as in other NWs and the a-Si shells play a primary role in enhancing the light absorption in the coaxial NWs as compared to in the c-Si NWs of same geometric dimensions. It has been further shown that the light absorption in the coaxial NWs can be optimized by tuning the dimensions of the coaxial NWs. For the single coaxial NWs, the light absorption is optimized when the thickness of the a-Si shell is about as twice as the radius of the Si core, at which the ultimate photocurrent can be enhanced up to 560 % in comparison with that in the single c-Si NWs. As for the coaxial NW arrays, on the other hand, the optimal absorption occurs when the coaxial NWs are about half-filled and the pitch is about 600 nm, at which it can be increased up to 14 % with the 1 μm long NWs considered, as compared to that in the c-Si NW arrays. The effective light-trapping scheme in the coaxial NWs together with the absorption optimization results can provide opportunities for designing not only functional power sources of nanoelectronic devices but also next generation PV cells.

Acknowledgments This work was supported by the National Major Basic Research Project of 2012CB934302 and the Natural Science Foundation of China under contracts 11074169, 11174202, and 61234005.

References

1. Nelson, J.: The Physics of Solar Cells. Imperial College Press, London (2003)
2. Dale, B., Rudenberg, H.G.: High efficiency silicon solar cells. In: Proceedings of the 14th Annual Power Sources Conference, p 22 (1960)
3. Pillai, S., Catchpole, K.R., Trupke, T., Green, M.A.: Surface plasmon enhanced silicon solar cells. J. Appl. Phys. **101**, 093105 (2007)
4. Beck, F.J., Polman, A., Catchpole, K.R.: Tunable light trapping for solar cells using localized surface plasmons. J. Appl. Phys. **105**, 114310 (2009)
5. Ferry, V.E., Verschuuren, M.A., Li, H.B.B.T., Schropp, R.E.I., Atwater, H.A., Polman, A.: Improved red-response in thin film a-Si:H solar cells with soft-imprinted plasmonic back reflectors. Appl. Phys. Lett. **95**, 183503 (2009)
6. Atwater, H.A., Polman, A.: Plasmonics for improved photovoltaic devices. Nat. Mater. **9**, 205–213 (2010)

7. Ferry, V.E., Verschuuren, M.A., Li, H.B.B.T., Verhagen, E., Walters, R.J., Schropp, R.E.I., Atwater, H.A., Polman, A.: Light trapping in ultrathin plasmonic solar cells. Opt. Express **18**, A237–A245 (2010)
8. Chen, X., Jia, B.H., Saha, J.K., Cai, B.Y., Stokes, N., Qiao, Q., Wang, Y.Q., Shi, Z.R., Gu, M.: Broadband enhancement in thin-film amorphous silicon solar cells enabled by nucleated silver nanoparticles. Nano Lett. **12**, 2187–2192 (2012)
9. Nunomura, S., Minowa, A., Sai, H., Kondo, M.: Mie scattering enhanced near-infrared light response of thin-film silicon solar cells. Appl. Phys. Lett. **97**, 063507 (2010)
10. Muskens, O.L., Rivas, J.G., Algra, R.E., Bakkers, E.P.A.M., Lagendijk, A.: Design of light scattering in nanowire materials for photovoltaic applications. Nano Lett. **8**, 2638–2642 (2008)
11. Cao, L.Y., White, J.S., Park, J.S., Schuller, J.A., Clemens, B.M., Brongersma, M.L.: Engineering light absorption in semiconductor nanowire devices. Nat. Mater. **8**, 643–647 (2009)
12. Pei, Z.W., Chang, S.T., Liu, C.W., Chen, Y.C.: Numerical simulation on the photovoltaic behavior of an amorphous-silicon nanowire-array solar cell. IEEE. Electron Device Lett. **30**, 1305–1307 (2009)
13. Liu, W.F., Oh, J.I., Shen, W.Z.: Light trapping in single coaxial nanowires for photovoltaic applications. IEEE. Electron Device Lett. **32**, 45–47 (2010)
14. Diedenhofen, S.L., Janssen, O.T.A., Grzela, G., Bakkers, E.P.A.M., Rivas, J.G.: Strong geometrical dependence of the absorption of light in arrays of semiconductor nanowires. ACS Nano. **5**, 2316–2323 (2011)
15. Tian, B.Z., Zheng, X.L., Kempa, T.J., Fang, Y., Yu, N.F., Yu, G.H., Huang, J.L., Lieber, C.M.: Coaxial silicon nanowires as solar cells and nanoelectronic power sources. Nature **449**, 885–890 (2007)
16. Kempa, T.J., Tian, B.Z., Kim, D.R., Hu, J.S., Zheng, X.L., Lieber, C.M.: Single and tandem axial p-i-n nanowire photovoltaic devices. Nano Lett. **8**, 3456–3460 (2008)
17. Tian, B.Z., Kempa, T.J., Lieber, C.M.: Single nanowire photovoltaics. Chem. Soc. Rev. **38**, 16–24 (2009)
18. Kim, S.K., Day, R.W., Cahoon, J.F., Kempa, T.J., Song, K.D., Park, H.G., Lieber, C.M.: Tuning light absorption in core/shell silicon nanowire photovoltaic devices through morphological design. Nano Lett. **12**, 4971–4976 (2012)
19. Hu, L., Chen, G.: Analysis of optical absorption in silicon nanowire arrays for photovoltaic applications. Nano Lett. **7**, 3249–3252 (2007)
20. Lin, C.X., Povinelli, M.L.: Optical absorption enhancement in silicon nanowire arrays with a large lattice constant for photovoltaic applications. Opt. Express **17**, 19371–19381 (2009)
21. Li, J.S., Yu, H.Y., Wong, S.M., Li, X.C., Zhang, G., Lo, P.G.Q., Kwong, D.L.: Design guidelines of periodic Si nanowire arrays for solar cell application. Appl. Phys. Lett. **95**, 243113 (2009)
22. Tsakalakos, L., Balch, J., Fronheiser, J., Korevaar, B.A., Sulima, O., Rand, J.: Silicon nanowire solar cells. Appl. Phys. Lett. **91**, 233117 (2007)
23. Stelzner, T., Pietsch, M., Andrä, G., Falk, F., Ose, E., Christiansen, S.: Silicon nanowire-based solar cells. Nanotechnology **19**, 295203 (2008)
24. Fang, H., Li, X.D., Song, S., Xu, Y., Zhu, J.: Fabrication of slantingly-aligned silicon nanowire arrays for solar cell applications. Nanotechnology **19**, 255703 (2008)
25. Garnett, E.C., Yang, P.D.: Silicon nanowire radial p-n junction solar cells. J. Am. Chem. Soc. **130**, 9224–9225 (2008)
26. Gunawan, O., Guha, S.: Characteristics of vapor-liquid-solid grown silicon nanowire solar cells. Sol. Energy Mater. Sol. Cells **93**, 1388–1393 (2009)
27. Bao, H., Ruan, X.L.: Optical absorption enhancement in disordered vertical silicon nanowire arrays for photovoltaic applications. Opt. Lett. **35**, 3378–3380 (2010)
28. Garnett, E., Yang, P.D.: Light trapping in silicon nanowiresolar cells. Nano Lett. **10**, 1082–1087 (2010)
29. Adachi, M.M., Anantram, M.P., Karim, K.S.: Optical properties of crystalline-amorphous core-shell silicon nanowires. Nano Lett. **10**, 4093–4098 (2010)
30. Chen, C., Jia, R., Yue, H.H., Li, H.F., Liu, X.Y., Wu, D.Q., Ding, W.C., Ye, T.C., Kasai, S., Tamotsu, H., Chu, J.H., Wang, S.L.: Silicon nanowire-array-textured solar cells for photovoltaic application. J. Appl. Phys. **108**, 094318 (2010)

31. Xie, W.Q., Oh, J.I., Shen, W.Z.: Realization of effective light trapping and omnidirectional antireflection in smooth surface silicon nanowirearrays. Nanotechnology **22**, 065704 (2011)
32. Cao, L.Y., Fan, P.Y., Vasudev, A.P., White, J.S., Yu, Z.F., Cai, W.S., Schuller, J.A., Fan, S.H., Brongersma, M.L.: Semiconductor nanowireoptical antenna solar absorbers. Nano Lett. **10**, 439–445 (2010)
33. Cao, L.Y., Fan, P.Y., Brongersma, M.L.: Optical coupling of deep-subwavelength semiconductor nanowires. Nano Lett. **11**, 1463–1468 (2011)
34. Wang, B., Leu, P.W.: Tunable and selective resonant absorption in vertical nanowires. Opt. Lett. **37**, 3756–3758 (2012)
35. Kelzenberg, M.D., Boettcher, S.W., Petykiewicz, J.A., Turner-Evans, D.B., Putnam, M.C., Warren, E.L., Spurgeon, J.M., Briggs, R.M., Lewis, N.S., Atwater, H.A.: Enhanced absorption and carrier collection in Si wire arrays for photovoltaic applications. Nat. Mater. **9**, 239–244 (2010)
36. Yablonovitch, E.: Statistical ray optics. J. Opt. Soc. Am. **72**, 899–907 (1982)
37. Tiedje, T., Yablonovitch, E., Cody, G.D., Brooks, B.G.: Limiting efficiency of silicon solarcells. IEEE Trans. Electron Devices **31**, 711–716 (1984)
38. Kosten, E.D., Warren, E.L., Atwater, H.A.: Ray optical light trapping in silicon microwires: exceeding the 2 n^2 intensity limit. Opt. Express **19**, 3316–3331 (2011)
39. Palik, E.D.: Handbook of optical constants of solids. Academic Press, London (1985)
40. Shah, A.V., Schade, H., Vanecek, M., Meier, J., Vallat-Sauvain, E., Wyrsch, N., Kroll, U., Droz, C., Bailat, J.: Thin-film silicon solar cell technology. Prog. Photovolt. **12**, 113–142 (2004)
41. Dong, Y.J., Yu, G.H., McAlpine, M.C., Lu, W., Lieber, C.M.: Si/a-Si core/shell nanowires as nonvolatile crossbar switches. Nano Lett. **8**, 386–391 (2008)
42. Taguchi, M., Kawamoto, K., Tsuge, S., Baba, T., Sakata, H., Morizane, M., Uchihashi, K., Nakamura, N., Kiyama, S., Oota, O.: HIT^{TM} cells-high-efficiency crystalline Si cells with novel structure. Prog. Photovolt. **8**, 503–513 (2000)
43. Bohren, C.F., Huffman, D.R.: Absorption and scattering of light by small particles. Wiley, New York (1998)
44. Moharam, M.G., Gaylord, T.K.: Rigorous coupled-wave analysis of planar-grating diffraction. J. Opt. Soc. Am. **71**, 811–818 (1981)
45. Moharam, M.G., Grann, E.B., Pommet, D.A., Gaylord, T.K.: Formulation for stable and efficient implementation of the rigorous coupled-wave analysis of binary gratings. J. Opt. Soc. Am. A **12**, 1068–1076 (1995)
46. Moharam, M.G., Pommet, D.A., Grann, E.B., Gaylord, T.K.: Stable implementation of the rigorous coupled-wave analysis for surface-relief gratings: enhanced transmittance matrix approach. J. Opt. Soc. Am. A **12**, 1077–1086 (1995)
47. Brönstrup, G., Jahr, N., Leiterer, C., Csáki, A., Fritzsche, W., Christiansen, S.: Optical properties of individual silicon nanowires for photonic devices. ACS Nano **4**, 7113–7122 (2010)
48. Zhu, J., Yu, Z.F., Burkhard, G.F., Hsu, C.M., Connor, S.T., Xu, Y.Q., Wang, Q., McGehee, M., Fan, S.H., Cui, Y.: Optical absorption enhancement in amorphous silicon nanowire and nanocone arrays. Nano Lett. **9**, 279–282 (2009)
49. Lee, Y.J., Ruby, D.S., Peters, D.W., McKenzie, B.B., Hsu, J.W.P.: ZnO nanostructures as efficient antireflection layers in solar cells. Nano Lett. **8**, 1501–1505 (2008)
50. Chiu, C.H., Yu, P.C., Kuo, H.C., Chen, C.C., Lu, T.C., Wang, S.C., Hsu, S.H., Cheng, Y.J., Chang, Y.C.: Broadband and omnidirectional antireflection employing disordered GaN nanopillars. Opt. Express **16**, 8748–8754 (2008)
51. Khoury, C.G., Norton, S.J., Vo-Dinh, T.: Investigating the plasmonics of a dipole-excited silver nanoshell: Mie theory versus finite element method. Nanotechnology **21**, 315203 (2010)
52. Air Mass 1.5 Spectra. American society for testing and materials (http://rredc.nrel.gov/solar/spectra/am1.5/) (2012)
53. Liu, W.F., Oh, J.I., Shen, W.Z.: Light absorption mechanism in single c-Si (core)/a-Si (shell) coaxial nanowires. Nanotechnology **22**, 125705 (2011)
54. Xie, W.Q., Liu, W.F., Oh, J.I., Shen, W.Z.: Optical absorption in c-Si/a-Si:H core/shell nanowire arrays for photovoltaic applications. Appl. Phys. Lett. **99**, 033107 (2011)

Chapter 4
Applications of Ordered Si Nanowire Array to Solar Energy Harvesting and NEMS

Yuerui Lu and Amit Lal

Abstract Nanostructured silicon thin-film solar cells are promising, due to the strongly enhanced light trapping, high carrier collection efficiency, and potential low cost. Ordered nanostructure arrays, with large-area controllable spacing, orientation, and size, are critical for reliable light-trapping and high-efficiency solar cells. Available top–down lithography approaches to fabricate large-area ordered nanostructure arrays are challenging due to the requirement of both high lithography resolution and high throughput. Here, a novel ordered silicon nano-conical-frustum array structure, exhibiting an impressive absorbance of ~99 % (upper bound) over wavelengths 400–1100 nm by a thickness of only 5 μm, is realized by our recently reported technique self-powered parallel electron lithography that has high throughput and high resolution. High-efficiency (up to 10.8 %) solar cells are demonstrated, using these ordered ultrathin silicon nano-conical-frustum arrays. Moreover, these ordered nano-structures have been successfully integrated into nano-electro-mechanical system (NEMS), enabling high-efficiency and broad-band optical actuation for NEMS devices. The first-ever nanopillar membrane acoustic speaker, using nano-scale photonic crystal optical absorbers for thermo-mechanical excitation of speaker membrane, is demonstrated.

4.1 Introduction

A solar cell [1] (or photovoltaic cell) is a device that converts the energy of light directly into electricity by the photovoltaic effect. Assemblies of solar cells are able to capture energy from sunlight. The photovoltaic effect is the creation of voltage or

Y. Lu (✉) · A. Lal
Sonic MEMS Laboratory, School of Electrical and Computer Engineering,
Cornell University, Ithaca, NY 08540, USA
e-mail: yl676@cornell.edu

A. Lal
e-mail: amit.lal@cornell.edu

H. Li et al. (eds.), *Silicon-based Nanomaterials*,
Springer Series in Materials Science 187, DOI: 10.1007/978-1-4614-8169-0_4,

© Springer Science+Business Media New York 2013

electric current in a material upon exposure to light. In a photovoltaic effect process, electrons are transferred between different bands (i.e., from the valence to conduction bands) within material, leading to the buildup of voltage between two electrodes. For example, a p–n junction solar cell works in three steps. First, incident sunlight is absorbed by semiconducting materials, such as silicon. Second, the absorbed photons will generate a large number of free electrons and holes, which will diffuse inside the materials. Third, the built-in electric field of the depletion region of the p–n junction will separate the generated free electrons and holes, leading to built-up voltage [1].

Currently, the big challenge of solar sell application is how to further reduce its cost. Solar cells with lower material cost are of interest to decrease the cost of solar energy. By far, the mostly used bulk material for solar cells is crystalline silicon (c–Si). Manufacturers of wafer-based cells have responded to high silicon prices in 2004–2008 with rapid reductions in silicon consumption. Therefore, thin-film solar cell that can reduce material cost, is becoming next generation technology to enable low-cost solar energy harvesting.

Nano-structured silicon thin-film solar cells are promising, due to the strongly enhanced light trapping, high carrier collection efficiency, and potential low cost. Solar cells with Si nanowire (Si NW) arrays incorporating radial p–n junctions not only have antireflective, broad-band optical absorption properties but also offer the advantage of orthogonalizing the light absorption and carrier collection processes, which is the challenging problem for bulk Si solar cells [2–5]. Previously, single Si NW radial p–n junction solar cells, using vapor–liquid–solid NW growth, were reported [6]. This single NW solar cell had a low open circuit voltage (V_{OC}), possibly due to the p-i-n structure, resulting in low overall solar cell efficiency of less than 3.4 %. Large-area solar cells based on Si NW were reported using Au catalytic chemical vapor deposited (CVD) p-type Si NW, followed with a plasma-enhanced CVD conformal n-type amorphous silicon (a-Si:H) layer to create the photoactive p–n junction [7]. These solar cells showed very low conversion efficiency (~0.1 %), particularly due to the geometry of the NWs with random and uncontrolled NW orientation and varied NW diameter distribution. Recently, large-area solar cells from arrays of Si NW with radial p–n junctions were shown [8, 9], using an aqueous suspension of silica beads to get a close-packed monolayer, followed by deep reactive ion etching to form NW and boron diffusion to form the radial p–n junction. By reducing surface roughness and improving the NW diameter and density control, the solar cell efficiency was enhanced significantly from initially 0.5 % to as high as 5.3 %. But its overall efficiency is still not high for practical application, probably due to the imperfect NW spacing, and absence of any surface passivation. Calculations have showed that with optimized wire spacing, diameter, and wire height, vertical Si NW array solar cells with photonic crystal architecture could offer conversion efficiency as high as 24 % or more [10], due to photonic crystal optical absorption enhancement effect with the presence of optical guided resonance modes [10–12]. Moreover, theoretical study also shows that ordered Si NW structures with small variations in the radius of the wires could create additional resonant modes at other frequencies not present in the perfectly ordered structures, leading to broad-band solar-energy harvesting [13]. Amorphous silicon nanocone arrays were demonstrated to have enhanced optical absorption in the 400–700 nm wavelength range [14], but they are hard to apply to high-performance thin-film solar cells due to the random nanocone spacing and wide cone size distribution.

The fabrication of large-area ordered controllable Si nanostructure arrays needs top-down planar lithography with both high throughput and high resolution. Conventional optical lithography has high throughput, but its critical dimension is limited to a fraction of the wavelength. Traditional electron beam lithography has the highest resolution <10 nm, but it suffers from high cost and low throughput due to the required electron beam raster scanning serial exposure. Nanoimprint lithography could be used to achieve nano-structured arrays, but the prospect of mask mechanical contact to substrate leads to a large number of defects and short mask life [15]. Our recently reported nanolithography system RIPEL [16, 17], using large-area planar radioactive beta-electron thin-film emitters to parallel expose e-beam resist through a stencil mask, demonstrated sub-35 nm resolution. Using naturally emitted high-energy beta particles, the RIPEL system can be compact as the electron focusing column needed in electron beam lithography is no longer needed. Elimination of vacuum in RIPEL will significantly simplify the overall lithography system and greatly reduce the cost, while enabling large area massively parallel high-throughput electron lithography with high resolution. Here, we successfully used RIPEL to fabricate a novel ordered Si nano-conical-frustum (NCF) array structure, which exhibits an impressive absorbance of ~99 % (upper bound) over wavelengths 400–1100 nm with a thickness of only 5 μm. Moreover, high-efficiency (up to 10.8 %) solar cells have been demonstrated, by using these ordered Si NCF arrays. Besides its application to solar energy harvesting, these Si nano-structures have been integrated into NEMS devices [18–21]. For example, the first-ever nano-pillar membrane acoustic speaker [19], using nano-scale photonic crystal optical absorbers for thermomechanical excitation of speaker membrane, is demonstrated.

4.2 Background for Photonic Crystal

Photonic crystals are composed of periodic dielectric or metallodielectric nanostructures that affect the propagation of electromagnetic waves, which is in the similar way as the periodic potential in a semiconductor crystal affects the electron motion by defining allowed and forbidden electronic energy bands [22]. Photonic crystals contain periodic internal regions of high and low dielectric constant, periodic along one, two, or three dimensions. Photons propagate through this structure—or not—depending on their wavelength, leading to guiding of light. The groups of light wavelengths that are allowed to travel, form bands spectrum in the photonic crystal, whole disallowed bands of wavelengths are called photonic band gaps.

Photonic crystal Si nanowire (NW) arrays vertically aligned to the substrate demonstrate broad-band optical absorption due to strong light trapping by multiple scattering of the incident light [3]. The optical properties of the photonic crystal Si NW structures are highly dependent on the typical geometric factors, including wire diameter, length, and periodicity. In particular, photonic crystal nanocone (NC) array (tapered nanowires) further suppresses the Fresnel reflection over a broad wavelength range at various angles of incidence, due to the gradual transition in the effective refractive indexes across the air–wire interface [14, 23].

4.3 Ordered Si Nano-structure Array Fabricated by RIPEL

Figure 4.1 shows the schematic illumination of the fabrication processes for ordered NCF arrays. A thin layer of plasma-enhanced chemical vapor deposition (PECVD) silicon dioxide (200 nm thick) was deposited on a silicon on insulator (SOI) (Si(5 μm)/SiO_2(2 μm)/Si stack) wafer substrate. RIPEL was used to pattern the e-beam resist, and the pattern was transferred to SiO_2, using SiO_2 reactive ion etch-

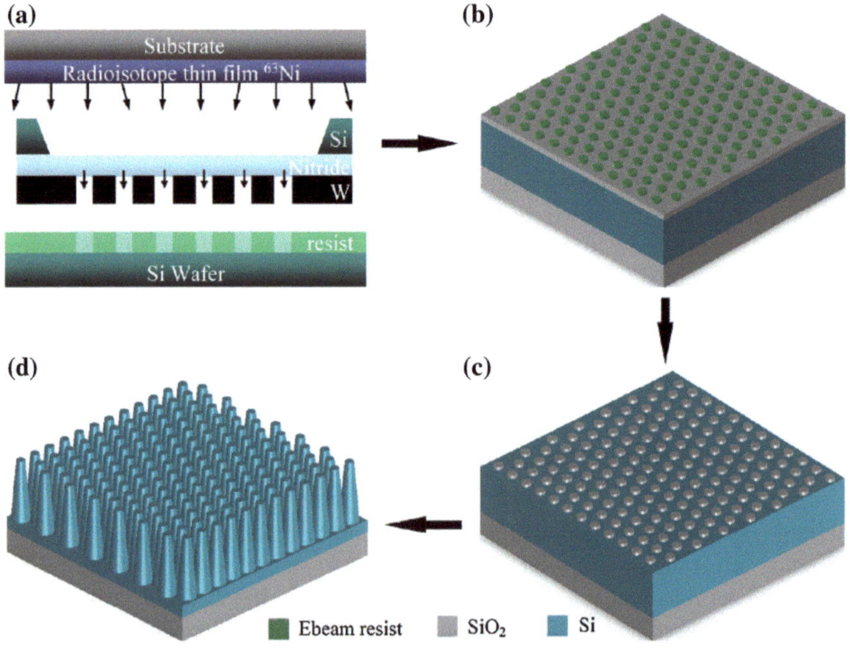

Fig. 4.1 Schematic drawings for ordered Si NCF arrays, fabricated by RIPEL. **a** *Cross section* schematic drawing for RIPEL experimental setting up, using large-area planar radioactive beta-electron thin-film emitters (*top*) to parallel expose e-beam resist (*bottom*) through a stencil mask (*middle*). Using nature's high-energy particle, RIPEL potentially enables large-area massively parallel high throughput electron lithography with low cost, while maintaining sub-35 nm resolution. **b** Electron resist is patterned on stack substrate SiO_2(0.2 μm)/Si(5 μm)/SiO_2(2 μm)/Si, by RIPEL. **c** Pattern is transferred to SiO_2 layer by oxide RIE etching, using e-beam resist as the etching mask. **d** Ordered Si NCF arrays are formed by Si RIE etching, using SiO_2 as the etching mask. The bottom Si substrate of the SOI wafer is not shown here

ing (RIE). Ordered Si NCF arrays were formed, using silicon angled RIE etching recipe, with SiO_2 as the etching mask. Ordered Si NCF arrays with different slant angles can be obtained, depending on the reactive ion etching conditions (Supporting Information). Figure 4.2 shows the scanning electron microscopy (SEM) images of two types of NCF arrays, with different slant angles, using the same ordered SiO_2 post arrays as the etching mask. The first NCF arrays (Fig. 4.2a, b) have 170 nm of top radius (r_1), 400 nm of base radius (r_2), 3.5 μm of height (h), 3.8° of slant angle (θ), and 800 nm of lattice constant. The second (Fig. 4.2c, d) type has 170 nm of equal top and base radii, height of 3.5 μm, 0° slant angle, and 800 nm lattice constant, which we name ordered quasi-nanowire (QNW) arrays. In experiments, the top radii and the lattice constant were controlled by RIPEL mask design. The slant angle and height are determined by the silicon RIE etching condition and etching time, respectively. The bases of all the adjacent NCFs were connected (Fig. 4.2b right), while the bases of the adjacent QNWs were apart from each other (Fig. 4.2d).

4.4 High-Efficiency Light-Trapping for Nano-structure Array

The transmitted diffraction pattern of the ordered Si NCF arrays (800 nm lattice constant) using $\lambda = 609$ nm laser (Fig. 4.3a), shows the high periodicity and uniformity of the nanostructures. The reflection and transmission spectra are shown in parts c and d of Fig. 4.3, respectively. In order to get the membrane window for transmission measurements, the SOI wafer was backside etched through, by Si deep reactive ion etching (DRIE), using 4 μm of PECVD SiO_2 as the etching mask and the buried oxide (2 μm) as an etch stop. The window membranes consist of 3.5 μm of ordered Si NCF arrays (or QNW arrays), 1.5 μm of planar Si layer underneath, and 2 μm of the SOI buried oxide. The control sample made of bare silicon membrane has a thickness of 5 μm planar Si and 2 μm of the SOI buried oxide. We obtained absorption (A) spectra (Fig. 4.3e) after acquisition of reflection (R) and transmission (T) spectra, by $A = 100\,\% - R - T$. The ordered Si NCF arrays exhibit maximal absorbance of ~99 % (upper bound) over wavelengths 400–1100 nm. A small part of diffusely reflected light was not included in the reflection measurements, due to the experimental setup limitation. Therefore, the absorption spectra represent an upper limit.

In order to investigate the light trapping properties as a function of lattice constants, we fabricated three kinds of ordered NCF arrays with the same height (3.5 μm) and constant slant angle ($\theta = 3.8°$), but with different lattice constants, 600 nm ($r_1 = 69$ nm, $r_2 = 300$ nm), 800 nm ($r_1 = 170$ nm, $r_2 = 400$ nm), and 1000 nm ($r_1 = 273$ nm, $r_2 = 500$ nm) on the same SOI wafer. The reflectance spectra are shown in Fig. 4.4a. The ordered NCF arrays with 600 nm lattice constant give the lowest reflectance (<0.5 %), due to the smallest top radius for minimal reflectance, which are consistent with expectations from theoretical predictions. All these three NCF samples show strong antireflective properties throughout the spectral range, which are consistent with the "black surface" (Fig. 4.3b). Both the ordered NCF with lattice

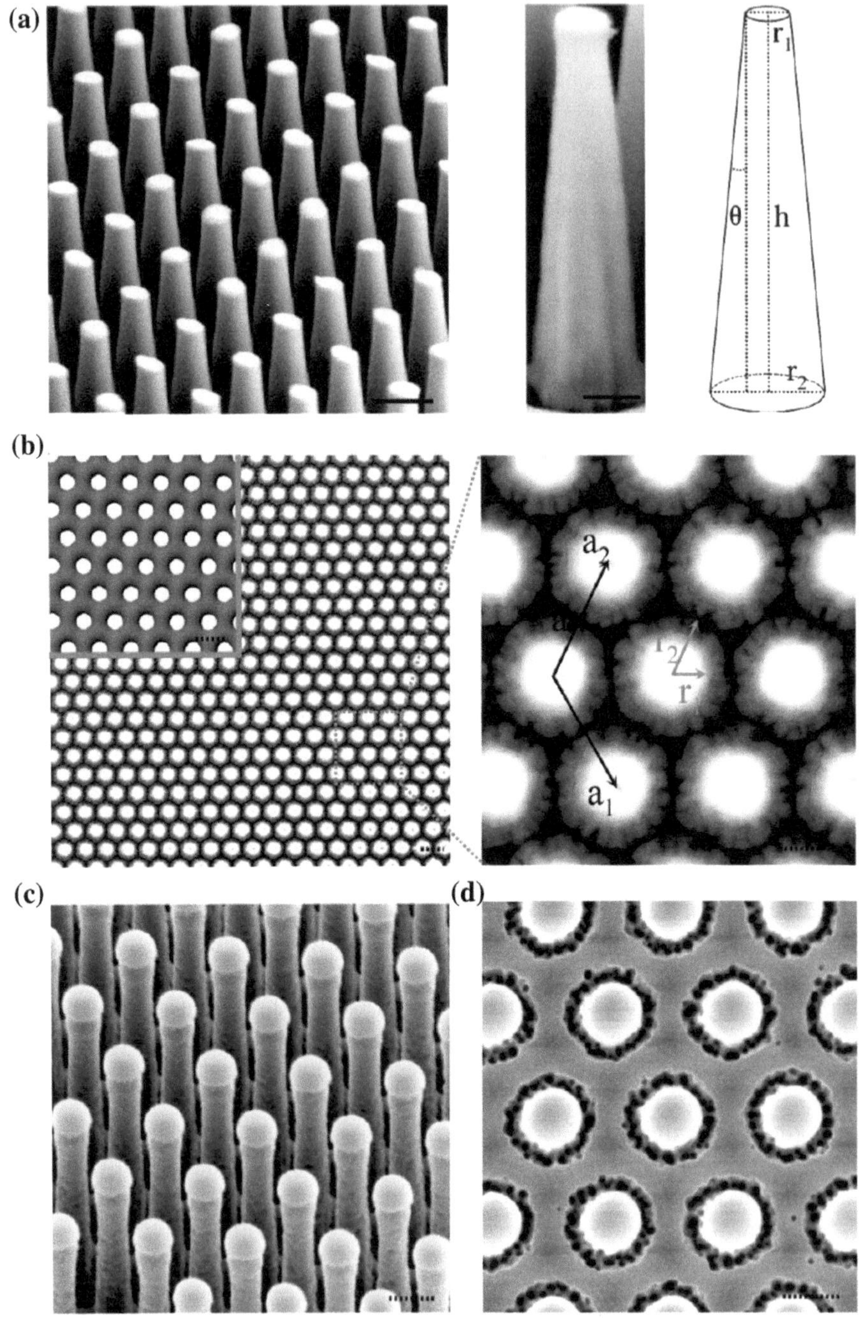

Fig. 4.2 (continued)

◀ **Fig. 4.2** Microscopy images of ordered Si NCF arrays and the ordered quasi-nanowire (QNW) arrays. **a** *Left part* shows the *side view* (45°) SEM image of the ordered Si NCF arrays, with scale bar 800 nm. The zoom in SEM image (45° *side view*) of a single NCF is shown on *right*, with 170 nm of *top* radius (r_1), 400 nm of *base* radius (r_2), 3.5 μm of height (h), and 3.8° of slant angle (θ). The scale bar is 400 nm. **b** *Top view* SEM images of the ordered Si NCF arrays (*left*, 800 nm scale bar) and the zoom in SEM image (*right*, 400 nm scale bar). *Inset* shows the SEM image of the corresponding etching mask SiO_2 post arrays (800 nm scale bar). a_1 and a_2 are the lattice vectors, with lattice constant a. The bases of the adjacent NCFs are connected, with 800 nm of lattice constant and 400 nm of NCF base radius. **c** *Side view* (45°) SEM image of the ordered Si QNW arrays, with scale bar 400 nm. **d** *Top view* SEM image of the ordered Si QNW arrays (d), with scale bar 400 nm

constants of 800 and of 1000 nm show ultralow transmittance spectra (Fig. 4.4b) throughout the spectral range 400–1100 nm, while the ordered NCF arrays with 600 nm lattice constant show slightly increasing transmittance for the wavelength near the band gap region 800–1100 nm.

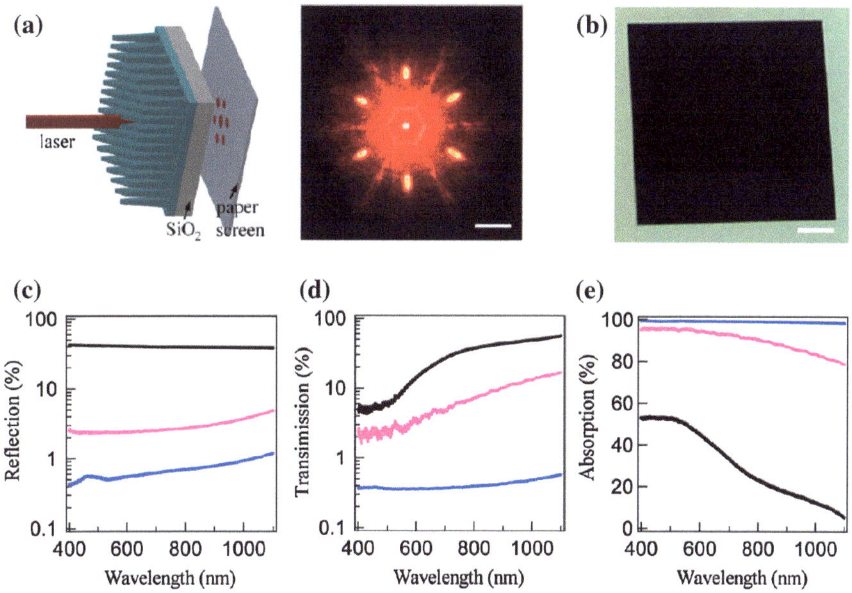

Fig. 4.3 Light trapping measurements. **a** Transmitted diffraction pattern of the NCF arrays (800 nm lattice constant) using a $\lambda = 609$ nm laser, projected onto a white paper screen and taken by digital camera. The paper screen is 4 mm away from the SiO_2 substrate. Scale bar is 3 mm. The window membranes consist of 3.5 μm of ordered Si NCF arrays, 1.5 μm of planar Si layer underneath and 2 μm of the SOI buried oxide. **b** Photograph of the ordered Si NCF arrays, with scale bar 1 mm. Parts **c**, **d**, and **e** show the measured reflection, transmission, and absorption spectra, respectively, for light normally incident on the ordered Si NCF arrays (*blue*), ordered Si QNW arrays (*pink*), and control bare silicon (*black*). The window membranes consist of 3.5 μm of ordered Si NCF (or QNW) arrays, 1.5 μm of planar Si layer underneath and 2 μm of the SOI buried oxide. The control bare silicon has a thickness of 5 μm

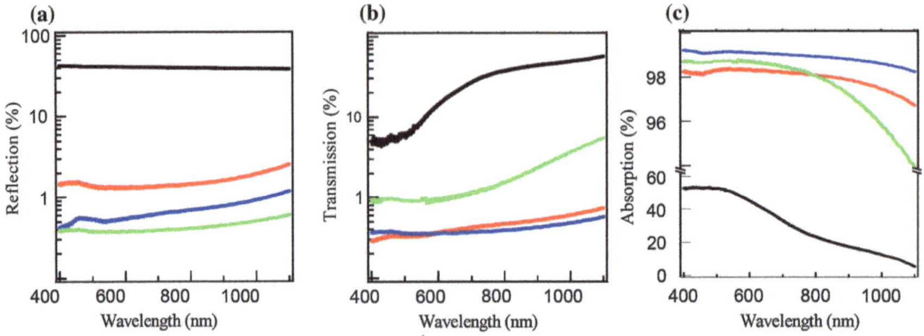

Fig. 4.4 Experimentally measured reflection (**a**), transmission (**b**) and absorption (**c**) spectra for light normally incident on the ordered Si NCF arrays. Four types of samples were measured: bare silicon (5 μm thick) (*black*), NCF arrays with lattice constants of 600 nm ($r_1 = 69$ nm, $r_2 = 300$ nm, *green*), 800 nm ($r_1 = 170$ nm, $r_2 = 400$ nm, *blue*), and 1 μm ($r_1 = 273$ nm, $r_2 = 500$ nm, *red*). All the window membranes consist of 3.5 μm of ordered Si NCF arrays, 1.5 μm of planar Si layer underneath, and 2 μm of the SOI buried oxide

4.5 High-Efficiency Nano-structured Solar Cell

Due to the maximal absorption throughout the spectra range, the ordered Si NCF arrays with 800 nm lattice constant were selected for solar cells fabrication, to see the solar energy conversion efficiency. The radial p–n junction was formed via boron ion implantation ($E = 10$ keV), starting with n^+-type ordered Si NCF arrays fabricated using SOI top silicon (resistivity 0.015 $\Omega \cdot$cm, doping concentration 2×10^{18} cm^{-3}). Ion implantation offers the advantage of precise control over the total dose of dopants, depth profile, and most importantly works well also for high doping levels of the order of 10^{19}–10^{21} cm^{-3}. Considering the slant angle (3.8°) of the NCF, the sidewall implantation depth was estimated to be 20 nm, using the lateral straggling depth value calculated from SRIM ion implantation simulator. The initial doping concentration on the NCF sidewall surface was controlled to be 1×10^{20} cm^{-3} during the ion implantation. Rapid thermal annealing (RTA) at 900 °C for 60 s was used directly after implantation to activate the dopants and remove the implantation damage. In order to minimize the loss from the surface recombination coming from the large total surface area increased by NCF array geometry, a thin layer (~20 nm thick) of SiO_2 wet oxide was formed as the outside passivation layer for the p–n junction, at 900 °C for 6 min. During the RTA and wet oxidation high temperature processes, the boron dopants diffused toward the core by approximately 80 nm more, estimated from the literature [24], offering a final junction depth of 100 nm and final boron doping concentration of 2×10^{19} cm^{-3} after the diffusion. From outer surface to the core center, the radial p–n junction consists of 20 nm of outside passivation SiO_2, 100 nm of quasi-neutral boron-doped p^+ type shell region, 25 nm of depletion inter shell region, and 45 nm top radius (275 nm base radius) of quasi-neutral n^+ type Si core region (in radii). It is important to use both heavily doped n^+ and p^+ quasi-neutral region to make a small depletion region in each NCF, so that the NCF is not

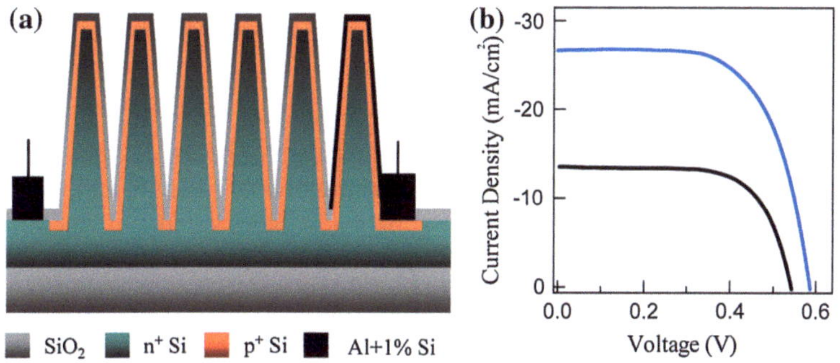

Fig. 4.5 Solar cell output characteristics. **a** *Cross section* schematic drawing of the solar cell structures. **b** Current density versus voltage curves for the solar cells consisting of bare silicon (*black*), Si NCF arrays with 800 nm (*blue*), under AM 1.5 G normal illumination

fully depleted, since the quasi-neutral core region should exist to conduct carriers down to the bottom contact. The schematic solar cell structure plot is shown in Fig. 4.5a.

Figure 4.5b shows representative current density versus voltage measurements of the individual solar cells, under AM 1.5 G normal illumination of 100 mW/cm^2 at room temperature. For the solar cells with only controlled bare Si and with ordered Si NCF arrays (800 nm lattice constant), they had short circuit current densities (J_{sc}) of 13.6 and 26.4 mA/cm^2, open circuit voltages (V_{oc}) of 0.54 and 0.59 V, fill factors (FF) of 0.71 and 0.69, and efficiencies (η) of 5.2 and 10.8 %, respectively. The high energy conversion efficiency of the ordered Si NCF arrays is due to the strong light trapping and absorption properties throughout the spectra range, with the presence of both photonic crystal enhancement effect and radii-variation-induced broad-band absorption. Although these NCF structure arrays have high light absorption efficiency, the solar cell efficiency is still less than half of that for bulk crystalline Si solar cells [25] and the open circuit voltage and fill factor are lower too. This is probably due to the high surface recombination losses, which are introduced by the highly enhanced surface area from the nanostructures, although a thin passivation oxide layer was used.

In summary, a novel large-area ordered silicon NCF array structure, exhibiting an impressive absorbance of ~99 % (upper bound) over wavelengths 400–1100 nm with a thickness of only 5 μm, is demonstrated by using the RIPEL technique that has the potential for very high lithography throughput, high resolution, and low cost. High-efficiency solar cells have been achieved using these ordered Si NCF arrays. Moreover, this novel high light-trapping structure array and the RIPEL fabrication technique can be also used for low-cost substrate solar cells, using poly-silicon or other semiconductors [26].

4.6 Photonic Crystal-Based All-Optical Pressure Sensor

An all optical pressure sensor has several advantages of being able to be used in harsh environments where wiring and electronics cannot be used. Optical pressure sensors can be interrogated at large distances using optical imaging and waveguide. To date, optical pressure sensors use air-gaps made using sacrificial etching under poly-silicon or dielectric films [27, 28]. Such sensors result in a gap that is also a function of temperature and has dynamic range limited by the gap. Most importantly, they are designed to be used with a specific wavelength of interrogating light.

Here, we realize an all-optical pressure sensor by fabricating controllable vertical silicon nanowire arrays on a Si/SiO_2 membrane [20]. Vertical nanowire array acts as a photonic crystal and can affect the propagation of electromagnetic waves in the same way as the periodic potential in a semiconductor crystal affects the electron motion by defining allowed and forbidden electronic energy bands [29]. In our design, the light absorbing photonic crystal is formed laterally on a membrane with ordered vertical nanowires array. This eliminates the need for under-etching, and very high accuracy array can be formed using RIPEL. The photonic-crystal absorbs light over a wide wavelength range enabling white-light interrogation. Higher dynamic range is realized as the nanowires can be moved out as the membrane is pressurized.

We can use RIPEL to pattern the vertical silicon nanowire arrays, on a SOI wafer, by silicon reactive ion etching (RIE) using SiO_2 nano-posts as the etching mask. The vertical silicon nanowire array has 100 nm diameter wires, 1 μm pitch and 1.5 μm height (Figs. 4.6 and 4.7). These nanowire arrays show uniform wire diameter, controllable wire orientation and wire pitch spacing. Optical lithography with backside alignment was used to pattern the backside circular cavity (800 μm diameter), followed by silicon DRIE to etch through the silicon substrate using the SOI buried oxide as the stopping layer. The oxide layer on the silicon membrane has a built in compressive stress which when released after etch, results in a slightly wrinkled membrane as shown in Fig. 4.8. The pressure sensor was mounted on a glass tube that was pressurized, applying pressure onto the membrane. By modulating the pressure, the photonic crystal nanowire structures reflect different patterns with spatially varying colors (Fig. 4.9). Hence, instead of the displacement, the resulting color pattern is the pressure indicator. Our visual pressure sensor showed high sensitivity between 0 up to 8×10^4 Pa with 1 Pa resolution determined by the cross-correlation with the CMOS imager used to measure the displacement. Given that the CMOS imagers have a sensitivity of $I_{\mathrm{read}}[\lambda]$, we can estimate $I_{\mathrm{read}}[P]$ as 0.5 Å/Pa. With a current noise of 0.5 Å/pixel, we can estimate the pressure sensitivity of this sensor to be 1 Pa.

For a uniform circular membrane geometry, with a uniform pressure load p, the plate's deflection $z(r, p)$ in z direction as a function of radial position, r, plate radius, r_0, and the plate material's flexural rigidity, D, can be expressed as [30]:

$$z(r, p) = \frac{p \cdot r_0^4}{64 \cdot D}\left(1 - \frac{r^2}{r_0^2}\right)^2 \tag{4.1}$$

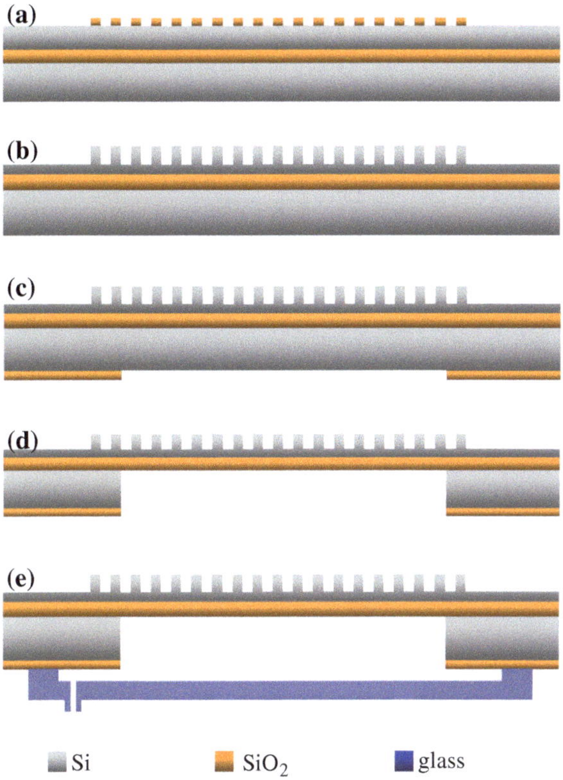

Fig. 4.6 Photonic crystal-based all-optical pressure sensor fabrication process schematic drawings. **a** A thin layer (~300 nm) of PECVD SiO_2 was deposited on SOI wafer (3 μm Si/2 μm SiO_2). The PECVD SiO_2 layer was patterned by electron beam lithography, followed by oxide RIE etching. **b** Vertical Si nanowire arrays were formed by Si RIE etching, using SiO_2 as the etching mask. **c** A thick layer (~4 μm) PECVD SiO_2 was deposited on back side of the SOI wafer. Then it was patterned by optical lithography EV620 using backside alignment, followed by SiO_2 RIE etching. **d** The SOI wafer Si substrate was etched through by Si deep RIE, using the back PECVD SiO_2 layer as the etching mask and the SOI buried oxide as the stopping layer. **e** A glass pressure controller adapter was mounted to the back by superglue

Flexural rigidity is given by [31], where t is the plate effective thickness, and E and υ are the plate material's effective Young's modulus and Poisson ratio, respectively.

$$D = \frac{E \cdot t^3}{12 \cdot (1 - \upsilon^2)} \tag{4.2}$$

From equation (4.1), we see that the maximum plate deflection, which occurs at the plate's center ($r = 0$), can be expressed as:

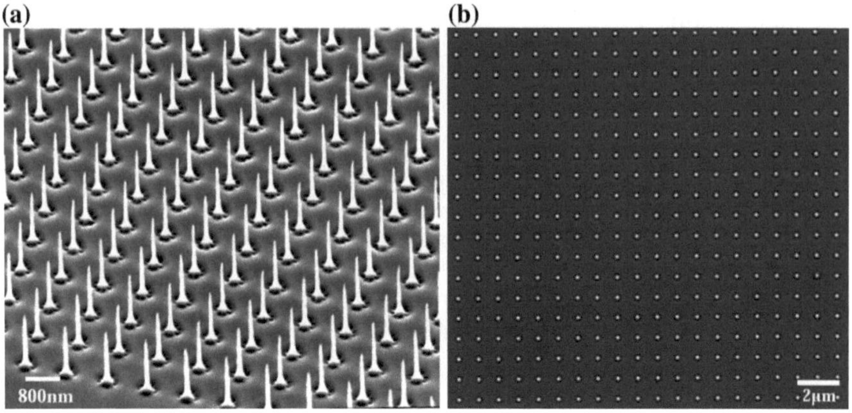

Fig. 4.7 **a** SEM image (45° *side view*) of the vertical Si nanowire arrays, with 100 nm of wire diameter, 1 μm of nanowire pitch and 1.5 μm of nanowire height. **b** The *top view* SEM image of the SiO_2 posts arrays in Fig. 6.5a

$$z(p)_{\max} = \frac{p \cdot r_0^4}{64 \cdot D}. \tag{4.3}$$

As the pressure changes, the membrane is deformed, and the nanowire deflection angle are changed leading to a nanowire reflectance modulation. The reflection wavelength shift $\Delta\lambda$ should be a function of the radial position, r, and the pressure load, p. It can be given by,

$$\begin{aligned}\Delta\lambda(r,p) &= a \cdot \sin\left[\tan^{-1}\left(-\frac{\partial(z(r,p))}{\partial r}\right)\right] \\ &= a \cdot \sin\left[\tan^{-1}\left(\frac{p \cdot r_0^2 \cdot r}{32 \cdot D}\left(1 - \frac{r^2}{r_0^2}\right)\right)\right]\end{aligned} \tag{4.4}$$

where a is the lattice constant of the photonic crystal nanowire arrays, and p is the membrane pressure load.

4.7 High-Efficiency Opto-Thermo-Mechanical MEMS Excitation

The pulses of light, incident on a sample of matter, can be absorbed and converted to heat, leading to local thermal excitation and detectable acoustic waves. This kind of photo-acoustic effect continues to be applied in an increasing number of fields, such as photo-acoustic imaging in biomedicine [32], photo-acoustic spectroscopy in trace gas monitoring [33], optical actuation for MEMS devices [34], etc. Optical actuation

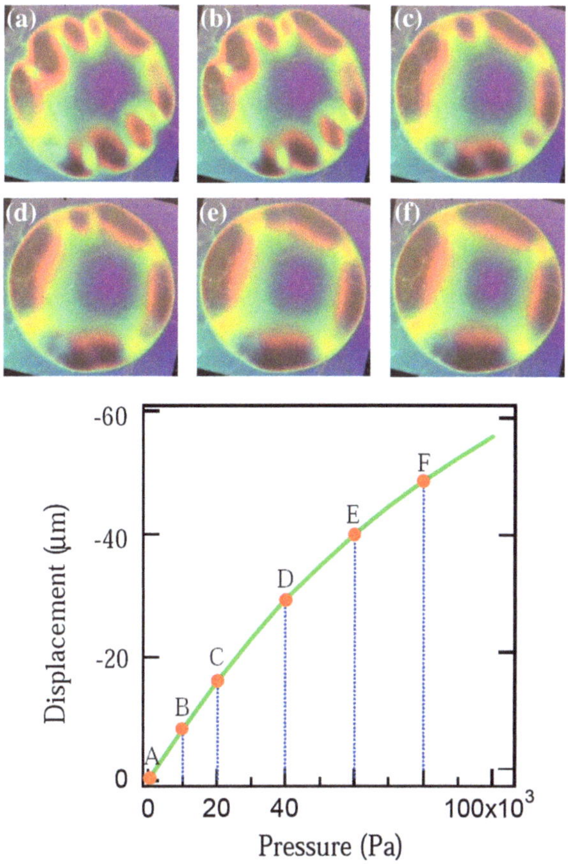

Fig. 4.8 Visual pressure sensor experiment test and simulation. The *circular* membrane has 800 µm diameter and its load pressure is defined as the pressure difference between the outside pressure and the cavity pressure. **a–f** show the *top view* photograph images of the nanowire membrane color change as increasing the membrane load pressure from 0 up to 8×10^4 Pa. The displacement corresponds to the maximum displacement of the *center* point of the *circular* membrane. The initial pattern is due to compressive stress in the Si/SiO_2 bilayer

for MEMS takes advantage of directly coupling photonic energy into selected device areas without any electrical interconnects as required in integrated electrostatic and piezoelectric actuation [35]. In opto-thermo MEMS actuation, light incident onto the structures is absorbed and converted to heat via photon absorption. Hence, the optical actuation efficiency is highly dependent on the material optical absorption efficiency. Different materials and molecules will absorb photons to different extents at different optical wavelengths. For example, silicon has reasonably high optical absorption efficiency in the visible light spectrum, while it is almost transparent to infrared (IR) light. Therefore, a system that could enhance light absorption at broad-band wavelengths is required for high-efficiency broad-band optical MEMS

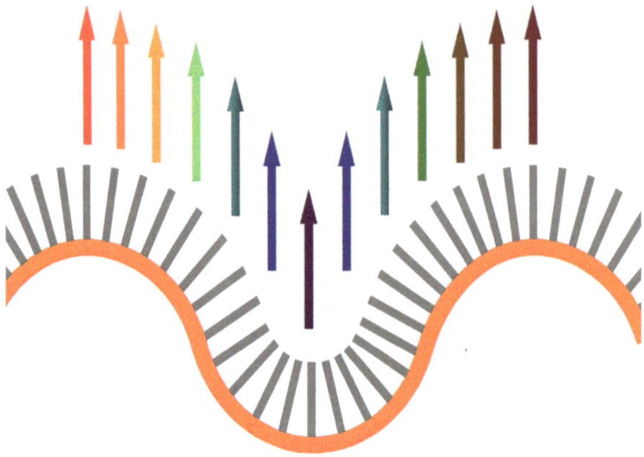

Fig. 4.9 Schematic mechanism explanation for the light reflection from these photonic crystal vertical nanowire structures on deformed membrane. The bending of the membrane modulates the view angle of the nanowire arrays, resulting in wavelength dependent reflection. The *arrows with different color* correspond to the reflecting visible lights with different wavelengths

actuation. Thin-film vertical silicon nano-pillar arrays with photonic crystal structure has been demonstrated to have highly enhanced light trapping and broad-band (from UV to near IR) optical absorption efficiency [14, 36]. Here, we present the first-ever opto-thermal-mechanical MEMS actuation for membrane resonator, using photonic crystal nanopillar arrays on top of membrane as the optical absorbers. This optical actuation for MEMS takes advantage of directly coupling of energy into selected device areas without any electrical interconnects. More importantly, we can use various lasers with different wavelength (from UV to near IR), or even modulated white light with continuous spectra, to actuate our resonator, since the vertical silicon nanopillar arrays with photonic crystal structures and wavelength comparable lattice constant, have been demonstrated to have highly enhanced broad-band optical absorption efficiency. Therefore, this realizes a high-efficiency broad-band optical actuation. Most interestingly, a nanopillar membrane acoustic speaker was demonstrated, by using a closely located microphone to detect the vibration response in the frequency range from 1 to 20 kHz [19].

Fabrication and experiments: Electron beam lithography was used to pattern the vertical silicon nanopillar arrays, on a SOI wafer, followed by silicon RIE etching, using SiO_2 nanoposts as the etching mask. The vertical silicon nanopillar array has conical pillars with 1.5 μm height, 250 nm base diameter, 100 nm top diameter, at a 600 nm pitch (Figs. 4.10 and 4.11). As the photons enter the nanopillar array they encounter an infinitum of reflective boundary conditions forming equivalent optical resonators at different wavelengths, resulting in light trapping. The measured optical absorption efficiency on these ordered nanopillar arrays were more than 90 % over broad-band wavelengths 400–1000 nm, consistent with the "fully black" surface (Fig. 4.3b). Optical lithography with backside alignment was used to pattern the backside circular cavity (Fig. 4.10, 2.5 mm diameter), followed by silicon DRIE to

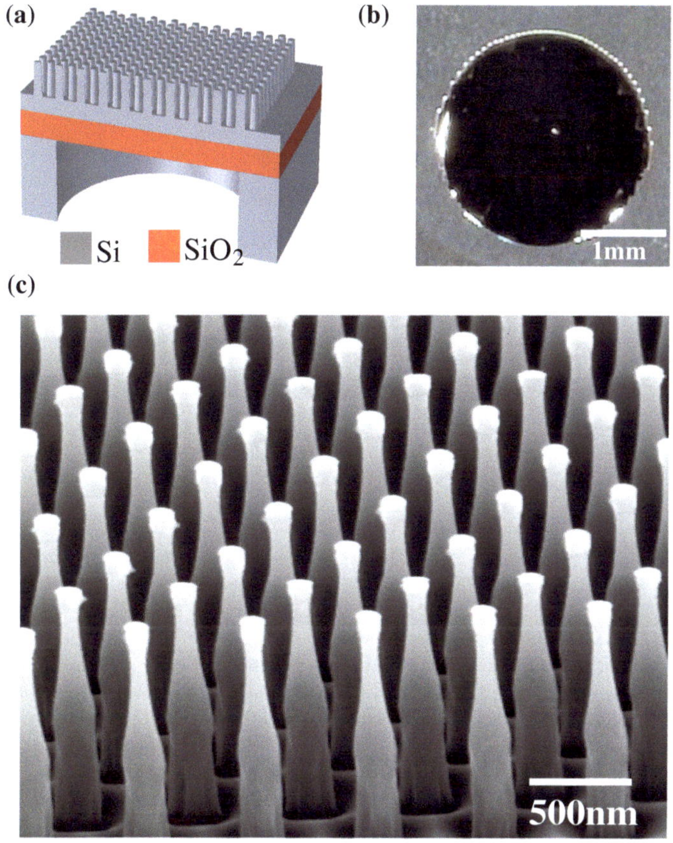

Fig. 4.10 **a** Schematic of the Si/SiO_2 (1.5 μm/2 μm) membrane with vertical Si nanopillar array. **b** Photo image (*top view*) of the *backside circular* cavity. **c** SEM image (60° view) of the nanopillar arrays, with 1.5 μm height, 250 nm base diameter, 100 nm top diameter, at a 600 nm pitch

etch through the silicon substrate using the SOI buried oxide as the stopping layer. A thin 50 nm layer of nitride was deposited to compensate the compressive stress in the SiO_2 layer.

For optical excitation, the irradiation of the modulated laser diode was focused onto the center of the membrane with nanopillar arrays. The local heating from the incident modulated laser leads to Si/SiO_2 bending movements (Fig. 4.12b, c). Since the top Si layer was thermally isolated by the SiO_2, convection losses are negligible. In order to verify the optical excitation, we used piezoelectric actuation, by mounting a PZT plate to the backside of the device membrane resonator was measured by PolyTec^{TM} interferometer with sub-1-nm amplitude resolution. Measured resonance spectra response of the membrane resonator from optical actuation matched very well with the spectra from PZT excitation. These data show linear optical actuation. The standing wave vibration amplitude 2D mapping patterns of the modes were

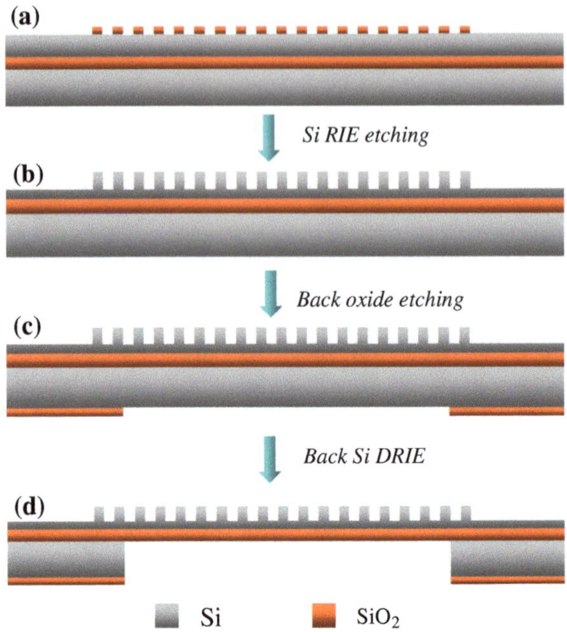

Fig. 4.11 Photonic crystal-based all-optical pressure sensor fabrication process schematic drawings. **a** A thin layer (~300 nm) of PECVD SiO_2 was deposited on SOI wafer (3 μm Si/2 μm SiO_2). The PECVD SiO_2 layer was patterned by electron beam lithography, followed by oxide RIE etching. **b** Vertical Si nanowire arrays were formed by Si RIE etching, using SiO_2 as the etching mask. **c** A thick layer (~4 μm) PECVD SiO_2 was deposited on back side of the SOI wafer. Then it was patterned by optical lithography EV620 using backside alignment, followed by SiO_2 RIE etching. **d** The SOI wafer Si substrate was etched through by Si deep RIE, using the back PECVD SiO_2 layer as the etching mask and the SOI buried oxide as the stopping layer

measured (Fig. 4.13). During mapping, the optical actuation frequency was fixed at the resonance mode and the interferometer laser gun position was controlled by a 2-axis stage controller with sub-0.5 μm lateral resolution.

Most interestingly, a nanopillar membrane speaker was demonstrated, by using a closely located microphone to detect the vibration response in the frequency range from 1 to 20 kHz (Fig. 4.14). The acoustic emission could be measured using B&K 2205 Sound Pressure Level Meter, corresponding to sound pressure levels of 50–60 dB. When the actuation was fixed at a specific resonance frequency of the membrane 11.04 kHz, real-time microphone output could be recorded (Fig. 4.14b). When the optical actuation was switched between "on" and "off" back and forth, the output was tuned accordingly.

Model and Simulation: For our Si/SiO_2 bilayer membrane, we could simply model the system to be a flexural rigidity-dominated elastic unimorph plate with silicon and silicon-dioxide layers. With opto-thermal excitation, the local heating from the incident modulated laser leads to Si/SiO_2 bending moments (Fig. 4.12c). This

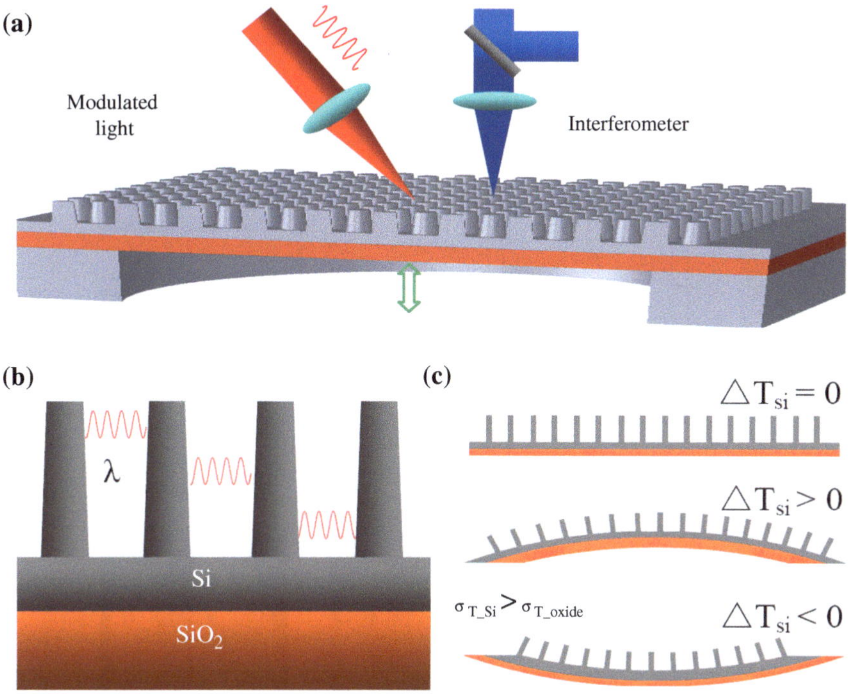

Fig. 4.12 **a** Schematic for experimental setup in vacuum. The vibration response of the membrane resonator was measured using PolyTec interferometer. The laser diode was controlled by laser diode driver, with modulation input from a function generator. **b** Schematic of light trapping and absorption in the photonic crystal nanopillar arrays. The light is absorbed and converted to heat. **c** Schematic for Si/SiO_2 bending moments, due to the local heating from the modulated laser

opto-thermal excitation could be modeled as a transverse loading per unit area $f(r, \theta, t)$. By considering damping, the nonlinear partial differential equation governing the membrane can be obtained as equation [31]:

$$\rho h \frac{\partial^2 Z}{\partial t^2} + D\nabla^4 Z + \zeta \frac{\partial Z}{\partial t} = f(r, \theta, t) \tag{4.5}$$

where $Z(r, \theta, t)$ is the transverse displacement, ζ is the damping coefficient, h is the effective membrane thickness, ρ is the effective mass density, and D is the effective flexural rigidity of the membrane, and ∇^2 is the Laplacian. By solving this equation for a circular membrane with radius r_0, the resonant frequencies and their corresponding transverse displacement profiles at different mode shapes are obtained as follows [37]:

$$f_{mn} = \frac{1}{2\pi}\sqrt{\frac{\gamma_{mn}^4 D}{\rho h}} \tag{4.6}$$

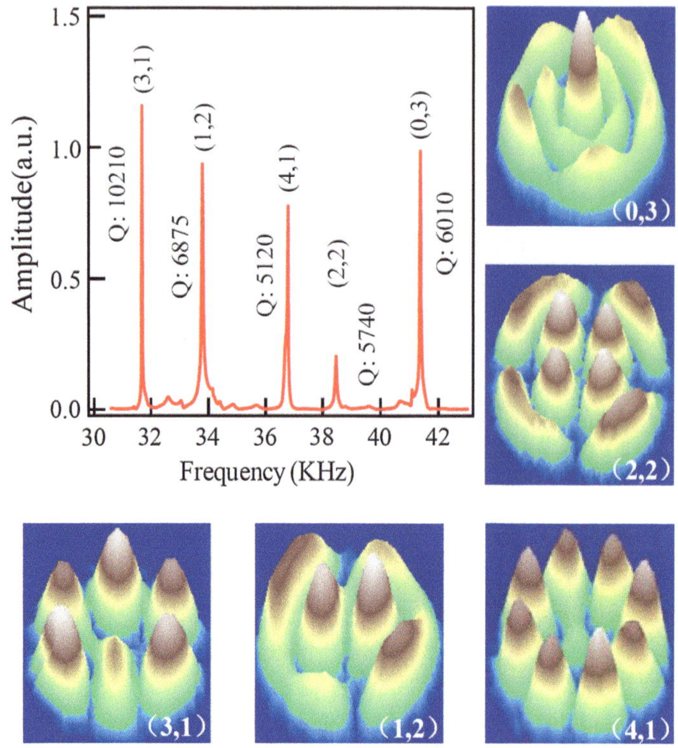

Fig. 4.13 Measured resonances spectra response for optical actuation by the irradiation of the modulated laser diode, using interferometer (Fig. 4.12a). Our interferometer system has a limited minimum frequency response of 20 kHz. The modes are named by (n, m), where n is circular node and m is diametrical node. The m and n values are determined by experimental mode amplitude mapping and simulation. Surrounding image plots show the measured amplitude mapping patterns of those corresponding modes. When mapping, the optical actuation was fixed at the resonance mode and the interferometer laser gun was controlled by a stage controller for lateral movement with sub-1 μm lateral resolution. These experimental data matched well with the theoretical vibration modes

$$Z_{mn}(r,\theta) = \left[J_m(\gamma_{mn}r) - \frac{J_m(\gamma_{mn}r_0)}{I_m(\gamma_{mn}r_0)} I_m(\gamma_{mn}r) \right] \cdot [A_{1m}\sin(m\theta) + A_{2m}\cos(m\theta)] \tag{4.7}$$

where the functions $J_m(\gamma_{mn}r)$ and $I_m(\gamma_{mn}r)$ are the Bessel function and the modified Bessel function of the first kind, γ_{mn} are the eigenvalues and the subscripts m and n represent the number of nodal diameters and nodal circles for the vibration mode, respectively. The final opto-thermo-mechanical actuation efficiency should be a combination of opto-thermo efficiency $\eta_{\text{op-th}}$ and thermal-mechanical efficiency $\eta_{\text{th-me}}$:

$$\eta_{\text{final}} = \eta_{\text{op-th}} \cdot \eta_{\text{th-me}} \tag{4.8}$$

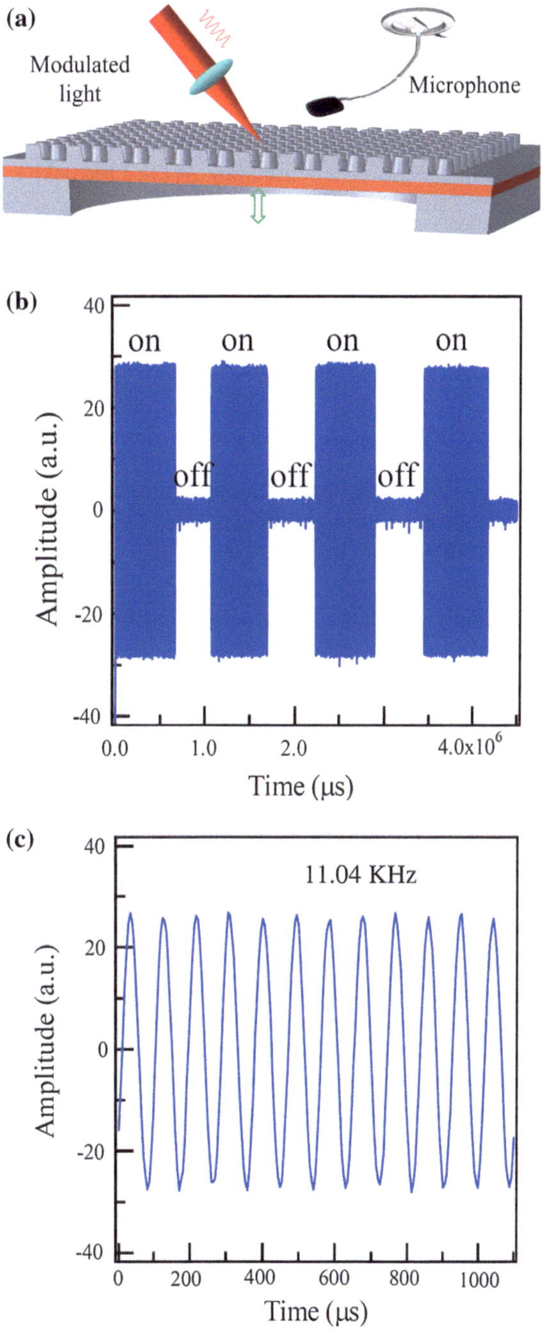

Fig. 4.14 Nanopillar membrane opto-thermo-mechanical actuated acoustic speaker demonstration. **a** Schematic for experimental setup in air. A microphone, which was placed closely to the membrane, was used to record the vibration response of the membrane resonator in the frequency range from 1 to 20 kHz. **b** Real-time microphone recorded output, in which the optical actuation was switched "on" and "off" back and forth. The actuation was fixed at a specific resonance frequency of the membrane, 11.04 kHz. **c** Zoom in plot of (**b**) at "on" state

The thermal-mechanical efficiency depends on the strain and temperature distributions in the resonator. Here, we assume that we have the same thermal-mechanical efficiency. Therefore, the final actuation efficiency only depends on the opto-thermal efficiency. For laser heating actuation, light incident onto the resonator is absorbed and converted to heat via photon absorption. The opto-thermo efficiency is related to the light absorption efficiency $A(\lambda)$ of the device.

$$\eta_{\mathrm{op-th}} \propto A(\lambda) \tag{4.9}$$

Therefore, we could calculate the efficiency enhancement factor F for our device, compared with the planar Si membrane devices.

$$F = \frac{\eta_{\mathrm{final\_NW}}}{\eta_{\mathrm{final\_planarSi}}} = \frac{\eta_{\mathrm{op-th\_NW}}}{\eta_{\mathrm{op-th\_planarSi}}} = \frac{A_{\mathrm{NW}}(\lambda)}{A_{\mathrm{planarSi}}(\lambda)} \tag{4.10}$$

where λ is the wavelength of the driving light.

For the resonant mode detected at 11 kHz, the sound pressure level was measured to be 55 dB, indicating a net acoustic power of 1.9 nW, using 30 % acoustic power collection efficiency by the microphone. The amount of optical energy being dissipated is 10 mW of AC power. Thus the net efficiency of optical to acoustics is 0.2 ppm.

References

1. Gevorkian, P.: Sustainable Energy Systems Engineering: The Complete Green Building Design Resource. McGraw-Hill Professional, New York (2007)
2. Huang, Y.-F., et al.: Improved broadband and quasi-omnidirectional anti-reflection properties with biomimetic silicon nanostructures. Nat. Nanotechnol. **2**, 770–774 (2007)
3. Tsakalakos, L., et al.: Strong broadband optical absorption in silicon nanowire films. J. Nanophotonics **1**, 013552 (2007)
4. Sivakov, V., et al.: Silicon nanowire-based solar cells on glass: synthesis, optical properties, and cell parameters. Nano Lett. **9**, 1549–1554 (2009)
5. Stelzner, T., et al.: Silicon nanowire-based solar cells. Nanotechnology **19**, 295203 (2008)
6. Tian, B., et al.: Coaxial silicon nanowires as solar cells and nanoelectronic power sources. Nature **449**, 885–889 (2007)
7. Tsakalakos, L., et al.: Silicon nanowire solar cells. Appl. Phys. Lett. **91**, 233117 (2007)
8. Garnett, E.C., Yang, P.: Silicon nanowire radial p–n junction solar cells. J. Am. Chem. Soc. **130**, 9224–9225 (2008)
9. Garnett, E., Yang, P.: Light trapping in silicon nanowire solar cells. Nano Lett. **10**, 1082–1087 (2010)
10. Lin, C., Povinelli, M.L.: Optical absorption enhancement in silicon nanowire arrays with a large lattice constant for photovoltaic applications. Opt. Express **17**, 19371–19381 (2009)
11. Hu, L., Chen, G.: Analysis of optical absorption in silicon nanowire arrays for photovoltaic applications. Nano Lett. **7**, 3249–3252 (2007)
12. Yu, Z., Raman, A., Fan, S.: Fundamental limit of nanophotonic light trapping in solar cells. Proc. Nat. Acad. Sci. **107**, 17491–17496 (2010)

13. Chutinan, A., John, S.: Light trapping and absorption optimization in certain thin-film photonic crystal architectures. Phys. Rev. A **78**, 023825 (2008)
14. Zhu, J., et al.: Optical absorption enhancement in amorphous silicon nanowire and nanocone arrays. Nano Lett. **9**, 279–282 (2008)
15. Costner, E.A., Lin, M.W., Jen, W.-L., Willson, C.G.: Nanoimprint lithography materials development for semiconductor device fabrication. Annu. Rev. Mater. Res. **39**, 155–180 (2009)
16. Lu, Y., Lal, A.: Vacuum-free self-powered parallel electron lithography with sub-35-nm resolution. Nano Lett. **10**, 2197–2201 (2010)
17. Lu, Y., Yoshimizu, N., Lal, A.: Self-powered near field electron lithography. J. Vac. Sci. Technol. B **27**, 2537–2541 (2009)
18. Lu, Y., Peng, S., Luo, D., Lal, A.: Low-concentration mechanical biosensor based on a photonic crystal nanowire array. Nature Commun. **2**, 578 (2011)
19. Lu, Y., Lal, A.: Acoustic speaker based on high-efficiency broadband nano-pillar photonic crystal opto-thermo-mechanical mems excitation. In: Proceedings of the 16th International Conference on Solid-State Sensors, Actuators and Microsystems, Beijing, China, pp. 2678–2681 (2011)
20. Lu, Y., Lal, A.: Photonic crystal based all-optical pressure sensor. In: Proceedings of the 24th International Conference on Micro Electro Mechanical Systems, Cancun, Mexico, pp. 621–624 (2011)
21. Lu, Y., Peng, S., Luo, D., Lal, A.: Femtomolar sensitivity DNA photonic crystal nanowire array ultrasonic mass sensor. In: Proceedings of the 25th IEEE International Conference on Micro Electro Mechanical Systems, Paris, France, pp. 88–91 (2012)
22. John, S.: Strong localization of photons in certain disordered dielectric superlattices. Phys. Rev. Lett. **58**, 2486–2489 (1987)
23. Diedenhofen, S.L., et al.: Broad-band and omnidirectional antireflection coatings based on semiconductor nanorods. Adv. Mater. **21**, 973–978 (2009)
24. Garnett, E.C., et al.: Dopant profiling and surface analysis of silicon nanowires using capacitance-voltage measurements. Nat. Nanotechnol. **4**, 311–314 (2009)
25. Zhao, J., Wang, A., Altermatt, P., Green, M.A.: Twenty-four percent efficient silicon solar cells with double layer antireflection coatings and reduced resistance loss. Appl. Phys. Lett. **66**, 3636–3638 (1995)
26. Fan, Z., et al.: Challenges and prospects of nanopillar-based solar cells. Nano Res. **2**, 829–843 (2009)
27. Kim, Y., Neikirk, D.P.: Micromachined fabry-perot cavity pressure transducer. IEEE Photonics Technol. Lett. **7**, 1471–1473 (1995)
28. Xiao, G.Z., Adnet, A., Zhang, Z., Lu, Z., Grover, C.P.: Fiber-optic fabry-perot interferometric gas-pressure sensors embedded in pressure fittings. Microwave Opt. Technol. Lett. **42**, 486–489 (2004)
29. Ohtaka, K.: Energy band of photons and low-energy photon diffraction. Phys. Rev. B **19**, 5057–5067 (1979)
30. Lohfink, A., Eccardt, P.C.: Linear and nonlinear equivalent circuit modeling of cmuts. IEEE Trans. Ultrason. Ferroelectr. Freq. Control **52**, 2163–2172 (2005)
31. Timoshenko, S.P., Woinowsky-Kreiger, S.: Theory of Plates and Shells, 2nd edn. McGraw-Hill Higher Education, New York (1964)
32. Xu, M., Wang, L.V.: Photoacoustic imaging in biomedicine. Rev. Sci. Instrum. **77**, 041101 (2006)
33. Harren, F.J.M., Cotti, G., Oomens, J., Hekkert, S.T.L.: Encyclopedia of Analytical Chemistry. Wiley, Chichester (2006)
34. Ilic, B., Krylov, S., Aubin, K., Reichenbach, R., Craighead, H.G.: Optical excitation of nanoelectromechanical oscillators. Appl. Phys. Lett. **86**, 193114 (2005)
35. Klaasse, G., Puers, R., Tilmans, H.A.C.: Piezoelectric versus electrostatic actuation for a capacitive rf-mems switch. In: Proceedings of the SeSens, p. 631 (2002)
36. Lu, Y., Lal, A.: High-efficiency ordered silicon nano-conical-frustum array solar cells by self-powered parallel electron lithography. Nano Lett. **10**, 4651–4656 (2010)

37. Olfatnia, M., Xu, T., Ong, L.S., Miao, J.M., Wang, Z.H.: Investigation of residual stress and its effects on the vibrational characteristics of piezoelectric-based multilayered microdiaphragms. J. Micromech. Microeng. **20**, 015007 (2010)

Chapter 5
Synchrotron-Excited Photoluminescence Spectroscopy of Silicon- and Carbon-Containing Quantum Dots in Low Dimensional SiO_2 Matrices

Anatoly F. Zatsepin and Evgeny A. Buntov

Abstract A comprehensive method to study semiconductor nanoparticles in thin film SiO_2 matrices has been developed. Selective and high-intensity synchrotron excitation allows the investigation of the nanoparticles energy structure. It is shown that the interference fringes affecting the optical excitation spectra of thin films may be neutralized by means of a special numerical technique. The spectral and kinetic properties of the Si, C, and SiC quantum dots (QD) formed by ion implantation in thin silica films were studied in details. Photoluminescence thermal quenching is shown to contain two stages and is dominated by Street law at low temperatures. Several indirect QD excitation mechanisms are realized, involving point defects, free, and self-trapped SiO_2 matrix excitons. An exciton-assisted mechanism is dominating at helium temperatures. A resonant energy transfer mechanism taking place in the silica matrix reveals average defect-QD distance of 6–9 nm. A direct excitation channel is found only for carbon nanoclusters. An overall scheme of energy levels and optical transitions in the "matrix-cluster" system is proposed.

5.1 Introduction

Intensive studies of silicon nanocrystals in the fused silica [1] have begun shortly after the first demonstration of relatively bright visible luminescence from porous silicon [2]. The 750 nm photoluminescence band emerging in such structures presents significant opportunities for the Si/SiO_2 nanocomposites to be applied in full-optical circuits, light-emitting devices, volatile memory [3, 4], and other fields of micro-, optoelectronics and photonics [5, 6]. A high optical gain of silicon nanocrystals containing media, comparable to that of direct-gap semiconductors, points out a possibility of quantum dot (QD) lasers creation [7]. Additional impurity doping can

A. F. Zatsepin (✉) · E. A. Buntov
Institute of Physics and Technology, Ural Federal University, Mira street 19, Ekaterinburg 620002, Russia
e-mail: a.f.zatsepin@ustu.ru

H. Li et al. (eds.), *Silicon-based Nanomaterials*,
Springer Series in Materials Science 187, DOI: 10.1007/978-1-4614-8169-0_5,

© Springer Science+Business Media New York 2013

serve to modify the emission wavelength. Particularly, the introduction of erbium leads to spectral conversion from 800 nm to the industrially important area of 1.5 μm [8].

In the last decades, the silicon carbide nanostructures became popular due to unique optical properties and advanced thermal stability. Both these features are extremely useful in the high-energy electronics [9]. Predominantly, it is the wide band gap (2.23 eV for 3C-SiC modification) that makes SiC nanocrystals a good candidate for blue–ultraviolet light sources thanks to size effects. Moreover, the size decrease down to several nanometers might suppress the nonradiative transitions due to quantum confinement [10–12].

Since the mid1990s, there have been attempts to obtain the blue emission of SiC nanocrystals formed by C^+ ion implantation in silicon matrix [13], by co-implantation of C^+ and Si^+ ions into SiO_2 [14] and the C_{60} implantation of porous Si [15]. It has been found that the luminescent properties of SiC nanoclusters strongly depend on synthesis conditions. Resulting emission band had occupied a wide range from 400 to 500 nm. However, the quantum confinement effects have not been directly observed until recently. The main reasons were the too high nanoclusters diameter and a complex system of surface and defect states dominating in the luminescence spectrum [16].

Despite the relative success in synthesis of the silicon and silicon carbide nanocrystals, there is no profound understanding of the mechanisms governing the optical excitation and relaxation processes. Even the luminescence of silicon nanoparticles is still controversial. It might be explained by recombination of confined excitons, intracenter relaxation of surface (cluster–matrix interface), or implantation-induced defects. Many authors are leaning toward the radiative recombination of excitons on the interface traps [17, 18]. Some tend to other opinion: minor energy shifts observed in PL spectra are caused by cluster–cluster interactions [18].

The lack of nanoclusters nature understanding mainly stems from the unknown energy structure of the quantum dots. A detailed study of the excited levels is impossible when the popular laser excitation is applied. The details of nanocluster–matrix interaction also cannot be studied by such a method. In this chapter, we describe a new technique of synchrotron-assisted selective photoluminescence spectroscopy which features selective excitation, temperature, and time-resolved measurements. We show how this technique helps to shed the light on the energy structure and dynamics of silicon, carbon, and silicon carbide nanoparticles formed by ion implantation in thin SiO_2 films. We believe that such an approach is the most efficient way to obtain optimal characteristics of functional materials for optoelectronics and photonics.

5.2 Ion-Beam Synthesis and Characterization of Quantum Dots in SiO_2 Matrix

The ion implantation is the bombardment of a material with ions accelerated by electric field. Ion energy may be varied in the wide range from several eV to tenth of MeV. The high energy implantation is applicable to a broad range of elements and host matrices and allows achieving arbitrary ion concentration profile in the subsurface layers, regardless of thermodynamic conditions [19]. A high-dose implantation creates a supersaturated solid solution of embedded atoms in the host matrix. The excess of implanted ions may lead to the nanoparticles precipitation. Subsequent thermal annealing results in nanocluster growth due to Ostwald ripening process. For instance, it is known that for silicon nanocrystals in SiO_2 matrix an optimum regime would consist of 900–1100 °C annealing during 1–2 h in dry nitrogen atmosphere [20].

Thus, the ion-beam treatment is a most universal method of solid-state nanostructuring. A simple calculation of the concentration profile allows developing advanced synthesis techniques, including co-implantation and complex nanoclusters creation. In this chapter, we will show the application of the method to synthesize Si, C, and SiC nanoparticles within the amorphous SiO_2 film.

Amorphous SiO_2 films with a thickness of 800 nm and 300 nm, thermally grown at 1100 °C in wet oxygen atmosphere on p-type Si substrates were used as an original material. The samples were consequently implanted with Si^+ (100 keV energy, $7 \times 10^{16}\,cm^{-2}$ fluence) and C^+ (50 keV) ions. Implantation profiles are presented in Fig. 5.1. The fluence of carbon ions was varied within the range of $7\times10^{15}-1.5\times10^{17}\,cm^{-2}$. Two series of samples were annealed in dry N_2 atmosphere (1100 °C, 2 hours) in order to assemble the nanoclusters of the ions implanted: (a) once after all implantations; (b) twice: before and after the carbon introduction. Both the transmission electron microscopy and the selected area electron diffraction show that the dominant kind of particles formed after ion implantation and thermal treatments is the 3–5 nm sized amorphous nanoclusters (nc). More detailed description of the samples preparation and characterization procedures is presented elsewhere [21].

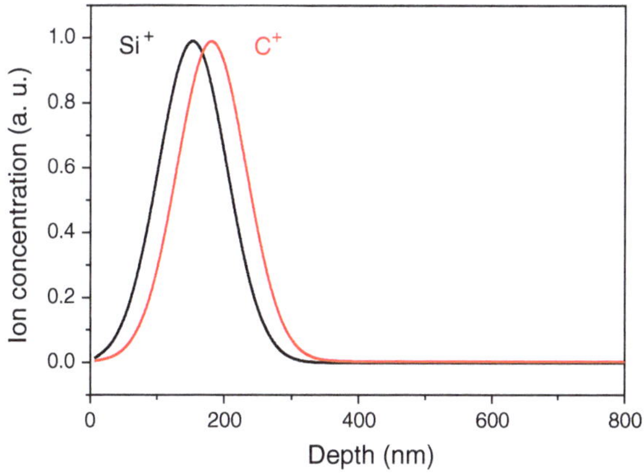

Fig. 5.1 Implantation profiles simulated by SRIM 2012 [22] software for Si^+ and C^+ ions (100 keV and 50 keV energies, respectively) in the 800 nm SiO_2 film

5.3 QD Optical Excitation Techniques

The conventional photoluminescence spectroscopy method may involve various types of optical excitation. The most popular are monochromatic (lasers, gas discharge lamps) and continuous (tungsten and deuterium lamps), steady state and pulsed light sources. Having a coherent and high intensity radiation, lasers are able to excite the broad range of emission centers due to multiphoton processes. They provide pulsed excitation with the highest possible time resolution. In case of photonics-oriented quantum dots, the lasers may offer the operating conditions which are identical to future applications. Thus, the lasers are ideal for applied research and photonics devices design. However, such type of excitation lacks spectral resolution and therefore is not suitable for the sophisticated investigations of quantum dots energy structure. On the other hand, the standard laboratory continuous sources provide low intensity, limited time resolution, and wavelength range.

In contrast, a synchrotron facility typically provides very high light flux intensity, broadest wavelength range and a comprehensive time resolution, which is determined by the electron bunch width and interpulse delay. Propagating through a vacuum channel along a closed curved trajectory, the bunch of electrons emits electromagnetic waves (synchrotron radiation, SR) of Visible-X-Ray spectral range (Fig. 5.2). In order to study the photoluminescence of a solid sample, one must use a monochromator (M1) to pick out an excitation beam with a single energy $h\nu_{exc}$. Being exposed to the beam, a specimen emits the luminescence radiation which is further monochromated by M2 and registered by the photomultiplier tube (PMT) in order to get the emission intensity at a single energy $h\nu_{em}$. By varying both energies on may obtain photoluminescence (PL) and excitation (PLE) spectra.

Beside the steady-state PL and PLE spectra, one can also get time-resolved ones: those collected in the adjustable time windows Δt_1 and Δt_2, delayed from the SR

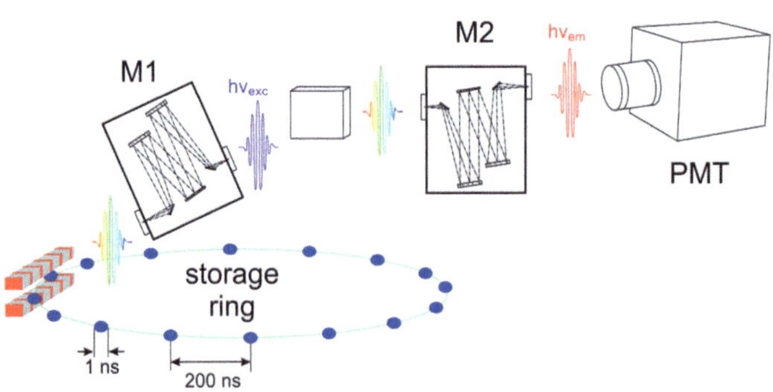

Fig. 5.2 Time-resolved spectroscopy scheme: electromagnetic waves of a wide Visible-X-Ray spectral range are generated by the electron bunches moving along a curved path through the synchrotron storage ring and being guided by bending magnets. The first monochromator M1 selects a radiation with energy $h\nu_{exc}$ which is then focused to a sample. A broad PL emission, further monochromated by M2 is registered with a photomultiplier tube

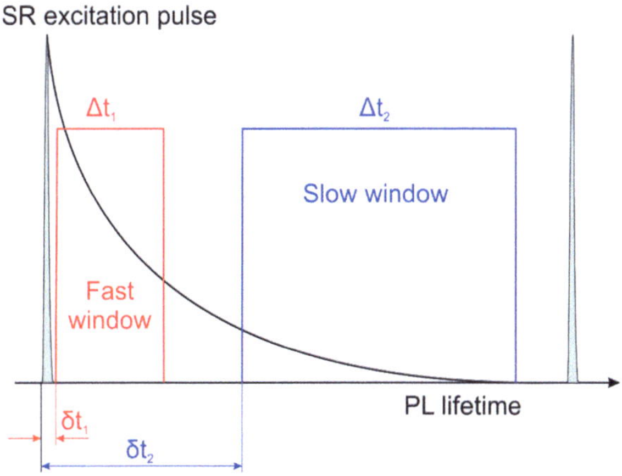

Fig. 5.3 A PMT detector may be gated in order to get the emission in the Δt-wide time window shifted with a delay of δt from the SR excitation pulse. The maximum window size is limited by a 200 ns interpulse distance

excitation pulse by δt_1 and δt_2, respectively (Fig. 5.3). The most detailed information on the PL decay may be obtained by kinetic curve measurements at fixed wavelengths. All the noted experiments might be performed in the broad range of temperatures.

For all the results in this chapter, the steady-state and time-resolved photoluminescence spectra were registered under pulse synchrotron radiation within the 3.7–16 eV excitation energy range at SUPERLUMI station, Hamburg (Beam-line I, HASYLAB, DESY [23]). The scheme of experiment is sketched on Fig. 5.2. The excitation channel incorporated a 2 m vacuum monochromator (M1) equipped with aluminum grating (25×10^{-3} eV spectral resolution). Photoluminescence excitation (PLE) spectra were normalized with a reference spectrum of sodium salicylate. A 0.3 m ARC Spectra Pro-300i monochromator (M2), R6358P photoelectron multiplier (Hamamatsu), and a CCD-camera were used for PL spectra acquisition. By means of this monochromator, the luminescence excitation spectra (PLE) were measured with a wavelength resolution of $\Delta\lambda = 3.2$ Å. Synchrotron radiation is pulsed with pulse duration (bunch width) about 1 ns. The interpulse distance of 200 ns limits the possibility of long PL decay measurements. The temporal resolution of the whole system is about 1 ns. A deconvolution method is applied in order to retrieve fast components of PL decay curves with an uncertainty that does not exceed 2 %.

5.4 Interference Effects

The question of incident light interference is especially important for a particular type of light sources—photoluminescent quantum dots, nonuniformly distributed inside the plane-parallel thin films [24]. Being valid even for laser excitation, this phenomenon deserves special attention when a selective synchrotron radiation (SR) excitation with a variable wavelength is applied. It may cause periodic oscillations of the photoluminescence excitation (PLE) spectra thus preventing the adequate analysis of PLE spectra, and therefore hinders to reveal the nature of the luminescent centers.

During the photoluminescent studies of the SiO_2 films implanted with Si, C, S, and Sn ions carried out at SUPERLUMI station, the unusual excitation spectra containing a series of regularly spaced bands (oscillations) of similar widths were observed [24–26]. It was hypothesized that the origin of these bands is related to interference effects which may occur for the incident electromagnetic waves in the film with a thickness close to the wavelength. The influence of the interference effects on the photoluminescence spectra of thin silica films with silicon nanoclusters was described in [27], where IMD software [28] was successfully applied to model the phenomenon. However, possible effects taking place in the PL excitation spectra were not addressed. The main purpose of the present work is to establish a new numerical technique and corresponding software to model the phenomenon of excitation radiation interference within thin films and to correct the shape of distorted PLE spectra.

5.4.1 Numerical Technique

IMD is an IDL language-based application developed by D.L. Windt and aimed to calculate specular and diffuse optical functions of an arbitrary multilayer structure [28]. Specular optical functions, particularly the electric field intensities, are computed using an algorithm that is based on recursive application of the Fresnel equations [29], modified to include interfacial roughness and/or diffuseness [30, 31]. In order to compute, in addition to the optical functions, the electric field intensity as a function of depth in a multilayer stack, a slightly different formalism, comprising Maxwell equations solution, is applied.

In our case, the calculation of the electric field intensities was performed for a 500 nm SiO_2 film on top of a semi-infinite silicon substrate. The excitation radiation with an incidence angle of 17° and quanta energies of 4–10 eV was assumed in the full accordance with the experimental conditions of the SUPERLUMI station. The obtained two-dimensional map of field intensity I in its dependence on film depth z and excitation energy $h\nu$ is presented in Fig. 5.4a. It reveals a set of interference fringes which may cause the PLE spectra oscillations. In order to model the interference pattern that distorts the PLE spectrum, one should take into account the depth distribution of the luminescent centers, in our case either point defects or nanoclusters. Both of them have a concentration proportional to the initial ion implantation profile.

The implanted ions distribution C(z) was calculated by means of SRIM-2012 software package [22]. The depth profile for the 150 keV Si^+ ions is shown in Fig. 5.4c.

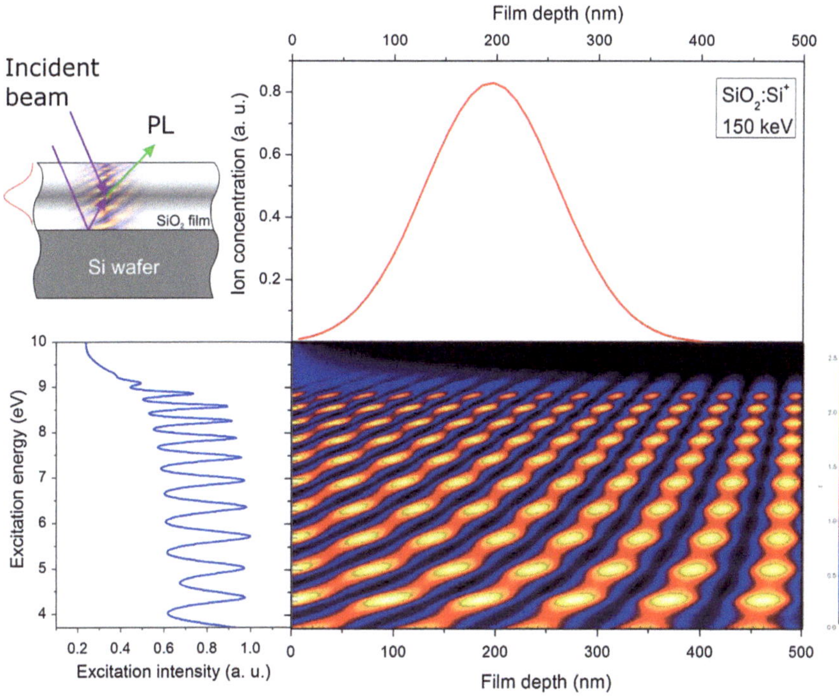

Fig. 5.4 Excitation light interference within a thin ion-implanted SiO_2 film on a silicon substrate: **a** Beam propagation scheme; **b** The depth distribution of the implanted ions concentration; **c** Excitation radiation spectral profile with interference fringes; **d** Electric field intensity map in its dependence on the depth and photon energy

Then the electric field intensity map $I(z, h\nu)$ was integrated with respect to depth z using the ion distribution as a weighting function:

$$I_{\mathrm{int}}(h\nu) = \int_0^d I(z, h\nu) \cdot C(z) \cdot A(z) dz, \tag{5.1}$$

where d—film thickness, $C(z)$—ion implantation profile, $A(z)$—a coefficient accounting for the optical absorption of the luminescence radiation. The resulting interference pattern $I_{\mathrm{int}}(h\nu)$, shown in Fig. 5.4b, is proportional to the excitation light intensity (photons number) at the specific excitation energy and is valid for the specified film thickness, light incidence angle, and luminescent centers distribution. It may be used regardless of emission centers physical nature unless the centers concentration is high enough to alternate the optical properties of the host matrix. The latter case may be modeled by a specially arranged multilayer structure with a variable refraction index which is beyond the scope of the present article and would not be considered here. The third term $A(z)$ is important when the thin film material has a considerable optical absorption at the emission wavelength. If the luminescence observed lies within the transparency gap of the host material, this term might be neglected.

An interference pattern may affect the intrinsic excitation spectrum (energy dependence of the photoluminescence quantum yield) $\eta_{PLE}(h\nu)$, so that the instrument registers a distorted variant $\eta^*_{PLE}(h\nu)$:

$$\eta^*_{PLE}(h\nu) = I_{int}{}^{(h\nu)} \cdot \eta_{PLE}{}^{(h\nu)}. \tag{5.2}$$

Thus, the main goal is to retrieve the undistorted initial spectrum using the model approximation of $I_{int}(h\nu)$. The most straightforward way to achieve the goal is to divide back by $I_{int}(h\nu)$. However, the restoration of the experimental spectrum is actually complicated due to the following reasons:

- The roughness of the film interfaces affects the field intensity map and in most cases is not known exactly;
- Though the film thickness is usually measured or calculated, it always contains an uncertainty;
- Light incidence angle may be altered due to inaccurate specimen positioning;
- SRIM-calculated ions distributions introduce an error of 4–6 % even if all parameters are specified properly;
- The equations used in IMD are most of all phenomenological and also contribute to the overall error.

From the Fig. 5.3, one may notice the way in which the most important parameters (film thickness, light incidence angle, and luminescent centers distribution) influence the calculated interference profile. The thickness changes lead to the scaling of the oscillations period: the thinner the film, the bigger the distance between interference fringes. The growth of the incidence angle results in the energy shift of oscillations. A narrow Gaussian distribution (12 nm-wide peak at 60 nm depth) of the luminescent centers gives the highest excitation amplitude while a wide one (40 nm-wide maximum at 213 nm depth) gives the lowest, almost coinciding with the uniform distribution. In order to compensate the uncertainties of these parameters, one must propose a more comprehensive method of correction. We suggest the following linear extension of the equation (5.2) introducing four additional factors:

$$\eta_{PLE}(h\nu) = \frac{\eta^*_{PLE}(h\nu)}{a_2 \cdot I_{int}(a_1 \cdot h\nu + b_1) + b_2}, \tag{5.3}$$

where a_1, a_2—scaling factors for the energy and intensity, respectively; b_1, b_2—shifts for the same quantities. The b_1 parameter is aimed to compensate the incidence angle error, a_1 accounts for film thickness uncertainty, a_2 and b_2 may correct the PL centers distribution accuracy. Numerical optimization taken on these parameters gives a better approximation of the initial undistorted PLE spectrum.

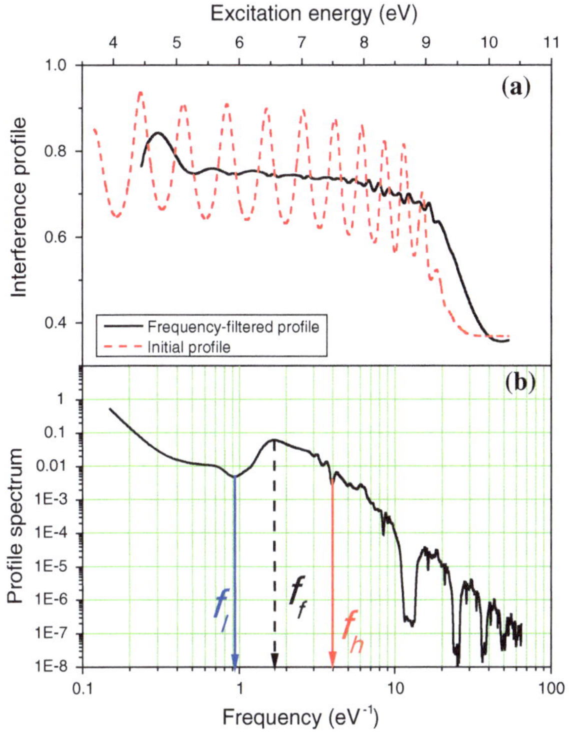

Fig. 5.5 Interference profile frequency filtering allows eliminating of the fringes: **a** the calculated profile before (*dashed*) and after (*solid*) third-order Butterworth filtering; **b** Frequency spectrum of the initial profile lets us to calculate the filter's higher f_h and lower f_l frequencies on the basis of fundamental frequency f_f

5.4.1.1 Correction Criterion

The most essential criterion for such optimization is the absence or minimal amplitude of interference-related fringes and oscillations in the corrected spectrum. In order to distinguish and remove these oscillations we try to use the frequency filtering, namely the third-order band-stop Butterworth filter [32] (Fig. 5.5a). The lower f_l and higher f_h cut-off frequencies are taken from the spectrum of the $I_{\text{int}}(h\nu)$ interference profile (Fig. 5.5b) by multiplying its fundamental frequency f_f by the factors of 0.5 and 2, respectively. The noted factors values give adequate bandwidth for the filter though they may be argued and changed for the particular cases. However, the values should be kept constant for all the spectra corrected with a single profile.

Being filtered with such Butterworth filter, the $I_{\text{int}}(h\nu)$ pattern lacks the majority of its oscillations and becomes smooth (see a black solid line in the Fig. 5.5a). Similar effect is expected to be achieved for the PLE spectrum obtained from equation (5.3) when the optimized parameters values are used. The result of equation (5.3) evaluation is subjected to the corresponding band-pass filter with the same frequencies. The root-mean-square value of the filtered signal, containing mostly the interference-related oscillations, is then minimized in order to get the optimum parameters values.

5.4.1.2 Model-based Testing

A freely available software module named iCorrector [33] was developed by us to automate the optimization procedure described. We performed three model-based tests where the algorithm was applied to correct the Gaussians deliberately divided by the interference profile shown in Fig. 5.4b. The technique proposed restores successfully the line shape for the peaks whose width is four and two times wider than the period of oscillations, respectively. An effective correction of the complex spectrum containing two wide bands was demonstrated. In case of a narrow band or a maximum with a fine structure, having a width comparable or less than the oscillations period, it is hardly possible to restore the shape automatically.

5.4.2 Correction of the Experimental Spectra

One of the most pronounced interference effects was observed for ion-implanted and N2-annealed silica films on a silicon substrate [25]. The spectrum is shown by a black solid line in the Fig. 5.6a. One may notice that the high-energy part ($h\nu$ >7.5 eV) of the spectrum is not affected by oscillations while the low-energy part demonstrates almost equally spaced maxima similar to those found in the interference pattern (Fig. 5.4b). The modeled profile does not predict at all such a sharp boundary at 7.5 eV. It can be suggested that the two parts of the PLE spectrum originate from different luminescent centers, some of which feel the interference effects and others do not. Taking into account bands spectral positions the 7–10 eV part may be related to some silica excitons while the 4–7.5 eV region is usually connected with point defects excitation. An extra evidence for their different nature is nanosecond and microsecond-order decay kinetics observed for the low-energy and high-energy sides of the spectrum [25].

A direct application of the algorithm proposed to the PLE spectrum of $SiO_2:Si^+$ films leads to the unwanted oscillations appearing in the high energy region. In the full conformance with the physical interpretation of the bands, we have approximated the 7.5–10 eV wing of the spectrum with an auxiliary Gaussian function ($h\nu_{max} = 7.6$ eV, FWHM = 3 eV, dashed line in Fig. 5.6a). Then this function was subtracted from the initial spectrum to get its interference-affected part (red line on Fig. 5.6a). The latter part was corrected by the iCorrector module and then added to the 7.6 eV band. The resulting spectrum is shown on the Fig. 5.6b. It still has a complex shape which may be deconvoluted to six elementary components. The spectroscopic features allow attributing the 5.1, 7.09 and 7.68 eV PLE peaks (Fig. 5.6b) to the singlet excited

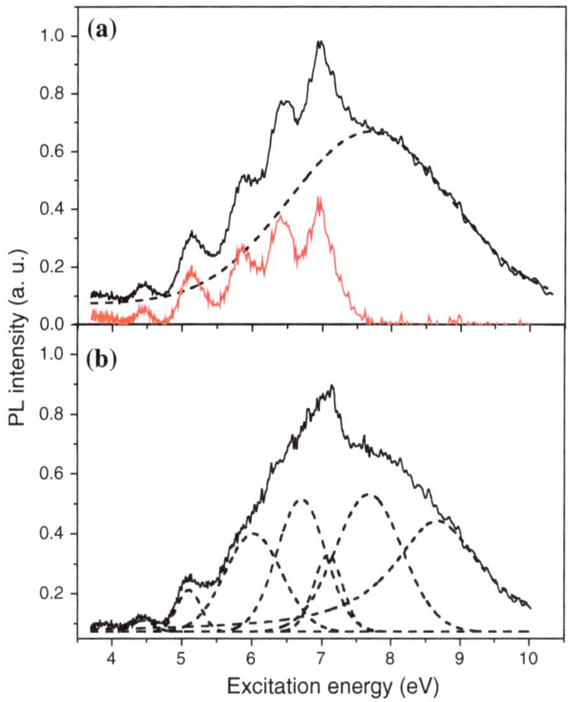

Fig. 5.6 The correction of the PL excitation spectrum allows further spectroscopic analysis: **a** Experimental PLE spectrum (*black solid line*) of the 2.18 eV PL band for SiO_2:Si^+ sample [25] containing undistorted (*dashed line*) and distorted (*red solid line*) parts. **b** Corrected spectrum lacks oscillations but keeps a complex shape. Deconvoluted Gaussian components are denoted by *dashed lines*

states of ODC-type defects [34]. The 6 eV maximum is close to the known optical absorption band of the E'$_S$-center. A band with the highest energy of 8.7 eV has a Lorentzian shape and is connected with the SiO_2 matrix exciton subsystem.

Thus, the developed interference modeling and correction technique together with corresponding software allows an automated restoration of the initial PLE spectra demonstrating pronounced regular oscillations. Resulting spectra provide meaningful information about the luminescent center nature and can be further analyzed by comprehensive spectroscopic methods including the decomposition into elementary components. However, a less contrast interference pattern may be observed in the excitation spectra when a different film thickness or a different luminescent center is considered. In order to test the applicability of the correction techniques in case of low-contrast fringes, the ion-implanted SiO_2:Si^+:C^+ [24] films were used. Samples have a wide PL excitation spectra containing multiple bands (Fig. 5.7a). The correction procedures performed have allowed to reduce significantly the interference fringes contribution and to remove false inflection points within the 5.5–8 eV spectral range (Fig. 5.7b).

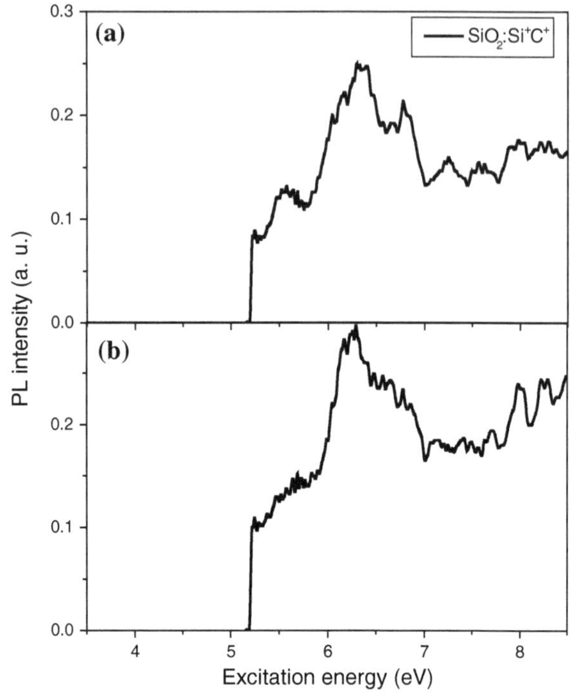

Fig. 5.7 The correction procedure performed on the $SiO_2:Si^+:C^+$ [24] samples: **a** initial spectra; **b** corrected spectra

5.5 Photoluminescence of Silicon and Carbon Quantum Dots

5.5.1 SiO_2: Si^+ System

The PL spectra of the implanted films are dominated at room temperature by the luminescence of the known ODCs with the radiative transitions at 3.1 eV (a triplet-singlet transition, $\tau = 100\,\mu s$) and 4.5 eV (a singlet-singlet transition, $\tau = 20\,ns$). They have a characteristic PL excitation spectrum (the maximum near 5.0–5.1 eV). Postimplantation annealing does not change parameters of the observed PL of ODCs. The shape of the PL spectra of the samples annealed up to 900 °C changed considerably at low temperatures. Steady-state PL spectra obtained at different energies of excitation photons are shown in Fig. 5.8. The nonelementary 4.5 eV PL band corresponding to the singlet luminescence of α-ODC (4.3 eV) and β-ODC (4.5 eV) [34], as well as an unidentified peak near 3.6 eV are observed under 5.1 eV excitation.

The growth of the luminescence intensity near 3.1 eV is assumed to be due to the presence of the ODC triplet radiation. The intensity of the triplet PL bands is quite low because of the activation barrier for the transfer of the excitation energy between the singlet and triplet states. Also, some new peaks appear in the PL spectrum in the long-wavelength region at energies of 1.34, 1.72, 1.9, 2.18, and 2.58 eV. The shape of the PL spectrum depends on the energy of excitation photons. As regards their

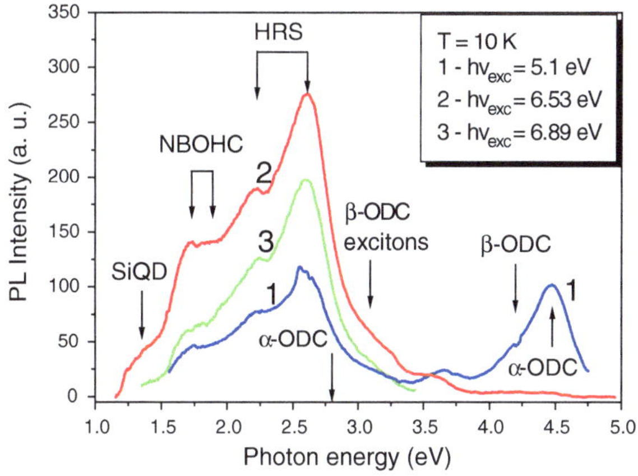

Fig. 5.8 PL spectra of a $SiO_2:Si^+$ film at Eexc = 5.10 eV (*1*), 6.53 eV (*2*) and 6.89 eV (*3*) recorded by a CCD camera

location, the bands at 1.7 and 1.9 eV can be referred to surface and bulk types of nonbridging oxygen hole centers (NBOHCs) [35]. The maxima at 2.18 and 2.58 eV can presumably be related to either the luminescence of silicon nanoclusters and hydrogen-related species [36].

Figure 5.9 presents time-resolved excitation spectrum of 2.81 eV band. The excitation spectra have a complicated "stepped" shape and include a set of relatively sharp bands in the low-energy region ($h\nu_{exc} = 4.45$; 5.14; 5.88; 6.45; 6.96 eV, FWHM is about 0.5 eV) and a wider band (FWHM near 2 eV) in the region of 8–10 eV. All these peaks cannot be attributed to some type of ODCs. Moreover, the wide band at 7.75 eV corresponds to the region of the exciton generation and the edge of the

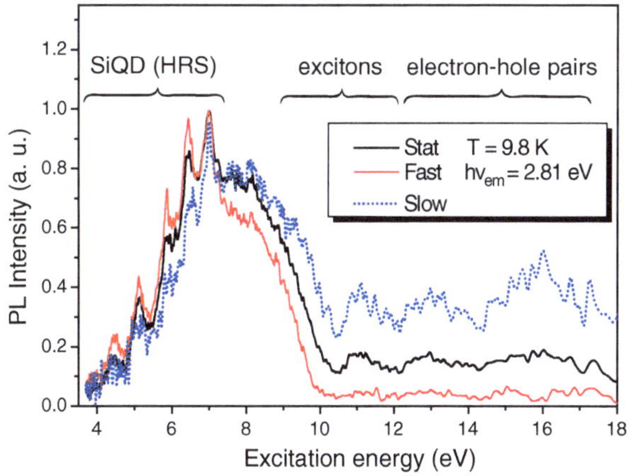

Fig. 5.9 Time-resolved PL excitation spectra of the 2.58 eV emission band in a $SiO_2:Si^+$ film

Table 5.1 Kinetic parameters of the nanoclusters' radiative transitions under excitation through the SiO_2 matrix point defects

Donor	Acceptor	Equation (5.4)			Equation (5.5)		
(PLE)	(PL)	τ (ns)	β	Q	S	R_0* (nm)	τ_0 (ns)
ODC(II) (7.0 eV)	nc-Si (1.82 eV)	3.5	0.58	2.4/1.4	6/8	6.9/6.2	29/16
ODC(II) (7.0 eV)	nc-C (2.48 eV)	2.9	0.58	1.3	6	5.6	13.5
ODC(II) (7.0 eV)	nc-SiC (3.31 eV)	1.3	0.46	5.1	6	8.8	63

interband transitions region for the SiO_2 matrix. The positions of the maximum, FWHM, and the relative integral intensity of all the bands are almost the same for each of the three PL peaks, see Table 5.1.

To investigate the nature of the discovered luminescence, we performed a set of kinetic measurements. Time-resolved luminescence excitation spectra over a wide spectral range and a series of decay kinetics curves for different PL bands at different excitation energies were recorded (Fig. 5.10). The time-resolved spectra demonstrate different radiative relaxation kinetics for various regions of the excitation energy. The normalized fast component dominates over the slow component on the low-energy side of the spectrum, while their ratio reverses at 7–20 eV. The difference in the PL decay kinetics points to the corresponding difference in PL excitation mechanisms for these centers. This conclusion agrees well with results of direct kinetic measurements, see Fig. 5.10. The decay kinetics of the PL band at 2.53 eV can be approximated well by two exponents with the average decay times of 4.5 and 35 ns, respectively.

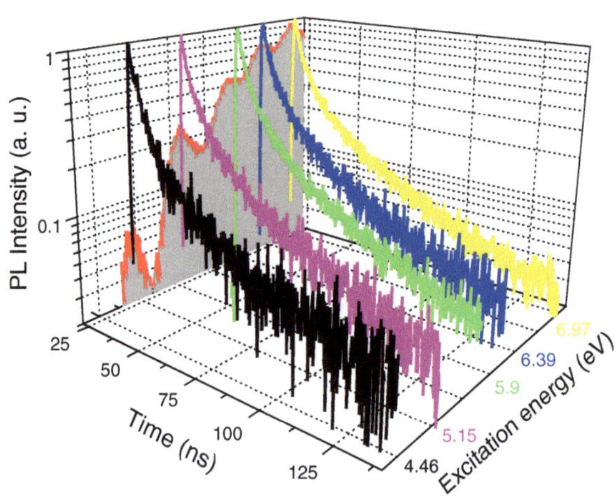

Fig. 5.10 PL decay kinetics of the 2.53 eV emission band in a SiO_2:Si^+ sample on the semilogarithmic scale. The PL excitation spectra (the *red-dashed line*) are given for reference

Lifetimes almost do not depend on excitation energy. At the same time, emission band strongly influences decay kinetics. The 2.14 eV band has the same order of the decay time: 6.5 and 50 ns. At the same time, the PL kinetics in both bands slow down at high excitation energies near 8 eV (6 and 43 ns for $h\nu_{em} = 2.53$ eV; 7 and 55 ns for $h\nu_{em} = 2.14$ eV), respectively. The third PL band at 1.72 eV has the maximum decay time: 8, 50 ns and components of the μs range.

The obtained results suggest that in addition to the well-known ODCs, a specific type of defect centers is present in the SiO_2:Si^+ films under study. Parameters of the aforementioned defects are close to those of defects in nanosized modifications of SiO_2 [36], respectively. Their specific features include an abnormally wide spectral region and a "stepped" shape of excitation spectra. Some investigators relate similar centers to a specific type of defects (so-called Hydrogen-Related Species—HRS) observed in amorphous SiO_2 nanoparticles [36].

5.5.2 SiO_2:Si^+:C^+ System

Figure 5.11 shows the typical PL spectra of the once and twice annealed SiO_2:Si:C samples. It can be seen that the sequential implantation of silicon and carbon ions leads to the manifestation of a set of luminescence bands in the visible range. The figure demonstrates that even for different synthesis and excitation conditions the spectra contain three selective peaks. They are observed at helium temperatures near 1.8, 2.45, and 3.05 eV (noted here and further as R, G and B, respectively). There is also an additional peak appearing in high carbon content samples at 2.1–2.2 eV.

The G band dominates the PL spectra for all the samples under study, whereas the ratio of R and B peaks depends on the thermal treatment regime: additional

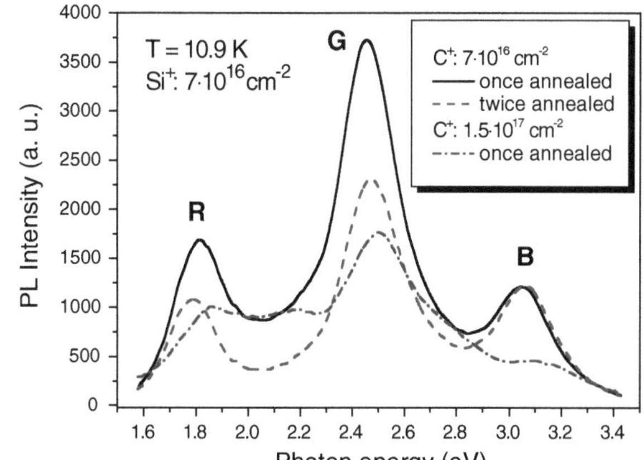

Fig. 5.11 Low-temperature PL spectra of the 800 nm SiO_2:Si:C films recorded under different excitation energies: 11.3 eV (*solid line*), 13.3 eV (*dash*), 6.39 эВ (*dash-dot*)

intermediate annealing results in the relative growths of the B maximum. Besides, spectral positions of the noted peaks depend on the thermal processing and the implantation fluence. Intermediate annealing causes the 0.05 eV red shift of R band together with a tiny (0.02 eV) blue shift of G band. The former is explained by the silicon clusters growth during the annealing. The latter may be caused by the peculiarities of carbon oxidation processes in case of the lack of interstitial silicon atoms which were assembled into Si nanoclusters during the intermediate annealing, and therefore could not participate in SiC formation. Ion fluence increase leads to the 0.05 and 0.08 eV shifts toward higher energies for G and B peaks, respectively. Figure 5.12 demonstrates the PL spectra temperature evolution for R and G maxima. One can notice that the PL intensity decreases gradually with the growing temperature while the overall shape of the spectra and the bands halfwidths persist over the whole temperature range from 14 to 300 K. Generally, such behavior is characteristic for the highly disordered systems due to spectral bands inhomogeneous broadening caused by static disorder [37].

The investigation of the luminescence mechanisms of an ion-implanted material should inevitably include the stage of luminescent centers identification. In the particular case of SiO_2:Si:C system, the PL emission observed could be caused by either confined excitons recombination or intracenter relaxation of the Si–SiO_2 interface defects which may persist at the nanoparticle-oxide phase boundary even after a high temperature annealing [11, 38].

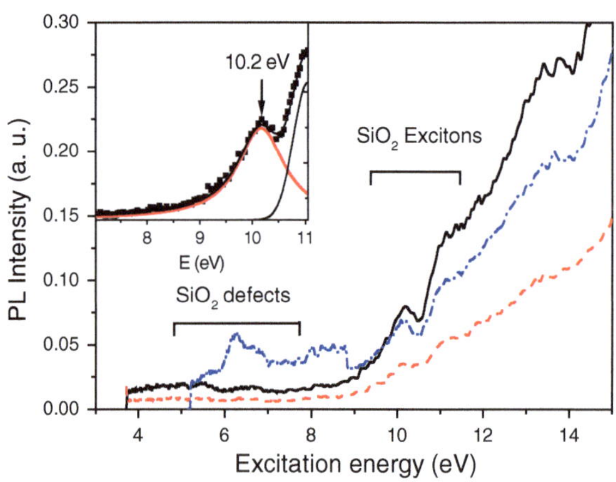

Fig. 5.12 The PLE spectra for G (2.47 eV, *solid* and *dash-dot lines*) and B (3.05 eV, *dashed line*) bands of once annealed 800 nm SiO_2:Si:C films, recorded at 10.9 K. The *solid* and *dash-dot lines* have different carbon fluences: 7×10^{16} and $1.5 \times 10^{17}\,cm^{-2}$, respectively. Inset graph shows the Lorentzian approximation of a 10.2 eV PLE band of the solid spectrum. Agreement with experimental data is very close on the left band wing, whereas the right one is distorted by band-to-band transitions described by an auxiliary Gaussian function. Experimental dots also demonstrate the noise level of the excitation spectra

The set of experimental data obtained points out a relation between the luminescent properties of the SiO_2:Si:C system and the presence of the implanted atoms clusters. Electron microscopy images also evidence the occurrence of 3–5 nm sized inclusions [21]. Both the high resolution TEM and the selected area electron diffraction show a predominantly amorphous structure of the particles assembled. The R band (1.8 eV) was described by many authors as appearing in case of silicon-implanted silica films and was mostly ascribed to the photoluminescence of amorphous Si clusters [39–42]. The minor shift of this peak after an additional annealing is due to the clusters growth.

The most intensive G band (2.45 eV, Fig. 5.11) is close by its position to the self-trapped excitons luminescence (STEPL) known in amorphous SiO_2 [43]. However the appearance of the nanosecond-range decay together with the low-energy excitation within the silica transparency gap undoubtedly disproves the STEPL explanation, drawing attention to its cluster-related nature. Furthermore, both the coincidence of the G-band transition energy with an amorphous carbon films band gap [44] and the fluence dependence of the 4.4 eV PLE band could serve as the extra arguments for the interpretation proposed even though in literature there are numerous spectral assignments of carbon-related PL to different emission bands, ranging from near infrared to blue PL bands [45, 46]. According to [47, 48], the blue PL band (B) could be due to the SiC clusters emission though sometimes the G and B are attributed to the $Si_yC_{1-y}O_x$ complexes ($x < 2$) [49].

Thus, the results presented in accordance with the literature data [11, 50] allow us to interpret the R, G and B bands of SiO_2:Si:C films as a luminescence of the nanometer-sized silicon, carbon and binary (SiC) clusters, respectively. All of them seem to have an amorphous structure.

The excitation spectra (Fig. 5.12) for the PL bands observed have a complex shape and strongly depend on the emission wavelength. For example the G and B peaks reach the maximum intensity under high energy excitation, corresponding to the SiO_2 interband transitions ($h\nu > 11$ eV). A slightly lower maximum is located at 10.2 eV and probably corresponds to the matrix exciton subsystem.

As it is seen from Fig. 5.12, the G band has multiple excitation maxima inside the SiO_2 optical transparency region (4–8 eV). These peaks reach the maximum intensity under the highest carbon fluence of $1.5 \times 10^{17}\,cm^{-2}$. Therefore, we can presumably attribute this PLE maximum to the intrinsic excited states of either C or SiC nanoclusters.

As opposed to a laser excitation, the monochromatic synchrotron radiation allows to study all the different excitation channels present with the quanta energies up to 40 eV. It is seen from the Fig. 5.12 that the direct excitation (4.4 eV) for the G band is the least efficient, as compared to high energy channels. The same could be said for the R and B peaks. In this connection, we will further consider the indirect excitation mechanisms involving the host matrix states.

The matrix–cluster optical excitation phenomena have been observed already for the silicon-implanted silica films [25]. It was shown that the excitation is still efficient at the photon energies comparable or even higher than SiO_2 band gap width. According to D.L.Griscom [51], the silica photon irradiation with $h\nu > 9$ eV could generally produce a large amount of self-trapped holes (STH). It could therefore be suggested that the interband transitions powered excitation should comprise the stages of hole trapping and subsequent free electron recombination on the holes trapped. A detailed consideration of such a mechanism requires a special investigation and could be a matter of a separate publication. The present article focuses on the two concurrent channels of the indirect excitation related to the matrix excitons and point defects.

The current understanding of the silica optical absorption band just above 10 eV is still controversial. Due to the quartz crystal to silica glass transition it may shift from 10.2 to 10.4 eV [52, 53]. Some authors [52] resolve several overlapping maxima corresponding to different exciton types within a single maximum, depending on their mobility. However, the recent studies of the VUV optical reflection spectra for SiO_2 glass (Suprasil F300, type IV) [53] suggest an elementary Lorentz-like shape of the 10.4 eV absorption band throughout the wide temperature range. In our case, similar to the data of Messina et al [53], the shape of the 10.2 eV exciton band is fitted well by a single Lorentz function (see inset graph on Fig. 5.12). According to the theory, proposed by Toyozawa [54], the lorentzian line shape suggests a weak exciton–phonon interaction providing high exciton mobility. The distributed network of both the defects and the nanoclusters formed after the implantation together with high exciton mobility create favorable conditions for the excitation energy transfer. The possible explanation of such a contradiction may come from the thin film structure being different from that of bulk glass. The excitonic transfer mechanism may be especially efficient in case of the ion-implanted layers.

Along with the excitonic channel there is a possibility of energy transfer from the silica point defects such as ODC and E'-centers. The PLE spectra structure as appeared on the Fig. 5.4 in the 5–8 eV range could be partially distorted by the SR interference effects. Nevertheless, the very occurrence of the PL excitation within the noted spectral region is an evidence of energy transfer from SiO_2 point defects to the clusters of the implanted ions. Indeed, the PLE maxima energies (Fig. 5.12) match those of the optical absorption bands of bulk (ODC(I), ODC(II) and E'_{γ}) and surface (E'_{s}) defects. It is known that the E'-centers are not the luminescent defects but could play a role of the effective donors for a relatively long-range nonradiative energy transfer. Zatsepin et al. have shown [55] that the electron emission quantum yield does not exceed the 20 % value for the E'-centers while the rest of them relax nonradiatively. Therefore, there must be the favorable conditions for energy transfer from an excited E'-center.

5.6 Energy Transfer and Optical Transitions

5.6.1 PL Decay Kinetics

The PL decay kinetics for the R, G, and B bands under 7 eV excitation (Fig. 5.13) have a nonelementary shape, so the experimental curves could not be fit by a single

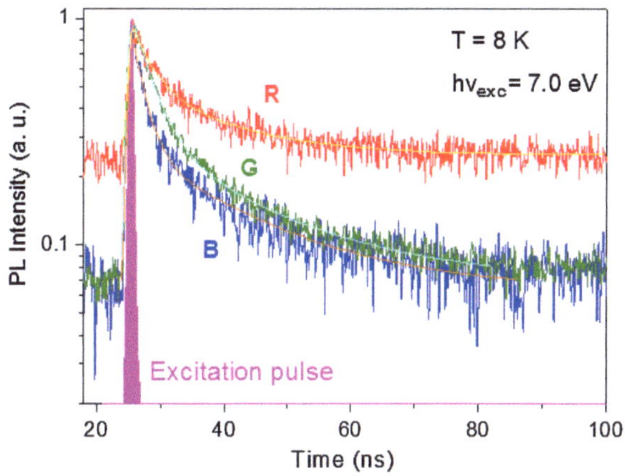

Fig. 5.13 PL decay curves recorded under 7 eV excitation for the R, G and B bands ($h\nu = 1.82, 2.48$ and 3.31 eV, respectively) of the once annealed 300 nm SiO_2:Si:C film (C^+ fluence $1.5 \times 10^{17}\,cm^{-2}$)

exponential. One can notice that the slow components of a microsecond range are present throughout the whole spectral range, appearing as a flat background in Fig. 5.13. A short SR pulse period of 200 ns does not allow to evaluate quantitatively so long PL lifetimes.

As a first approach to the nonelementary shape problem the two-exponential approximation was applied. The method reveals a fast (2–3 ns) and a slow (13–18 ns) components for the all three curves in Fig. 5.13. The longer exponential is almost the same for the G and B bands (18 ns), while the R band has a quite different lifetime (13 ns). Contrary, the steeper exponentials are identical for R and G (3 ns), but comprise a distinct value for B (2 ns).

A detailed analysis of the luminescence kinetics requires a more reasonable model for the relaxation processes and will be performed in the discussion. Nevertheless, the nonelementary shape of PL decay curves may indicate the complexity of a multistage excitation energy transfer process or the significant influence of the QD and matrix structural disorder on the optical excitation and emission mechanisms.

A relatively slow PL kinetics of microsecond order is observed when the excitonic mechanism is realized. Such a decay deceleration should be related with the intermediate stages occurring during the energy transport process. The silicon dioxide is known to have both the mobile and localized excitons at helium temperatures. The former, in turn, could either undergo localization at the lattice defects and imperfections or get over the self-trapping barrier. The stage of exciton migration and subsequent localization is able to decelerate the whole process of an excited state relaxation.

Contrary, the point defect-related excitation produces the relatively fast and nonelementary decay of nanosecond order (Fig. 5.13). The decay curve shape deviation from the single exponential could be due to the peculiarities of emission center energy structure or center–matrix interactions. All the named phenomena require different data analysis approaches.

On the one hand, the presence of a system with several tightly spaced energy levels can produce various alternative relaxation channels demonstrating different recombination rates. The tiny splitting between the levels together with high emission line widths may result in quasi-Gaussian PL spectra. In this case, the decay curve can be represented by a convolution of the exciting SR pulse with two or more exponentials. However, the occurrence of an identical energy structure with similar lifetimes is hardly possible for the three types of clusters observed.

The PL decay curves for the silicon nanocrystals are often described by a stretched exponential [42, 43]:

$$I(t) = I_0 \cdot \exp\left[-\left(\frac{t}{\tau}\right)^{\beta}\right] \tag{5.4}$$

where I_0 is the PL intensity at $t = 0$; τ—characteristic lifetime; β—dispersion factor. The expression (5.2) allows a high-quality approximation of the kinetic curves for the disordered systems within a wide temperature range.

The dispersion of the luminescent centers spectral and kinetic parameters may be related to the nanocomposite features, such as a nanoparticle size distribution, a nanocluster structural disorder, or a difference in their local surroundings. For instance, the nonelementary decay of silicon nanocrystals is generally assigned to an anomalous dispersion of the nanocrystallites kinetic parameters [56, 57] and therefore the stretched exponential (5.4) is used. The dispersion factor β, other conditions being equal, depends on the lifetime distribution width for a set of luminescent centers.

The left part of the Table 5.1 presents the parameters derived from Fig. 5.13 data using the equation (5.4). One can notice that the characteristic lifetime is close to the fast component of the two-exponential analysis (2–3 ns) while the contribution of the slow component is well compensated by a proper selection of the β parameter. The β values for the elementary Si and C clusters are identical and lie close to those for the porous silicon and the SiO_2:Si^+ implanted films [58]. Kobitsky et al [46] have studied the β parameter temperature dependence for the silicon nanocrystals. According to their results, the β factor reaches the value of 0.5 near T = 10 K, which is rather close to the observed values for R, G, and B bands. The lifetime τ, being obtained from the nanosecond part of PL kinetics, is significantly lower than the microsecond-order values obtained under laser excitation [58]. Binary nanoclusters show the fastest kinetics and the lowest dispersion factor value, the latter being an evidence of the wider cluster lifetime distribution.

The experiments performed show that the change of excitation energy (from 7 eV to 6.2–6.4 eV) does not vary the β value but leads to the lifetime increase up to 4–5 ns. Such a deceleration can be caused by a change of energy transfer mechanism. Particularly, the exchange mechanism involving the overlapped wave functions of the donor and acceptor electrons is the most favorable for the surface (interface) centers of E'$_s$ type absorbing at 6.3 eV.

In order to evaluate quantitatively, the parameters of energy transport between the defects and the nanoclusters we used the Inokuti–Hirayama model [59] that is often applied in case of a single-step transfer. Assuming a random distribution of luminescent centers within amorphous and glass-like systems one can write down:

$$I(t) = I_0 \exp\left[-\frac{t}{\tau_0} - Q\left(\frac{t}{\tau_0}\right)^{3/S}\right] \tag{5.5}$$

where τ_0 is the lifetime not taking into account the energy transfer, Q—energy transport intensity, S—multipolarity factor being equal to 6, 8, or 10 for dipole–dipole, dipole–quadrupole and quadrupole–quadrupole interactions, respectively. The Q parameter depends on the donor–acceptor distance: $Q = 4\pi/3 \cdot \Gamma(1 - 3/S)N_0R_0^3$. Here Γ—gamma-function, N_0—acceptor concentration, R_0—critical donor–acceptor distance at which the radiative and nonradiative relaxation probabilities become equal.

All the kinetic curves are fitted well by the Eq. (5.5) assuming the dipole–dipole interaction ($S = 6$, see the right part of the Table 5.1). At the same time, the silicon nanoparticles PL decay could be also approximated with $S = 8$. In the latter case, the Q and τ_0 parameters are decreased by a factor of 2. The uppermost transfer efficiency as well as the maximum lifetime is realized for the binary nanoclusters. An estimative calculation gives the average nanoparticles concentration of about 10^{18} cm^{-3} for the 7×10^{16} cm^{-2} fluence. Keeping in mind the concentration, we were able to calculate the critical distance R_0 (Table 5.1) for all the PL bands using the Q value taken from the approximation. The quite large distances of several tens of angstroms point out a resonant transfer mechanism that may be realized in our case.

5.6.2 PL Thermal Quenching

Both approaches used provide good approximations of the experimental PL decay curves. Most probably the nonelementary kinetics is caused by the energy transfer effects as well as by lifetime distribution. The latter indicates a high degree of structural disorder of the objects under study. It is known [60] that a disorder-related distribution of the PL thermal quenching activation energies of the luminescent centers leads to the appearance of the empirical Street law [61] employing the so-called anomalous Berthelot exponential [62]:

$$I_{PL}(T) = \frac{I_0}{1 + p_s \cdot e^{\frac{T}{T_0}}} \tag{5.6}$$

where p_s—pre-exponential factor; T_0—characteristic temperature.

In order to investigate the temperature dependence of the nonradiative processes, the 1.5–2.5 eV PL energy region comprising the silicon (R) and carbon (G) clusters

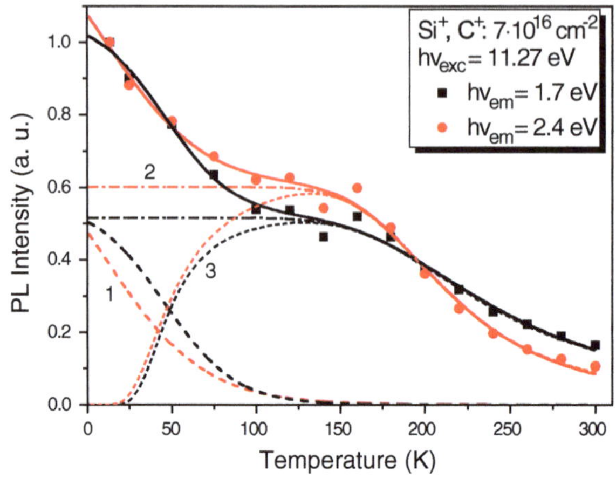

Fig. 5.14 The PL quenching curves for the R and G bands of the once annealed SiO_2:Si:C film. The *dashed, dash-dot* and *dotted lines* denote the Street [61] (*1*), Mott [62] (*2*), and Wang [63] (*3*) approximation functions, respectively

emission was chosen. The first quenching stage for the R and G maxima (10–75 K, Fig. 5.14) is described well by the expression (5.6). The second stage (100–300 K) may be approximated by different functions depending on the mechanism assumed.

One of the possible relaxation channels is the excitation energy transfer to the defects of an interfacial layer on the cluster–matrix boundary. Then the PL temperature dependence should be determined by the acceptor defect energy structure. The point defect intracenter PL quenching is often being described by a Mott function [63] that includes usual Arrhenius exponential:

$$I_{\mathrm{PL}}(T) = \frac{I_0}{1 + p_m \cdot e^{-\frac{E_a}{kT}}} \tag{5.7}$$

where I_0—PL intensity at $T \rightarrow 0$ K; k—Boltzmann constant; p_m—frequency factor; E_a—quenching activation energy.

Within the framework of the first approach, the experimental data are fitted by a sum of (5.6) and (5.7) functions (Fig. 5.14, curves 1 and 2). The numerical values for the both PL maxima are presented in the Table 5.2 and significantly differ from the known characteristics of both surface and bulk SiO_2 defects for which the quenching is performed at lower temperatures (50 ÷ 150 E). Thus, even though the Mott law nominally fits the experimental data, its application is not well founded and the parameters do not give valuable information concerning the luminescent centers nature.

An alternative emission mechanism could be an immediate recombination of the quantum-confined excitons within the implantation-produced quantum dots. In general, considering the intracenter relaxation processes for the molecular type PL

Table 5.2 The temperature quenching characteristics for the R (1.7 eV) and G (2.4 eV) maxima

Function	Parameter	PL band	
		2.4 eV	1.7 eV
Street	T_0, K	28 ± 1	20 ± 1
Equation (5.6)	p_s	0.65 ± 0.05	0.11 ± 0.02
Mott	E_a, eV	0.12 ± 0.01	0.10 ± 0.01
Equation (5.7)	p_m	637 ± 5	120 ± 3
Wang	E_a, eV	0.012 ± 0.005	0.016 ± 0.005
Equation (5.8)	E_{nr}, eV	0.115 ± 0.005	0.098 ± 0.005
	A	0.092 ± 0.005	0.026 ± 0.005
	B	549 ± 7	115 ± 5

centers one must take into account the different spin states, namely the singlet and triplet ones. The same approach can be applied to the excitons. The lowest excited state is commonly the triplet one. An activation barrier between the singlet and triplet states causes the triplet–singlet ($T_1 \rightarrow S_0$) PL intensity increase with growing temperature. A three-level model of a luminescent center can be used to describe the excitation and relaxation processes observed [11]. This model must involve the radiative and nonradiative transitions, as well as singlet–triplet conversion. Transforming the equations of the formal kinetics one can derive a function similar to that used by Wang et al. [63]:

$$I_{\mathrm{PL}}\,(T) = \frac{I_{01}}{\left(1 + Ae^{\frac{E_r}{kT}}\right)\left(1 + Be^{-\frac{E_{nr}}{kT}}\right)} \tag{5.8}$$

where E_r—triplet luminescence activation energy; E_{nr}—nonradiative relaxation activation energy; A and B—pre-exponential factors.

Similar dependencies with a temperature-driven luminescence increase were already seen by us for the ODC-centers in SiO_2 [64]. A strict consideration of the formal kinetics equations for the singlet and triplet states clarifies the physical meaning of the A and B factors of equation (5.8):

$$A = P_{\mathrm{LS}}/p_0^C;\; B = p_0^{\mathrm{NT}}/P_{\mathrm{LT}} \tag{5.9}$$

Here P_{LS}—singlet luminescence probability, which is inversely proportional to the excited state lifetime measured at low temperatures; P_{LT}—triplet luminescence probability; p_0^C and p_0^{NT}—frequency factors of the singlet–triplet and the triplet-ground state nonradiative transitions, respectively.

The sum of (5.6) and (5.8) functions provides better approximation of the experimental dependence I_{PL}(T). The E_{a} activation energy values are close to those obtained for silicon nanoparticles, though the PL final quenching in our case is observed at lower temperatures than under Si nanocrystals laser excitation.

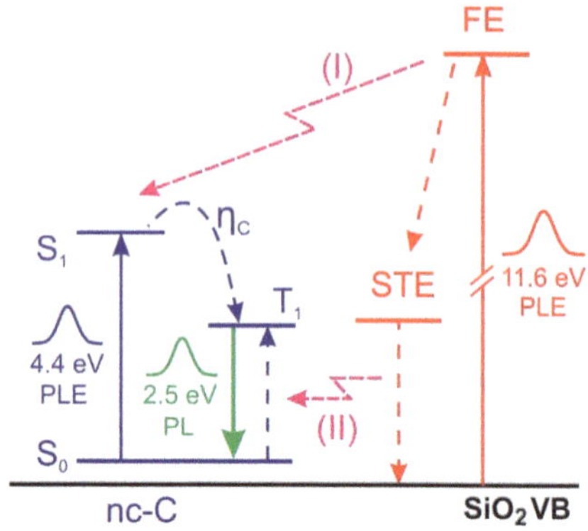

Fig. 5.15 A presumable scheme of the carbon nanocluster PL excitation–relaxation mechanism involving the free (FE) and self-trapped (STE) excitons taking place within the implanted SiO_2 films. S_0, S_1—singlet exciton states of the nanoclusters; T_1—the lowest triplet state

5.6.3 General Scheme of QD Excitation-Relaxation Processes

We suggest that the two quenching stages observed are due to the multiple excitation sources present. Indeed, a silicon dioxide film being exposed to a high energy excitation may contain both the nonrelaxed excitons (free, FE) and the self-trapped excitons (STE). The former cannot relax radiatively but are rather able to transfer their energy to the singlet excited states of the nanoparticles with subsequent singlet–triplet conversion and radiative recombination (channel I, Fig. 5.15):

$$FE \rightarrow S_1 \rightarrow T_1 \rightarrow S_0 + h\nu$$

A huge energy difference between the initial exciton and the quantum dot first excited states during the process I points out the participation of a higher excited state of the nanoclusters. After a fast thermalization, the confined excitons get over the $S_1 \rightarrow T_1$ conversion barrier causing the PL intensity increase at $T > 50\,K$.

Worthy of note, the second quenching stage occurs at the lower temperatures than those usually observed for silicon nanocrystals [65]. Such a peculiarity may be explained by either clusters amorphous structure or the decrease of energy transfer efficiency due to exciton–phonon interaction becoming stronger at T >200 K.

An amorphous silicon dioxide has the well-known self-trapped exciton photoluminescence (STEPL) band at 2.4 eV which resembles the G peak observed in our experiments. The STEPL band undergoes a Street-like thermal quenching [53] that is a characteristic of disordered systems having a wide activation energy distribution with a maximum close to zero [60]. Thus, the localized excitons created from the free ones during the relaxation can either recombine on their own or transfer their energy resonantly to the triplet state of a nearby cluster according to the following reaction (channel II, Fig. 5.15):

$$STE \rightarrow T_1 \rightarrow S_0 + h\nu$$

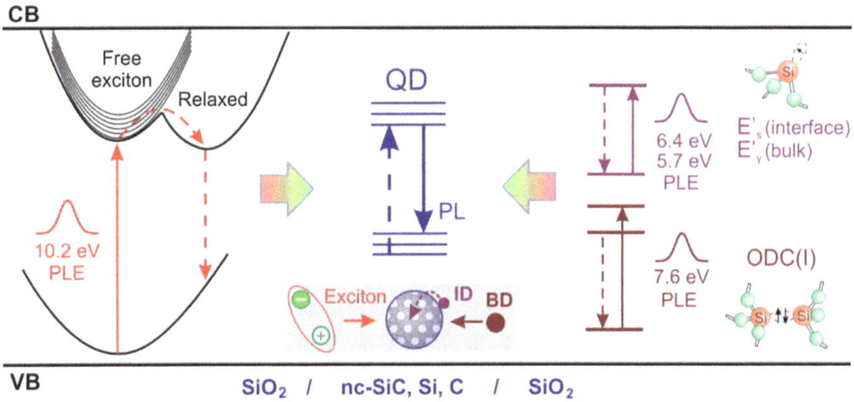

Fig. 5.16 A scheme of indirect PL excitation processes valid for the Si, C, and SiC quantum dots within the ion-implanted SiO_2 films. The *left part* of the figure contains the configuration curves for both free and relaxed excitons, while the *right part* presents the energy diagrams and structural models of surface (interface—ID) and bulk (BD) matrix defects. The only one type of ODC(I) is shown for clarity though the ODC(II) also take part in the energy transfer

Both processes noted must obey the Street law just as it was seen in the experiments. Figure 5.16 shows a more general scheme of the possible nonradiative energy transfer channels including the excitonic and the defect mechanisms of the luminescent quantum dots indirect excitation for SiO_2:Si:C films.

According to our data, the dominating excitation mechanisms are the exciton- and the electron-hole pairs-related ones which include a stage of energy transfer from the matrix relaxed and free excitons as well as from the oxide e-hole pairs to the quantum dots. A significant probability of energy transport from the bulk E'$_\gamma$-centres and ODCs appears with growing defect concentration. The resonant type of such a transfer is confirmed by relatively long critical distances R_0 (Table 5.2). To the contrary, the surface E'$_s$-centres may be located at the cluster–matrix interface region, and therefore an exchange mechanism could be realized which involves the electron wave functions overlapping.

5.7 Conclusion Remarks

In this chapter, we have considered the information capabilities of the synchrotron radiation excited luminescence spectroscopy for the investigation of optical properties of semiconductor quantum dots in a dielectric matrix. It was shown that the synchrotron radiation is an effective instrument to control the quantum dots energy structure, its formation and modification. These capabilities were demonstrated with the example of SiO_2/Si film structures with Si, C and SiC nanoclusters, formed by ion-beam synthesis. The method allows studying practically important spectral and kinetic properties during both optical excitation and relaxation processes of the quantum dots. Main results of the SiO_2:Si:C system investigation are the following:

1. Analysis of spectrally luminescent characteristics of the films with Si, C, and SiC nanodots evidences realization of the quantum confinement effect under UV-VUV excitation. The magnitude of the spectral shift of the emission band of quantum dots corresponds to the diameters of 3–5 nm range. A specific feature of the photoluminescence of thin film structures is a distortion of the experimental excitation spectra induced by interference effects which highly manifest itself for the SiO_2:Si system. Thus, the excitation spectra correction is a necessary step of experimental data analysis. This procedure is performed by means of special software taking into account dimensional factors and nanodots spatial distribution.
2. Using time-resolved techniques we divided the excitation spectrum into several parts corresponding to various QD excitation mechanisms. At low temperatures, the dominant mode of optical excitation of Si, C, and SiC clusters is the energy transfer with mobile and self-trapped SiO_2 excitons, as well as with participation of the matrix point defects. A resonant energy transfer mechanism from the matrix bulk defects (E' and ODC) is proved to be valid and is characterized quantitatively by means of a proper PL kinetics analysis. A direct excitation is realized only for carbon nanodots.
3. Based on the patterns of excitation and quenching of photoluminescence, an energy diagram of optical transitions in the "matrix-cluster" system was proposed. Scheme supposes the transfer of excitation energy to the singlet and triplet states of the spatially confined excitons localized in the quantum dots.

We believe that the results presented in the chapter point out the opportunities and application prospects of the time-resolved, synchrotron-excited photoluminescence spectroscopy to solve both the physical and technological problems in the field of silicon-related nanomaterials.

Acknowledgments We are grateful to our colleagues Prof. D.I. Tetelbaum, Prof. H.-J. Fitting and Prof. V.A. Pustovarov for useful collaboration during the acquisition of some results presented here.

References

1. Tiwari, S., Rana, F., Hanafi, H., et al.: A silicon nanocrystals based memory. Appl. Phys. Lett. **68**, 1377–1380 (1996)
2. Guo, L., Leobandung, E., Chou, S.Y.: A room-temperature silicon single-electron metal-oxide-semiconductor memory with nanoscale floating-gate and ultranarrow channel. Appl. Phys. Lett. **70**, 850–853 (1997)
3. Lalic, N., Linnros, J.: Light emitting diode structure based on Si nanocrystals formed by implantation into thermal oxide. J. Lumin. **80**, 263–267 (1999)
4. Normand, P., Tsoukalas, D., Kapetanakis, E., et al.: Formation of 2-D arrays of silicon nanocrystals in thin SiO2 films by very-low energy Si+ ion implantation. Electrochem. Solid State Lett. **1**, 88–90 (1998)

5. Pavesi, L., Dal Negro, L., Mazzoleni, C. et al.: Optical gain in silicon nanocrystals. Nature. **408**, 440–444 (2000)
6. Fujii, M., Yoshida, M., Kanzawa, Y., et al.: 1.54 μm photoluminescence of Er3+ doped into SiO2 films containing Si nanocrystals: evidence for energy transfer from Si nanocrystals to Er3+. Appl. Phys. Lett. **71**, 1198 (1997)
7. Brongersma, M.L., Polman, A., Min, K.S., Atwater, H.A.: Depth distribution of luminescent Si nanocrystals in Si implanted SiO2 films on Si. J. Appl. Phys. **86**, 759 (1999)
8. Brongersma, M.L., Polman, A., Min, K.S., et al.: Tuning the emission wavelength of Si nanocrystals in SiO2 by oxidation. Appl. Phys. Lett. **72**, 2577–2580 (1998)
9. Devaty, R.P., Choyke, W.J.: Optical characterization of silicon carbide polytypes. Phys. Stat. Sol. (a) **162**, 5–38 (1997)
10. Canham, L.T.: Silicon quantum wire array fabrication by electrochemical and chemical dissolution of wafers. Appl. Phys. Lett. **57**, 1046–1048 (1990)
11. Cullis, A.G., Canham, L.T., Calcott, D.J.: The structural and luminescence properties of porous silicon. J. Appl. Phys. **82**, 909–965 (1997)
12. Brus, L.E., Szajowski, P.F., Wilson, W.L., et al.: Electronic spectroscopy and photophysics of Si nanocrystals: relationship to bulk c-Si and porous Si. J. Am. Chem. Soc. **117**, 2915–2922 (1995)
13. Liao, L.S., Bao, X.M., Yang, Z.F., Min, N.B.: Intense blue emission from porous β-SiC formed on C^+-implanted silicon. Appl. Phys. Lett. **66**, 2382–2384 (1995)
14. Zhao, J., Mao, D.S., Lin, Z.X., et al.: Intense short-wavelength photoluminescence from thermal SiO2 films co-implanted with Si and C ions. Appl. Phys. Lett. **73**, 1838–1840 (1998)
15. Wu, X.L., Siu, G.G., Stokes, M.J., et al.: Blue-emitting β-SiC fabricated by annealing C60 coupled on porous silicon. Appl. Phys. Lett. **77**, 1292–1294 (2000)
16. Petrova-Koch, V., Sreseli, O., Polisski, G., et al.: Luminescence enhancement by electrochemical etching of SiC(6H). Thin Solid Films **255**, 107–110 (1995)
17. White, C.W., Withrow, S.P., Meldrum, A., et al.: Optical properties Of Si nanocrystals formed In SiO_2 by ion implantation. Mater. Res. Soc. Symp. Proc. **507**, 249 (1998)
18. Kovalev, D., Diener, J., Heckler, H., et al.: Optical absorption cross sections of Si nanocrystals. Phys. Rev. B. **61**, 4485–4487 (2000)
19. Zatsepin, A.F., Kaschieva, S., et al.: Formation and electron-beam annealing of implantation defects in a thin-film Si-SiO2 heterostructure. Tech. Phys. **54**, 323–326 (2009)
20. Salh, R., Fitting, L., Kolesnikova, E.V., et al.: Si and Ge nanocluster formation in silica matrix. Semiconductors **41**, 397–402 (2007)
21. Tetelbaum, D.I., Mikhaylov, A.N., Vasiliev, V.K., et al.: Effect of carbon implantation on visible luminescence and comosition of Si-implanted SiO_2 layers. Surf. Coat. Technol. **203**, 2658–2663 (2009)
22. Ziegler, J.F., Biersack, J.P.: SRIM—The Stopping and Range of Ions in Solids. SRIM Company, Chester (2012)
23. Zimmerer, G.: SUPERLUMI: a unique setup for luminescence spectroscopy with synchrotron radiation. Radiat. Meas. **42**, 859–864 (2007)
24. Zatsepin, A.F., Buntov, E.A., Kortov, V.S. et al.: Mechanism of quantum dot luminescence excitation within implanted SiO_2:Si:C films. J. Phys.: Condens. Matter **24**, 045301 (2012)
25. Zatsepin, A.F., Pustovarov, V.A., Kortov, V.S., et al.: Time-resolved photoluminescence of implanted $SiO_2:Si^+$ films. J. Non-Cryst. Sol. **355**, 1119–1122 (2009)
26. Zatsepin, A.F., Buntov, E.A., Kortov, V.S.: Low-temperature photoluminescence of ion-implanted $SiO_2:Sn^+$ films and glasses. J. Surf. Invest. X-ray, Synchrotron Neutron Techniques **6**, 668–672 (2012)
27. Smirani, R., Martin, F., Abel, G., et al.: The effect of size and depth profile of Si-nc imbedded in a SiO_2 layer on the photoluminescence spectra. J. Lumin. **115**, 62–68 (2005)
28. Windt, D.L.: IMD—Software for modeling the optical properties of multilayer films. Comput. Phys. **12**, 360–370 (1998)
29. Born, M., Wolf, E.: Principles of Optics, 6th edn. Pergamon Press, Oxford (1980)

30. Stearns, D.G.: The scattering of x rays from nonideal multilayer structures. J. Appl. Phys. **65**, 491–506 (1989)
31. Nevot, L., Croce, P.: Caractérisation des surfaces par réflexion rasante de rayons X. Application à l'étude du polissage de quelques verres silicates. Revue. Phys. Appl. **15**, 761–779 (1980)
32. Smith, S.: The Scientist and Engineer's Guide to Digital Signal Processing. California Technical Publishing, San Diego (1997)
33. Buntov, E.A.: ICorrector software. http://sourceforge.net/projects/icorrector/ (2012). Accessed 9 Nov 2012
34. Skuja, L.N.: Optically active oxygen-deficiency-related centers in amorphous silicon dioxide. J. Non-Cryst. Solids. **239**, 16–48 (1998)
35. Salh, R., von Czarnowski, A., Zamoryanskaya, M.V. et al.: Cathodoluminescence of SiOx under-stoichiometric silica layers. Phys. Stat. Sol. (a). **203**, 2049–2057 (2006)
36. Glinka, YuD, Lin, S.-H., Chen, Y.-T.: The photoluminescence from hydrogen-related species in composites of SiO_2 nanoparticles. Appl. Phys. Lett. **75**, 778–780 (1999)
37. Gan, F.H., Fuxi, G.: Optical and Spectroscopic Properties of Glass. Springer-Verlag, Berlin (1992)
38. Calcott, P.D.J., Nash, K.J., Canham, L.T. et al.: Identification of radiative transitions in highly porous silicon. J. Phys.: Condens. Matter **5**, L91 (1993)
39. Ghislotti, G., Nielsen, B., Asoka-Kumar, P., et al.: Effect of different preparation conditions on light emission from silicon implanted SiO_2 layers. J. Appl. Phys. **79**, 8660–8663 (1996)
40. Inokuma, T., Wakayama, Y., Muramoto, T., et al.: Optical properties of Si clusters and Si nanocrystallites in high-temperature annealed SiOx films. J. Appl. Phys. **83**, 2228–2234 (1998)
41. Yi, L.X., Heitmann, J., Scholz, R., Zacharias, M.: Si rings, Si clusters, and Si nanocrystals—different states of ultrathin SiOx layers. Appl. Phys. Lett. **81**, 4248–4250 (2002)
42. Serincan, U., Aygun, G., Turan, R.: Spatial distribution of light-emitting centers in Si-implanted SiO_2. J. Lumin. **113**, 229–234 (2005)
43. Trukhin, A.N.: Excitons, localized states in silicon dioxide and related crystals and glasses In: Pacchioni, G., Skuja, L., Griscom, D.L. (eds.) Defects in SiO_2 and Related Dielectrics: Science and Technology. Springer, New York (2002)
44. Mathioudakis, C., Kopidakis, G., Patsalas, P., Kelires, P.C.: Disorder and optical properties of amorphous carbon. Diam. Relat. Mater. **16**, 1788–1792 (2007)
45. Robertson, J.: Recombination and photoluminescence mechanism in hydrogenated amorphous carbon. Phys. Rev. B. **53**, 16302–16305 (1996)
46. Henley, S.J., Carey, J.D., Silva, S.R.: Room temperature photoluminescence from nanostructured amorphous carbon. Appl. Phys. Lett. **85**, 6236–6238 (2004)
47. Fan, J.Y., Wu, X.L., Chu, P.K.: Low-dimensional SiC nanostructures: fabrication, luminescence, and electrical properties. Prog. Mater. Sci. **51**, 983–1031 (2006)
48. Shimizu-Iwayama, T., Hole, D.E., Boyd, I.W.: Mechanism of photoluminescence of Si nanocrystals in SiO_2 fabricated by ion implantation: the role of interactions of nanocrystals and oxygen. J. Phys.: Condens. Matter **11**, 6595 (1999)
49. Rebohle, L., Gebel, T., Frob, H., et al.: Ion beam processing for Si/C-rich thermally grown SiO_2 layers: photoluminescence and microstructure. Appl. Surf. Sci. **184**, 156–160 (2001)
50. Zhuravlev, K.S., Gilinsky, A.M., Kobitsky, A.Y.: Mechanism of photoluminescence of Si nanocrystals fabricated in a SiO_2 matrix. Appl. Phys. Lett. **73**, 2962–2964 (1998)
51. Griscom, D.L.: Self-trapped holes in amorphous silicon dioxide. Phys. Rev. B. **40**, 4224–4227 (1989)
52. Trukhin, A., Poumellec, B.: Energy transport in silica to oxygen-deficient luminescence centers. Comparison with other luminescence centers in silica and α-quartz. Solid State Commun. **129**, 285–289 (2004)
53. Messina, F., Vella, E., Cannas, M., Boscaino, R.: Evidence of delocalized excitons in amorphous solids. Phys. Rev. Lett. **105**, 116401 (2010)
54. Toyozawa, Y.: Dynamics and instabilities of an exciton in the phonon field and the correlated absorption-emission spectra. Pure Appl. Chem. **69**, 1171–1178 (1997)

55. Zatsepin, A.F., Biryukov, D.Yu., Kortov, V.S.: Photoelectron spectroscopy of E′ centers in crystalline and glassy silicon dioxide. Phys. Solid State **48**, 245–254 (2006)
56. Dovrat, M., Goshen, Y., Jedrzejewski, J., et al.: Radiative versus nonradiative decay processes in silicon nanocrystals probed by time-resolved photoluminescence spectroscopy. Phys. Rev. B. **69**, 155311 (2004)
57. Mihalcescu, I., Vial, J.C., Romestain, R.J.: Carrier localization in porous silicon investigated by time-resolved luminescence analysis. J. Appl. Phys. **80**, 2404–2411 (1996)
58. Linnros, J., Lalic, N., Galeckas, A., Grivickas, V.: Analysis of the stretched exponential photoluminescence decay from nanometer-sized silicon crystals in SiO_2. J. Appl. Phys. **86**, 6128–6134 (1999)
59. Inokuti, M., Hirayama, F.: Influence of energy transfer by the exchange mechanism on donor luminescence. J. Chem. Phys. **43**, 1978 (1965)
60. Zatsepin, A.F., Buntov, E.A., Ageev, A.L.: The relation between static disorder and photoluminescence quenching law in glasses: a numerical technique. J. Lumin. **130**, 1721–1724 (2010)
61. Street, R.A.: Recombination in a-Si:H: defect luminescence. Adv. Phys. **25**, 5775–5784 (1976)
62. Mott, N.F., Davis, E.A.: Electronic Processes in Non-crystalline Materials. Oxford University Press, Oxford (1979)
63. Wang, J., Righini, M., Gnoli, A., et al.: Thermal activation energy of crystal and amorphous nano-silicon in SiO_2 matrix. Solid State Commun. **147**, 461–464 (2008)
64. Zatsepin, A.F.: Statics and dynamics of excited states of oxygen-deficient centers in SiO_2. Phys. Solid State **52**, 1176–1187 (2010)
65. Kapoor, M., Singh, V.A., Johri, G.K.: Origin of the anomalous temperature dependence of luminescence in semiconductor nanocrystallites. Phys. Rev. B. **61**, 1941 (2000)

Chapter 6
Silicon Nanoparticles-Based Light Emitting Capacitors

A. Morales Sánchez, J. Barreto, C. Domínguez Horna, M. Aceves Mijares, J. A. Luna López and L. Licea Jiménez

Abstract This chapter is focused on the study of the microstructural, optical, electrical, and electro-optical properties of Si-nanoparticles (Si-nps) embedded in a silica matrix for light emitting devices applications. Si-nps were created from silicon-rich oxide [SRO, (SiO_x, $x < 2$)] films which are deposited by low pressure chemical vapor deposition and followed by a thermal annealing at high temperature. The composition, microstructure, and optical properties of SRO films are analyzed as a function of the silicon excess and thermal annealing temperature. Once the properties of these materials are known, SRO films which exhibited the best photoluminescent (strongest PL) properties were chosen in order to analyze their electrical and electroluminescent (EL) properties. Simple Metal–Oxide–Semiconductor structures using the SRO films as the dielectric layer were fabricated for these studies. Blue and red EL was observed by changing the Si-np size from 1.5 to 2.7 nm embedded in the silica matrix, respectively. EL is ascribed to the charge injection into the Si-nps embedded in the SRO films through a balanced transport network. The

A. M. Sánchez (✉) · L. L. Jiménez
Centro de Investigación en Materiales Avanzados (CIMAV),
Alianza Norte 202 Parque de Investigación e Innovación Tecnológica,
Apodaca,66600 Nuevo León , México
e-mail: alfredo.morales@cimav.edu.mx

J. Barreto
Centre for Quantum Photonics, HH Wills Laboratory,
University of Bristol,Tyndall Avenue, Bs8, Bristol, UK

C. D. Horna
Instituto de Microelectrónica de Barcelona (CNM-CSIC),
Campus UAB, 08193 Bellaterra, España

M. A. Mijares
Electronics Department, INAOE ,
Apdo 51, 72000 Pue. Pue., México

J. A. L. López
IC-CIDS Benemérita Universidad Autónoma de Puebla,
Ed. 103 C or D, C.U., Col. San Manuel, 72570 Puebla, México

H. Li et al. (eds.), *Silicon-based Nanomaterials*,
Springer Series in Materials Science 187, DOI: 10.1007/978-1-4614-8169-0_6,
© Springer Science+Business Media New York 2013

EL emission is observed with the naked eye and in daylight conditions on the whole area of devices. Therefore, these results prove the feasibility to obtain LECs by using simple capacitors with SRO films as the active layer.

6.1 Introduction

Observation of luminescence in porous silicon [1] seemed to solve the physical inability of silicon (Si) to act as light emitter; however its poor chemical stability, robustness, and luminescence degradation [2] have hindered its development. Silicon nanoparticles (Si-nps) embedded in a dielectric matrix look as a better alternative for light emitting devices (LED's) due to their strong, stable luminescence, and chemical stability. One of the most studied materials has been the silicon rich oxide (SRO), because the silicon excess separates from the oxide matrix and forms nanometer-sized silicon grains dispersed throughout the oxide when the film is subjected to high temperature anneal. These materials have shown interesting properties in charge trapping [3], electroluminescence (EL) [4, 5] and they have been used with a PN junction to get a radiation sensor [6]. In order to advance in the photonics integration objective complementary metal oxide semiconductor (CMOS) compatible techniques must be used from the available ones in the fabrication of the nanostructured materials. Low pressure chemical vapor deposition (LPCVD) allows obtaining silicon-rich oxide layers using oxidant species like nitrous oxide (N_2O) and silicon compounds (SiH_4) as reactant gasses. Silicon excess is easily controlled by changing the partial pressure ratio (*R*o) between N_2O and SiH_4. Silicon excess as high as 17 at.% is obtained with Ro = 3, and stoichiometric oxide can be obtained for Ro = 100 [7–9].

For EL in Si-nps, intense investigations have been carried out to get an efficient light emitting device, including porous silicon [10–15]. Some of these results prove the feasibility of these nanostructured materials to be used in the fabrication of silicon-based light emitting devices. However, one of the main problems in the development of efficient LEDs based on Si nps is the carrier injection into the dielectric matrix. It is well-known that EL intensity is directly related with charge injection to the luminescent centers [10–14]; but also, it has been reported that when high current flows through the active layer, charge trapping or degradation of the matrix could occur, quenching the EL properties [15–17]. Nevertheless, the decay of the EL properties at DC is not necessarily related with those effects, instead, the annihilation of unstable Si-nps could be occurring. Moreover, the annihilation of such unstable Si-nps could change the way to obtain EL.

On the other hand, both DC and AC electroluminescence in Si-nps-based devices have been reported [18–26]. Nevertheless, most of the reported luminescent devices exhibit red EL [19–23]. Some other show dual color EL, either blue and red or violet and white [25, 26]. However, in devices emitting dual color, the blue electroluminescence is obtained just by the selection of a certain electric field. A slight variation in the electric field implies a combination of the color of emission. Moreover, strong enough EL to be seen by naked eye has been reported to appear as a small number of bright spots [18, 24, 27]. Recently, some authors have reported devices emitting EL observed with the naked eye and in the whole active area [25, 26]. This fact could be related with the characteristics of the active layer.

In this chapter, the microstructural, optical, electrical, and electro-optical properties of Si-nanoparticles (Si-nps) embedded in a silica matrix are studied and analyzed for their application in light emitting capacitors. Since it is expected that a strong PL will produce an intense EL, SRO films that exhibited the best photoluminescent (strongest PL) properties were chosen in order to analyze their electrical and electroluminescent (EL) properties. Simple Metal–Oxide–Semiconductor (MOS) structures using SRO films with 4 and 5.1 at.% of Si-excess as the dielectric layer were fabricated for these studies. Blue and red electroluminescence (EL) was observed by changing the silicon nanoparticle (Si-np) size from 1.5 to 2.7 nm (4 to 5.1 at.% of Si-excess) embedded in the silica matrix. Nevertheless, blue emission appears after the electrical annihilation of preferential conductive paths, that is, a switching from high conduction state (HCS) to a low conduction state (LCS). EL in the HCS was also observed, but as shining spots which disappear after the annihilation of those preferential conductive paths. EL in these devices is ascribed to the charge injection into the Si-nps embedded in the SRO films through a balanced transport network. The EL emission is observed with the naked eye and in daylight conditions on the whole area of devices. Therefore, these results prove the feasibility to obtain light emitting capacitors (LECs) by using simple capacitors with SRO films as the active layer.

6.2 Experimental Procedure

6.2.1 Silicon Nanoparticles Based Materials

In this work, SRO films were deposited in a homemade LPCVD hot wall reactor using pure nitrous oxide (N_2O) and 5%-nitrogen (N_2) diluted silane (SiH_4) as the reactant gasses. In this system, the pressure is controlled using mass flow controllers and a Baratron pressure sensor, the low pressure is obtained from a mechanical vacuum pump. Si excess in the deposited films was controlled by the ratio of partial pressure (*R*o) of the reactant gases of N_2O and SiH_4, defined by the Eq. 6.1. Deposition of SRO films was carried out on 2 inches N type Si wafers ((100)-oriented) at temperature of 720°C. SRO films with *R*o values of 10, 20, 30, and 40 were deposited with the parameters shown in Table 6.1. The expected thickness of the films was about 550 nm.

$$R_o = \frac{P\,(N_2O)}{P\,(SiH_4)} \tag{6.1}$$

After deposition, SRO films were thermally annealed at 1,100 and 1,250 °C for 180 min in a N_2 atmosphere in order to induce a Si and SiO_2 phase separation, and therefore the formation of silicon nanoparticles embedded in a SiO_2 matrix.

The thickness and refractive index of the deposited SRO films were measured using a Gärtner L117 ellipsometer with incident laser wavelength of 632.8 nm. A FTIR BRUCKER Vector 22 spectrometer was employed to study the composition and structure of the SRO films. The scanned range was between 350 cm^{-1} and 3,500 cm^{-1} with 2.5 cm^{-1} of resolution. The silicon excess in SRO films was measured with a PHI ESCA–5500 X-ray photoelectron spectrometer (XPS) using a monochromatic Al radiation source with energy 1,486 eV. Depth composition profiles inside the films were obtained measuring by XPS in the surface, then etching and measuring again. XPS analysis was done on thermally annealed SRO samples. In order to know the microstructure of SRO films and corroborate the Si-nanoparticles formed inside of these films after thermal annealing, some samples were analyzed by means of Energy Filtered Transmission Electron Microscopy technique (EFTEM). EFTEM images were obtained using a JEOL JEM 2010F electronic microscope. All the EFTEM images were measured using a Si plasmon of 17 eV and cross-section views. PL at room temperature was carried out with a Fluoromax 3 spectrofluorometer from Horiba Jobin Yvon, which is controlled with a computer. The samples were excited using a 250 nm radiation. PL was measured between 400 and 900 nm with a resolution of 1 nm.

6.2.2 Silicon Nanoparticles Based LECs

Once the properties of these materials are known, SRO films that exhibited the best photoluminescent (strongest PL) properties were chosen in order to analyze their electrical and electroluminescent properties. Two fabrication processes were done to study the electrical and electro-optical properties of SRO films. Simple Metal–Oxide–Semiconductor (MOS) structures using the SRO films as the dielectric layer were fabricated for these studies.

In the first process, 25 and 50 nm thick SRO films were deposited on P-type silicon wafers ((100)-oriented, resistivity 0.1–1.4 Ω cm) in a LPCVD using nitrous oxide (N_2O) and silane (SiH_4) as the reactant gasses at 720 °C. The deposition time was about 5.5 and 11 min for 25 and 50 nm thick SRO films, respectively. The total pressure and partial pressure of N_2O and SiH_4 was about 2.5, 1.5 and 1 torr, respectively, in order to get a *R*o value of 30 for this experiment. The silicon excess for SRO film used in these structures is ∼ 4 at.% as measured by XPS (see Table 6.2).

Table 6.1 Parameters of deposited SRO – LPCVD films, T = 720 °C

*R*o =	Pressure of gases (Torr)		Time
N_2O/SiH_4	N_2O	SiH_4	(min)
10	0.60	1.20	75
20	1.06	1.06	120
30	1.40	0.93	140
40	1.40	0.70	115

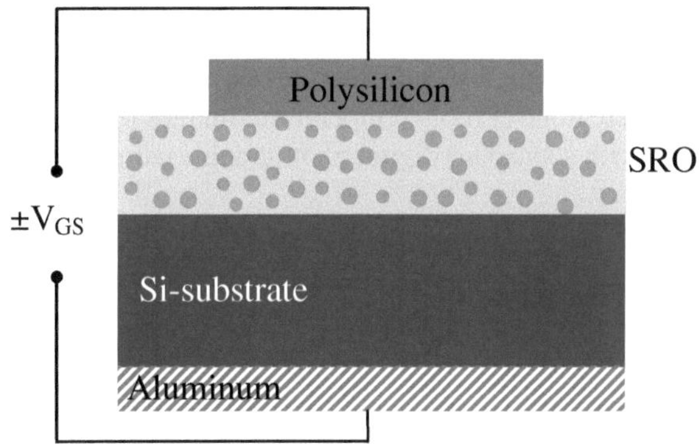

Fig. 6.1 Schematic diagram of the light emitting capacitors (LECs)

After deposition, a thermal annealing at 1,100° C in N_2 atmosphere was carried out for 180 min in order to induce silicon agglomeration. These devices feature 1 μm thick Al/Cu backside contacts and a 350 nm thick n^+ polysilicon gate. The area of gate electrode is $9.604 \times 10^{-3}cm^2$.

In the second process, SRO films were deposited on N-type silicon substrates ((100)-oriented) with resistivity of 5–10 Ω-cm by LPCVD using the flow ratio (*R*o) between nitrous oxide (N_2O) and silane (SiH_4) at 720 °C. *R*o values of 20 and 30 were used for this experiment, which allows to obtain SRO films with 5.1 and 4.0 at.% of silicon excess, respectively. After deposition, SRO films were thermally annealed at 1,100 °C for 3 h. The thickness of annealed SRO films with $Ro = 20$ and 30 is, respectively, 119.3 ± 4.9 and 70.1 ± 2.3 nm, as measured with a Gaertner L117 ellipsometer (@632.8 nm). A ~400 nm thick semitransparent n^+ polycrystalline silicon (Poly) gate was then deposited onto the SRO film surface by LPCVD. After a lithography process step, square shaped gates of 4 mm^2 area were defined. Finally, the backside contact was formed with 1 μm thick aluminium layer by evaporation. A general schematic diagram of the final devices is shown in Fig. 6.1.

Current–voltage (I–V) measurements were done with a computer-controlled Keithley 2,400 source meter. Frequency capacitance–voltage (C–V) measurements were performed from inversion to accumulation at 100 kHz (sine wave with 30 mV of amplitude) using a computer-controlled HP4192A impedance analyzer. The EL spectra were measured by biasing the devices with a constant DC voltage. Electrical stress was applied between the gate electrode and the back contact. Light emission

Table 6.2 Silicon excess of thermally annealed SRO films obtained by XPS

*R*o = N_2O/SiH_4	Si excess (at.%)	Concentración (%) O	Si	N	x = O/Si
10	12.7	54.00	46.00	0.00	1.17
20	5.1	60.81	38.46	1.00	1.58
30	4.0	62.01	37.35	0.62	1.66
40	2.2	64.50	35.50	0.00	1.81

was collected with an optical fiber, which was located right onto the surface of devices and connected to a Fluoromax 3 spectrofluorometer and analyzed with a computer. All measurements were done at room temperature. Images from electroluminescent devices were taken with a standard digital photographic camera.

6.3 Results and Discussion

6.3.1 *Silicon Nanoparticles Based Materials*

The study of composition and microstructure of SRO films before and after applying a thermal annealing play an important role in order to understand their luminescence, electrical, and electro-optical properties. In this sense, several techniques such as null ellipsometry, Fourier transform infra-red spectroscopy, X-ray photoelectron spectroscopy, and energy filtered transmission electron microscopy have been used.

An easy way to verify the presence of Si excess in SRO films is measuring their refractive index (n). The refractive index of SiO_2 and Si is about 1.46 and 3.8 (for the real part), respectively. It is well-known that n values between 1.46 and 3.8 are due to silicon excess into the film, if no other element or impurity is in the SRO films. Figure 6.2 shows the refractive index of SRO films before and after annealing. In as-deposited films when $Ro \leq 30$ the refractive index is higher than 1.46 indicating

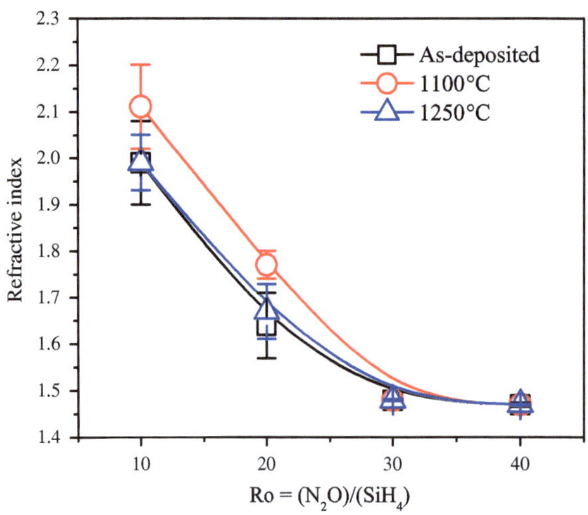

Fig. 6.2 Refractive index from as deposited and annealed SRO–LPCVD films

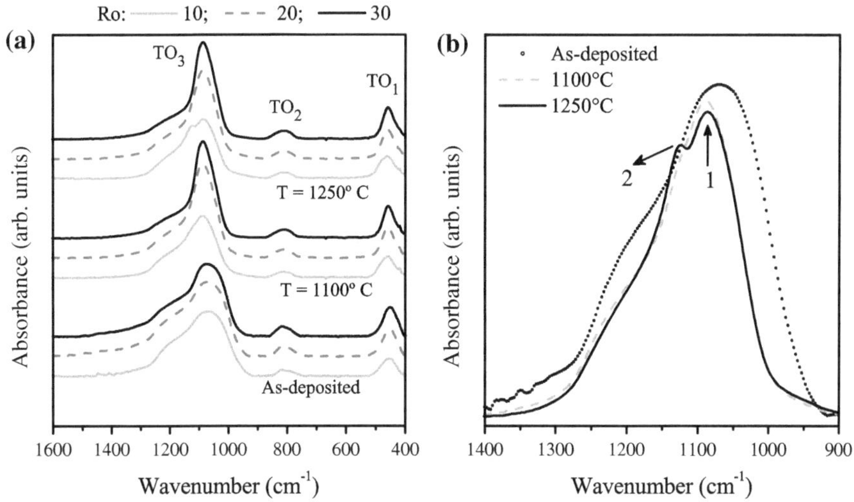

Fig. 6.3 Infrared absorption spectra of **a** SRO films before and after thermal annealing and SRO films with **b** *R*o = 10 before and after thermal annealing in the 900–1500 cm^{-1} range

the presence of silicon excess into the films. As *R*o becomes higher, the refractive index tends to the SiO_2 value, as expected by the lower value of (SiH_4) pressure listed in Table 6.1. Thickness of SRO films was reduced (not shown), if compared to the as-deposited films, after thermal annealing (1,100 and 1,250 °C), which is due mainly to a microstructural re-ordering and densification of the material. By the contrary, the refractive index was increased, being more evident for samples with higher silicon excess (see Fig. 6.2).

Figure 6.3 shows the IR absorbance spectra obtained from SRO films before and after annealing. It has been reported that SRO films deposited by the mixture of N_2O and SiH_4 display absorption bands associated with SiN, NH, SiH, and OH vibrations in addition to the three characteristic bands related to the Si–O–Si bonding arrangement [28]. However, the presence of some nitrogen and hydrogen characteristic peaks was not observed in the IR spectra of the films. Only three vibration bands related to Si–O–Si rocking (TO_1), Si–O–Si bending (TO_2) and Si–O–Si stretching (TO_3) appeared [28, 29].

It is well-known that the dominant feature in the IR spectrum of SRO films is associated to the stretching motion of the oxygen atoms TO_3. A shifting in the frequency of the TO_3 mode is associated with the presence of silicon excess and with a change in the microstructure of the material [29, 30]. For a thermally grown 400 nm thick SiO_2 film, the TO_3 mode and its full width at half maximum (FWHM) are about 1088 and 85 cm^{-1}, respectively. [31]. Figure 6.4 exhibits the TO_3 vibration band frequency and its full width at half maximum (FWHM) as a function of the flow ratio *R*o. For as-deposited SRO films, the TO_3 band slightly shifted toward higher frequencies and its FWHM was reduced as *R*o was changed from 10 to 40 corresponding to a larger oxygen concentration and to the presence of different

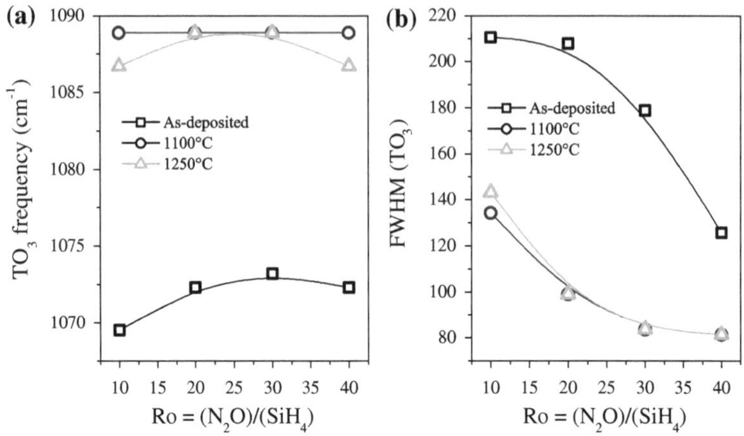

Fig. 6.4 **a** TO_3 vibration band frequency and **b** its FWHM as a function of the flow ratio *R*o

bonding arrangement (microstructural disorder), respectively [28–30]. This effect is directly related to the decrease in the refractive index (*n*) of the as-deposited SRO films where *n* tends to the SiO_2 value as *R*o is varied from 10 to 40. Another shift and width (FWHM) reduction of the TO_3 vibration band was obtained when the SRO films were subjected to a thermal annealing at both temperatures of 1,100 or 1,250 °C indicating a strong arrangement on the microstructure of the films. The TO_3 vibration band was placed at about 1,088 cm^{-1} for all films, independently of the flow ratio (*R*o). In addition, the FWHM decreases as *R*o increases (lower silicon excess). Due to the silicon excess inside these films, these changes are ascribed to a phase separation of Si and SiO_2 in the SRO films.

An important feature is observed in the IR-TO_3 mode for SRO films with $Ro = 10$. Figure 6.3b shows the evolution of the TO_3 mode at different temperatures of anneals. For as-deposited film, a broad peak is observed with a maximum at 1,069 cm^{-1}. However, as mentioned before, the FWHM reduces after thermal annealing and the maximum peak shifts toward higher wavenumber. Both, microstructural arrangement and high silicon excess in the SRO film induced a splitting of the TO3 in two peaks when the temperature of anneal was 1,250 °C. The first one placed at ~1,088 cm^{-1} and the second one at ~1,124 cm^{-1}, as shown in Fig. 6.3b. It has been reported that the shoulder located on the high wave numbers side of the TO_3 mode is the result of the overlapping of higher frequency bands attributed to three symmetric stretching modes: two longitudinal optical (LO_2 and LO_1) modes at 1,150 and 1,250 cm^{-1} and one transverse optical TO_4 mode at 1,200 cm^{-1} [30]. Then, the absorption peak at 1,124 cm^{-1} could be a result of larger phase separation between Si and SiO_2.

The silicon (Si) and oxygen (O) concentration in SRO films as measured by XPS are depicted in Fig. 6.5a,b, respectively for different *R*o values. As it was expected, the silicon excess varies depending on the flow ratio (*R*o). For SRO films with $Ro = 40$, the silicon profile stays about 35.5 at.% and it increases until 46 at.% as *R*o value

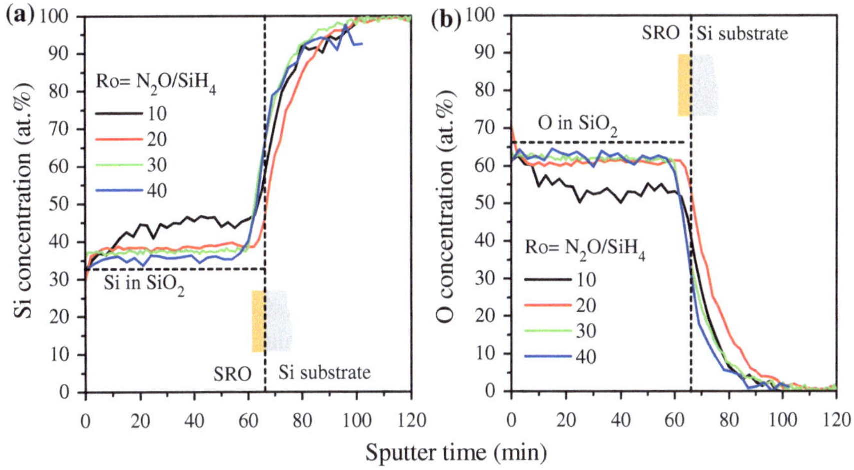

Fig. 6.5 **a** Silicon and **b** Oxygen concentration in SRO films with different flow ratio (*R*o) values as measured by XPS

reduces to 10. A thin silicon dioxide layer is formed at the SRO surface for low *R*o values; which can be explained by the silicon excess oxidation with air.

A nitrogen profile was also observed, however, is very low (around 0.8 at.%) as compared to other works reported for SRO-PECVD films where the N incorporation is about 10 at.% [32]. Table 6.2 summarizes the silicon excess in the SRO films. It is observed that excess is ranging between 2.2 and 12.7 at.% when the ratio changes between *R*o 40 to 10.

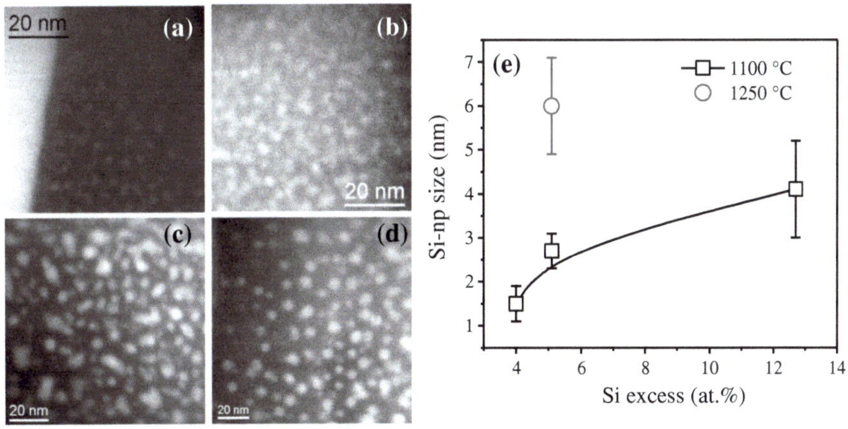

Fig. 6.6 Cross-view EFTEM images of SRO films with **a** 4, **b** 5.1 and **c** 12.7 at.% of silicon excess annealed at 1, 100 °C. The EFTEM image of SRO film with **d** 5.1 at.% of silicon excess annealed at 1, 250 °C and the **e** Si-np size as a function of silicon excess are also reported

Figure 6.6 shows the cross-view EFTEM images of Si-nanoparticles for the SRO films after thermal annealing at 1,100 °C and the Si-np size as a function of silicon excess. In all of these images, the bright zones are associated to the presence of silicon in the film. The EFTEM image for the SRO film with 4 at.% of silicon excess shown in Fig. 6.6a indicates some slightly bright zones. In spite of low contrast and sharpness, the contour of some of the Si nanoparticles is observed. As expected, a phase separation between Si and SiO_2 becomes more evident when the silicon excess increases. Figure 6.6b shows the EFTEM image for the SRO with 5.1 at.% of silicon excess, where more well-defined Si-nanoclusters were observed. The largest Si-clusters were observed in the SRO film with the highest silicon excess [Fig. 6.6c]. While some show an elliptical shape, others have an enlarged one, which could be a result of the agglomeration of neighbored Si-clusters. The silicon agglomeration in SRO film with 5.1 at.% of silicon excess was enhanced when the temperature of annealing increased to 1,250 °C, forming Si-nps of larger size, see Fig. 6.6d,e. The mean size Si-nps extended from 1.5 up to 4.1 nm when the silicon excess in SRO films, annealed at 1,100 °C, was increased from 4.1 at.% to 12.7 at.%, as indicated in Fig. 6.6e.

Figure 6.7a shows the PL spectra for the SRO films thermally annealed at 1,100 °C. All the films exhibited a PL band in the 1.4–2.1 eV range. A dependence of the PL peak energy and its intensity on the silicon excess is observed.

The SRO film with the highest (12.7 at.%) silicon excess emitted a PL peak at 1.62 eV, but with very low intensity, as observed in Fig. 6.7a,b. As the silicon excess reduces to 5.1 at.%, the PL intensity increases up to 20 times. In spite of a difference in the mean size of the Si-nps between these two films, 4.1 nm for 12.7 at.% and 2.7 nm for 5.1 at.% of silicon excess, the PL peak remains at the same energy, as shown in Fig. 6.7b. Moreover, then it still remains at the same energy as the Si-nps

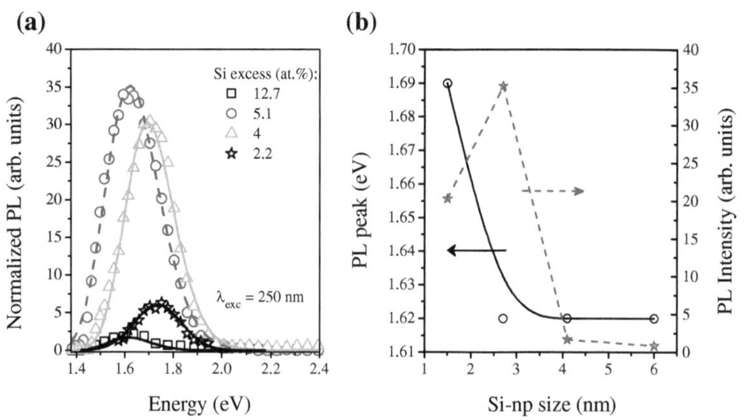

Fig. 6.7 PL spectra from **a** SRO films annealed at 1,100 ° C and **b** the PL intensity and peak energy of emission as a function of the silicon nanoparticle size. *Symbols* are experimental data and *lines* are the Gaussian fits

size is increased to 6 nm. On the other hand, the SRO film with 4 at.% of silicon excess showed a PL peak at 1.69 eV; that is, a redshift took place when the silicon excess was increased from 4 to 5.1 at.%. The maximum emission was observed with a silicon excess around 4 and 5.1 at.%. The energy of the emission peak does not seem to depend on the size of the silicon nanoparticle (silicon excess), contrary to this, the PL intensity seems to depend on the size of the Si-nps of the Si-nps, as observed in Fig. 6.7b.

PL intensity of about 20 a.u. was observed in the SRO film with embedded Si-nps of 1.5 nm in size; when the Si-nps size enlarged to 2.7 nm, the PL increased to about 35 a.u.; however, PL was dramatically reduced by more than 40 times when the Si-np size was larger than 4.1 nm. All the SRO films annealed at 1,250 °C exhibited a PL of very low intensity. The Si-np size in these films is expected to be larger than the annealed at 1,100 °C, such as depicted by EFTEM images. Then, there exist a direct relation between the Si-np size with the PL intensity rather than with the emission energy.

6.3.2 Silicon Nanoparticles Based LECs

For clarity, we will call P-devices and N-devices to those LECs fabricated on P-type and N-type silicon wafers, respectively. Figure 6.8a exhibits the typical I–V curve measured in as-fabricated P-devices which have the SRO film with 4at.% of Si-excess (Si-np mean size of 1.5 nm). A high current conduction is measured at low voltages, at both negative (forward bias, FB) and positive biases (reverse bias, RB). However, at certain voltage (V_{drop}) the current drops (regardless on the sweep rate) from a high conduction state (HCS) to a low conduction state (LCS). Afterwards, the current enters in a high electric field conduction regime. At FB, four different regions can be identified as *A, B, C,* and *D* corresponding to [0 to ∼−21 V], [∼−21 to ∼−32 V], [∼−32 to ∼−44 V] and [∼−44 to ∼−50 V], respectively. The *B* zone is the region where the current remains at some pA (LCS) after dropping from 10^{-3} A. Additionally, in *C* region, a steady current increase behavior with bias is observed, which is continued on *D* with a slightly different slope. At RB, the *A, C* and *D* zones are also present. Furthermore, in reverse bias, the current drop occurs at higher voltages compared to that at forward bias. In the *A* zone (low electric fields), the current exhibits a strong dependence on the gate voltage. In this region, tunneling mechanisms such as Fowler–Nordheim (F–N) [32, 33] or Pool–Frenkel (P–F) [16, 34] should be excluded because of the SRO thickness. Therefore, other conduction mechanism should be responsible for that current observed; the Trap-Assisted Tunneling (TAT) [35] being the most probable. Meanwhile, in *C* and *D* zones the electric field inside the oxide is strong enough (>6 MV/cm) to cause a strong band bending, so electrons tunnel through a triangular barrier from the polysilicon gate toward the silicon substrate. After the current drop and at FB, a subsequent I–V measurement, on the same device, exhibits a very low-level current of about 10^{-12} A in both *A* and *B* zones. This effect could be explained by the annihilation

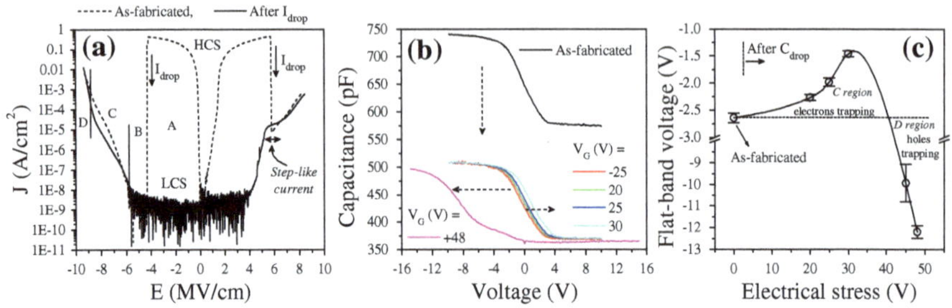

Fig. 6.8 **a** I–V characteristic of the P-devices which have the SRO film with 4at.% of Si-excess (Si-np mean size of 1.5 nm) before and after the current drop, **b** C–V curves and **c** flat-band voltage measured before and after the current drop. All measures were done from inversion to accumulation at 100 kHz with a sweep rate of 0.1 Vs^{-1}

of conductive paths in the SRO film, which connect the silicon substrate with the polysilicon gate. As the gate voltage becomes more negative, *C* and *D* regions are obtained.

Meanwhile at RB, the current oscillates at a low level ($< 10^{-10}$ A) and at low electric fields (<3.75 MV/cm). For gate voltages $V_G \geq 20$ V, the current increases until a step-like behavior is observed at about 5.43 MV/cm. This step-like current is almost constant (as marked by the arrow in Fig. 6.8a until entering into the high field conduction regime. Similar current behaviors have been reported [16, 19] and ascribed to charge trapping in a Si-ncs layer through tunneling and a consequent Coulomb blockade effect on the charges trying to tunnel behind. Then, in these SRO films charges could be getting trapped during the sweep voltage. In fact, I–V curve measured from similar devices which present the same behavior but with gate area of 2.304×10^{-3} cm^2 exhibit a clear staircase-like behavior between 16 and 28 V [36]. This quantum phenomenon is due to Coulomb blockade (CB) effects as a result of single electron trapping in the silicon nanoparticles embedded in the SRO film [37, 38].

In order to estimate the electrical influence on the Si-nps embedded in the SRO films, C–V hysteresis curves were performed (not shown) with the voltage swept in both directions, that is, from inversion to accumulation and back to inversion. However, no flat-band voltage shifting was observed as one should expect from an oxide layer with nanoparticles [32, 39]. Rather than being trapped charges seem to flow through the SRO films. Then, C–V measurements were repeated in pursue of a charge trapping effect. This time a constant gate voltage $V_G = V_{drop}$ (the voltage value when current drops) was applied on as-fabricated devices. Figure 6.8b,c displays, respectively, the C–V curves and the flat-band voltage before and after electrical stressing P-devices. Measurements done on devices without any electrical stress showed a maximum capacitance of 708 ± 26 pF. After stressing the devices with a gate voltage of $V_G = V_{drop}$, the capacitance drops, homogeneously, to a lower value (495 ± 7 pF), as shown in Fig. 6.8b. Hence, current and capacitance dropping are directly correlated.

Si nanoparticles in SRO films are produced by the diffusion and agglomeration of Si-excess at high temperature. That is, when SRO films are thermally annealed, the silicon atoms diffuse creating silicon clusters [nanoparticles (Si_n, $n \geq 3$)] at nucleation sites. Annealed SRO films with Si-excess of 4 at.% contain embedded Si-nps with a mean size of 1.5 nm. However, due to the low silicon excess and assuming a dispersion in the Si-nps' size, other very small Si-nps (with few Si atoms bonded, <1 nm) could exist. Then, in order to understand our results, we differentiate nanostructures in: small ones (<1 nm) "clusters" and big ones (≥1 nm) "nanoparticles". It is expected that all of these Si-nps and Si-clusters (Si-cls) could be randomly distributed; such as a mix in which some of the Si-cls could be located between Si-nps. In this model, we assume that the tunneling current is not uniformly distributed through the whole capacitor area, instead of this; it flows through narrow conductive paths within the oxide.

Therefore, in as-fabricated devices, if the distance between Si-nps and Si-cls is quite small, they can act as nodes for electrical conduction paths between the polysilicon gate and the Si-substrate making possible the high current conduction at low voltages. Therefore, the only possible reason for the capacitance and current reduction is a change in the conductivity of the films by the annihilation of such conductive paths. A particular case may be depicted by the E′ centers formation in SRO films [34]. E′ centers can be created in the SRO films when charges are injected into the film breaking-off Si–Si bonds during a voltage sweep or an electrical stress. It has been reported that most of E′ centers in SiO_2 are neutral [40]. This would explain why the capacitance drops without a noticeable Flat-Band voltage (V_{FB}) shifting.

Although the capacitance reduction might be related to a charge trapping phenomenon, no changes have been measured on the V_{FB} value. Similar effects have been observed by other authors [41] ascribing them mainly to charge trapping in Si-ncs. However, if charge gets trapped in a SiO_2 matrix with embedded Si-nps, it should produce a V_{FB} shifting in the C–V curves [32, 39]. In these devices, a V_{FB} shifting was obtained only when the devices were biased right after the current had dropped. Then, after the capacitance drops, see Fig 6.8b, the same device was biased at several gate voltages and two trends can be clearly observed: First, the C–V curve exhibits a positive shift for electrical stresses of $V_G = 20$ to 30 V, indicating electrons

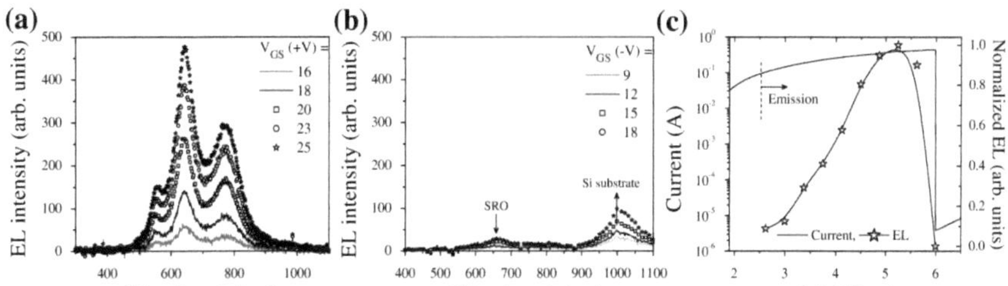

Fig. 6.9 EL spectra measured from as-fabricated devices at **a** reverse bias and **b** forward bias; **c** correlation between current and EL in P-devices

trapping. Second, the C–V curve shifts toward negative voltages when a higher electrical stress (V_G = +48 V) is applied, indicating that holes are trapped. Charge trapping can be correlated with C and D zones, as marked in Fig. 6.8c. Both electrons' and holes' trapping seem to occur in Si-nps which remain after the current drops. The size of these remaining Si-nps in similar SRO films from this work was estimated from the stair width of the staircase current behavior observed after the current and capacitance drop [38]. It was found that the size of Si-nps is about 1.21 ± 0.06 nm, which is a value close to the observed by EFTEM (1.5 nm) measurements.

Light emission was observed in the as-fabricated P-devices in the HCS. The EL spectra as a function of the gate voltage at RB and FB are shown in Fig. 6.9a,b, respectively. At RB, Fig. 6.9a, a broad visible spectrum with various peaks between 450 and 900 nm is observed for all gate voltages. This behavior of several peaks has been mainly related to the transmittance spectrum of the top polycrystalline silicon electrode [42]. Nevertheless, SRO films with 4 at.% of silicon excess show a photoluminescent peak with very low intensity at ~470 nm [43]. It is well-known that neutral oxygen vacancy (NOV) defects emit at this wavelength [11, 44]. Due to the low silicon excess, it is possible that these luminescent centers exist in the SRO film. In fact, cathodoluminescence (CL) studies on similar SRO films exhibit an emission band with a maximum peak at about 460 nm and a tail at longer wavelengths [44, 45]. EL emission in these devices at that region is not observed due to the zero transmittance of polysilicon at these wavelengths [46], but EL emission with low intensity at 550 nm is observed. Then, it is expected that these films also emit in the blue region. On the other hand, EL is observed with the naked eye as shining spots on the surface of the P-devices [36]. The EL intensity increases with the gate voltage. At FB, Fig. 6.8b, two peaks at about 650 and 1,000 nm are observed, but they are very weak. The emission at 650 nm could be produced by the recombination of e-h pairs inside the SRO film; meanwhile the emission at 1,000 nm could be due to the Si substrate. The emission at 1,000 nm has been reported as a silicon emission [47, 48] in spite of its indirect band. Nevertheless, further studies need to be done in order to clarify this emission.

Figure 6.9c shows the reverse I–V characteristic and the EL peak intensity at 650 nm as a function of gate voltage normalized to the highest emission measured from as-fabricated devices. As can be seen, when the voltage increases the current through the SRO film increases and at a certain gate voltage, where the current reaches ~1 mA, the emission starts. As the voltage increases even more, the current and the EL intensity increase. Nevertheless, after a certain gate voltage the current drops quenching the EL. Therefore, the current decay from the HCS to the LCS is directly related to the EL quenching.

Figure 6.10 exhibits the current density as a function of the applied electric field (J-E) of N-devices. N-devices with SRO films with the same silicon excess of 4at.% showed the same current drop behavior. Light emitting capacitors (LECs) with embedded Si-nps of 1.5 nm in size show a high conduction state (HCS) at low voltages when they are forward biased (positive voltage applied to the gate respect to the substrate). Nevertheless, when a threshold voltage is reached, the current drops (I_{drop}) abruptly to ~79 μA/cm^2 (almost four orders of magnitude). Again, as in case

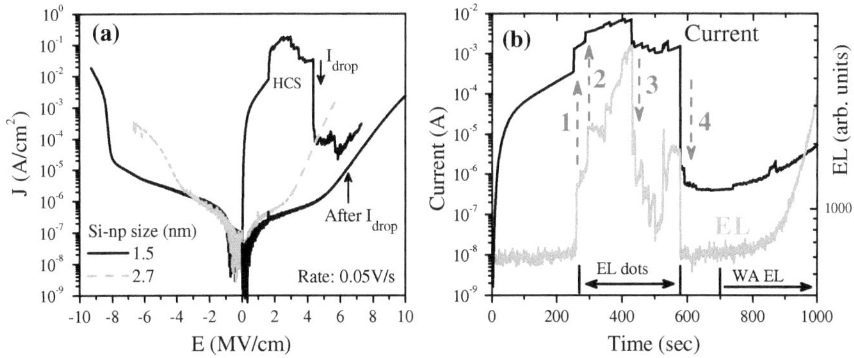

Fig. 6.10 **a** I–V curve for N-devices with embedded Si-nps of 1.5 nm and 2.7 nm and **b** correlation between EL and the current that flows through the LEC with embedded Si-nps of 1.5 nm

of P-devices, once the current drop occurs, subsequent I–V measurements exhibit a low conduction state at low voltages and the HCS is not observed anymore, as shown in Fig. 6.10a. As mentioned above, the change of conductivity is related to the annihilation of conductive paths created by adjacent stable Si-nps and unstable silicon nanoclusters (Si-ncls) and the possible creation of E′ defects.

A clear increase in the current density at low electric fields (E) is observed by increasing the Si-nps' size embedded in the SRO layer, as observed by the dashed line in Fig. 6.10a. The behavior of HCS in these LECs was not observed and it could be related to the Si-nps density. The Si-nps density in these devices is $\sim 2.46 \times 10^{12}$ cm^{-2}, about twice the Si-nps density in devices with a mean size of 1.5 nm, $\sim 1.1 \times 10^{12}$ cm^{-2}. Therefore, as the Si-nps density increases an uniform network of conductive paths becomes possible allowing uniform charge flows through the whole capacitor area. Meanwhile as the Si-nps density decreases, the distance between them increases reducing the amount of available paths, with a resulting set of discrete and preferential conductive paths within the oxide.

Light emission in these LECs was observed when they were biased with a constant DC voltage, at both, forward and reverse biases. N-devices with embedded Si-nps of 1.5 nm in size also exhibited shining spots when the LECs are operating within a HCS region. Figure 6.10b shows a correlation between current and EL for as-fabricated devices with embedded Si-nps of 1.5 nm. A clear correlation between current jumps (arrows 1 and 2) and drops (arrows 3 and 4) with the increase and decrease of the integrated luminescence is observed. Those luminescence jumps and drops are due to, respectively, the appearing and disappearing of luminescent spots on the surface of devices. After the current dropping (arrow 4), EL dots totally disappear and EL in the whole area (WA) is obtained.

Figure 6.11a shows the normalized WA-EL spectra from LECs with embedded Si-nps of 1.5 and 2.7 nm in size forwardly biased with an electric field of 7.4 and 4.4 MV/cm, respectively. Devices with Si-nps of 1.5 nm in size exhibit a maximum EL peak at ~466 nm with an additional peak of low intensity at about 600 nm, as shown

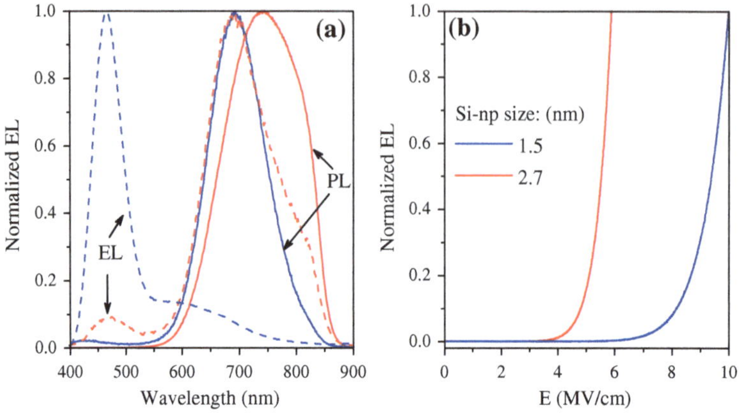

Fig. 6.11 **a** EL and PL spectra and **b** Normalized EL intensity from LEC devices with SRO films

in Fig. 6.11a. When the Si-nps size increases to 2.7 nm, the maximum EL peak is shifted to the red region. EL spectra show the main emission peak at ~685 nm, but with an additional peak of low intensity at about 470 nm, as observed in Fig. 6.11a. Reverse-biased devices present similar EL spectra (not shown). For both Si-nps sizes the photoluminescence (PL) peak is located at the red region as shown in Fig. 6.11a. However, cathodoluminescence (CL) studies on SRO films with the same silicon excess present two emission bands: at the blue/green and at the red region [45]. This fact indicates the luminescence depends on the excitation method. The PL mechanism in SRO-LPCVD films has been recently related to defects surrounding the Si-nps [49]. Yet, in this case, since charge is not generated directly within the matrix, but has to be injected, the EL mechanism could be due to a combination of quantum confinement effects and defects surrounding the Si-nps.

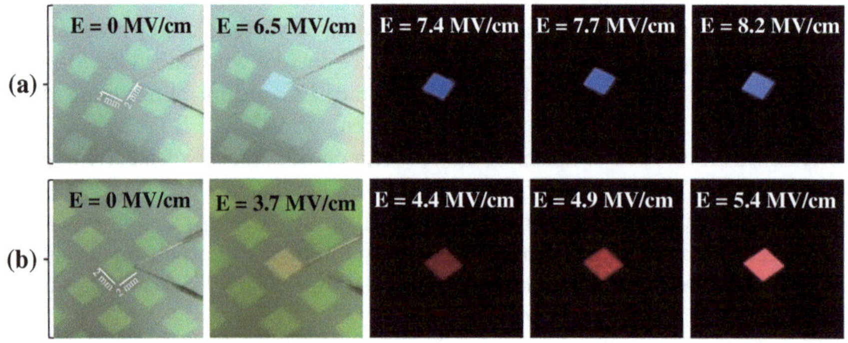

Fig. 6.12 Images of the LEC devices with embedded Si-np size of **a** 1.5 and **b** 2.7 nm biased with different electric fields. From Morales-Sánchez et al. [50]

Since current drops have been ascribed to the annihilation of preferential conductive paths, the whole area light emission could be related to an improvement in the carrier injection. Within a balanced transport network carriers can reach the radiative centers without establishing preferential paths. This, in turns, reduces the current density per path resulting in a much stabler and homogeneous network [36]. As the electric field is increased, the band bending is stronger, thus more charges are injected into the SRO film and the EL emission becomes more intense, as shown in Fig. 6.11b. For bigger Si-nps (2.7 nm), the energy bandgap is smaller resulting in a lower current injection barrier (more efficient) and a notable red shift of the luminescence. Yet, both blue and red luminescence spectra, emitted by the LECs reported in this work, remain unchanged as the driving electric field increases (see Fig. 6.12). These results prove the feasibility to obtain light emitting devices by using simple capacitors with SRO films as the active layer.

6.4 Conclusion

SRO films deposited by LPCVD with low silicon excess exhibit more intense PL than that with high silicon excess. The optimum conditions where the strongest PL is obtained are 1,100 °C for annealing with a silicon excess about 4 and 5 at. %. Silicon nanoparticles with a mean size of 1.5, 2.7 and 4.1 were obtained in SRO films annealed at 1,100 °C and for Si-excess of 4, 5.1 and 12.7 at.%, respectively. Simple Metal–Oxide–Semiconductor (MOS) structures using SRO films with 4 and 5.1 at.% of Si-excess, which exhibited the strongest PL properties, as the dielectric layer were fabricated for these studies. Reduction of current and capacitance was observed at specific voltage in LECs with embedded Si-nps of 1.5 nm in size. A clear correlation between the reduction in capacitance and current was observed. This phenomenon results of an electrical anneal of the SRO layers, and was produced by the annihilation of conductive paths. Moreover, this anneal was related to the creation of E′ defects in the SRO layer. In these devices, charge trapping was only observed after the capacitance reduction. A broad electroluminescent emission spectrum (from shining spots) with maximum peaks was observed in devices with SRO films with a Si-excess of 4 at.%. A maximum and then an EL quenching were observed. The quenching EL was analyzed and correlated to a current drop, which results in the annihilation of conductive paths during the charge injection in the SRO films. Blue whole area emission appeared after the electrical annihilation of preferential conductive paths in these devices, that is, after the switching from high conduction state (HCS) to a low conduction state (LCS). Red whole area EL was obtained by changing the Si-nps size to 2.7 nm. EL in these devices is ascribed to the charge injection into the Si-nps embedded in the SRO films through a balanced transport network. The EL emission is observed with the naked eye and in daylight conditions on the whole area of devices. Therefore, these results prove the feasibility to obtain light emitting capacitors (LECs) by using simple capacitors with SRO films as the active layer.

Acknowledgments Dr. Morales acknowledges the support received from project CONACyT-180992.

References

1. Canham, L.T.: Silicon quantum wire array fabrication by electrochemical and chemical dissolution of wafers. Appl. Phys. Lett. **57**, 1046–1048 (1990)
2. Chang, I.M., Pan, S.C., Chen, Y.F.: Light-induced degradation on porous silicon. Phys. Rev. B **48**, 8747–8750 (1993)
3. Kapetanakis, E., Normand, P., Tsoukalas, D., Beltsios, K., Stoemenos, J., Zhang, S., Van den Berg, J.: Charge storage and interface states effects in Si-nanocrystal memory obtained using low-energy Si^+ implantation and annealing. Appl. Phys. Lett. **77**(21), 3450–3452 (2000)
4. Jambois, O., Rinnert, H., Devaux, X., Vergnat, M.: Photoluminescence and electroluminescence of size-controlled silicon nanocrystallites embedded in SiO2 thin films. J. Appl. Phys. **98**, 046105 (2005)
5. Gong-Ru, L., Chun-Jung, L.: Improved blue-green electroluminescence of metal-oxide-semiconductor diode fabricated on multirecipe Si-implanted and annealed SiO2/Si substrate. J. Appl. Phys. **95**(12), 8484–8486 (2004)
6. Aceves, M., Carrillo, J., Carranza, J., Calleja, W., Falcony, C., Rosales, P.: Duality metal oxide semiconductor-PN junction in the Al/silicon rich oxide/Si structure as a radiation sensor. Thin Solid Films **373**, 134–136 (2000)
7. DiMaria, D.J., Kirtley, J.R., Pakulis, E.J., Dong, D.W., Kuan, T.S., Pesavento, F.L., Theis, T.N., Cutro, J.A., Brorson, S.D.: Electroluminescence studies in silicon dioxide films containing tiny silicon islands. J. Appl. Phys. **56**(2), 401–416 (1984)
8. Aceves, M., Falcony, C., Reynoso, A., Calleja, W., Torres, A.: The conduction properties of the silicon/off-stoichiometry-SiO_2 diode. Solid-State Electronics **39**, 637–644 (1996)
9. Morales, A., Domínguez, C., Barreto, J., Riera, M., Aceves, M., Luna, J.A., Yu, Z., Kiebach, R.: Spectroscopical analysis of luminescent silicon rich oxide films. Rev. Mex. Fís. **53**, 279–282 (2007)
10. Lin, ChJ: Lin, G.R.: Defect-enhanced visible electroluminescence of multi-energy silicon-implanted silicon dioxide film. IEEE J. Quantum Electron. **41**, 441–447 (2005)
11. Song, H.Z., Bao, X.M., Li, N.S., Zhang, J.Y.: Relation between electroluminescence and photoluminescence of Si^+-implanted SiO_2. J. Appl. Phys. **82**, 4028–4032 (1997)
12. Lin, G.R., Lin, ChJ, Lin, ChK, Chou, L.J., Chueh, Y.L.: Oxygen defect and Si nanocrystal dependent white-light and near-infrared electroluminescence of Si-implanted and plasma-enhanced chemical-vapor deposition-grown Si-rich SiO_2. J. Appl. Phys. **97**, 094306 (2005)
13. Lin, G.R.: Lin, Ch.J: Improved blue-green electroluminescence of metal-oxide-semiconductor diode fabricated on multirecipe Si-implanted and annealed SiO_2/Si substrate. J. Appl. Phys. **95**, 8484–8486 (2004)
14. Walters, R.J., Carreras, J., Feng, T., Bell, L.D., Atwater, H.A.: Silicon Nanocrystal Field-Effect Light-Emitting Devices. IEEE J. Sel. Top. Quantum Electron. **12**, 1647–1656 (2006)
15. Linnros, J., Lalic, N., Knápek, P., Luterová, K., Kocka, J., Fejfar, A., Pelant, I.: Instabilities in electroluminescent porous silicon diodes. Appl. Phys. Lett. **69**, 833–835 (1996)
16. Irrera, A., Iacona, F., Crupi, I., Presti, C.D., Franzò, G., Bongiorno, C., Sanfilippo, D., Di Stefano, G., Piana, A., Fallica, P.G., Canino, A., Priolo, F.: Electroluminescence and transport properties in amorphous silicon nanostructures. Nanotechnology **17**, 1428–1436 (2006)
17. Gebel, T., Rebohle, L., Sun, J., Skopura, W., Nazarov, A.N., Osiyuk, I.: Correlation of charge trapping and electroluminescence in highly efficient Si-based light emitters. Physica E **16**, 499–504 (2003)
18. Valenta, J., Lalic, N.: J. Linnros.: Electroluminescence microscopy and spectroscopy of silicon nanocrystals in thin SiO_2 layers. Opt. Mater. **17**, 45–50 (2001)

19. Franzò, G., Irrera, A., Moreira, E.C., Miritello, M., Iacona, F., Sanfilippo, D., Di Stefano, G., Fallica, P.G., Priolo, F.: Electroluminescence of silicon nanocrystals in MOS structures. Appl. Phys. A **74**, 1–5 (2002)
20. Walters, R.J., Bourianoff, G.I., Atwater, H.A.: Field-effect electroluminescence in silicon nanocrystals. Nature Mater. **4**, 143–146 (2005)
21. Cho, K.S., Park, N., Kim, T., Kim, K., Sung, G.Y., Shin, J.H.: High efficiency visible electroluminescence from silicon nanocrystals embedded in silicon nitride using a transparent doping layer. Appl. Phys. Lett. **86**, 071909 (2005)
22. Barreto, J., Perálvarez, M., Rodriguez, J.A., Morales, A., Riera, M., López, M., Garrido, B., Lechuga, L.M., Domínguez, C.: Pulsed electroluminescence in silicon nanocrystals-based devices fabricated by PECVD. Physica E **38**, 193–196 (2007)
23. Perálvarez, M., Carreras, J., Barreto, J., Morales, A., Dominguez, C., Garrido, B.: Efficiency and reliability enhancement of silicon nanocrystal field-effect luminescence from nitride-oxide gate stacks. Appl. Phys. Lett. **92**, 241104 (2008)
24. Lacombe, A., Kadari, B., Beaudoin, F., Barba, D., Martin, F., Ross, G.G.: Electroluminescence microspectroscopy of silicon nanocrystals obtained by Si+ ion implantation in SiO_2. Nanotechnology **19**, 465702 (2008)
25. Fernández, González: A. A., Aceves Mijares, M., Morales Sánchez, A., Leyva, K. M.: Intense whole area electroluminescence from low pressure chemical vapor deposition-silicon-rich oxide based light emitting capacitors. J. Appl. Phys. **108**, 043105 (2010)
26. Cen, Z.H., Chen, T.P., Liu, Z., Liu, Y., Ding, L., Yang, M., Wong, J.I., Yu, S.F., Goh, W.P.: Electrically tunable white-color electroluminescence from Si-implanted silicon nitride thin film. Opt. Express **18**, 20439–20444 (2010)
27. Cheong, H.J., Tanaka, A., Hippo, D., Usami, K., Suchiya, Y., Mizuta, H., Oda, S.: Visible Electroluminescence from spherical-shaped silicon nanocrystals. Jpn. J. Appl. Phys. **47**, 8137–8140 (2008)
28. Pai, P.G., Chao, S.S., Takagi, Y.: Infrared spectroscopic study of SiOx films produced by plasma enhanced chemical vapor deposition. J. Vac. Sci. Technol. A **4**(3), 689–694 (1986)
29. Ay, F., Aydinli, A.: Comparative investigation of hydrogen bonding in silicon based PECVD grown dielectrics for optical waveguides. Opt. Mater. **26**, 33–46 (2004)
30. Fazio, E., Barletta, E., Barreca, F., Neri, F.: Investigation of nanocrystalline silicon phase embedded in SiOx thin films grown by pulsed laser deposition. J. Vac. Sci. Technol. B **23**(2), 519–524 (2005)
31. Martinet, C., Devine, R.A.B.: Analysis of the vibrational mode spectra of amorphous SiO2 films. J. Appl. Phys. **77**(9), 4343–4348 (1995)
32. Kouvatsos, D.N., Ioannou-Sougleridis, V., Nassiopoulou, A.G.: Charging effects in silicon nanocrystals embedded in SiO_2 films. Mater. Sci. and Eng. B **101**, 270–274 (2003)
33. Kameda, E., Matsuda, T., Emura, Y., Ohzone, T.: Fowler-Nordheim tunneling in MOS capacitors with Si-implanted SiO_2. Solid-State Electronics **42**(11), 2105–2111 (1998)
34. Lelis, A.J., Oldhan, T.R.: Time dependence of switching oxide traps. IEEE Transact. Nuc. Sci. **41**(6), 1835–1843 (1994)
35. Perera, R., Ikeda, A., Hattori, R., Kuroki, Y.: Trap assisted leakage current conduction in thin silicon oxynitride films grown by rapid thermal oxidation combined microwave excited plasma nitridation. Microelectron. Eng. **65**, 357–370 (2003)
36. Morales-Sánchez, A., Barreto, J., Domínguez, C., Aceves-Mijares, M., Perálvarez, M., Garrido, B., Luna-López, J.: A.: DC and AC electroluminescence in silicon nanoparticles embedded in silicon-rich oxide films. Nanotechnology **21**, 085710 (2010)
37. Yu, Z., Aceves, M., Ipiña, M.A.: Single electron charging and transport in silicon rich oxide. Nanotechnology **17**, 3962–3967 (2006)
38. Morales-Sánchez, A., Barreto, J., Domínguez, C., Aceves, M., Yu, Z., Luna-López, J.A.: Coulomb blockade effects in silicon nanoparticles embedded in thin silicon-rich oxide films. Nanotechnology **19**, 165401 (2008)
39. Salonidou, A., Nassiopoulou, A.G., Giannakopoulos, K., Travlos, A., Ioannou-Sougleridis, V., Tsoi, E.: Growth of two-dimensional arrays of silicon nanocrystals in thin SiO_2 layers by low

pressure chemical vapour deposition and high temperature annealing/oxidation. Investigation of their charging properties. Nanotechnology **15**, 1233–1239 (2004)
40. Afanas'ev, V. V., Stesmans, A.: Charge state of paramagnetic E' centre in thermal SiO_2 layers on silicon. J. Phys.: Condens. Matter**12**, 2285–2290 (2000)
41. Ng, C.Y., Liu, Y., Chen, T.P., Tse, M.S.: Charging/discharging of silicon nanocrystals embedded in an SiO_2 matrix inducing reduction/recovery in the total capacitance and tunneling current. Smart Mater. Struct. **15**, S43–S46 (2006)
42. Perálvarez, M., Barreto, J., Carreras, J., Morales, A., Navarro, D., Lebour, Y., Domínguez, C., Garrido, B.: Si-nanocrystal-based LEDs fabricated by ion implantation and plasma-enhanced chemical vapour deposition. Nanotechnology **20**, 405201 (2009)
43. Luna-López, J.A., Morales-Sánchez, A., Aceves-Mijares, M., Yu, Z., Domínguez, C.: Analysis of surface roughness and its relationship with photoluminescence properties of silicon-rich oxide films. J. Vac. Sci. Technol. A **27**(1), 57–62 (2009)
44. Flores, F.F., Aceves, M., Carrillo, J., Domínguez, J., Falcony, C.: Photoluminescence and cathodoluminescence characteristics of SiO_2 and SRO films implanted with Si. Superficies y Vacío **18**, 7–13 (2005)
45. Lopez, R., Aceves, M., Yu Z., Falcony, C., Cathodoluminescence of Silicon Rich Oxide with nitrogen incorporated. 4th International Conference on Electrical and Electronics, Engineering, pp. 341–344 (2007)
46. Perálvarez, M., García, C., López, M., Garrido, B., Barreto, J., Domínguez, C., Rodríguez, J.A.: Field effect luminescence from Si nanocrystals obtained by plasma-enhanced chemical vapor deposition. Appl. Phys. Lett. **89**, 051112 (2006)
47. Lin, ChF: Liu, C.W., Chen, M.J., Lee, M.H., Lin, I.C.: Electroluminescence at Si band gap energy based on metal-oxide-silicon structures. J of. Appl. Phys. **87**(12), 8793–8795 (2000)
48. Lin, ChF: Liu, C.W., Chen, M.J., Lee, M.H., Lin, I.C.: Infrared electroluminescence from metal-oxide-semiconductor structures on silicon. J. Phys. Condens. Matter **12**, L205–L210 (2000)
49. Ray, M., Hossain, S.M., Klie, R.F., Banerjee, K., Ghosh, S.: Free standing luminescent silicon quantum dots: evidence of quantum confinement and defect related transitions. Nanotechnology **21**, 505602 (2010)
50. Morales-Sánchez, A., Aceves-Mijares, M., Monfil-Leyva, K., González, A.A., Luna-López, J.A., Carrillo, J., Domínguez, C., Barreto, J., Flores-Gracia, F.J.: Strong blue and red luminescence in silicon nanoparticles based light emitting capacitors. Applied Physics Letters 99, 171102 (2011)

Chapter 7
Electronic and Optical Properties of Silicon Carbide Nanostructures

Hung-Chung Hsueh, Guang-Yu Guo and Steven G. Louie

Abstract The electronic and optical properties of quasi-one-dimensional single-walled zigzag/armchair silicon-carbide nantotubes (SiC-NTs) as well as a two-dimensional SiC monolayer are investigated by *ab initio* methods. In order to elucidate many-electron effects on SiC nanosystems, we apply the *ab initio* many-body Green's function approach to calculate the quasiparticle and optical properties of SiC nanostructures. The significant band gap correction, more than 1 eV, to the Kohn-Sham gap of density functional theory within the local density approximation of semiconducting SiC-NTs and a SiC monolayer is mainly due to the many-electron interaction effect, which is included in the GW approximation for the electron self energy. Furthermore, taking into account electron-hole interaction, the optical spectra of SiC-NTs are calculated by solving the Bethe-Salpeter equation (BSE) for the electron-hole amplitudes. Our GW+BSE calculations reveal the presence of excitons with a large binding energy as well as strong anisotropy in the optical properties in the low-dimensional SiC systems. The characteristics of the strongly bound electron-hole pairs or excitons in SiC nanostructures are also discussed in terms of the corresponding excitonic wavefunctions.

H.-C. Hsueh (✉)
Department of Physics, Tamkang University, Tamsui, 25137 New Taipei, Taiwan
e-mail: hchsueh@mail.tku.edu.tw

G.-Y. Guo
Department of Physics, National Taiwan University, 10617 Taipei, Taiwan

S. G. Louie
Department of Physics, University of California at Berkeley, Berkeley, CA 94720, USA

Materials Sciences, Lawrence Berkeley National Laboratory, Berkeley, CA 94720, USA

H. Li et al. (eds.), *Silicon-based Nanomaterials*,
Springer Series in Materials Science 187, DOI: 10.1007/978-1-4614-8169-0_7,

© Springer Science+Business Media New York 2013

7.1 Introduction

The most recent revolution of modern materials science is triggered by the advance of nanotechnology. Many exceptional and fabulous characteristics have been revealed with the dimensions of materials go down to the nanoscale. As a milestone in materials science research, the discoveries of carbon nanotubes (CNT) and graphene open a new gate into the wonderland of the nanoworld. Especially, the novel properties of these low-dimensional nanostructures, including supreme mechanical stiffness and elasticity, extremely high mobility of electrical and thermal carriers, and extraordinary magnetic phenomena [1–3], not only deepen our understanding of nanomaterials but also enable more promising applications in different, versatile fields. In fact, the recent breakthroughs in studies of CNT and graphene stimulate fruitful research on other geometrical equivalent compounds such as BN, BC_2N, GaN, SiC, MoS_2, and many others, and inspire more possibilities of functional devices [4, 5].

Silicon carbide (SiC) crystallizes in either a cubic or a hexagonal form, and exhibits interesting polytypism [6, 7]. The polytypes are made of identical hexagonal layers with different stacking sequences. These polytypes are semiconductors with a range of band gaps, from 2.39 eV in the zincblende polytype (3C) to 3.33 eV in the wurtzite polytype (2H) [6, 8]. The distinctive properties of 3C and 6H SiC [9] offer possible applications for high temperature, high-power, and high-frequency devices [10–13]. Meanwhile, 6H SiC with a band gap of 2.86 eV is a useful material for blue light-emitting diode applications [8]. Recently, SiC nanotubes (NTs) were synthesized by reaction of carbon nanotubes (CNTs) with SiO at various temperatures [14]. Like other tubular materials such as C [1, 15, 16], BN [17, 18], AlN [19], and GaN [20] nanotubes, SiC-NTs display very interesting properties different from their bulks. In particular, based on density-functional calculations, Miyamoto and Yu [21] predicted a lower strain energies of SiC-NTs comparing with CNTs and a chirality dependence of direct/indirect fundamental band gaps of SiC-NTs. Also, using both tight-binding molecular dynamics and *ab initio* methods, Menon and coworkers [22] showed that single-walled SiC-NTs are very stable with a large band gap.

Like BN-NTs (see, e.g., [23–25]), the polar nature of SiC-NTs may induce unusual physical properties in contrast to CNTs. For example, zigzag SiC-NTs may become piezoelectric, and also show second-order nonlinear optical response. A knowledge of the optical properties of SiC-NTs is crucial for their optical and electro-optical applications. In fact, recent density functional theory (DFT) calculations within a local density approximation (LDA) [26, 27] found that, as in BN nanotubes, [28] SiC nanotubes are all semiconducting irrespective to tubular diameter and chirality with exceptions of the ultrasmall (3, 0) and (4, 0) zigzag tubes, which may be regarded as the thinnest conducting SiC nanowires [26]. Furthermore, the DFT results indicate that both zigzag and chiral SiC-NTs have pronounced second-harmonic generation (SHG) and linear electro-optical coefficients, which are significantly larger than that of bulk SiC polytypes and BN-NTs [27]. All these suggest that SiC-NTs are promising materials for nonlinear optical and optoelectric applications in, for example, sum-frequency generation and electro-optical switches.

Nevertheless, theoretical studies of the electronic and optical properties of SiC-NTs mentioned above were mainly based on the independent-particle approximation within the DFT-LDA framework [26, 27] or on semi-empirical calculations [22]. However, many-body interactions are known to play an important role in low-dimensional systems such as nanotubes and graphene [29–31], due to reduced charge screening and enhanced electron–electron correlation. From the geometrical connection between SiC monolayer and SiC-NTs, understanding the optical properties of an isolated SiC sheet would also help learn the many-body effects on the optical responses of SiC-NTs. Furthermore, the optical properties of the two-dimensional (2D) SiC monolayer are interesting on their own account, and, in particular, a single SiC sheet bears many similarities to graphene which exhibits many fascinating properties. [32, 33] Therefore, motivated by the need for more fundamental understanding of the science and the prospects of optoelectronic device applications of SiC monolayer and SiC-NTs, we perform *ab initio* calculations to study the many-body effects on the quasiparticle band gaps and optical spectra of SiC monolayer and SiC-NTs using the GW [34, 35] and GW+BSE [36, 37] methods. Indeed, we find pronounced GW quasiparticle corrections of more than 1.0 eV to the LDA band gaps in both SiC monolayer and SiC-NTs. Remarkably, the calculated optical absorption spectra from the SiC nanostructures (atomic layer and tubes) are all dominated by the discrete exciton peaks due to strongly bound excitons with a large binding energy of up to 2.0 eV. The formation of strongly bound excitons is attributed to the enhanced electron-hole interaction in these low-dimensional systems. Moreover, a different character exciton, attributed to the specific chirality, is observed in the armchair SiC-NTs in which the bound exciton wavefunctions are more localized and also strongly noncylindrical symmetric. The large excitation energy of $\sim$3.0 eV of the first bright exciton in the armchair SiC-NTs with no dark exciton below it, suggests that the small-radius armchair SiC-NTs can be useful for optical devices working in the UV regime. On the other hand, the zigzag SiC-NTs have many dark excitons below the first bright exciton and hence may have potential applications in tunable optoelectric devices ranging from infrared to UV frequencies by external perturbations.

7.2 Theory and Computational Methodologies

In this work, ab-initio calculations of the ground and excited-state properties are accomplished within the BerkeleyGW package [38] in three main steps: first, we calculate the electronic ground-state properties within DFT using appropriate approximations to the exchange-correlation potential ($V_{\mathrm{xc}}^{\mathrm{DFT}}$) [39]. Second, we obtain the quasiparticle energies ($E_{n\mathbf{k}}^{\mathrm{QP}}$) within the GW approximation for the electron self-energy Σ [34, 35, 40]. Finally, the coupled electron-hole excitation energies and optical spectra can be evaluated by solving the Bethe-Salpeter equation (BSE) of the two-particle Green's function. [36, 37] Details of our computations are described below.

7.2.1 Ground-State Electronic Properties

An appropriate mean-field description of the ground-state properties of solids is essential to perform quasiparticle calculations within many-body perturbation theory. For the conventional covalent and ionic-bonded materials, ground-state DFT eigen wavefunctions and eigen energies have been shown to be a good starting point for the many-body perturbation calculations [35, 37]. In many DFT packages [41–43], mean-field electronic orbitals ($\phi_{n\mathbf{k}}^{\mathrm{DFT}}$) and eigenvalues ($E_{n\mathbf{k}}^{\mathrm{DFT}}$) are obtained by solving Kohn-Sham equation [39] self-consistently (in atomic units)

$$H^{KS}\phi_{n\mathbf{k}}^{\mathrm{DFT}} = \left[-\frac{1}{2}\nabla^2 + V_H + V_{\mathrm{ion}} + V_{\mathrm{xc}}^{\mathrm{DFT}}\right]\phi_{n\mathbf{k}}^{\mathrm{DFT}} = E_{n\mathbf{k}}^{\mathrm{DFT}}\phi_{n\mathbf{k}}^{\mathrm{DFT}} \tag{7.1}$$

In our studies, the LDA is employed to describe the $V_{\mathrm{xc}}^{\mathrm{DFT}}$ in Kohn-Sham Hamiltonian (H^{KS}) for all the calculations of SiC monolayer and SiC-NTs. The electronic configurations of Si($3s^2, 3p^2$) and C($2s^2, 2p^2$) are treated as active valence states when generating norm-conserving pseudopotentials [44]. A 40 Ry cutoff energy for the plane-wave basis and a $10\times10\times1$ ($1\times1\times32$) k-point grid for the Brillouin zone (BZ) integrals [45] in the calculations of the SiC monolayer (SiC-NTs) are adopted to ensure converged LDA results. A supercell scheme is also used to simulate low-dimensional nanosystems [46]. Meanwhile, to avoid spurious interaction between replicas in neighboring supercells, a sufficient interlayer/intertube separation (more than 10 Å) is introduced in nonperiodic directions. Moreover, the underlying equilibrium atomic positions and lattice constants are taken from Ref. [26] where the theoretical structures were obtained by the conjugate gradient method with the atomic forces and the stress calculated from first-principles.

7.2.2 Quasiparticle Self-Energy Calculation

In the language of many-body perturbation theory, $V_{\mathrm{xc}}^{\mathrm{DFT}}$ can be viewed as a independent-particle mean-field approximation of the electronic self-energy operator and the quasiparticle energies ($E_{n\mathbf{k}}^{\mathrm{QP}}$) and wavefucntions ($\phi_{n\mathbf{k}}^{\mathrm{QP}}$) can be computed from their mean-field counterparts ($\phi_{n\mathbf{k}}^{\mathrm{DFT}}$, $E_{n\mathbf{k}}^{\mathrm{DFT}}$) under the perturbation of many-body self energy. Therefore, the quasiparticle band structure characterized by the many-body self-energy operator ($\Sigma(E_{n\mathbf{k}}^{\mathrm{QP}})$) can be evaluated by solving the following Dyson equation [34, 35, 40]

$$\left[H^{KS} + \left(\Sigma(E_{n\mathbf{k}}^{\mathrm{QP}}) - V_{\mathrm{xc}}^{\mathrm{DFT}}\right)\right]\phi_{n\mathbf{k}}^{\mathrm{QP}} = E_{n\mathbf{k}}^{\mathrm{QP}}\phi_{n\mathbf{k}}^{\mathrm{QP}} \tag{7.2}$$

The GW approximation to the electron self energy corresponds to a series expansion of Sigma in terms of a series in the screened Coulomb interaction W

and the interacting single-particle Green's function G. In practice, compared with experiments, a very high accuracy for $\Sigma(E_{n\mathbf{k}}^{\mathrm{QP}})$ in most conventional/novel insulating and metallic systems is achieved with considering only a first-order perturbation on the Dyson equation within the GW approximation in the so called G^0W^0 approximation [35, 47–49]. In this approach, the irreducible polarizibility P^0 is a convolution of the mean-field Green's function G^0 (constructed from $\phi_{n\mathbf{k}}^{\mathrm{DFT}}$ and $E_{n\mathbf{k}}^{\mathrm{DFT}}$), i.e., $P^0 = -iG^0G^0$. This gives rise to a dynamical dielectric matrix within the random phase approximation (RPA) scheme. Here we further use a generalized plasmon-pole (GPP) model [35] to describe the frequency-dependent part of the dielectric matrix. The screened Coulomb potential W^0 is obtained from the bare Coulomb interaction v through the equation $W^0 = \upsilon/\left[1 - \upsilon P^0\right]$. Finally, the quasiparticle self-energy operator is given by $\Sigma = iG^0W^0$. In this chapter, we adopt the G^0W^0 approach to calculate the quasiparticle properties of the single SiC sheet and the SiC-NTs. Here a rather dense k-point sampling for the SiC sheet ($18 \times 18 \times 1$) and also for the SiC-NTs ($1 \times 1 \times 32$) is used to ensure converged self-energy.

Particularly, a proper description of the screening properties reflecting the nature of highly intrinsic anisotropic screening interactions plays an important role in obtaining reliably quasiparticle properties of low-dimensional nanosystems. Since a supercell is used, we also employ an efficient truncation scheme [50, 51] to the Coulomb interaction (in which a step function at the boundary of the supercell) to remove the long-range Coulomb interaction between the structure of interest and its images in the neighboring supercells.

7.2.3 Electron-Hole Excitation and Optical Spectrum Calculation

The optical properties of a crystal are associated with interband transitions within an independent quasiparticle picture. However, for the systems with significant electron-hole interaction, excitons which are composed of correlated electron-hole pairs are a principle issue that affects all optical processes involving interaction between light and matter. The connection of the exciton energies E^S and corresponding electron-hole amplitudes $A_{vc\mathbf{k}}^S$ of the correlated electron-hole excitations S is governed by an effective two-particle equation, the Bethe-Salpeter equation [37],

$$(E_{c\mathbf{k}}^{\mathrm{QP}} - E_{v\mathbf{k}}^{\mathrm{QP}})A_{vc\mathbf{k}}^S + \sum_{v'c'\mathbf{k}'} \langle vc\mathbf{k}|K^{ph}|v'c'\mathbf{k}'\rangle A_{vc\mathbf{k}}^S = E^S A_{vc\mathbf{k}}^S \tag{7.3}$$

where v and c denote the quasiparticle valence and conduction bands, respectively, and $A_{vc\mathbf{k}}^S$ is the wavefunction amplitude of the excited state S. Within the Tamm-Dancoff approximation [52], only transition from valence to conduction bands is considered in the electron-hole interaction kernel K^{eh}, which may be evaluated knowing the electron self energy operator [37]. To obtain the converged optical spectra, in the present work, the kernel K^{eh} of a SiC monolayer (SiC-NTs) is evaluated on a sparse k-point grid of $18\times18\times1$ ($1\times1\times32$) and then interpolated [36] onto a denser k-point grid of $36\times36\times1$ ($1\times1\times64$).

Instructively, the electron-hole amplitude in real space may be visualized by expanding in the quasi-electron/hole basis $\{\phi_{c\mathbf{k}}(\mathbf{r}_e), \phi_{v\mathbf{k}}(\mathbf{r}_h)\}$,

$$\Phi_S(\mathbf{r}_e, \mathbf{r}_h) = \sum_{\mathbf{k}} \sum_{c,v} A^S_{vc\mathbf{k}} \phi_{c\mathbf{k}}(\mathbf{r}_e) \phi_{v\mathbf{k}}(\mathbf{r}_h) \tag{7.4}$$

Because of the complexity of the six-dimensional exciton wavefunction, a simpler distribution of the electron amplitude square with the hole position fixed, i.e., $|\Phi_S(\mathbf{r}_e, \mathbf{r}_h = 0)|^2$, is usually used to reveal the essence of the electron-hole correlation of the exciton [29, 30].

Considering the optical spectrum of nanosystems, the imaginary part of frequency-dependent macroscopic dielectric function $\varepsilon_2(\omega)$ may be calculated using the Fermi Golden rule. Since the incident photons carry negligible momentum, ignoring electron-hole interaction, the optical excitations associated with vertical interband transitions from independent hole states to electron ones are given by

$$\varepsilon_2(\omega) = \frac{16\pi^2 e^2}{\Omega\omega^2} \sum_{c,v} |\lambda \cdot \langle v|v|c\rangle|^2 \delta(\omega - (E_c - E_v)) \tag{7.5}$$

where $\boldsymbol{\lambda}$ and $\boldsymbol{v}$ indicate the polarization vector of the light and single-particle velocity operator, respectively. The photon energy is denoted by ω and Ω is the unit cell volume. The oscillator strength of a specific transition can be obtained from the optical transition matrix elements ($\langle v|v|c\rangle$) derived from the electronic ground and excited states involved. The transition energy associated with a peak in the optical absorption spectrum can be simply estimated by knowing the interband joint density of states given by $E_c^{\mathrm{LDA}} - E_v^{\mathrm{LDA}}$ or $E_c^{\mathrm{QP}} - E_v^{\mathrm{QP}}$ corresponding to a mean-field (LDA) or a quasiparticle (GW+RPA) level of approximations, respectively.

However, in the GW+RPA approach, the dielectric function gives the interband optical absorption spectrum beyond the LDA, but still without including electron-hole interaction (a excitonic) effects. A more realistic optical spectrum can be obtained from GW+BSE calculations which yield an $\varepsilon_2(\omega)$ which includes quasi-particle self-energy corrections, the local-field corrections and excitonic effects:

$$\varepsilon_2(\omega) = \frac{16\pi^2 e^2}{\Omega\omega^2} \sum_{S} |\lambda \cdot \langle 0|v|S\rangle|^2 \delta(\omega - E^S), \tag{7.6}$$

where S is the exciton states from equation 7.3.

Finally, the arbitrary vacuum space in a supercell approach for reduced dimensional systems will generate an arbitrary volume effect on the dielectric function computed. In order to compare directly to experimental measurements, we calculate the imaginary part of polarizability per unit area for the SiC monolayer, $\alpha_2(\omega)$ which is derived from calculated dielectric susceptibility of the supercell arrangement, $\chi = (\varepsilon - 1)/4\pi$, multiplied by the distance between neighboring SiC monolayers [31]. Thus, the corresponding optical absorbance can be evaluated from $Abs(\omega) = \frac{4\pi\omega}{c}\alpha_2(\omega)$. For the one-dimensional (1D) nanotubes, the measured optical spectrum is determined by the imaginary part of the optical polarizability per single tube, α_2. This α_2 is equal to the imaginary part of the calculated dielectric function, $\varepsilon_2^{\mathrm{calc}}$, multiplied by the cross-sectional area of the supercell perpendicular to the tubular axis ($\Lambda_{sc}^{\perp}$), i.e., $\alpha_2 = \Lambda_{sc}^{\perp}\left(\varepsilon_2^{\mathrm{calc}}\right)/4\pi$.

7.3 Electronic and Optical Properties of SiC Monolayer

The equilibrium atomic geometry of 2D SiC honeycomb monolayer can be determined from the structural coordinates of the atoms with respect to the minimum of the calculated total energy on the Born-Oppenheimer surface. Without including thermal fluctuation which could induce a deformation of the perfect planar structure, such as corrugations and intrinsic ripples observed in graphene [53], a stable and nonbuckled 2D SiC honeycomb sheet was found in DFT-LDA calculations. Meanwhile, a peculiar buckled configuration has been recently predicted and revealed in silicene and germanene [54, 55], the graphene equivalent of silicon and germanium, which is attributed to the nature of sp^2/sp^3-like orbital hybridization in group-IV elements. It would be interesting to understand the buckling effects on the structural stability of 2D SiC layers. Our total-energy calculations of SiC monolayers in Fig. 7.1a show a planar sheet with a optimized lattice constant ($a = 3.07$Å) is the

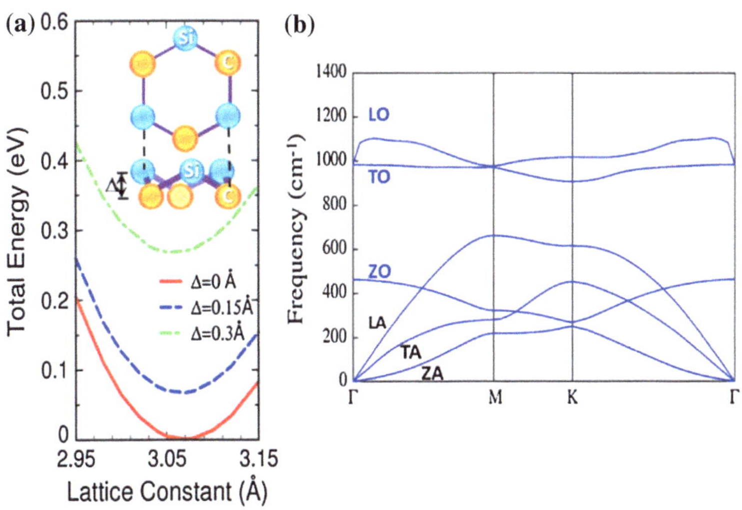

Fig. 7.1 **a** Total energy as a function of lattice constant for differently buckled SiC monolayers (characterized by a buckled distortion Δ as shown in the side-view of a monolayer in the *inset*) and **b** phonon dispersions for a flat SiC monolayer (Δ = 0 Å) with the optimized lattice constant of $a = 3.07$Å

most stable 2D structure compared with buckled ones characterized by the atomic distortion Δ (as indicated in the side-view of the SiC monolayer in Fig. 7.1a). Furthermore, to confirm the stability of this flat atomic arrangement, we have performed *ab-initio* phonon calculation of the planar SiC sheet within the scheme of density functional perturbation theory [42, 43]. The calculated phonon dispersion relations of three optical (longitudinal (LO), transverse (TO), and out-of-plane (ZO)) and three acoustic (LA, TA, and ZA) branches in Brillouin Zone are shown in Fig. 7.1b. The calculated phonon frequency of LO (and TO) modes at Γ point is 984 cm^{-1} which is in good agreement with early force-constant method calculation [56]. Especially, the out-of-plane acoustic ZA mode is usually utilized as a signature of instability of 2D planar structure. Our calculated nonimaginary frequency of ZA mode near Γ point (460 cm^{-1}) provides a convincing evidence of structural stability of planar SiC monolayer. Note that the sharp drop in frequency for the LO mode near Γ point is due to an artificial LO-TO splitting effect within the supercell framework of DFPT calculations in a plane-wave DFT code [57].

Figure 7.2 shows the LDA band structure (left panel) and also the GW corrections to the LDA eigenvalues (right panel) of the 2D SiC sheet. Since C and Si atoms are different, the massless Dirac fermion feature near the Fermi energy in graphene is no longer present in the SiC monolayer. In the LDA approximation, this heteropolar character gives a direct Kohn-Sham energy gap of 2.56 eV at the *K* point of the

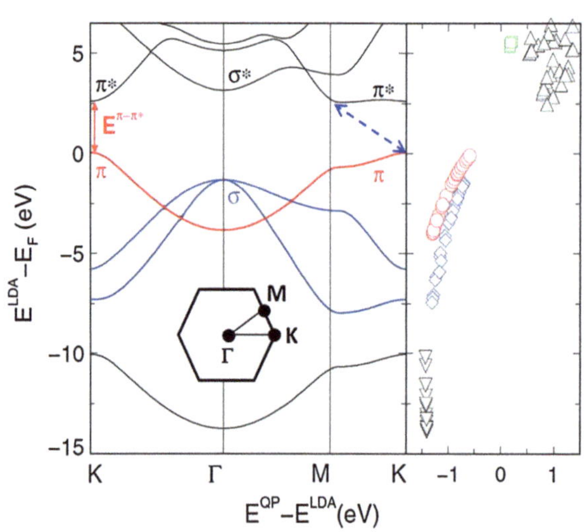

Fig. 7.2 The LDA band structure (*left panel*) and GW corrections to the LDA eigenvalues (*right panel*) of the SiC monolayer. The π and σ bands (*left panel*) are denoted as *red* and *blue lines*, respectively. In the *right panel*, the GW corrections for the π and σ bands are represented by *red circles* and *blue diamonds*, respectively, whereas the GW corrections of the NFE states are represented by *green squares*. The LDA indirect fundamental band gap and direct $E^{\pi-\pi*}$ gap are marked respectively as an *blue* (*dashed*) and *red* (*solid*) *arrows* in the *left panel*. The valence band maximum is aligned at 0 eV (After Ref. [59]). High-symmetry *k*-points and lines inside the 1st BZ are indicated in the inset of the band structure

hexagonal Brillouin zone, as shown by the red arrow in the left panel of Fig. 7.2. Meanwhile, the LDA fundamental band gap is indirect (indicated by the dashed arrow in Fig. 7.2) with a slightly smaller gap of 2.51 eV. The quasiparticle corrections to the 2D SiC monolayer state are obtained through the GW calculations with special care on Coulomb truncation to account for the finite interlayer vacuum space in the supercell geometry as described in Sect. 7.2. The calculated quasiparticle eigenvalues show a rather significant GW corrections to the LDA fundamental band gap (from 2.51 to 4.42 eV) as compared with the GW correction to the LDA band gap of bulk 2H SiC (about 1.0 eV). [58, 59] Strikingly, the GW corrections transform the indirect LDA band gap ($K \rightarrow M$) into a direct band gap ($K \rightarrow K$) for the SiC monolayer.

The nature of the direct band gap in the SiC monolayer indicates that the lowest energy exciton is optically active (bright). Furthermore, based on the simple zone-folding approximation, we could expect no momentum-mismatch induced optically inactive (dark) excitons below the lowest energy bright exciton in the SiC-NTs [60]. This anticipation is indeed confirmed by our GW+BSE calculations for both the SiC sheet and armchair SiC-NTs presented below. However, the breakdown of such oversimplified argument occurs in the very small zigzag SiC-NTs when electron-electron/electron-hole interactions are included in our GW+BSE calculations, as will be discussed in Sect. 7.5.

Indeed, the quasiparticle effect depends sensitively on the character of the electron orbitals, which can be strongly affected by the anisotropy in low-dimensional systems. For instance, the nearly free electron (NFE) states which mainly lie more than a few Ås away from the SiC monolayer are characterized by weakly bound nature and much less sensitive to many-body screening effects in the GW approximation(green squares in Fig. 7.2). These sophisticated screening effects enforce the notion that the quasiparticle calculations rather than the simple scissor-operator approximation are necessary for low-dimensional systems such as the 2D SiC monolayer.

The influence of the degree of geometrical anisotropy is substantial not only on the electronic structures but also on optical properties of low-dimensional systems. Such geometry effects can be apparently illustrated by comparing DFT-LDA $\varepsilon_2(\omega)$ of bulk $2H$-SiC and SiC monolayer. For the wurtzite $2H$-SiC crystal, spectral region is mainly distributed from 2 to 10 eV in Fig. 7.3a. Clearly, both transitions corresponding to the two electric field polarizations, $\mathbf{E}//\mathbf{a}$ and $\mathbf{E}//\mathbf{c}$, along the different crystallographic a- and c-axes have considerable contributions but with distinct features. However, Fig. 7.3b shows that the optical properties of SiC monolayer can be divided into two spectral regions. The spectrum within the low-energy region from 2 to 5 eV is dominated by the interband optical transitions for the electric field polarization parallel to the monolayer plane ($\mathbf{E}//\mathbf{a}$). Optical absorption corresponding to light polarization perpendicular to the plane ($\mathbf{E}//\mathbf{c}$) can only be detected in high-frequency range (from 6 to 10 eV) and thus render a monolayer transparent for photon energy below 5 eV. This anisotropic optical phenomenon, so-called depolarization effect, is originated from induced polarization inside low-dimensional materials under the application of an external field and has also been observed in CNTs [61]. Due to the huge depolarization effect, the optical spectrum

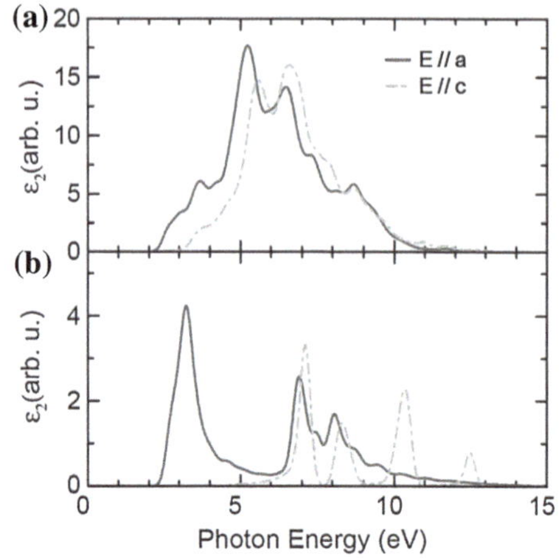

Fig. 7.3 The imaginary part of dielectric function of (**a**) wurtzite $2H$-SiC and (**b**) SiC monolayer for different light polarizations as indicated

of SiC monolayer for light polarization parallel to the plane is therefore focused in the remainder of the present work.

The excitonic effects on the optical properties of a single SiC sheet can be examined by comparing the calculations with (GW+BSE) and without (GW+RPA) electron-hole interactions. Figure 7.4 shows the optical polarizability (a) and absorbance (b) from both GW+RPA and GW+BSE calculations as well as the $E_1^{\pi-\pi*}$ exciton wavefunction (c) of the SiC monolayer. First, the GW+BSE calculated

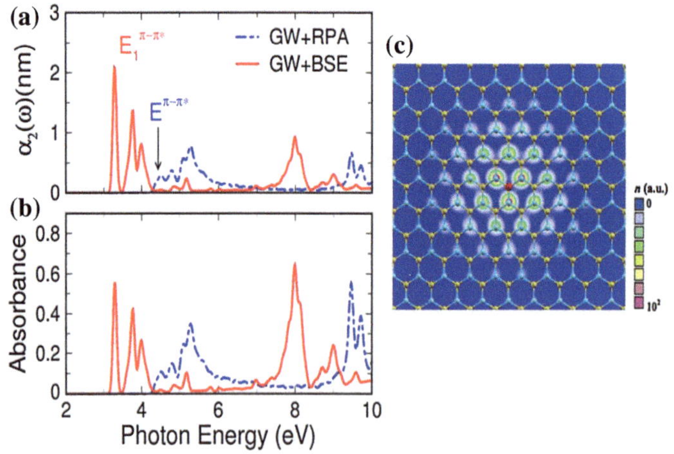

Fig. 7.4 **a** Optical polarizability per unit area and **b** absorbance of SiC monolayer with (GW+BSE) and without (GW+RPA) excitonic effects included. **c** $E_1^{\pi-\pi*}$ exciton wavefunction of the SiC sheet. In **a** and **b**, the theoretical spectra are broadened with a Gaussian smearing width of 0.10 eV. In **c**, the hole position (*red sphere*) is fixed at the top of a C atom (*yellow sphere*) and the squared electron amplitude is mainly distributed on the Si atoms (*cyan spheres*) next to the hole (After Ref. [59])

optical absorption spectrum of the SiC sheet $\alpha_2(\omega)$ indicates that the electron-hole interaction not only triggers a red-shift of the onset optical transition energies but also modifies their relative absorption intensities from those of the GW+RPA calculations. Second, the absorption features in low-energy range (from 3.0 to 5.0 eV) and high-energy one (from 5.0 to 8.0 eV) in Fig. 7.4a are dominated by the optical transitions involving mainly π bands and σ bands, respectively. For example, $\pi \rightarrow \pi^*$ transitions near the K point at the zone edge (Fig. 7.2) contribute pronounced absorption peaks within 3.0–4.0 eV, whereas a rather minor optical feature at 5.2 eV is mainly due to the $\sigma \rightarrow \sigma^*$ transition at the zone center. Particularly, the first prominent peak located at $E_1^{\pi-\pi*} = 3.25\,\text{eV}$ in the GW+BSE absorption spectrum comes from a bright exciton due to the excitation between the π and π^* states near the K point [see Fig. 7.2 (left panel)]. This is a strongly bound exciton with a large binding energy of $E_B = 1.17\,\text{eV}$ as measured by the difference between the $E_1^{\pi-\pi*}$ and the onset energy of the GW corrected electron-hole continuum ($E^{\pi-\pi*} = 4.42\,\text{eV}$). In contrast, the theoretical exciton binding energy in bulk 2H SiC is only 0.1 eV [59]. This manifests clearly that the reduced dimensionality of a SiC monolayer strongly confines the quasiparticles, and this significantly enhances the overlap between the electron and hole wave functions and hence the electron-hole interaction. Moreover, in addition to this quantum confinement, the presence of the vacuum region for isolating a monolayer reduces the screening and hence provides extra contribution to the large excitonic effect in 2D SiC sheet. Furthermore, because the GW corrections transform the indirect LDA band gap of the SiC sheet into the direct band gap, as already mentioned, no dark exciton was found in the optical spectrum until the photon energy up to 4.2 eV which is above the lowest bright exciton ($E_1^{\pi-\pi*}$). Finally, an analysis of $\int |\Phi_S(\mathbf{r}_e, \mathbf{r}_h)|^2 d\mathbf{r}_e$ and $\int |\Phi_S(\mathbf{r}_e, \mathbf{r}_h)|^2 d\mathbf{r}_h$ shows that the hole is mainly resided in the C-sublattice where as the excited electron is on the neighboring Si-sublattice. In order to elucidate the extent of the $E_1^{\pi-\pi*}$ exciton wavefunction, we plot the electron amplitude square with the hole position fixed at the position slightly above the C atom (see Fig. 7.4c). Clearly, the electron orbital is of the p_z character, and distributes mainly on the nearest neighbor Si atoms around the hole to form a fairly localized exciton.

7.4 Orbital Rehybridization in SiC Nanotubes

The ideal SiC-NTs with different diameters and chiralities (n, m) are constructed by simply rolling up a SiC monolayer strip along specific lattice vectors into corresponding nanocylinders. The atomic positions and lattice constants were then fully relaxed by the conjugate gradient (CG) method in DFT-LDA approximation. Studies on CNT [62, 63] show that tubule curvature induces a significant strain energy in the small-radius single-wall CNT. The curvature energies E_c defined as total energy relative to that of single SiC monolayer per atom, also known as the strain energy, of the SiC-NTs in zigzag $(n, 0)$, armchair (n, n) and chiral (n, m) as a function of optimized tube diameters are compiled in Fig. 7.5. Interestingly, E_c of the SiC-NTs

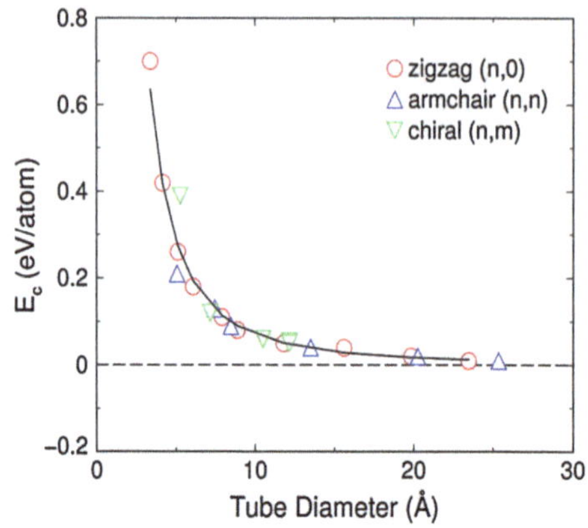

Fig. 7.5 Curvature energy (E_c) of zigzag $(n, 0)$, armchair (n, n), and chiral (n, m) of SiC-NTs as a function of tube diameters. The *solid line* between the symbols is a guide to the eye

can be very well fitted with $E_c = \alpha/R^2$ with $\alpha = 2.004\,\text{eV}\cdot\text{Å}^2$/atom which is almost identical to that of the CNTs but larger than that (1.248 eV Å^2/atom) of the BN-NTs [64]. This behavior indicates that the classical elastic theory works well even down to such a small length scale.

The geometrical connection between a single-wall nanotube and the isolated monolayer has inspired the application of the zone-folding method to construct the zeroth-order electronic band structure and phonon dispersions for single-wall nanotubes [15]. However, the zone-folding scheme alone cannot address the significant orbital rehybridization due to the curvature effect and also the many-body interaction effects (e.g., screening effect, phase-space effect, etc.) in the nanotubes. To include these effects, we have calculated the diameter-dependence of the fundamental band gap of selected SiC-NTs with different chiralities in both DFT-LDA and quasiparticle GW calculations, as shown in Fig. 7.6. First of all, we note that all the SiC-NTs are semiconductors (except metallic ultrasmall-diameter (3, 0) and (4, 0) nanotubes [26]), and this may be attributed to the different potentials of the Si and C atoms, being similar to the case of the BN nanotubes [17]. Second, an asymptotic convergence of band gaps of the zigzag and armchair SiC-NTs as a function of tube size toward that of the isolated SiC monolayer is observed in the LDA calculations.

Interestingly, the mechanism of the reciprocal radius dependence of fundamental band gap in some single-walled CNTs was recognized as curvature-induced mixture of electronic orbitals [62]. However, we find a peculiar band gap reduction with the decrease of tube size for both zigzag and armchair SiC-NTs. This band gap reduction of SiC-NTs with the increasing tubular curvature can be attributed to the curvature-induced hybridization between the π^* and σ^* orbitals near the conduction band minimum (CBM) in the small radius SiC-NTs similar to that found in the very small diameter CNTs [65]. To illustrate this notion, we display the charge density contours of the CBM of the zigzag $(n, 0)$ and armchair (n, n) nanotubes at the 1D Brillouin zone center (Γ point) and zone boundary (Z point), respectively, in Fig. 7.7. It shows that in the small zigzag $(n, 0)$ SiC-NTs such as (5, 0) and (6, 0) nanotubes, a strong mixture of the π^* and σ^* states in a SiC monolayer now exists inside the tube to form

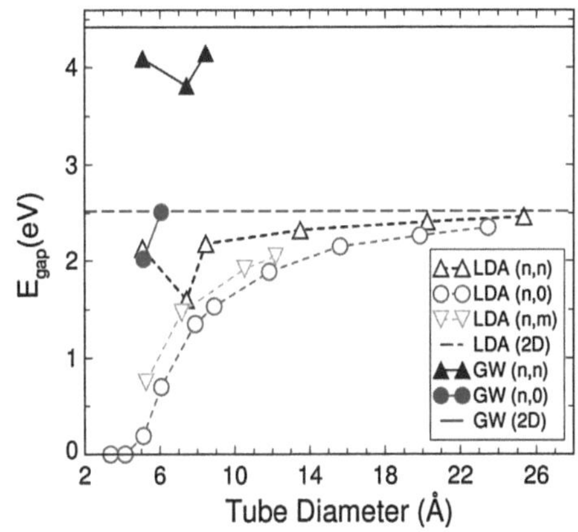

Fig. 7.6 The fundamental band gaps of some zigzag $(n, 0)$ and armchair (n, n) SiC nanotubes as a function of the tube diameter from the GW (*solid symbols*) and LDA [26] (*open symbols*) calculations. The band gap of the isolated SiC sheet obtained from the GW and LDA calculations is also displayed as the *solid* and *dashed line*, respectively. The LDA results for a few (n, m) SiC nanotubes are also included for comparison. *Dotted lines* between the symbols are a guide to the eye only

a ring-like charge distribution, whereas the feature of the p_z state is floating above the Si and C atoms outside the SiC-NT even under considerable structural distortion. Therefore, a substantial band gap reduction upon increasing the tube curvature can be expected. Indeed, this is confirmed by our LDA and quasiparticle calculations (see circle symbols in Fig. 7.6). In the limit cases, both (3, 0) and (4, 0) SiC-NTs with extremely small diameters can be regarded as the thinnest conducting SiC nanowires to give a zero band-gap character at least within DFT-LDA. In contrast, a much smaller degree of the orbital hybridization in the armchair SiC-NTs is induced by the folding-curvature, as revealed by the cylindrically symmetric distribution of the Si p_z component of the CBM at the Brillouin zone boundary (Z point) in the (5, 5) nanotube (Fig. 7.7). This gives rise to a minor size dependence of the band gap ("triangle up" symbols in Fig. 7.6) for the armchair SiC-NTs. Nevertheless, a spectacular curvature effect in terms of strongly enhanced $\pi^* - \sigma^*$ hybridization does occur in the ultrasmall armchair (3, 3) SiC-NT(see Fig. 7.7). This unusual orbital rehybridization effects on the band gap have also been recognized before in other ionic group III-V nanotubes [18, 60].

7.5 Huge Excitonic Effects in SiC Nanostructures

In quasi-one dimensional nanotube systems, the wavefunctions of quasielectrons and quasiholes can overlap significantly because of quantum confinement, and therefore, electron-hole interaction may be considerably enhanced. At the same time, the

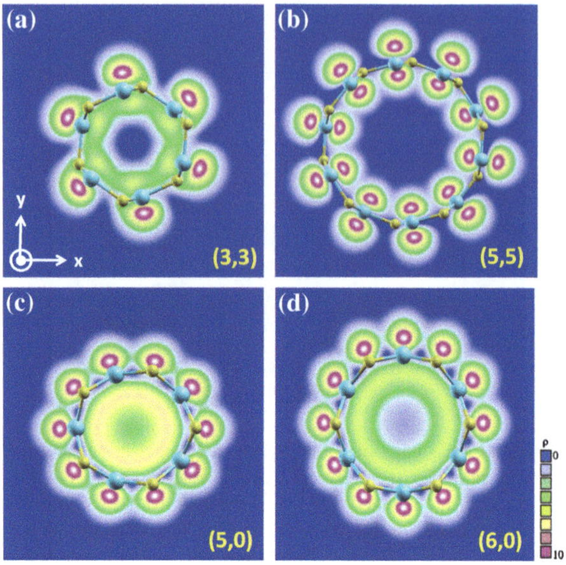

Fig. 7.7 The charge density contours of the states at the conduction band minimum (CBM) at the 1D Brillouin zone boundary (Z point) of the (3, 3) and (5, 5) SiC-NTs and also at the zone center (Γ point) of the (5, 0) and (6, 0) SiC-NTs. The Si and C atoms are represented by the *cyan* and *yellow spheres*, respectively (After Ref. [59])

screening among the quasiparticles can be systematically changed because curvature and geometrical reduced dimensionality effects. For example, the phenomenon of anti-screening was discovered recently in the CNTs [66].Therefore, the absorption features associated with the exciton formation can become prominent in the optical spectra in SiC-NTs. To study the excitonic effects on the optical properties of SiC-NTs, we have performed GW+BSE calculations by taking into account the electron-hole interaction kernel for several SiC-NTs of different sizes and different chiralities.

In Fig. 7.8a and b, the quasiparticle band structure and the imaginary part of the optical polarizability for the armchair (3, 3) SiC-NTs are presented. Because of the strong depolarization effects in nanotubes [67], we only consider the optical response for light polarized along the tubular axis. For comparison, the optical absorption spectrum obtained without the electron-hole interaction in the GW+RPA scheme is also displayed as a blue-dashed curve in Fig. 7.8b. The onset energy (4.86 eV) of this independent-particle optical spectrum corresponds to the lowest-energy dipole-allowed vertical transition (E^{I_1}) from the valence to conduction bands near 2/3 of the $\Gamma - Z$ line in the Brillouin zone which corresponds to the K point in the hexagonal BZ of the SiC sheet (see Fig. 7.8a). When electron-hole interaction is included, excitons with specific excitation energies (E^S) are formed to give rise to the prominent photoabsorption peaks. The binding energy of the lowest-energy exciton

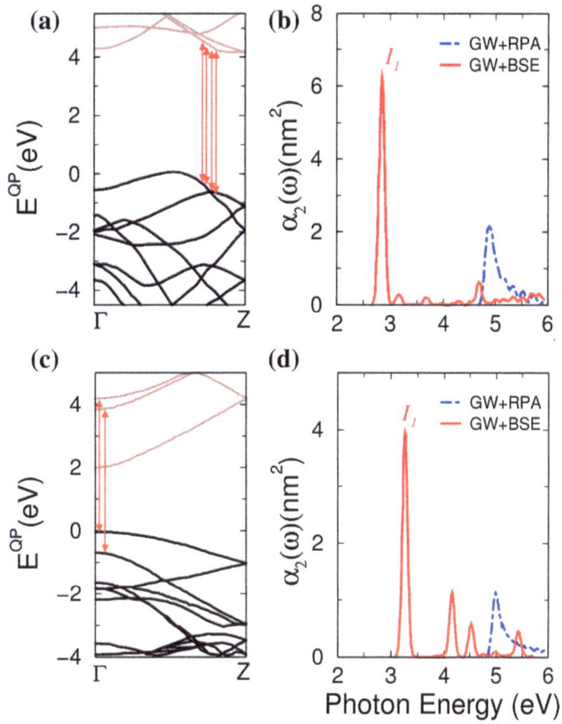

Fig. 7.8 The quasiparticle band structure and the optical absorption spectra of the armchair (3, 3) (**a** and **b**) and zigzag (5, 0) (**c** and **d**) SiC-NTs. Valence (conduction) bands are shown as *black* (*brown*) *curves* and the valence band maximum is aligned at 0 eV in band structures (**a** and **c**). In both **b** and **d**, the optical spectra from GW+RPA and GW+BSE calculations are denoted as the *blue-dashed* and *red-solid curves*, respectively. The theoretical absorption spectra were broadened with a Gaussian of 0.10 eV. The corresponding lowest-energy bright exciton (I_1) is formed by the mixture of four interband transitions as indicated by the *red arrows* in **a** and **c** for (3, 3) and (5, 0) SiC-NTs, respectively

is defined as the energy difference between the continuum onset and corresponding excitation energy.

For the armchair (3, 3) SiC-NT, the optical absorption spectra from the GW+BSE calculations are dominated by the discrete exciton peaks [see the red curves in Fig. 7.8b]. The first bright exciton (I_1) appearing at 2.86 eV with a large binding energy ($E_B^{I_1}$) of ∼ 2.0 eV is formed by a mixture of four pair of interband transitions, as indicated by the arrows in Fig. 7.8a. These four different interband transitions couple strongly to make the electron amplitude more localized and highly asymmetric around the hole position located outside the tube but near the C atom, as displayed in Fig. 7.9a, thereby resulting in a strongly bound character of exciton I_1. At the same time, the large electron-hole overlap in the (3, 3) SiC-NT lifts the double degeneracy of the $E_1^{\pi-\pi*}$ exciton in the 2D SiC monolayer to create the bright I_1 exciton and a dark exciton with a higher (by 0.02 eV) excitation energy. Similar excitonic effects are also found in the larger armchair (5, 5) SiC-NT. However, in Fig. 7.9b, a reduced electron-hole overlap of exciton I_1 in (5, 5) SiC-NT is demonstrated by a rather delocalized profile of electron-hole amplitude due to the decreased curvature. Moreover, apart from the bright bound and resonant excitons shown as red peaks in Fig. 7.8b, there also exist many dark excitons among the bright ones. However, we find no dark exciton below the I_1 exciton in the armchair SiC-NT. As mentioned before, this can

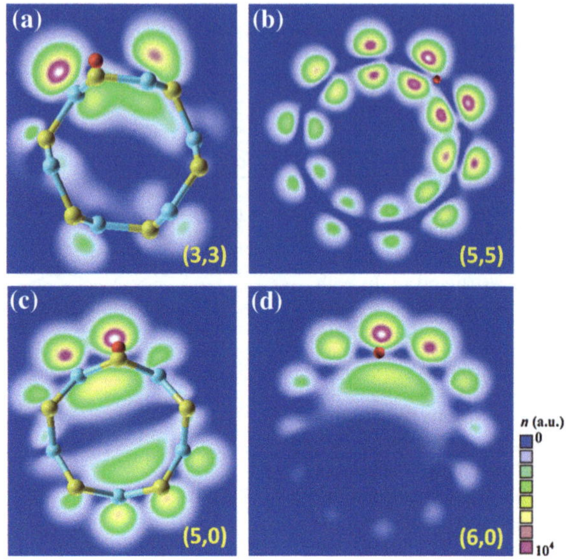

Fig. 7.9 The electron charge density contours of the first bright exciton of **a** (3, 3), **b** (5, 5), **c** (5, 0), and **d** (6, 0) on a cross-sectional tubular plane with the hole fixed at the position of the *red sphere* (see text) are displayed

be anticipated from the fact that the isolated SiC layer has a direct band gap at the K-point. In this case, there is no momentum mismatch when the SiC monolayer is rolled up to form an armchair SiC-NT, and hence the possibility of the formation of the optically inactive lowest-energy exciton diminishes [60].

Next, to study the excitonic effects on the optical absorption in the small radius zigzag SiC-NTs, we performed GW+RPA and GW+BSE calculations of the zigzag (5, 0) SiC-NT with a tube radius comparable to that of the armchair (3, 3) SiC-NT. The quasiparticle band structure and optical polarizability α_2 of the (5, 0) SiC-NT are displayed in Fig. 7.8c and d, respectively. Fig. 7.9c is a contour plot of the electron charge density distribution of the I_1 exciton of the (5, 0) SiC-NT with respect to a fixed hole position. Like the (3, 3) SiC-NT, the photoexcited spectrum of the (5, 0) SiC-NT is characterized by the discrete exciton peaks. The first bright exciton I_1 is the most prominent feature in the spectrum, and is consisted of the different interband transitions near the zone-center Γ-point (represented by the solid arrows). As a result, the charge density distribution is not cylindrically symmetric. Similar results are found in the (6, 0) SiC-NT (as shown in Fig. 7.9d). Interestingly, although the (5, 0) and (3, 3) SiC-NTs have a similar curvature, the relatively smaller binding energy (1.71 eV) of the I_1 exciton in the (5, 0) SiC-NT indicates a rather delocalized exciton. This pronounced chirality effect can be attributed to the considerable orbital rehybridization in the zigzag SiC-NTs emphasized above. This orbital rehybridization induces a significant band-gap reduction giving rise to an increase in the effective screening, and hence reduces the binding energy. Furthermore, according to the zone-folding scheme for nanotubes, the lowest energy optically active interband transition at the K point of the 2D monolayer can be folded into the Γ point to form a bright exciton with the lowest excitation energy in zigzag nanotubes.

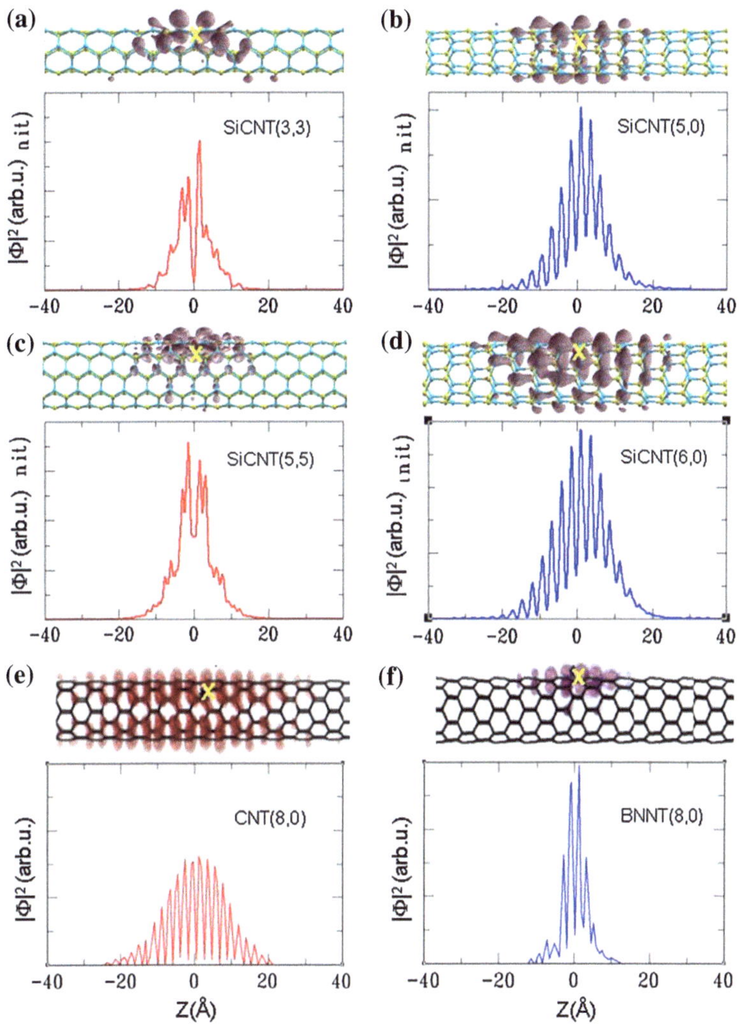

Fig. 7.10 Isosurface plots of the electron distribution $|\Phi|^2$ of the I_1 exciton (see Fig. 7.8) with the fixed hole position (indicated by a *yellow cross*) in the armchair (**a**) (3, 3) and (**c**) (5, 5) as well as the zigzag (**b**) (5, 0) and (**d**) (6, 0) SiC-NTs. The corresponding integrated intensities by averaging out the coordinates perpendicular to the tube axis, are also shown. Here the hole position is set at 0. The Si and C atoms are represented by the *cyan* and *yellow spheres*, respectively. Similar isosurface plots and integrated intensity of electron distribution along tubular axis of **e** (8, 0) CNT and (**f**) (8, 0) BN-NT are also including for comparison (After Ref. [30, 59])

However, in the very small-radius zigzag SiC-NTs such as the (5, 0) nanotube, the strong curvature-induced orbital rehybridization has brought a higher conduction band down to the bulk GW-band gap (0.0 – 4.0 eV) region (see Fig. 7.8c). Since the optical transitions from the top valence bands to this conduction band are dipole-forbidden, some low energy dark excitons appear in the small-radius zigzag SiC-NTs [59] The existence of localized dark excitons makes the zigzag SiC-NTs candidates for tunable optical devices for temperature or external field sensors.

Finally, we explore the extent of the exciton wavefunction localization in all directions in the different SiC-NTs by comparing the intensity of their electron-hole wavefunctions in real space. In Fig. 7.10, the isosurface plots of the electron density with the fixed hole position (the yellow crosses in the figure) of the first bright exciton (I_1) of the armchair SiC-NTs suggest a significant localized distribution in the direction along the circumference of the tube. Mixing of excitations to different subband pairs in the small-radius SiC-NTs is responsible for this tightly bound character. The spatial localization of the exciton wavefunction along the tubular direction is revealed by integrating out the electron coordinates in the perpendicular plane, as presented in Fig. 7.10a and c. In contrast to the carbon nanotubes (shown in Fig. 7.10e), the deviation from the one-dimensional nature of the low-energy exciton is mainly attributed to the strongly mixing of the interband transitions. A even more tightly bound behavior along the circumference direction of the lowest-energy bright exciton has been discovered in the BN-NTs (displayed in Fig. 7.10f) [30]. In the zigzag SiC-NTs, we find the similar anisotropic bound nature of the first bright exciton. The slightly broadened distribution along the tubular direction of the integrated wavefunction intensity (see the blue curves in Fig. 7.10b and d) suggests a less bound character in the zigzag SiC-NTs. The distinct spiky features in the curves of the zigzag SiC-NTs are located on the cation (Si atom) planes, and they indicate a charge transfer from an anion (hole) to its neighboring cations by photoexcitation. On the other hand, the spiky peak structures are not observed in the armchair SiC-NT because of the specific atomic arrangement dictated by its particular chirality. Furthermore, the asymmetric distribution of $|\Phi|^2$ with respect to the fixed hole position, is a consequence of the different nearest-neighbor cation-anion distances caused by rolling up the ideal hexagonal symmetric SiC monolayer. This curvature-induced spatial symmetry breaking is also responsible for the special nodal-like feature near the hole position in the smallest armchair (3, 3) SiC-NT, as shown in Fig. 7.10a. Nonetheless, by comparing the distribution of the exciton wavefunction of the (3, 3) SiC-NT with the larger (5, 5) nanotube (shown in Fig. 7.10c), we note that this symmetry breaking will decrease gradually as the curvature decreases or the tube diameter increases.

7.6 Conclusion

In summary, we have employed both mean-field (Kohn-Sham energies of DFT) and state-of-the-art many-body Green's-function methodologies to study the electronic band structure and optical properties of SiC nanostructures (including monolayer and SiC-NTs). First of all, our GW band structure calculations show that rather complicated orbital-dependent self-energy corrections are needed in obtaining accurate quasiparticle properties for a SiC monolayer. The profile of the optical spectra of a

SiC monolayer is modified dramatically when electron-hole interaction is included. In particular, a strongly bound bright exciton with a large binding energy (1.17 eV) is found to dominate the optical spectrum, because of the enhanced overlap between the electron and hole orbitals due to reduced-dimensionality effects and also the existence of vacuum region, which reduces the effective screening in the SiC monolayer.

Second, the quasiparticle band structure of the small-radius armchair and zigzag SiC-NTs is systematically studied within the GW approximation. Our detailed analysis on the charge density distributions reveals that a curvature-induced orbital rehybridization plays a vital role in determining the band gap of the small-radius SiC-NTs with different chiralities at both the LDA and GW levels. In particular, the quasiparticle band gaps as a function of tube diameter behave very differently, depending on the chirality of the SiC-NTs.

Finally, the calculated photoexcited spectra of the SiC-NTs are composed of discrete exciton peaks, thereby indicating strong excitonic effects in both the armchair and zigzag SiC-NTs. The optical absorption spectra of the small radius armchair and zigzag SiC-NTs are dominated by the first bright bound exciton I_1 with a significant binding energy up to ~2.0 eV. The highly asymmetric charge distribution of the exciton I_1 in the (3, 3) SiC-NT is found to be a consequence of the strong coupling of the four lowest optically allowed inter-subband transitions. Also, our GW+BSE calculations demonstrate that the simple zone-folding approach fails in predicting the low energy exciton characters in very small zigzag SiC-NTs, mainly because of the strong curvature-induced orbital rehybridization in these nanotubes. We believe that the large excitation energy of ~3.0 eV of the first bright exciton, with no dark exciton below it, may make the small-radius armchair SiC-NTs candidates for optical devices working in the UV regime. In contrast, the numerous dark excitons below I_1 in the zigzag SiC-NTs may lead to potential applications in tunable optoelectric devices ranging from infrared to UV frequencies by external perturbations. Moreover, the similarity between graphene and SiC monolayer in many aspects except for the electronic fundamental gap suggests a promising 2D-based sandwich-type heterostructures by tailoring various combinations of individual layers.

Acknowledgments We thank the National Science Council and NCTS of Taiwan ROC for support, and also NCHC and IAMS (Academia Sinica) of Taiwan ROC for CPU time. HCH acknowledges support under Grant No. NSC 101-2112-M-032-001-MY3. SGL acknowledges support by the Director, Office of Science, Office of Basic Energy Sciences, Materials Sciences and Engineering Division, US Department of Energy, under Contract No. DE-AC02-05CH11231.

References

1. Iijima, S.: Nature **354**, 56 (1991)
2. Charlier, J.-C., Blase, X., Roche, S.: Rev. Mod. Phys. **79**, 677 (2007)
3. Castro Neto, A.H., Guinea, F., Peres, N.M.R., Novoselov, K.S., Geim, A.K.: Rev. Mod. Phys. **81**, 109 (2009)
4. Cohen, Marvin L., Zettl, Alex: Phys. Today **63**, 34 (2010)
5. Novoselov, K.S., Fal'ko, V.I., Colombo, L., Gellert, P.R., Schwab, M.G., Kim, K.: Nature **490**, 192 (2012)

6. Hellwege, K.H., Madelung, O. (eds.): Semiconductors: Physics of Group IV Elements and III-V Compounds, Landolt-Börnstein, New Series, Group IV and III-V, vol. 17, Pt. A. Springe, Berlin (1982)
7. Wyckoff, R.W.G.: Crystal Structures. Wiley, New York (1963)
8. Ivanov, P.A., Chelnokov, V.E.: Semicond. Sci. Technol. **7**, 863 (1992)
9. Harris, G.L. (ed.): Properties of Silicon Carbide. INSPEC, Institution of Electrical Engineers, London (1995)
10. Persson, C., Lindefelt, U.: J. App. Phys. **82**, 5496 (1997)
11. Wang, R., Zhang, D., Liu, C.: Chemical. Phys. Lett. **411**, 333 (2005)
12. Rurali, R., Godigonon, P., Rebollo, J., Hernandez, E., Ordejon, P.: Appl. Phys. Lett. **82**, 4298 (2003)
13. Park, C.H., Cheong, B.H., Lee, K.H., Chang, K.J.: Phys. Rev. B **49**, 4485 (1994)
14. Sun, X.-H., Li, C.-P., Wong, W.-K., Wong, N.-B., Lee, C.-S., Lee, S.-T., Teo, B.-K.: J. Am. Chem. Soc. **124**, 14464 (2002)
15. Saito, R., Dresselhaus, G., Dresselhaus, M.S.: Physical Properties of Carbon Nanotubes. Imperial College Press, London (1998)
16. Guo, G.Y., Chu, K.C., Wang, D.S., Duan, C.G.: Phys. Rev. B **69**, 205416 (2004)
17. Rubio, A., Corkill, J.L., Cohen, M.L.: Phys. Rev. B **49**, 5081 (1994)
18. Guo, G.Y., Lin, J.C.: Phys. Rev. B **71**, 165402 (2005)
19. Zhao, M., Xia, Y.Y., Zhang, D.J., Mei, L.M.: Phys. Rev. B **68**, 235415 (2003)
20. Lee, S.M., Lee, Y.H., Hwang, Y.G., Elsner, J., Porezag, D., Frauenheim, Th: Phys. Rev. B **60**, 7788 (1999)
21. Miyamoto, Y., Yu, B.D.: Appl. Phys. Lett. **80**, 586 (2002)
22. Menon, M., Richter, E., Mavrandonakis, A., Froudakis, G., Andriotis, A.N.: Phys. Rev. B **69**, 115322 (2004)
23. Guo, G.Y., Lin, J.C.: Phys. Rev. B **72**, 075416 (2005)
24. Guo, G.Y., Lin, J.C.: Phys. Rev. B **77**, 049901 (E) (2008)
25. Guo, G.Y., Ishibashi, S., Tamura, T., Terakura, K.: Phys. Rev. B **75**, 245403 (2007)
26. Wu, I.J., Guo, G.Y.: Phys. Rev. B **76**, 035343 (2007)
27. Wu, I.J., Guo, G.Y.: Phys. Rev. B **78**, 035447 (2008)
28. Blase, X., Rubio, A., Louie, S.G., Cohen, M.L.: Europhys. Lett. **28**, 335 (1994)
29. Spataru, C.D., Ismail-Beigi, S., Benedict, L.X., Louie, S.G.: Phys. Rev. Lett. **92**, 077402 (2004)
30. Park, C.-H., Spataru, C.D., Louie, S.G.: Phys. Rev. Lett. **96**, 126105 (2006)
31. Yang, L., Deslippe, J., Park, C.-H., Cohen, M.L., Louie, S.G.: Phys. Rev. Lett. **103**, 186802 (2009)
32. Novoselov, K.S., Geim, A.K., Morozov, S.V., Jiang, D., Zhang, Y., Dubonos, S.V., Grigorieva, I.V., Firsov, A.A.: Science **306**, 666 (2004)
33. Castro Neto, A.H., Guinea, F., Peres, N.M.R., Novoselov, K.S., Geim, A.K.: Rev. Mod. Phys. **81**, 109 (2009)
34. Hybertsen, M.S., Louie, S.G.: Phys. Rev. Lett. **55**, 1418 (1985)
35. Hybertsen, M.S., Louie, S.G.: Phys. Rev. B **34**, 5390 (1986)
36. Rohlfing, M., Louie, S.G.: Phys. Rev. Lett. **81**, 2312 (1998)
37. Rohlfing, M., Louie, S.G.: Phys. Rev. B **62**, 4927 (2000)
38. Deslippe, J., Samsonidze, G., Strubbe, D.A., Jain, M., Cohen, M.L., Louie, S.G.: Comp. Phys. Commun. **183**, 1269 (2012). (http://berkeleygw.org/)
39. Kohn, W., Sham, L.J.: Phys. Rev. **140**, A1133 (1965)
40. Hedin, L.: Phys. Rev. **139**, A796 (1965)
41. Pfrommer, B.G., Cote, M., Louie, S.G., Cohen, M.L.: J. Comput. Phys. **131**, 233 (1997)
42. Gonze, X., Lee, C.: Phys. Rev. B **55**, 10355 (1997)
43. Baroni, S., de Gironcoli, S., Dal Corso, A., Giannozzi, P.: Rev. Mod. Phys. **73**, 515 (2001). (http://www.abinit.org/)
44. Kleinman, L., Bylander, D.M.: Phys. Rev. Lett. **48**, 1425 (1982)
45. Monkhorst, H.J., Pack, J.D.: Phys. Rev. B **13**, 5188 (1976)
46. Cohen, M.L., Schluter, M., Chelikowsky, J.R., Louie, S.G.: Phys. Rev. B **12**, 5575 (1975)

47. Aulbur, W.G., Jonsson, L., Wilkins, J.W.: Solid State Physics: Advances in Research and Applications vol. 54, p. 1. Academic, New York (2000)
48. Spataru, C.D., Ismail-Beigi, S., Capaz, R.B., Louie, S.G.: Carbon Nanotubes: Advanced Topics in the Synthesis, Structure, Properties and Applications vol. 111, p. 195. Springer, Heidelberg (2008)
49. Neaton, J.B., Hybertsen, M.S., Louie, S.G.: Phys. Rev. Lett. **97**, 216405 (2006)
50. Spataru, C.D., Ismail-Beigi, S., Benedict, L.X., Louie, S.G.: Appl. Phys. A **78**, 1129 (2004)
51. Ismail-Beigi, S.: Phys. Rev. B **73**, 233103 (2006)
52. Fetter, A.L., Walecka, J.D.: Quantum Theory of Many-Particle Systems. McGrawHill, New York (1971)
53. Meyer, J.C., Geim, A.K., Katsnelson, M.I., Novoselov, K.S., Booth, T.J., Roth, S.: Nature **446**, 60 (2007)
54. Cahangirov, S., Topsakal, M., Akt*ü*rk, E., Sahin, H., Ciraci, S.: Phys. Rev. Lett. **102**, 236804 (2009)
55. Vogt, Patrick, De Padova, Paola, Quaresima, Claudio, Avila, Jose, Frantzeskakis, Emmanouil, Asensio, M.C., Resta, A., Ealet, B., Lay, G.L.: Phys. Rev. Lett. **108**, 155501 (2012)
56. Sahin, H., Cahangirov, S., Topsakal, M., Bekaroglu, E., Akturk, E., Senger, R.T., Ciraci, S.: Phys. Rev. B **80**, 155453 (2009)
57. Wirtz, L., Rubio, A., de la Concha, R.A., Loiseau, A.: Phys. Rev. B **68**, 045425 (2003)
58. Ummels, R.T.M., Bobbert, P.A., van Haeringen, W.: Phys. Rev. B **58**, 6595 (1998)
59. Hsueh, H.C., Guo, G.Y., Louie, S.G.: Phys. Rev. B **84**, 085404 (2011)
60. Ismail-Beigi, S.: Phys. Rev. B **77**, 035306 (2008)
61. Marinopoulos, A.G., Reining, L., Rubio, A., Vast, N.: Phys. Rev. Lett. **91**, 046402 (2003)
62. Kane, C.L., Mele, E.J.: Phys. Rev. Lett. **78**, 1932 (1997)
63. Gulseren, O., Yildirim, T., Ciraci, S.: Phys. Rev. B **65**, 153405 (2002)
64. Guo, G.Y., Lin, J.C.: Phys. Rev. B **71**, 165402 (2005)
65. Blase, X., Benedict, L.X., Shirley, E.L., Louie, S.G.: Phys. Rev. Lett. **72**, 1878 (1994)
66. Deslippe, Jack, Dipoppa, Mario, Prendergast, David, Moutinho, M.V.O., Capaz, R.B., Louie, S.G.: Nano Lett. **9**, 1330 (2009)
67. Ajiki, H., Ando, T.: Physica (Amsterdam) **201B**, 349 (1994)

Chapter 8
Plasma Enabled Fabrication of Silicon Carbide Nanostructures

Jinghua Fang, Igor Levchenko, Morteza Aramesh, Amanda E. Rider, Steven Prawer and Kostya (Ken) Ostrikov

Abstract Silicon carbide is one of the promising materials for the fabrication of various one- and two-dimensional nanostructures. In this chapter, we discuss experimental and theoretical studies of the plasma-enabled fabrication of silicon carbide quantum dots, nanowires, and nanorods. The discussed fabrication methods include plasma-assisted growth with and without anodic aluminium oxide membranes and with or without silane as a source of silicon. In the silane-free experiments, quartz was used as a source of silicon to synthesize the silicon carbide nanostructures in an environmentally friendly process. The mechanism of the formation of nanowires and nanorods is also discussed.

8.1 Introduction

Due to its unique properties such as very wide electronic bandgap ($\sim$3 eV) and superior chemical, mechanical and thermal stability, silicon carbide (SiC) has attracted considerable attention as one of the most promising materials for the fabrication of nanostructures [1, 2]. Being used in nanoelectronic devices operated at high power or high frequency, SiC is able to ensure efficient current multiplication owing to its excellent thermal conductivity and high electric field [3, 4]. Compared to the bulk SiC counterparts, crystalline SiC nanostructures have superior mechanical and field emission properties [5, 6]. Recently, negative index properties of SiC have led to it being used in a number of novel applications [7].

SiC nanostructures can be fabricated by: decomposition of organic Si compounds [8], high-temperature reactions between carbon nanotubes and silicon oxide (SiO_2)

J. Fang (✉) · I. Levchenko · A. E. Rider · K. (Ken) Ostrikov
CSIRO Materials Science and Engineering, 218, West Lindfield, , NSW 2070 , Australia
e-mail: Jinghua.Fang@csiro.au

M. Aramesh · S. Prawer
University of Melbourne, Melbourne, VIC 3010, Australia

H. Li et al. (eds.), *Silicon-based Nanomaterials*,
Springer Series in Materials Science 187, DOI: 10.1007/978-1-4614-8169-0_8,
© Springer Science+Business Media New York 2013

[9], arc discharge [10, 11] and hot filament chemical vapor deposition [12]. However, these methods are very complex, slow and environmentally unfriendly. In addition, they are not able to effectively fabricate nanomaterials with precisely-controlled internal structure [13]. Plasma-based fabrication techniques have recently attracted considerable interest due to their controllability [14]. They are also able to control the nanostructure size and enhance the nucleation at lower processing temperatures [15].

High-density, low-frequency inductively coupled plasma chemical vapour deposition (ICP-CVD) is a promising method for the fabrication of SiC nanostructures due to high selectivity and excellent reproducibility of this process. Another popular method is a microwave plasma-enhanced chemical vapour deposition (MWPE-CVD) technique. The main advantage of the MWPE-CVD process compared with ICP-CVD is the ability of the microwave plasma to heat up the substrate to the temperature required for the optimal deposition. Here, we will discuss some aspects of the bottom-up fabrication of SiC nanostructures using both ICP-CVD and MWPE-CVD techniques.

Template-based nanofabrication is a novel method for growing highly controllable arrays of various nanostructured materials. Porous anodic aluminum oxide membranes (AAOM) can be used as a template. AAOMs are able to support arrays of carbon nanotubes [16, 17] and nanowires [18, 19], with the dimensions and density of the nanostructures precisely controlled by the length, width and separation of the template channels. By using highly ordered AAOM as a template, ordered arrays of SiC nanowires were successfully produced [20]. However, this process requires high temperatures, a separate source of silicon, and long (more than 48 h) treatment times. Moreover, it was found that quartz could be etched by MWPE-CVD, and an environmentally friendly approach for the fabrication of SiC nanostructures without toxic silane (SiH_4) gas has been developed based on the use of quartz substrates as a source of Si in MWPE-CVD process [13]. This technique will also be discussed in this chapter.

In this chapter, various techniques of the SiC nanostructure fabrication are discussed. Specifically, we will consider SiC quantum dots (QDs), nanowires (NWs) and nanorods (NRs). The fabrication methods include plasma-assisted growth with and without AAOM, and with or without SiH_4 as a source of Si. The chapter is organized as follows. In Sect. 8.2 we describe the simulation of the controlled growth of SiC quantum dots in a plasma environment. Then, experiments on the fabrication of SiC and related materials using SiH_4 as a source of Si (with and without AAOM) using plasma-based techniques (ICP-CVD and magnetron sputtering) will be discussed in Sect. 8.3. In Sect. 8.4, we will discuss the fabrication of SiC nanostructures without SiH_4, with and without AAOM. In Sect. 8.5, we will conclude with a brief recap of the relative merits of plasma-only and combined plasma and AAOM methods for the fabrication of SiC nanostructures and related nanomaterials.

8.2 Computational Studies: SiC QD Growth in a Plasma Environment

The properties of SiC nanostructures are controlled by size, shape, and elemental composition (i.e., x in $Si_{1-x}C_x$). These properties determine how the nanomaterials may be used in nanodevices. Thus, it is important to understand the influence of the growth environment on self-assembly of quantum dots, and to control their size, shape and elemental composition during the growth. In this section we will discuss a hybrid model of the epitaxial growth of self-assembled SiC quantum dots, focused mainly on determining the conditions required to achieve stoichiometric (i.e. SiC = 1:1) and core-shell structure. The effects of a plasma-based growth environment will be demonstrated through a computational study which examined the following structures:

1. Core–multishell QDs;
2. Compositionally graded QDs: $A_{1-x}B_x$, with x varying;
3. QDs of selected composition: $A_{1-x}B_x$, with constant x throughout the entire structure.

The bottom-up, supply-limited growth of SiC nanomaterials proceeds via surface diffusion and is strongly influenced by the surface temperature and particle influx (the effect of a plasma environment on these parameters will be discussed). Surface processes to take into account include the arrival of Si and C atoms to the surface, surface diffusion of Si and C adatoms, addition of adatoms to growing QDs, atom evaporation to the 2D surface vapour and to the 3D external vapour (see Fig. 8.1a). Levchenko et al. [21] considered a series of adatom balance equations as follows:

$$\frac{\partial \eta_{Si(C)}}{\partial t} = \Psi^{+}_{Si(C)} - \Psi^{-}_{Si(C)} + \sum\nolimits_{i=1}^{N} \eta_i \left(\vec{\Gamma}_{Si(C),i} - \overleftarrow{\Gamma}_{Si(C),i} \right), \tag{8.1}$$

where $\partial \eta_{Si(C)}/\partial t$ is the change in adatom concentration (Si for silicon, C for carbon) on the surface with time, Ψ^{+} is the incoming flux of atoms, Ψ^{-} is the flux of atoms leaving the surface, *N* is the number of atoms that make up the QD, η_i is the density of i-atom QDs, $\vec{\Gamma}$ and $\overleftarrow{\Gamma}$ are atom fluxes to and from the QD, respectively. These fluxes are given by:

$$\vec{\Gamma}_{Si(C),i} - \overleftarrow{\Gamma}_{Si(C),i} = \upsilon_{d,Si(C)} \sigma_i \eta_{Si(C)} - n_i \mu_{Si(C),i}, \tag{8.2}$$

where $\upsilon_{d,Si(C)}$ is the rate of surface diffusion, $\mu_{Si(C),i}$, is the frequency of atom re-evaporation to the 2D surface vapour, σ_i is the cross section of a i-atom QD. A measure of the composition of the QD is given by (8.3):

$$\wp_{Si-C}(t) = \int_0^t \bar{\Gamma}_{Si-C}(\tau) d\tau, \tag{8.3}$$

where $\bar{\Gamma}_{Si-C} = (\vec{\Gamma}_{Si} - \overleftarrow{\Gamma}_{Si})/(\vec{\Gamma}_{C} - \overleftarrow{\Gamma}_{C})$.

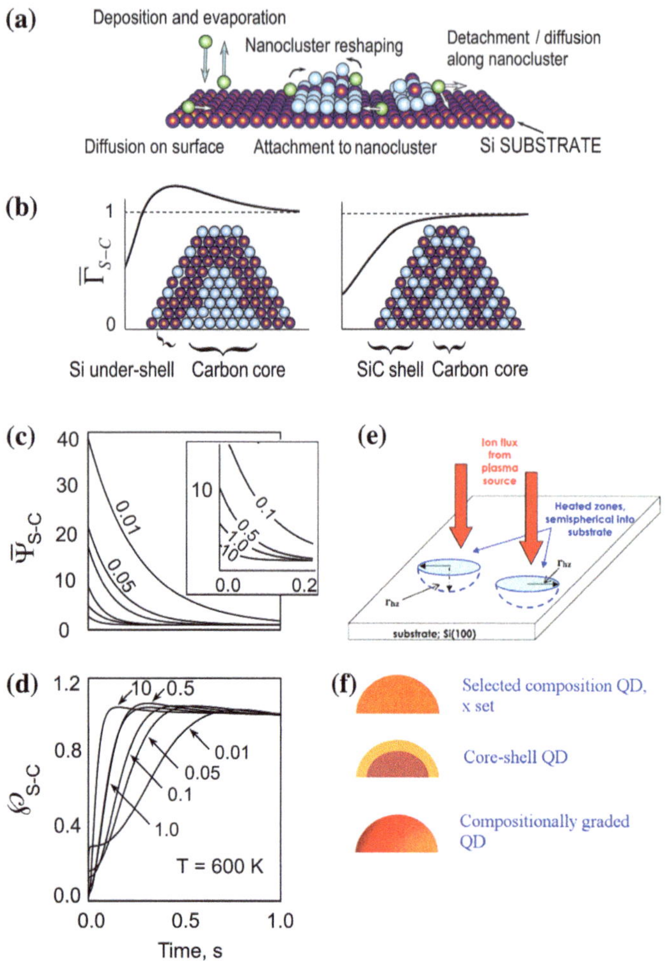

Fig. 8.1 **a** Processes considered in the model, **b** Schematic of the core-shell structure formation, **c** Time-variable influxes, **d** Change of elemental composition with time for a 600 K, **e**, **f** Variety of QD structures considered in Ref. [24]. Please note: **a**–**d** reprinted with permission from [21], copyrights 2007, American Institute of Physics, **e** reproduced from [24] by permission of IOP Publishing. All rights reserved, copyright 2008

When using a fixed influx ratio of Si and C, Rider et al. found that a stoichiometric composition was achieved quicker when higher substrate temperatures were used [22]. Levchenko et al. found that it was possible to obtain SiC core-shell QDs (Fig. 8.1b) by using a time variable Si/C influx ratio (Fig. 8.1c, d) [21].

By changing the substrate temperature and time dependence of Si/C fluxes, it is possible to control the internal structure as well as elemental composition of SiC QDs grown on a Si substrate. As shown in Fig. 8.1b [21], two different core-shell

SiC QDs are considered: (left) Carbon core and stoichiometric SiC outer shell [when influx is 0.1 ML/s and T = 600 K, where ML is monolayer], (right) carbon core, Si-enriched under-shell, and stoichiometric SiC outer shell [when influx is 0.1 ML/s and T = 800 K].

The effect of a plasma on QD growth was incorporated into the model by Rider et al. [24] by accounting for:

1. Local substrate heating (as shown in Fig. 8.1e)—this is advantageous as it widens the range of substrates that may be used for deposition.
2. A reduction in the surface diffusion activation energy, ε_d, which has a large effect on the rate of surface diffusion [24]:

$$v_{d,Si(C)} = \nu_0 \lambda_{Si} \exp[-\varepsilon_d / kT], \tag{8.4}$$

where ν_0 is the Debye frequency, λ_{Si} is the Si (100) substrate lattice parameter, ε_d is the adatom surface diffusion activation energy and k is Boltzmann's constant. By using these plasma-related effects as well as further tuning the time variable influx ratios, it was shown that it is possible to achieve selected-composition $Si_{1x}C_x$ dots (faster than for the non-plasma case), as well as core-shell and compositionally-graded QDs (Fig. 8.1f) [24]. This model is consistent with the experimental results where Cheng et al. [23, 26] studied the dependence of the gas flow ratio CH_4/SiH_4 on the resultant $Si_{1-x}C_x$ composition.

8.3 Experimental Studies: Growth of SiC Nanostructures in a Plasma Environment

Chemical vapour deposition is a method for depositing solid materials onto a surface, which usually includes chemical reactions between working gas precursors and the substrate. Plasma-enhanced CVD uses both neutral and ionized 'building units' (material the nanostructures are made from) and 'working units'(to prepare the surface for deposition). The benefits of using plasmas as a growth environment include, but are not limited to: a wider range of reactive species, lower growth temperatures (hence a wider range of substrates), higher size uniformity, control over the vertical alignment of nanostructures due to the electric field in a plasma-surface sheath [28]. In this section, we will be focusing on the use of inductively coupled plasma CVD to grow SiC nanostructures, specifically nanoislanded films and QDs, both with and without SiH_4.

Figure 8.2a schematically shows an ICP-CVD chamber, where power is generated by a RF power generator installed on the chamber top, and then induced into the chamber via a quartz window. A SiC disc is used as a source of material for the deposition. Ions formed in the plasma hit the target and disloge Si and C atoms, which deposit onto the substrate surface. The main parameters to be controlled during the process are RF and target power, flow rate of the working gases, substrate temperature

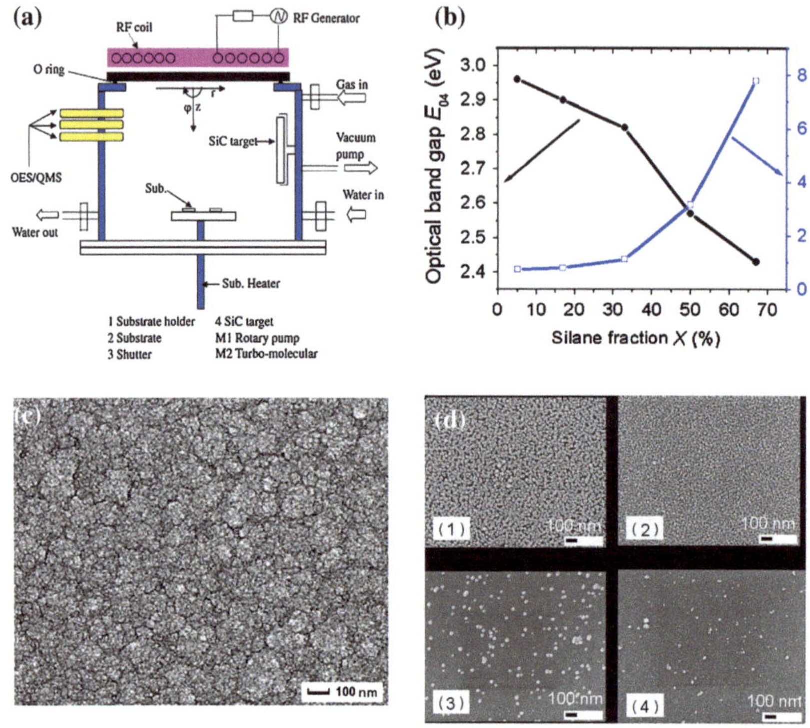

Fig. 8.2 **a** ICP-CVD chamber, reproduced from [25] by permission of IOP Publishing. All rights reserved, copyright 2007. **b** Growth rate of SiC films with SiH_4 fraction, reprinted with permission from [26] [1] copyrights 2006 and 2007, American Institute of Physics. **c** SiC film deposited at target power of 800 W, reproduced from [25] by permission of IOP Publishing. All rights reserved, copyright 2007. **d** SiC QDs deposited at different SiC target powers, reprinted with permission from [27], copyrights 2007, American Institute of Physics

and bias voltage, as well as pressure. One of the advantages of the ICP-CVD technique is that it provides a separate control over the plasma parameters and the rate of material deposition. Also, the substrate temperature is relatively low (< 600 °C) in an ICP-CVD process.

8.3.1 Fabrication of Nanocrystalline SiC Film by ICP-CVD Using Silane

Cheng et al. showed that homogeneous nanocrystalline 3C–SiC films can be synthesized in a process using a $SiH_4 + CH_4 + H_2$ environment. In this experiment, the flow rates of CH_4 and H_2 were kept constant (4 and 60 sccm, respectively), whereas SiH_4 flow was varied from 0.2 to 8 sccm (5–67 %). All other parameters such as pressure, inductive RF power (2,000 W), substrate temperature, and deposition time (100 min) were kept constant. These experiments have shown that higher SiH_4 fractions lead to the higher growth rates and lower optical bandgaps (Fig. 8.2b) [26]. The amount of silane defines the composition, growth rate, optical bandgap and binding energy of the film.

During the deposition, different bonds between the atoms can be established, such as Si–Si, C–C, Si–H, C–H, and Si–C. The Si–C bond is very strong, so it is the most favourable for 3C–SiC structure. High amount of hydrogen leads to etching the weak bonds in favour of stronger Si–C ones, and thus ensures the formation of a homogeneous and crystalline structure.

In similar study [29], different structure of SiC film was obtained by changing the amount of CH_4 (from 2 sccm to 96 sccm) in the plasma. Flow rates of SiH_4 and H_2 were kept constant (2 sccm and 50 sccm, respectively) at 500 °C, pressure of 30 m Torr and discharge power of 2,000 W. A CH_4/SiH_4 ratio defines the amount of the carbon in the resulted $Si_{1-x}C_x$ structure, i.e., more carbon in the film was found at the higher ratio. For example, for the ratios of 1, 2 and 48, the value of x was 0.17, 0.49, and 0.59, respectively. For $x = 0.17$ and $x = 0.59$, the structures were polycrystalline Si and amorphous SiC films, respectively. With $x = 0.49$, the structure was 3C–SiC film.

Hydrogenated amorphous $Si_{1-x}C_x$ structure can be fabricated using ICP-CVD at relatively low power densities ($\sim 0.025\,\mathrm{Wcm^{-3}}$) without hydrogen. In this process, methane and silane were used as precursors. Due to the dissociation in plasma, some hydrogen radicals were formed and then deposited during the growth. The carbon content in these structures can be varied from 0.09 to 0.71 by changing the methane/silane ratio. Silane plays an important role in terms of growth rate, and there was no growth in 100 % CH_4 (i.e., without silane). Electronic and optical properties of the structure are highly affected by the film composition.

8.3.2 Fabrication of Nanocrystalline SiC QDs by ICP-CVD Without Silane

High-density ICP ($Ar + H_2$) and RF magnetron sputtering of near-stoichiometric SiC targets were used to grow SiC films [25] and SiC QDs [27]. In this case the source of Si was the SiC target (not SiH_4 gas). A large amount of hydrogen ions in the plasma etched the weak bonds on the surface and thus resulted in the formation of the most stable bonding structure, i.e., crystalline SiC (3C-SiC) film or QDs.

An ICP with the discharge power of 800 W resulted in the formation of crystalline SiC films (Fig. 8.2c), whereas deposition without the ICP led to the formation of amorphous films. In this process, the surface temperature was about 400 °C. In this case, uniform amorphous particles with an average size of 18 nm covered the surface. On the other hand, when ICP power was applied to the system, much smaller highly crystalline particles with an average size of 3 nm were formed. In the ICP, Si and

C radicals were deposited onto the substrate, whereas other ions and radicals (such as hydrogen ions) caused etching of some bonds, and thus only the strongest bonds survived. Thus, ICP discharge leads to the formation of smaller crystalline particles. It was also found that the increase of the SiC target power from 100 to 250 W (Fig. 8.2d) leads to an increase in the growth rate of QDs from 10 to 58 Å/min. The increased power intensifies the sputtering of the target and hence, increases the rate of deposition. As a result, smaller and more size-uniform arrays of quantum dots are formed.

The working gas pressure changes the uniformity and size of QDs. Two different mechanisms control the structure at different pressure regimes:

1. At lower (less than 1 Pa) pressure, QD size increases with pressure and a uniform array of QDs is formed;
2. At higher (more than 1 Pa) pressure, QD size decreases with pressure and the QDs are non-uniform.

The dependence of the uniformity on the working pressure can be related to the mobility of adatoms on the surface. At higher pressures, the atoms evaporated from the target lose their kinetic energy in the plasma before reaching the substrate. Eventually, this results in the formation of non-uniform structure. At lower pressure, adatoms have more energy and the final structure is more homogenous, with smaller QDs.

8.4 Plasma Enabled Fabrication of SiC Nanostructures Using Quartz

As described in the introduction, MWPE-CVD can effectively etch Si-based materials. Thus, Si atoms can cause contamination during the MWPE-CVD fabrication of nanomaterials. However, whilst this is a drawback for the fabrication of nanodiamonds, it may be advantageous for the fabrication of other nanomaterials, since this process does not involve harmful silane (SiH_4). In this section, we will describe an efficient, simple, safe and SiH_4-free method which is based on the use of quartz as a Si source to fabricate SiC nanowires and nanorods with and without AAOM.

8.4.1 Fabrication of SiC Nanorods and Nanodiamonds Using Quartz Without AAOM

In these experiments, quartz substrates were chemically cleaned. Then, the cleaned surface was treated using a MWPE-CVD process for 1h in a plasma ignited in a mixture of H_2 (494 sccm) and CH_4 (3.5 sccm). The sample topography and structure has been studied using scanning electron microscopy (SEM), transmission electron microscope (TEM) and Raman spectroscopy techniques. Figure 8.3 illustrates a plain quartz surface after processing in H_2 and CH_4 plasmas in the MWPE-CVD system at 1,200 W for 30 min. It is clear that whilst nanodiamond particles were formed, the quartz surface was also etched and this can lead to the contamination of the nanodiamond particles.

Figure 8.4 shows the SEM images of the nanorods fabricated in the centre of the quartz substrates. During the treatment, quartz substrates were directly exposed to the plasma. It can be clearly seen that the diameters of the nanorods are around 100 nm, with the lengths ranging between 1 and 2 μm. The results of the Raman analysis shown in Fig. 8.5 confirmed the SiC structure of the nanorods, as evidenced by the presence of the typical SiC TO mode at 793 cm^{-1} and LO mode at 963 cm^{-1} in the spectrum.

Transmission electron microscopy was used to investigate the crystalline structure of the fabricated nanorods. Figure 8.6a–c are TEM images illustrating the nanorod

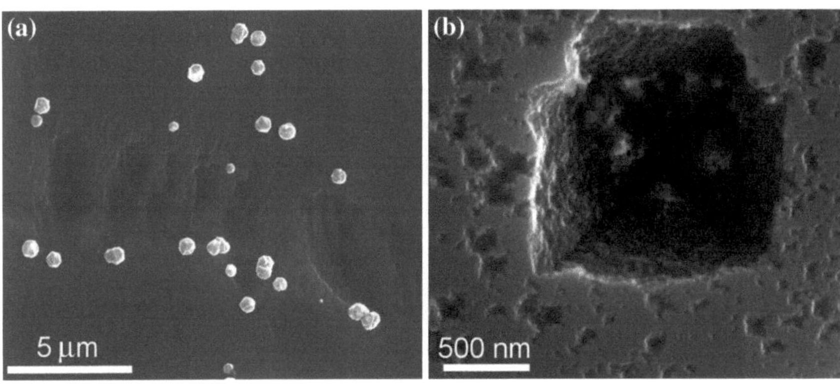

Fig. 8.3 SEM images of the diamond particles on the quartz (**a**) and etched quartz surfaces (**b**)

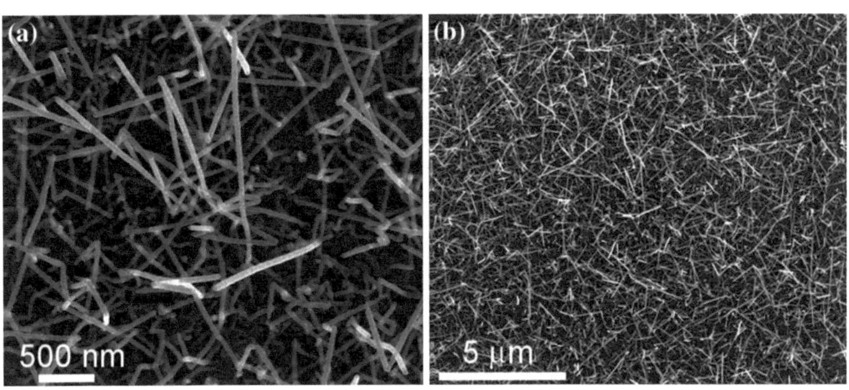

Fig. 8.4 SEM images of nanorods in the centre of quartz substrates at **a** high and **b** low magnifications

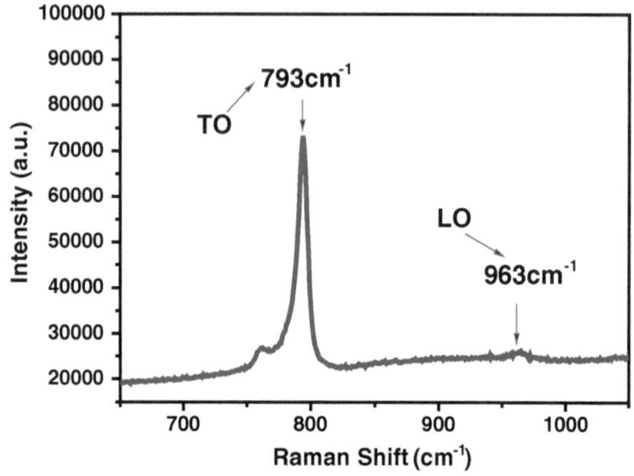

Fig. 8.5 Raman spectrum of the SiC nanorods

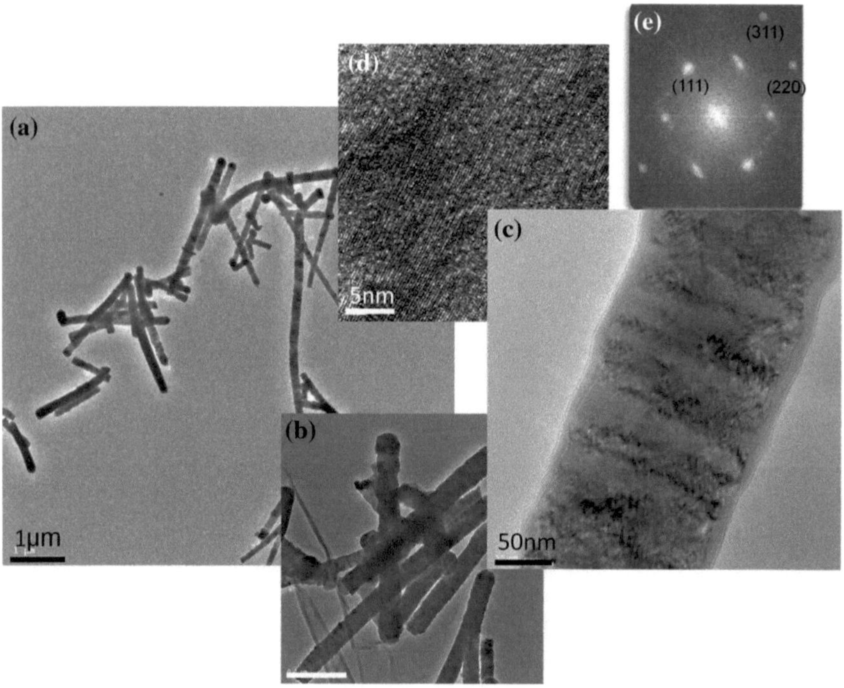

Fig. 8.6 **a–c** TEM images at different magnifications. **d** High-quality single-crystalline structure of the nanorod. **e** Fast Fourier Transform pattern taken from the sample shown in (**d**)

structure at different magnifications. It is clear that the nanorods have very uniform size with an average diameter of 100 nm. The nanorods have a high-quality single-crystalline structure (Fig. 8.6d). The lattice spacing is around 0.251 nm, corresponding to the (111) atomic plane of SiC. Figure 8.6e is the Fast Fourier Transform (FFT) pattern taken from the sample shown in Fig. 8.6d. This pattern evidences the presence of the (111), (220) and (311) lattice spacing of the SiC structure.

TEM images shown in Fig. 8.7a illustrate the cap on the top of nanorods. These caps indicate that the nanorods grow by an vapour-liquid-solid (VLS) mechanism. Diffraction patterns taken during TEM analysis indicate that these caps are a silica crystal (Fig. 8.7b). The results of the TEM analysis demonstrate that the SiC nanostructures can be synthesized without SiH_4, and the mechanism of the nanorod growth is related to VLS process where SiO_2 crystal acts as a catalyst. In order to confirm the role of SiO_2, an additional experiment was carried out to study the mechanism of the nanorod formation and investigate the influence of the time of plasma treatment on the structure.

Figure 8.8 shows nanostructures grown at three different plasma treatment times, namely 30, 45, and 60 min, respectively. Based on these experiments, the following growth mechanism may be proposed: first, quartz was etched by the ionized hydrogen, and a number of small SiO_2 particles was formed (Fig. 8.8a). These SiO_2 particles act as a catalyst enhancing the nanorod formation. Next, Si atoms together with the carbon radicals diffuse into SiO_2 and form the initial nanorods (Fig. 8.8b). It can be clearly seen that longer and denser nanorods are formed in 60 min (Fig. 8.8c).

This process is much simpler than conventional fabrication methods used to synthesize SiC nanostructures. In order to understand the role of AAO templates, the membranes were used to grow ultra-thin SiC nanowires under the same conditions.

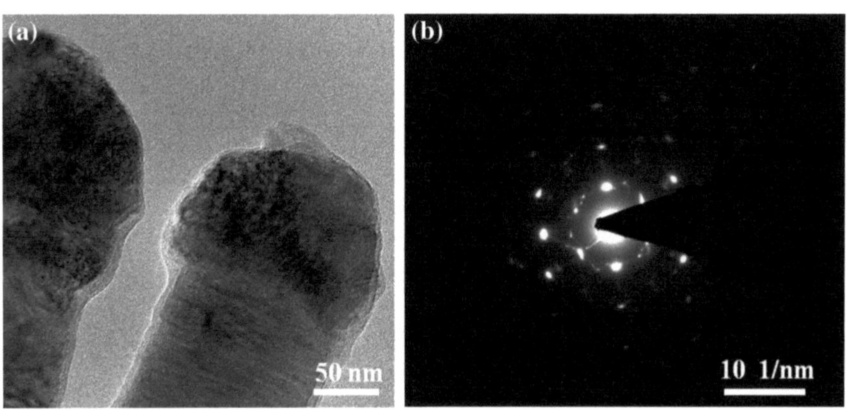

Fig. 8.7 **a** TEM image of the top of nanorods **b** diffraction pattern of SiC nanorods top

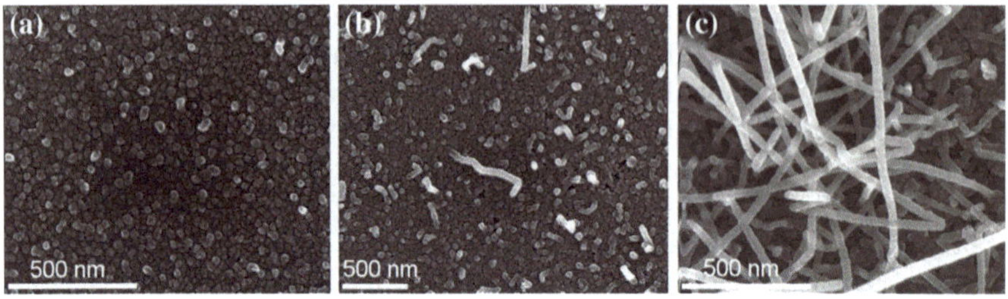

Fig. 8.8 SEM images of the array of nanorods grown for 30 min (**a**), 45 min (**b**), and 60 min (**c**), respectively

8.4.2 Fabrication of Ultra-Thin SiC Nanowires on AAO Template

As described above, the plasma-based techniques are able to effectively modify the surfaces, rather than deep channels of the AAO template. Here we show a very simple but efficient method of the SiC nanowires fabrication by using quartz as a source of Si. By exposing AAO templates on quartz to a $CH_4 + H_2$ plasma, long SiC nanowires with a core-shell structure were grown. The results of TEM, SEM, EDX and AFM have demonstrated that these nanowires have a single-crystalline SiC core in the center and a thin amorphous AlSiC shell. This is a fast, simple, environmentally friendly and SiH_4-free process without involving any pre-deposited metal catalysts.

To use quartz as a source of Si and increase the stability of the AAO template during plasma treatment, the templates were fabricated on quartz substrates. Initially, a quartz substrate was cleaned by acetone and ethanol for 5 and 10 min using ultrasonic bath, respectively. Then the sample was dried under nitrogen gas flow. A layer (1 μm) of high-purity (99.999 %) aluminum was deposited onto the cleaned quartz using e-beam evaporation. After that, an anodization process was used to fabricate the

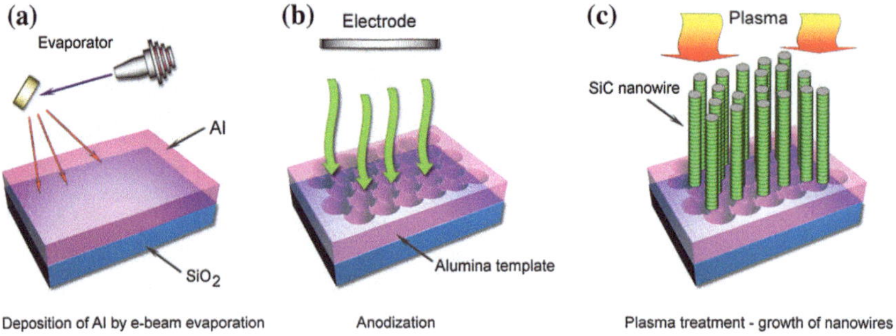

Fig. 8.9 Schematic depiction of the SiC nanowire synthesis process. **a** Deposition of Al by e beam evaporation onto quartz, **b** AAO fabrication using an electrochemical cell and **c** SiC nanowire growth on the top of AAO surface in the PECVD system. Reprinted with permission from [13]. Copyright (2012) American Chemical Society

templates [30]. The fabricated templates were then treated in a MWPE-CVD reactor for 30 min with gas supply rates of 3.5 (for H_2) and 494 (for CH_4) sccm, respectively. The substrate temperature was about 800 °C, and the microwave power was 1,200 W. No precursor was used during the growth. Due to the presence of the reductive H_2 and powerful plasma, Si atoms were easily extracted from the substrate and then reacted with the carbon radicals from CH_4 [31]. As a result, this reaction led to the formation of SiC. A schematic of this process is shown in Fig. 8.9.

Figure 8.10a is a SEM image of the tilted alumina template before the plasma treatment. It is clear that the nanopores are uniform, ordered, and parallel to each other. Figure 8.10b presents a FIB-cutted cross-sectional TEM image of the template. Black thin layers in this image are the protective gold and platinum films. It can be

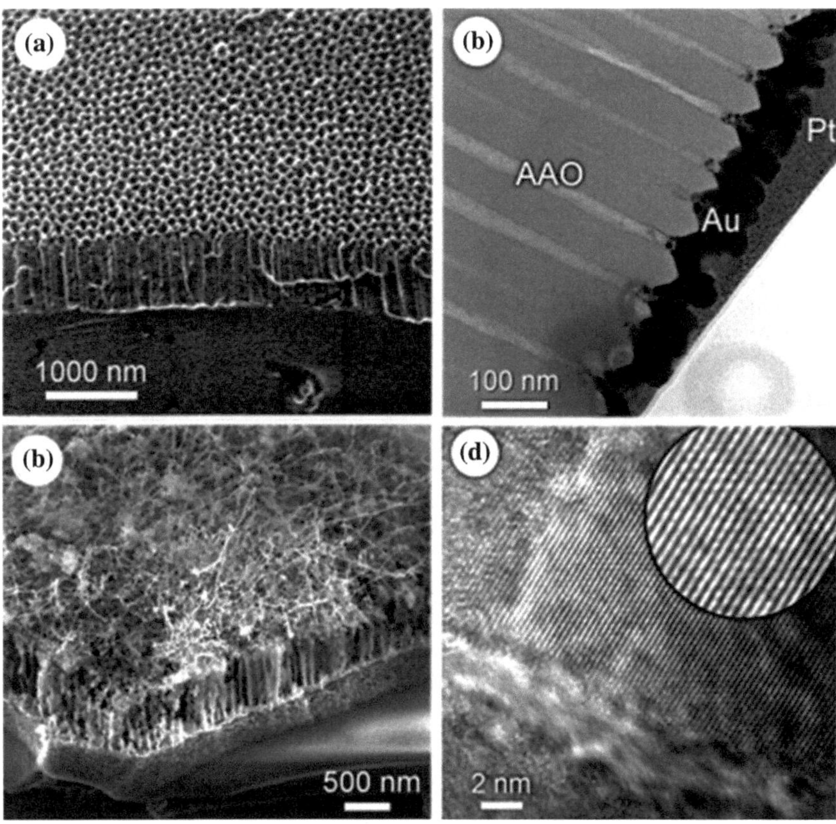

Fig. 8.10 **a** Representative SEM image of the as-prepared alumina template on quartz, **b** FIB-cut cross-section TEM image (*black thin layers* are gold and platinum layers deposited to prevent damage to the template during the FIB milling), **c** SEM image of the template with nanowires on the top surface after the deposition in the PECVD reactor for 30 min, **d** HRTEM image of the nanowire, inset shows a crystalline structure. Adapted with permission from [13]. Copyright (2012) American Chemical Society

clearly seen that the walls of channels are very thin (not exceeding 30 nm) and become even thinner near the top surface of the template. Significantly, the channel walls taper creating openings, and as a result, an ordered array of sharp nanocones was formed (periodic triangles in the cross-section). Figure 8.10c shows an SEM image of the top surface of the template after nanowire growth.

Interestingly, no nanowires were found either on the side or inside surfaces of the template, but only on a top face of the template which was in direct contact with the plasma. High-resolution TEM analysis shows that these nanowires have a crystalline structure (Fig. 8.10d).

To confirm that the nanowires were nucleated and grown on the upper surface of the template (not in the nanopores), we have partially removed the nanowires and then examined the obtained structure by SEM. It can be seen clearly in these SEM images that the nanopores are 'empty' and the nanowires did not grow from the inside of channels of AAO templates. Figure 8.11 are the low- and high-resolution SEM images of the alumina template surface after removal of most part of the nanowires. It is evident that the nanopores are empty and the edge of channels has been modified after the plasma treatment where nanowire grows.

In order to investigate the crystal structure of these nanowires, TEM analysis was performed. Figure 8.12a shows a TEM image of the entangled nanowires collected from the template. It can be seen that the length-to-radius ratio of the nanowires can reach 100:1. Also, the nanowire diameters are in a rather narrow distribution (see inset in Fig. 8.12a). Figure 8.12b is a selected area electron diffraction pattern taken from a number of nanowires. It shows two concentric circles consisting of individual diffraction spots, indicating the highly-crystalline structures. These two concentric circles identify (111) and (220) atomic planes with the spacing of 0.251 and 0.153 nm, respectively, thus confirming a crystalline structure of the nanowires. The arrangement of the concentric circles of individual diffraction spot is highlighted by red rings. Moreover, TEM image of the 100 nm section of nanowire confirms

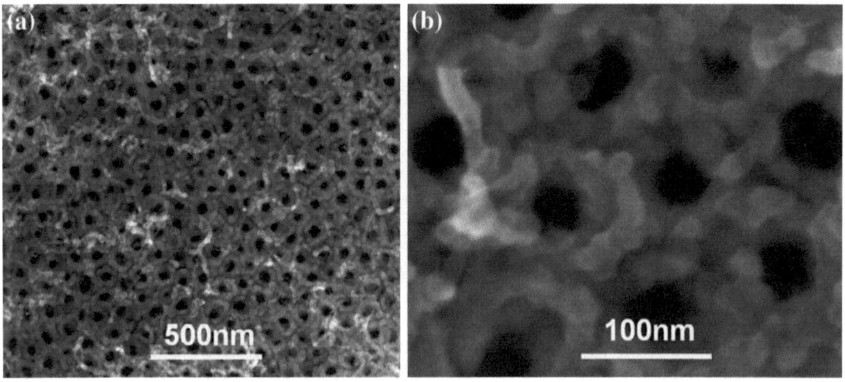

Fig. 8.11 Low- (**a**) and high- (**b**) resolution SEM images of the alumina template surface after removal of the most part of the nanowires

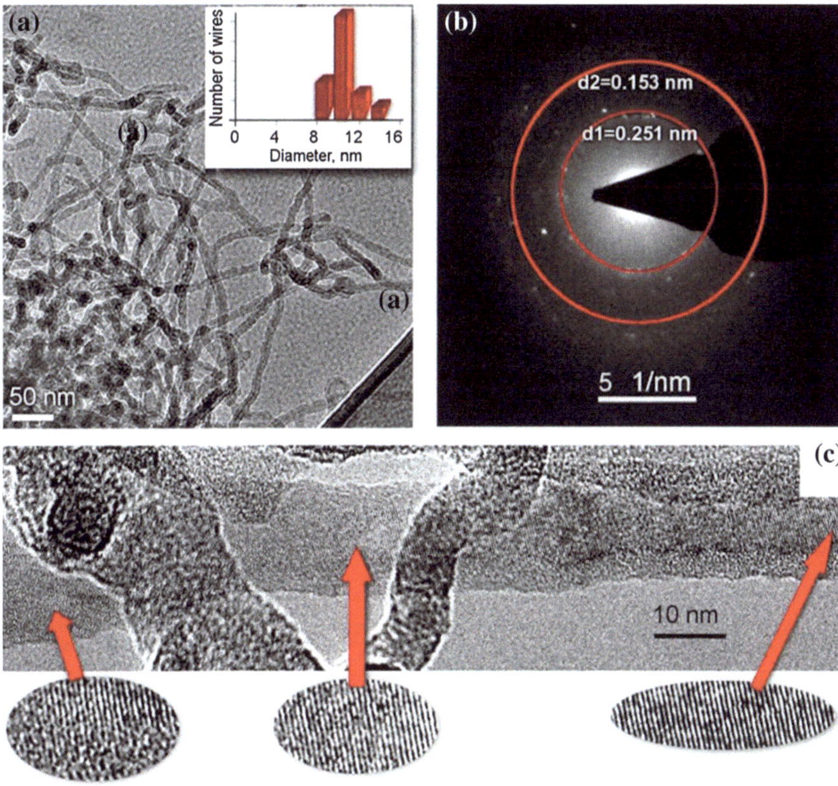

Fig. 8.12 **a** TEM image of the entangled net of nanowires (*inset* shows the distribution of the nanowire diameters), **b** Diffraction pattern recorded from the nanowire bundle. The *first* and the *second rings* correspond to (111) and (220) planes of –SiC, respectively, **c** TEM image of a long nanowire section showing the single-crystalline structure. Adapted with permission from [13]. Copyright (2012) American Chemical Society

that the nanowire has a high quality single-crystalline structure (Fig. 8.12c). X-ray diffraction (XRD) measurements were also used to confirm the structure of SiC. Figure 8.13 revealed that the fabricated nanowires exibit Al_4SiC_4 and SiC crystal structure.

To reveal the chemical composition and distribution of elements across the nanowires, EELS analysis was performed. The plot shown in Fig. 8.14b shows an EELS line scan recorded across the nanowire shown in Fig. 8.14a (yellow trace). This analysis reveals that both Si and carbon are present throughout the cross-section of the nanowire. Si concentration reaches the maximum near the outer shell, while the carbon concentration decreases. Furthermore, EELS analysis has shown that Si and carbon have approximately equal concentrations in the crystalline core, thus forming near-stoichiometric SiC. More interestingly, aluminium is concentrated on both sides of SiC nanowire. As a result, the HRTEM, EELS and other techniques demonstrated

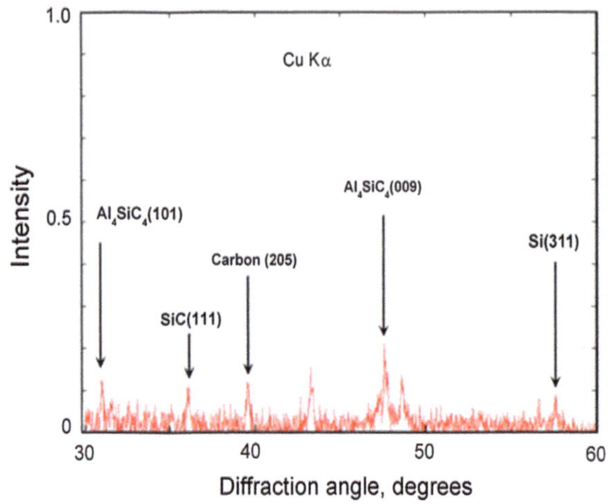

Fig. 8.13 XRD analysis of the nanowires. Adapted with permission from [13]. Copyright (2012) American Chemical Society

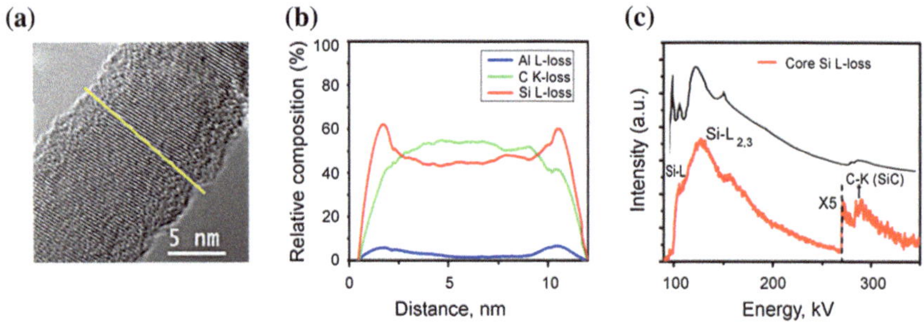

Fig. 8.14 **a** HRTEM image of the nanowire, revealing the <111> crystal planes of $\beta-SiC$. A core of crystalline nanowire is encapsulated in the amorphous shell with the thickness of about 1 nm, **b** An EELS line scan spectrum of the nanowire recorded along the yellow line shown in (**a**), **c** Si electron energy L-loss spectra recorded from the core of the nanowire. A typical SiC EELS spectrum (*black line*) is shown above the spectra. Reprinted with permission from [13]. Copyright (2012) American Chemical Society

that nanowires comprise crystalline SiC core and an amorphous AlSiC shell. Hence, the structure is proved to be SiC/AlSiC nanowire.

Based on SEM, HRTEM, EELS and EDX analysis of the fabricated nanowires, the following growth mechanism was proposed. First, both hydrogen and methane were ionized into radicals during the plasma-CVD growth phase. The charged hydrogen radicals react with the sharp cones of the alumina template resulting in the formation of a reduced aluminum phase, and possibly even small metallic aluminum droplets. Once aluminum droplets have been formed, Si ions and carbon radicals diffuse into the aluminum droplet. Once the concentrations of Si and carbon inside the metallic droplet reach saturation, SiC precipitates into a crystalline nanowire. This nanowire continues to grow until the aluminum droplet vanishes. During the growth of the wires, the presence of the aluminum is presumably from the alumina template, which forms the seed droplet. The aluminum vapor can condense along the sides of the nanowire, and consequently, form the amorphous sheath, as observed in the HRTEM images. Since the thickness of the shell is of the order of a nanometer, we assume this is a shell formed to reduce the surface energy of the nanowires.

8.5 Conclusion

In this chapter, we have discussed the fabrication of SiC films and various nanostructures by inductively coupled plasma CVD and microwave plasma-enhanced CVD techniques. Particularly, the controlled growth of SiC nanostructures with and without toxic silane was relieved. Moreover, we have considered the use of quartz substrates as a source of silicon in plasma-enhanced growth techniques. A separate section was devoted to the use of aluminium oxide templates for the formation of arrays of SiC nanodots and nanowires. The mechanism for the formation of SiC nanostructures was also discussed.

References

1. Hatty, V., Kahn, H., Trevino, J., Zorman, C.A., Mehregany, M., Ballarini, R., Heuer, A.H.: Fracture toughness of low-pressure chemical-vapor-deposited polycrystalline silicon carbide thin films. J. Appl. Phys. **99**, 013517 (2006)
2. Yonenaga, I., Hoshi, T., Usui, A.: High-temperature hardness of bulk single-crystal gallium nitride-in comparison with other wide-gap materials. J. Phys. Condens. Matter. **12**, 10319 (2000)
3. Ajayan, P.M., Ebbesen, T.W., Ichihashi, T., Iijima, S., Tanigaki, K., Hiura, H.: Opening carbon nanotubes with oxygen and implications for filling. Nature **362**, 522–525 (1993)
4. Fissel, A., Schroter, B., Richter, W.: The field emission properties of silicon carbide whiskers grown by CVD. Appl. Phys. Lett. **66**, 3182–3184 (1995)
5. Zhang, X.F., Sixta, M.E., De Jonghe, L.C.: Grain boundary evolution in hot pressed ABC SiC. J. Am. Ceram. Soc. **83**, 2813–2820 (2000)
6. Wong, E.W., Sheehan, P.E., Lieber, C.M.: Nanobeam mechanics: elasticity, strength, and toughness of nanorods and nanotubes. Science **277**, 1971–1975 (1997)
7. Cai, W., Chettiar, U.K., Kildishev, A.V., Shalaev, V.M.: Designs for optical cloaking with high-order transformations. Opt. Express **16**, 5444–5452 (2008)
8. Addamiano, A.: Preparation and properties of 2H SiC crystals. J. Cryst. Growth **58**, 617–622 (1982)
9. Zhou, D., Seraphin, S.: Production of silicon carbide whiskers from carbon nanoclusters. Chem. Phys. Lett. **222**, 233–238 (1994)
10. Volotskova, O., Fagan, J.A., Huh, J.Y., Phelan, F.R., Shashurin, A., Keidar, M.: Tailored distribution of single-wall carbon nanotubes from arc plasma synthesis using magnetic fields. ACS Nano **4**, 5187–5192 (2010)

11. Seeger, T., Kohler-Redlich, P.: Synthesis of nanometer-sized SiC whiskers in the arc-discharge. Adv. Mater. **12**, 279–282 (2000)
12. Kamlag, Y., Goossens, A., Colbeck, I., Schoonman, J.: Laser CVD of cubic SiC nanocrystals. Appl. Surf. Sci. **184**, 118–122 (2001)
13. Fang, J.-H., Aharonovich, I., Levchenko, I., et. al.: Plasma-enabled growth of single-crystalline SiC/AlSiC core-shell nanowires on porous alumina templates. Cryst. Growth Des. **12**, 2917–2922 (2012)
14. Ostrikov, K., Cvelbar, U., Murphy, A.B.: Plasma nanoscience: setting directions, tackling grand challenges. J. Phys. D Appl. Phys. **44**, 174001 (2011)
15. Ostrikov, K., Seo, D.H., Mehdipour, H., Cheng, Q., Kumar, S.: Plasma effects in semiconducting nanowire growth. Nanoscale (2012, in press). doi: 10.1039/C1NR10658A
16. Gao, H., Mu, C., Wang, F., Xu, D., Wu, K., Xie, Y., Liu, S., Wang, E., Xu, J., Yu, D.: Field emission of large-area and graphitized carbon nanotube array on anodic aluminum oxide template. J. Appl. Phys. **93**, 5602–5605 (2003)
17. Sui, Y.C., Acosta, D.R., Gonzalez-Leon, J.A., Bermudez, A., Feuchtwanger, J., Cui, B.Z., Flores, J.O., Saniger, J.M.: Structure, thermal stability, and deformation of multibranched carbon nanotubes synthesized by CVD in the AAO template. J. Phys. Chem. B **105**, 1523–1527 (2001)
18. Zong, R.L., Zhou, J., Li, B., Fu, M., Shi, S.K., Li, L.T.: Optical properties of transparent copper nanorod and nanowire arrays embedded in anodic alumina oxide. J. Chem. Phys. **123**, 094710 (2005)
19. Liu, C.H., Yiu, W.C., Au, F.C.K., Ding, J.X., Lee, C.S., Lee, S.T.: Electrical properties of zinc oxide nanowires and intramolecular junctions. Appl. Phys. Lett. **83**, 3168–3170 (2009)
20. Li, Z., Zhang, J., Meng, A., Guo, J.: Large-area highly-oriented SiC nanowire arrays: synthesis, raman, and photoluminescence properties. J. Phys. Chem. B. **110**, 22382–22386 (2006)
21. Levchenko, I., Rider, A. E., Ostrikov, K.: Plasma-aided nanofabrication: where is the cutting edge? J. Appl. Phys. **101**, 044306 (2007)
22. Rider, A.E., Levchenko, I., Ostrikov, K.: Plasma/ion-controlled metal catalyst saturation: enabling simultaneous growth of carbon nanotube/nanocone arrays. Appl. Phys. Lett. **90**(19), 193110 (2007)
23. Cheng, Q., Xu, S., Long, J.D., Ni, Z.H., Rider, A.E., Ostrikov, K.: High-rate, low-temperature synthesis of composition controlled hydrogenated amorphous silicon carbide films in low-frequency inductively coupled plasmas. J. Phys. D. Appl. Phys. **41**, 055406 (2008)
24. Rider, A. E., Ostrikov, K., Levchenko, I.: Tailoring the composition of self-assembled $Si_{1-x}C_x$ quantum dots: simulation of plasma/ion-related controls. Nanotechnology **19**(35), 355705 (2008)
25. Cheng, Q.J., Long, J.D., Xu, S.: Chemically active plasmas for deterministic assembly of nanocrystalline SiC film. J. Phys. D. Appl. Phys. **40**, 2304 (2007)
26. Cheng, Q., Xu, S.: Low-temperature synthesis of homogeneous nanocrystalline cubic silicon carbide films. J. Appl. Phys. **102**, 056101 (2007)
27. Cheng, Q.J., Long, J.D., Xu, S.: Growth dynamics and characterization of SiC quantum dots synthesized by low-frequency inductively coupled plasma-assisted RF magnetron sputtering. J. Appl. Phys. **101**, 094304 (2007)
28. Levchenko, I., Ostrikov, K., Keidar, M., Xu, S.: Deterministic nanoassembly: neutral or plasma route? Appl. Phys. Lett. **89**, 033109 (2006)
29. Cheng, Q., Xu, S., Long, J., Huang, S., Guo, J.: Homogeneous nanocrystalline cubic silicon carbide films prepared by inductively coupled plasma chemical vapor deposition. Nanotechnology **18**, 465601 (2007)
30. Fang, J.H., Spizzirri, P., Cimmino, A., Rubanov, S., Prawer, S.: Extremely high aspect ratio alumina transmission nanomasks: their fabrication and characterization using electron microscopy. Nanotechnology **20**, 065706 (2009)
31. Xu, S., Levchenko, I., Huang, S.Y., Ostrikov, K.: Silicon on silicon: self-organized nanotip arrays formed in reactive Ar+ H2 plasmas. Appl. Phys. Lett. **95**, 111505 (2009)

Chapter 9
Catalyst-Free Chemical Vapor Deposition for Synthesis of SiC Nanowires with Controlled Morphology

Jyoti Prakash, Sunil Kumar Ghosh and Dakshinamoorthy Sathiyamoorthy

Abstract Chemical vapor deposition hydrogen reduction of methyltrichlorosilane (MTS) is most prominent method for production of silicon carbide (SiC) nanowires with controlled morphology. In a typical SiC nanowire synthesis process the cracking of MTS is carried out in reducing atmosphere of hydrogen using chemical vapor deposition technique at high temperature and normal atmospheric pressure. Taguchi method is very useful to design experiments specially in the cases where large numbers of variables are to be considered. This statistical method has been used to design the experiments to get the optimum parameters for bulk production of silicon carbide wires of uniform diameter in nanometer range. Further the effect of different parameters on the morphology of SiC deposit has been discussed. XRD and SEM analysis showed the growth of crystalline β-SiC wires having different morphology. From the analysis of variance (ANOVA) of data it has been observed that growth temperature and hydrogen to MTS ratio in carrier gas are the two important parameters which decide the final growth morphology of SiC deposition. At higher temperature ($\geq$1400 °C), the SiC nuclei prefer to grow as SiC grains rather than wires. The optimum deposition conditions have been obtained by analyzing Signal to Noise (S/N) ratio corresponding to lowest deposition rate and minimum growth diameter of SiC wires. The optimum deposition conditions have been used for uniform diameter growth of SiC nanowires, smoothness of the surface, and homogeneous growth of SiC on the surface. It has been observed that the hydrogen to MTS flow rate ratio value should be above 20 for the growth of SiC wires of nanometer diameter. The deposition temperature for the growth of crystalline SiC wires should be 1100–1300 °C. The total flow rate of carrier gas comprising of argon and hydrogen should be in moderate range for particular hydrogen to MTS ratio. The effect of H_2/MTS

J. Prakash (✉) · D. Sathiyamoorthy
Powder Metallurgy Division, Bhabha Atomic Research Centre Trombay,
Mumbai 400085, India
e-mail: jprakash@barc.gov.in

S. K. Ghosh
Bio Organic Division, Bhabha Atomic Research Centre Trombay, Mumbai 400085, India

H. Li et al. (eds.), *Silicon-based Nanomaterials*,
Springer Series in Materials Science 187, DOI: 10.1007/978-1-4614-8169-0_9,
© Springer Science+Business Media New York 2013

mole ratio on morphology of SiC deposition by varying H_2/MTS mole ratio from 0 to ~80 has been discussed in detail. This detail process study has given a new perspective to produce SiC nanowires of high purity and homogeneous diameter by a simple atmospheric pressure CVD method without using a metallic catalyst. Even manipulation of growth parameters can be done to get desired morphology of SiC deposit.

9.1 Introduction

Since last decade one-dimensional nanostructures such as wires, rods, belts, and tubes have received continued interest as a result of their peculiar and fascinating properties, in comparison to their bulk counterparts [1, 2]. It is generally accepted that one dimensional nanostructures provide a good system to investigate the dependence of electrical and thermal transport or mechanical properties on dimensionally and size reduction. They are also expected to play an important role as both interconnects and functional units in fabricating electronics, optoelectronics, electrochemical, and electromechanical devices with nanoscale dimensions. Initially the development in the field of one dimensional nanostructures with respect to quantum dots and wells was slow due to difficulties associated with the synthesis of one-dimensional nanostructures with controlled morphology, purity and chemical composition. There were several issues related to the fabrication of such one-dimensional nanostructures. Since past decade, a variety of synthesis methods have been used for the synthesis of such nanostructures. In these methods several issues related to the synthesis of one-dimensional structure have been discussed. The strategies based on chemical synthesis routes [3] have advantages of diversity and low cost over the advance technique such as nanolithographics [4], proximal probe patterning, and X-ray or UV lithography [5, 6]. Thus, nanowires of elements, oxides, nitrides, carbides, and chalcogenides, [7, 8] have been generated by employing various chemical strategies. One of the crucial factors in the synthesis of nanowires is the control of composition, size, and crystallinity. Among the chemical methods employed, some are based on vapor phase techniques, while others are solution techniques. Thus, techniques involving chemical vapor deposition (CVD), precursor decomposition, solvothermal, hydrothermal, and carbothermal methods have been widely employed for synthesis of such one dimensional nanostructure.

The crystallinity [9] of one-dimensional nanostructure is the important aspect during its evolution from a vapor, a liquid, or a solid phase. Nucleation and growth are the two important process associated with the overall growth of one-dimensional nanostructure. As the concentration of the building units (atoms, ions, or molecules) of a solid becomes sufficiently high, they aggregate into small nuclei or clusters through homogeneous nucleation. These clusters serve as seeds for further growth to form larger clusters. These nuclei/clusters need to be supplied at a well-controlled rate in order to obtain crystals having a homogeneous composition and controlled morphology. While developing a synthesis strategy for controlled dimension nanostructure growth one should take care of the important issues like simultaneous control over dimensions, uniformity and shape. Based on the control over different growth parameters several synthetic strategies have been developed to grow one-dimensional nanostructure. Generally these synthesis strategies can be classified as: (i) anisotropic crystallographic structure facilitated growth of solid one-dimensional nanowire; (ii) reduction of symmetry of a seed nuclei by introduction of a solid–liquid interface; (iii) one dimensional morphological template based growth of one-dimensional nanostructures; (iv) use of supersaturation control to modify the growth habit of a seed nuclei; (v) use of capping agents to kinetically control the growth rates of the various facets of a seed; (vi) self-assembly of zero-dimensional (0D) nanostructures, and (vii) size reduction of one-dimensional microstructures.

Among these strategies the growth of nanowires from the isotropic medium (vapor or solution) is relatively simple and straight forward if the solid material has a highly anisotropic crystal structure. For many solids that have isotropic crystal structures (e.g. almost all metals have face centered cubic lattice), symmetry breaking is required in the nucleation step to induce anisotropic growth. In the other strategies (except first one) lowering the symmetry of seed/nuclei or the environment around the seed/nuclei is done to produce one-dimensional nanostructure. In principle these methods are suitable for all types of solid materials irrespective of their isotropic or anisotropic crystallographic structure. In addition the feed molecules also need to be supplied at a well-controlled rate in order to obtain crystals having a homogeneous composition and uniform morphology. We will briefly describe these methods based on its mechanism for developing effective synthesis strategy for one-dimensional nanostructure growth.

(i) **Anisotropic crystal based growth of one-dimensional nanostructure**
Anisotropic crystal structure of the solid material (Fig. 9.1) dictate the growth of solid materials in one-dimensional nanostructure [10–12]. Poly (sulphur nitride) is the best example in this class which natuarally grow as nanowires. Other materials, such as selenium [13, 14], tellurium [15] molybdenum chalcogenides [16, 17], biological systems such as cellulose [18] and collagen [19] are easily obtained as nanowires. They all have anisotropic bonding, which dictates the crystallization to occur along the c-axis, favoring the stronger covalent bonds over the relatively weak van der Waals forces between the chains.

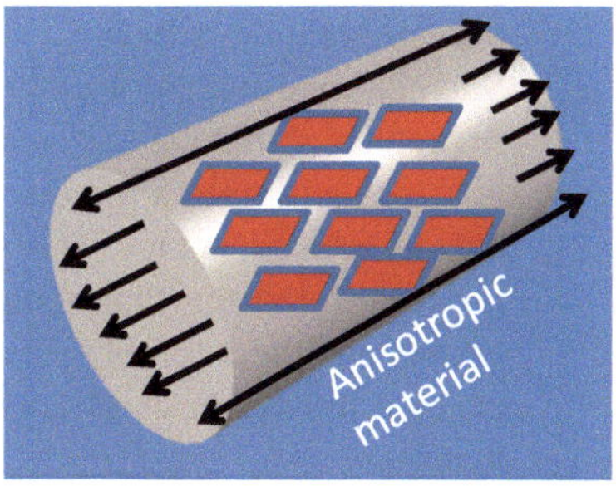

Fig. 9.1 Schematic representation of anisotropic crystal based one-dimensional crystal growth

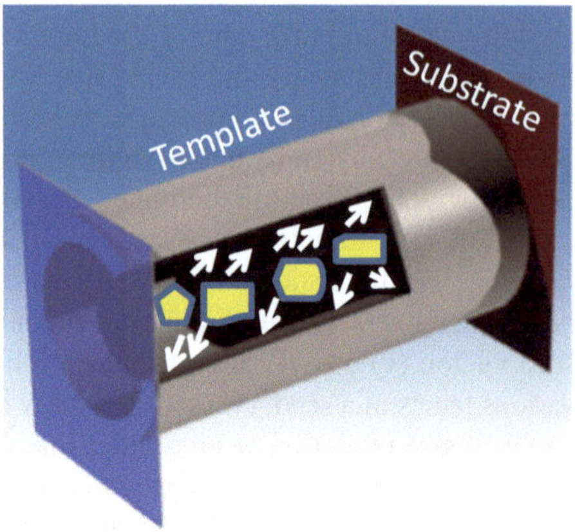

Fig. 9.2 Schematic representation of template based directional growth

(ii) **Templates having one-dimensional morphologies based growth of one-dimensional nanostructures**
Template-directed growth (Fig. 9.2) of one-dimensional nanostructure is a versatile technique. The template directed synthesis provides a simple, high yield, and economical procedure. There are different type of template methods which can be used for growth of one-dimensional nanostructure. Features/pattern on solid surface [20–22], channels in porous materials [23–28], mesophase structures self-assembled from surfactants [29, 30], and existing nanostructure template are the widely used methods [31–33]. The major drawbacks of template method are: (i) the nanostructures generated using template method are often polycrystalline, (ii) the quality of structures that can be produced in each run of synthesis is relatively limited.

(iii) **Solution–liquid–solid process**
Solution–liquid–solid (SLS) method (Fig. 9.3) was developed by Buhro and coworkers [34–37] for low temperature synthesis of crystalline nanowires of III–V semiconductors. In a typical procedure, a metal (e.g. In, Sn, Bi) with a low melting point is used as a catalyst, and the desired material generated through the decomposition of organometallic precursors. Nanowhiskers of InP, InAs, and GaAs have been prepared by low temperature (203 °C) solution phase reactions. The products obtained are generally single-crystalline.

(iv) **Solvothermal synthesis**
Solvothermal methodology [38, 39] is extensively employed as a solution route to produce semiconductor nanowires and nanorods. In this process, a solvent is mixed with metal precursors and crystal growth regulating or templating agents, such as amines. This solution mixture is placed in an autoclave maintained at relatively high temperatures and pressures to carry out the crystal growth and the assembly process. The methodology is quite versatile and has enabled the synthesis of crystalline nanowires of semiconductors and other materials.

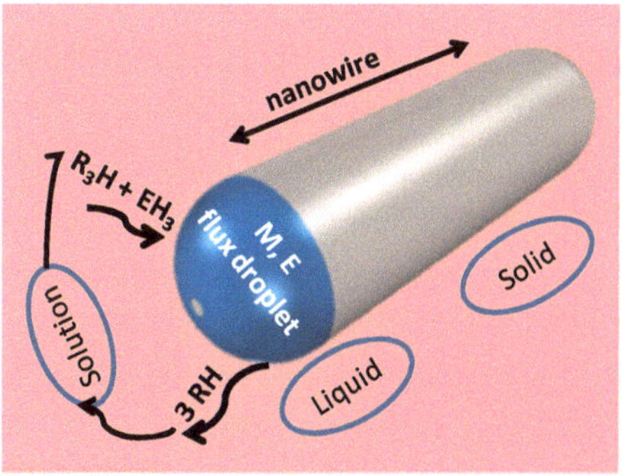

Fig. 9.3 Schematic representation of solution liquid solid based growth of one-dimensional nanostructure

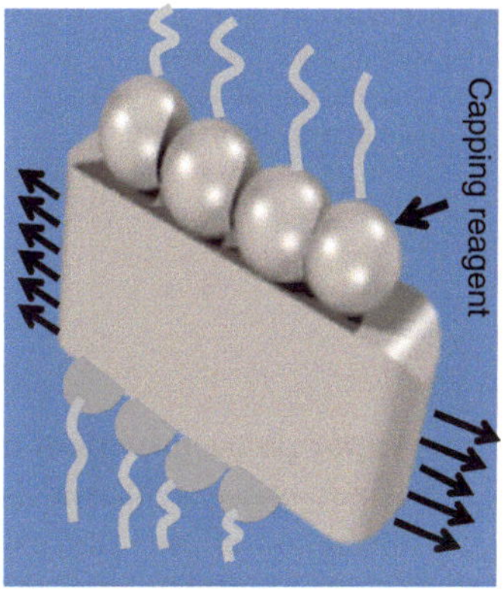

Fig. 9.4 Schematic representation of solution phase one-dimensional crystal growth based on capping reagent

(v) **Solution phase methods based on capping reagents**
One can control the final shape of the crystal by introducing a appropriate capping reagent (Fig. 9.4). The capping reagent is used to change the free energies of of the various crystallographic surfaces that would alter their growth rate. A range of compounds [40–42] have been evaluated as capping reagents to control the shape of colloidal particles synthesized using solution phase methods.

(vi) **Self-assembly of nanoparticles**
Self-assembly has been extensively explored for generating complex nanostructures on various scales [43, 44] (Fig. 9.5). Monodispersed colloids have been used as the building blocks for growth of nanowire structures through self or external manipulated assembly.

(vii) **Size Reduction**
The use of isotropic deformation of a polycrystalline or amorphous material, anisotropic etching of a single crystal and near field optical lithography with a phase shift mask are the most promising techniques to reduce the lateral dimension of one-dimensional structures from 1–10 μm to the regime $\leq$100 nm [45, 46] (Fig. 9.6).

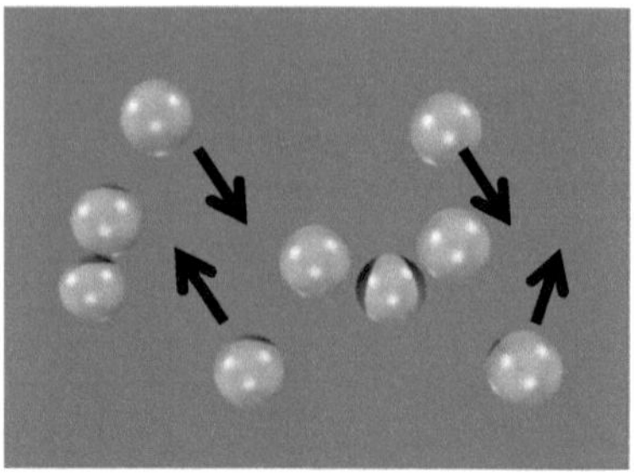

Fig. 9.5 Schematic representation of one-dimensional crystal growth by self-assembly of zero-dimensional nanostructures

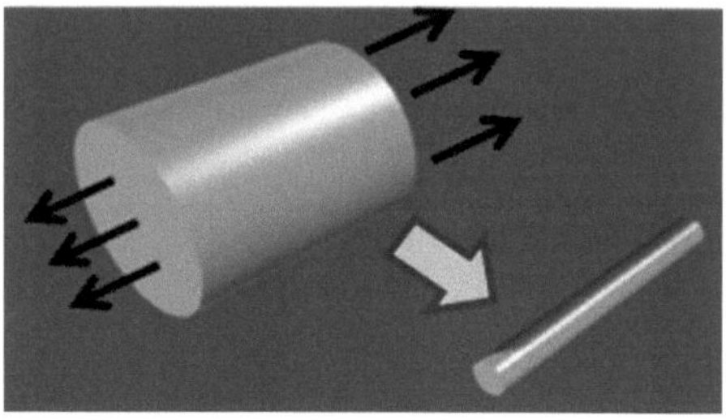

Fig. 9.6 Schematic representation of one-dimensional crystal growth by size reduction of one-dimensional microstructure

(viii) **Vapor phase growth of nanowires**

Vapor phase synthesis is the most extensively explored approach to the formation of one-dimensional nanostructures [47]. Starting with the simple evaporation technique in an appropriate atmosphere to produce nanowires, vapor–liquid–solid, vapor–solid and other processes are also made use of. In principle it is possible to process any solid material in to one-dimensional nanostructures by controlling the super saturation at relatively low level. It has been widely accepted that the control of super saturation is a prime consideration in obtaining one-dimensional nanostructures due to evidences that the degree of super saturation determines the growth morphology. It has been observed that a low supersaturation is required for whisker growth, a medium super saturation supports bulk crystal growth and at high super saturation powders are formed. The formation of powders occurred through homogeneous nucleation in the vapor phase. The lateral dimension of one-dimensional nanostructures can be varied by controlling a number of parameters that include the supersaturation factor, nucleation size, and growth time [48, 49].

The vapor phase synthesis of nanowires is broadly classified in three categories, direct vapor phase method, Indirect vapor phase method and vapor–liquid–solid method. In a typical direct vapor phase method the vapor species is generated by evaporation, chemical reduction and other types of gaseous reactions [50, 51]. The species are subsequently transported and condensed on the surface of a solid substrate placed in a zone with temperature lower than that of the source material. With proper control over supersaturation factor one could easily obtain one-dimensional nanostructures. In the indirect vapor phase

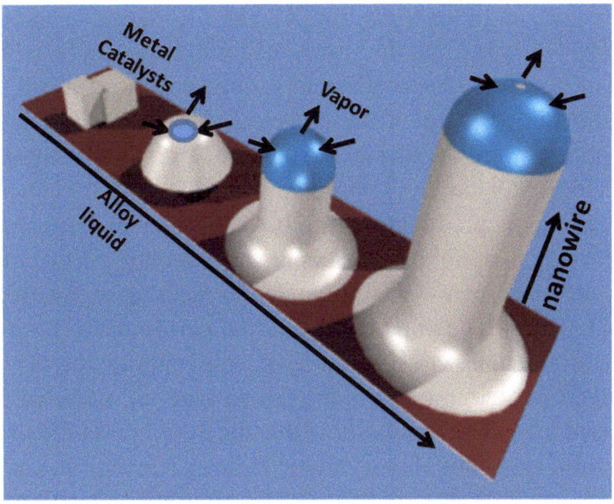

Fig. 9.7 Schematic representation of vapor phase one-dimensional crystal growth based on VLS mechanism

synthesis decomposition of compound in vapor phase and many other types of side reaction occurs before growth of final product [52, 53]. Vapor–liquid–solid (VLS) method (Fig. 9.7) is the most successful process among all vapor phase process for synthesizing nanowires with single crystalline structure and in relatively large quantity [54–58]. In a general VLS process the reaction starts with the dissolution of gaseous reactants in to nanosize liquid droplets of a catalyst metal, followed by nucleation and growth of single crystalline one-dimensional nanostructures. The liquid droplets which act as template plays crucial role for growth of individual nanowires. The major requirement for nanowire growth of desired material is that there should exist a good solubility of target material with liquid droplets. Ideally they should make eutectic compounds.

9.2 Background on Silicon Carbide Nanowire Growth

The discovery of carbon nanotubes in 90s has given the wide scope to the research in the area of nanomaterials [59]. Among nanomaterials one-dimensional nanomaterial has got attention in recent times. Nanotubes and nanowires of nitrides of boron, gallium and silicon, oxides of magnesium, calcium, silicon, carbides of silicon [60–69] have been widely reported in literature. These nanoscale one-dimensional structures can not only offer opportunities for fundamental research, but also be used in technological applications. CNT and the SiC nanowires/nanorods are the two most studied one-dimensional nanostructure. Their excellent mechanical properties have made them to use in composite materials to improve its property. The measurement of elasticity, strength, and toughness of CNT and SiC nanorods using atomic force microscopy (AFM) revealed that the carbon nanotubes and SiC nanorods have large elasticity values [70]. These large elasticity values obviously make these materials candidates for the reinforcing element in ceramic, metal, and polymer matrix composites. The reported strengths of the SiC nanorods were substantially greater than those found previously for larger SiC structures, and they approach theoretical values. SiC nanorods, despite their smaller elasticity values than carbon nanotubes, should be a better reinforcing material in some composite structures than carbon nanotubes. Therefore, SiC nanorods/nanowires are materials with great potential in mechanical applications [71, 72]. The good field emission performance, photoluminescence properties of such SiC nanowires widen up its application area [73–75]. Recently, study on hydrogen production and hydrogen storage on modified SiC nanowires have been reported [76]. The results show that the β-SiC nanowires can absorb visible light and exhibit excellent photocatalytic hydrogen evolution performance from pure water under visible light irradiation. The hydrogen evolution rate can reach more than 60 μL/g h. SiC nanowires by simple modification can greatly enhance the efficiency of H_2 production. Average H_2 production rate over the modified SiC is 76.1 % higher than that of unmodified SiC. The enhanced H_2 production is mainly due to stronger hydrophilic ability of the modified SiC. In addition, O_2 is detectable in the experiment, indicating that water has been decomposed into H_2 and O_2. These studies have shown that the SiC nanowires can act as an excellent support material in corrosive and high temperature environment. In addition, SiC can be a material of great technological interest for devices designed to operate at high temperature, high powers, high frequency, and harsh environments [77–79].

There are variety of established techniques in literature for the growth of β-SiC whiskers/nanowires/nanorods. The synthesis methods that have been used to synthesize SiC nanomaterials can be broadly classified into following categories: (a) carbothermal reduction of silica [80, 81]; (b) decomposition of organic silicon compounds/polymers [82]; (c) reaction between silicon halides and CCl_4;[83] (d) reacting carbon nanoclusters with SiO at elevated temperature [84, 85]; (e) metal catalyst reduction of silicon powder [86, 87] (f) High temperature catalytical reaction of SiC powder [88] (g) chemical vapor deposition (CVD) with the aid of catalysts through the vapor–liquid–solid (VLS) mechanism [89, 90] (h) CVD process hydrogen reduction of methyltrichlorosilane (CH_3SiCl_3) [91, 92].

We will briefly discuss all the above-mentioned methods. In the carbothermal reduction process, the SiC nanomaterials have been obtained by the carbothermal reduction of sol–gel-derived silica xerogels containing carbon source such as carbon nanoparticles/polymeric carbon compound. These carbothermal reductions are generally occurred at 1600–1800 °C under argon atmosphere. The synthesized SiC nanomaterials mostly accompanied by amorphous silica layers. In such method generally catalysts were used for the SiC nanowire growth. Growth of β-SiC nanowires in bulk quantities have been reported from a mixture of activated carbon and sol–gel derived silica embedded with Fe nanoparticles [75]. There were impurities like silica, catalyst were present in the grown SiC nanowires. The other method for synthesis of SiC nanowires like decomposition of organic compounds process for

SiC nanomaterial synthesis involves catalytic decomposition of organosilicon compounds at moderate temperature. The SiC nanowires were synthesized by decomposition of methylvinyldichlorosilane in presence of nickel catalyst at 800–1200 °C [93]. In this method SiC nanowires were synthesized at comparative lower temperature in comparison to other methods. Catalyst impurities were present in the grown SiC wires. Synthesis of SiC nanowires using chlorides of silicon and carbon as one of the silicon and carbon source is one of the oldest method used. This method involves the catalytical reaction of chlorides of silicon or carbon with corresponding other silicon or carbon source. Giovanni Attolini et al. [94] have prepared cubic SiC nanowires by chemical reaction of silicon powder in an open tube, with carbon tetrachloride (CCl_4) and nickel as the catalyst. This method is based on the reaction between chlorine and the silicon substrate surface in nitrogen atmosphere at a temperature of 1000 °C, where the carbon produced by CCl_4 thermal decomposition acts as a precursor for the formation of the silicon carbide nanowires. In such methods control of reactants concentration is the key factor to get stoichiometric SiC, i.e., free from free silicon and carbon. Generally deposition of free carbon or silicon occurs with SiC nanowires. The synthesis of SiC nanowires by reaction between silica as the silicon source and different carbon source such as carbon nanotubes, activated carbon black nanopowders, carbon monoxide is also one of the most studied methods. The main advantage of such method is that the final structure of SiC nanowires/nanotubes can be tailored by using suitable carbon source. There were several reports [84, 85] on the synthesis of SiC nanowires at high temperature reaction of SiO as silicon source and carbon nanotubes, activated carbon black nanopowders, carbon monoxide as carbon source. These are high temperature synthesis processes. Recently, there were several reports in which the synthesis temperature of above process has been drastically decreased by using sodium catalyst [95]. In this method SiC nanowires were synthesized from reaction between silicon oxide and amorphous active carbon with the assistance of metallic Na and elemental S. In-situ temperature monitoring showed that the reaction temperature could increase quickly to 850 °C, once the autoclave was heated to 100 °C. Such a rapid temperature increase was attributed to the intense exothermic reaction between Na and S, which assisted the formation of silicon carbide nanowires. The carbothermal reduction of WO3 by C provided reductive environment to synthesize crystalline SiC nanowires coated with carbon sheath in the growth temperature of 1000–1100 °C. The synthesis of SiC nanowires from commercially available SiC powders in a thermal evaporation process at 1700 °C with catalyst [88] is also one of the viable options. The as-grown nanowires have a high purity, a low length/diameter ratio and a good linearity of the curve. This method requires high temperature to grow SiC nanowires. SiC nanowire synthesis by thermal decomposition of Methyltrichlorosilane (MTS) is one of the most extensively studied methods. The thermal decomposition of MTS can be done in the presence and absence of catalyst. The chemical vapor deposition (CVD) process growth of SiC nanowires with the aid of catalysts follows the vapor–liquid–solid (VLS) mechanism [87, 88]. Catalyst plays the crucial role in deciding the final morphology of nanowire in this method. The non catalytic hydrogen reduction of MTS route for SiC nanowire growth has shown the potential to grow SiC nanowires of controlled morphology according to application [91, 92, 96, 97]. Whether it is matter of bulk production of SiC nanowire or growth of SiC nanowires on C/C composites, Silicon substrates for different application, hydrogen reduction of MTS method has shown its applicability in such kind of growth process. In a typical synthesis process MTS was decomposed in the hydrogen reducing environment. The reactive intermediates generated due to MTS decomposition use to get adsorbed on the substrate surface and further nucleation and growth of SiC nanowires occur. The concentration of reactive intermediates at the substrate surface is the key factor which decides final morphology of SiC deposit.

In most of the methods as described above, expensive raw materials, catalysts, and sophisticated techniques were used. These drawbacks may limit the massive fabrication and application of SiC nanowires. It is still a challenge for scientists and industrials to synthesize large-scale SiC nanowires by using a simple and rapid method. Researchers often focused on seeking some special techniques or route to realize the massive synthesis of SiC nanowires. For example, Wang et al. [98] has reported the SiC nanowires synthesis by heating a mixture of SiO powder and arc-discharge plasma pretreated carbon black with porous and loose structures in a high frequency induction furnace. The total heating-up and reaction time was less than 1 h, and more than 200 g products can be obtained per day. The formation of the loose and porous structures may be the key factor for the high yield of SiC nanowires. This method has shown the potential of high yield but used a complicated technique. List of most of the non catalytic route are tabulated (Table 9.1) which were used by researchers for SiC nanowires growth. In these techniques usual and natural precursors of silicon and carbon were used for SiC nanowire growth. Some of these techniques have reported the bulk production of SiC nanowires [98–102]. Although these methods reported large-scale synthesis but the issues like impurity, quality of SiC nanowires and the amount of product always remain there. Recently Jianjun Chen et al. [99] has demonstrated a simple catalyst free route for the bulk production of SiC nanowires. The SiC nanowires were synthesized by heating a special carbon source with porous and loose structures with silicon powder. They have used a special carbon source called expandable graphite (EG). Expandable graphite is a graphite intercalation compound in which some oxidants, such as sulfuric acid, potassium permanganate, are inserted between the carbon layers of the graphite. EG expands hundred times larger than its original volume when heated. Therefore, EG can form porous structure and has excellent absorptive capacity for reactant gas after expansion. They have used very simple technique but the yield of final SiC nanowires was less than expectation with the free silicon impurity. So still the challenges remain the same.

Among most of the reported methods which have been developed for producing SiC nanowires, there were several critical issues specially with the fabrication of SiC nanowires-reinforced ceramic matrix composites. There are several efforts being reported in the current literature on fabricating SiC nanowires-reinforced ceramic matrix composites but the challenges are still there [79]. This is possibly because following factors: First, sufficient amount of high quality SiC nanowires with uniform morphology (high crystallinity, pure Si–C chemistry, etc.) must be easily produced.

Table 9.1 List of most of the non catalytic synthesis route for SiC nanowires growth

Sr. No.	Authors	Synthesis method	Precursors	Title
1	G. W. Meng [81]	Carbothermal reduction of sol–gel derived silica xerogels containing	Silica xerogels	Preparation of β-SiC nanorods with and without amorphous SiO_2 wrapping layers
3	Weiqiang Han [64]	Vacuum heating	Silica and CNT	Direct synthesis of SiC nanowires by multiple reaction VS growth
6	Wen Yang [110]	CVI method	MTS and hydrogen	Fabrication in-situ SiC nanowires/SiC matrix composite by chemical vapor infiltration process
7	Jian Wei [111]	CVD method	silicon monoxide (SiO) and carbon monoxide (CO) vapor	Fabrication of composite structure of carbon fibers and high density SiC nanowires
8	Xu-Hui Sun [112]	Vacuum heating	SiO powders and MWCNT	Formation of silicon carbide nanotubes and nanowires via reaction of silicon (from disproportionation of silicon monoxide) with carbon nanotubes
9	Jianjun Chen [113]	Evaporation method (vacuum heating)	Silicon powder and graphite	Growth mechanism of twinned SiC nanowires synthesized by a simple thermal evaporation method
10	Y. L. Chiew [114]	Pyrolysis	Oil palm empty fruit bunch fibers infiltrated with tetraethyl orthosilicate	Growth of SiC nanowires using oil palm empty fruit bunch fibers infiltrated with tetraethylorthosilicate
11	Jian Wei [115]	CVD method	SiO_2 and graphite powder	Growth and morphology of one-dimensional SiC nanostructures without catalyst assistant
12	Zhaoqian Li [116]	Ablation of SiC-containing coating under oxyacetylene torch	Silicon carbide	Growth mechanism of silica nanowires without a metal catalyst via oxyacetylene torch ablation

(continued)

Table 9.1 (continued)

13	Rajnish Dhiman [117]	Vacuum heating	Wooden powder and TEOS	Growth of SiC nanowhiskers from wooden precursors, separation, and characterization
14	Weon-Ju Kim [91]	CVI method	(CH_3SiCl_3, MTS)	Growth of SiC nanowires within stacked SiC fiber fabrics by a noncatalytic chemical vapor infiltration technique
15	Jin-Seok Lee [118]	Vacuum heating	SiO_2 and carbon black	Improvement of porous silicon carbide filters by growth of silicon carbide nanowires using a modified carbothermal reduction process
16	Jin-Seok Lee [119]	Vacuum heating–Carbothermal reduction	SiO_2 powders and carbon black	In situ growth of SiC nanowires by carbothermal reduction using a mixture of low-purity SiO_2 and carbon
17	Wen Yang [92]	CVD	Methyltrichlorosilane (MTS)	In situ growth of SiC nanowires on RS-SiC substrate(s)
18	He Huang [120]	Pyrolysis	Anthracite coal and silicon powder	In situ growth of silicon carbide nanowires from anthracite surfaces
19	Wei Min Zhou [101]	Pyrolysis	SiO and activated carbon	Large-scale synthesis and characterization of SiC nanowires by high-frequency induction heating
20	Jian Wei [102]	Thermal reaction	Activated carbon and vaporized silicon	Large-scale synthesis and photoluminescence properties of hexagonal-shaped SiC nanowires
21	Yanhui Chu [121]	Vacuum heating	SiO_2, Si, and C powders	Microstructure and growth mechanism of SiC nanowires with periodically fluctuating hexagonal prisms by CVD
22	Lipeng Xin [122]	Sol–gel carbothermal reduction	TEOS and isopropanol	Morphological evolution of one-dimensional SiC nanomaterials controlled by sol–gel carbothermal reduction

(continued)

Table 9.1 (continued)

23	Jian Wei [123]	Vacuum heating	Silicon, activated carbon	Photoluminescence performance of SiC nanowires, whiskers, and agglomerated nanoparticles synthesized from activated carbon
24	Tomitsugu Taguchi [124]	Vacuum heating	Carbon nanotube and Si powder	Preparation and characterization of single-phase SiC nanotubes and C–SiC coaxial nanotubes
25	J. Z. Guo [125]	CVD method	Si and SiO_2 and gaseous CH_4	Preparation of SiC nanowires with fins by chemical vapor deposition
26	Xintong Li [126]	Carbothermal reduction	Novel resorcinol-formaldehyde (RF)/SiO_2 hybrid aerogel	Preparation of silicon carbide nanowires via a rapid heating process
27	Gong-yi Li [127]	Heating	Polycarbosilane and activated carbon	SiC nanowires grown on activated carbon in a polymer pyrolysis route
28	Sheng-Cheng Chiu [84]	Thermal pyrolysis	Carbonnanotube (CNT)/ Si–SiO_2	SiC nanowires in large quantities: synthesis, band gap characterization, and photoluminescence properties
29	Hongsheng Zhao [128]	High temperature treatment	Using silicon powder and phenolic resin	Silicon carbide nanowires synthesized with phenolic resin and silicon powders
30	Wasana Khongwong [129]	CVD method	Silicon powder and methane	Simple approach to fabricate SiC–SiO_2 composite nanowires and their oxidation resistance
31	Baosheng Li [130]	Pyrolysis	Silicon and MWCNT	Simultaneous growth of SiC nanowires, SiC nanotubes, and SiC/SiO_2 core–shell nanocables
33	Qian-Gang Fu [97]	CVD	MTS, hydrogen	Synthesis of silicon carbide nanowires by CVD without using a metallic catalyst

(continued)

Table 9.1 (continued)

34	Pengchao Kang [131]	High temperature treatment	Si,C nanopowders	Synthesis of SiO_2 covered SiC nanowires with milled Si,C nanopowders
35	G. Attolini [94]	Pyrolysis	Carbon monoxide and single crystal silicon	Synthesis and characterization of 3C-SiC nanowires
36	L. D. Zhang [132]	Carbothermal reduction	SiO_2 xerogel containing carbon nanoparticles	Synthesis and characterization of nanowires and nanocables
37	Jianjun Chen [133]	Heating	Thermal evaporation of silicon on the carbon fibers	Synthesis and photoluminescence of needle-shaped 3C-SiC nanowires on the substrate of PAN carbon fiber
38	Enlei Zhang [134]	Carbothermal reduction	Solid SiO powders, carbon powder	Synthesis and photoluminescence property of silicon carbon nanowires synthesized by the thermal evaporation method
39	Jian Wei [135]	Carbothermal reduction	Silicon and silicon dioxide powders and graphite sheet	Synthesis of centimeter-scale ultra-long SiC nanowires by simple catalyst-free chemical vapor deposition
40	Jun Jie Niu [136]	Carbothermal reduction	Silicon powder and CO_2	Synthesis of macroscopic SiC nanowires at the gram level and their electrochemical activity with Pt loadings
41	H. X. Zhang [137]	Pulse laser technique	SiC target	Synthesis of nanostructured SiC using the pulsed laser deposition technique
42	V. Raman [138]	Vacuum heating	Coal tar pitch blended with sol–gel derived silica.	Synthesis of silicon carbide nanofibers from pitch blended with sol–gel derived silica

Second, a compliant coating(s), such as carbon or BN, is necessary to be deposited on the nanowires with sufficient thickness and uniformity control before dispersing the nanowires in a ceramic matrix to guarantee a sound nanowire/ matrix interfacial bonding strength. when nanowires/whiskers are used as reinforcements in ceramic matrix composites, one of the key issues is the control of the interfacial bonding between the reinforcements and the matrix. It must be neither too strong (to allow interfacial debonding and deflection of matrix cracks by the reinforcement nanowires/whiskers during failure) nor too weak (to keep effective load transfer from the matrix to the nanowires/whiskers). Direct burying of SiC nanowires/whiskers into ceramic matrix, such as SiC, is likely to produce a strong interfacial bond, resulting in unnecessarily improved strength over the pure matrix. Third, a suitable method(s) has to be found to process the nanowires into the matrix with homogeneous space distribution. At the mean time, great care should be paid to protect the nanowires, as well as the compliant coating, from any significant damages during these processes. Summarizing all the methods as discussed above it can be concluded that Chemical vapor deposition (CVD) process hydrogen reduction of methyltrichlorosilane (CH_3SiCl_3) method has a potential to produce SiC in various shapes of thin films, powders, whiskers, and nanowires with control over morphology and deposition chemistry of reaction. This method can deal with most of the problems associated with SiC nanowires growth as discussed above.

9.3 Brief Introduction of Taguchi Method

Taguchi method of experimental design provides a systematic and efficient approach for conducting experiments to determine near optimum settings of process parameters for optimal performance, e.g., nanodimensional growth of SiC wires in present case. Taguchi's method for experimental design is straightforward and easy to apply to many engineering situations, making it a powerful yet simple tool. It can be used to quickly narrow the scope of a research project or to identify problems in a process from data already in existence. The Taguchi method utilizes orthogonal arrays to study a large number of variables with a small number of experiments. Using orthogonal arrays [103] significantly reduces the number of experimental configurations to be studied. The conclusions drawn from small-scale experiments are valid over the entire experimental region spanned by the control factors and their settings. This method can reduce research and development costs by simultaneously studying a large number of parameters.

The main disadvantage of the Taguchi method is that the results obtained are only relative and do not exactly indicate what parameter has the highest effect on the performance characteristic value. Also, since orthogonal arrays do not test all variable combinations, this method should not be used with all relationships between all variables. The Taguchi method has been criticized in the literature for its difficulty in accounting for interactions between parameters. Another limitation is that the Taguchi methods are offline, and therefore inappropriate for a dynamically changing process such as a simulation study. Furthermore, since the Taguchi methods deal with designing quality rather than correcting for poor quality, they are applied most effectively at early stages of process development.

In order to analyze the results, the Taguchi method uses a statistical measure of performance called signal-to-noise (S/N) ratio [104]. The S/N ratio takes both the mean and the variability into account. It also observes the effect of noise sources present in the experimental system. The S/N equation depends on the criterion for the quality characteristic to be optimized. Usually, there are three categories of the quality characteristic in the analysis of the S/N ratio, i.e., the lower-the-better, the larger-the-better, and the more-nominal-the-better. The S/N ratio for each level of process parameters is computed based on the S/N analysis. Regardless of the category of the quality characteristic, a larger S/N ratio corresponds to a better quality characteristic. Therefore, the optimal level of the process parameters is the level with the highest S/N ratio. Furthermore, a statistical analysis of variance is performed to see which process parameters are statistically significant. The optimal combination of the process parameters can then be predicted. Further statistical analysis is required to see whether the effect of parameters are statistically significant or not as well as to justify the adequacy of the models. The relative effect of different factors can be obtained by decomposition of total variation into its appropriate components, which is commonly called analysis of variance (ANOVA).

Using the Taguchi method for parameter design, the predicted optimum setting need not correspond to one of the rows of the matrix experiment. Therefore, an experimental confirmation is run using the predicted optimum levels for the control parameter is required. The purpose is to verify that the optimum conditions suggested by the matrix experiments do indeed give the projected improvement. If the observed and the projected improvements match, the suggested optimum conditions will be adopted. If not, it will be concluded that the additive model underlying the matrix experiment has failed, and ways need to be found to correct that problem. The corrective actions include finding better quality characteristics, or S/N ratios, or different control factors and levels, or studying a few specific interactions among the control factors.

9.4 SiC Nanowire Synthesis: Taguchi Method Optimization

Recently we have demonstrated a simple but effective synthesis route, CVD technique for the growth of homogeneous and smooth SiC wires and Taguchi method has been used to optimize the parameters for growth of nanodimensional SiC wires [105]. The SiC wires were grown on C/C composite (3.5 cm × 3.5 cm × 0.5 cm) using chemical vapor deposition technique. A high temperature vertical graphite reactor of dimension 400 mm length and diameter of 60 mm was used for the coating studies. Detail schematic diagram of the coating system is shown in Fig. 9.8. The Methyltrichlorosilane (MTS) was used as the precursor for SiC due to its stoichiometric

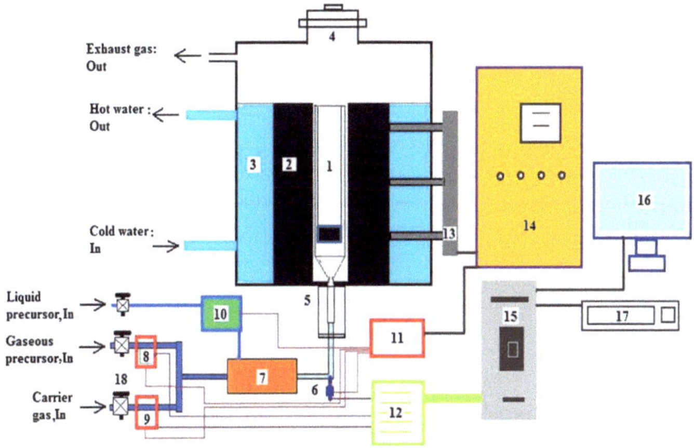

Fig. 9.8 Schematic diagram of chemical vapor deposition unit (1)Core graphite reactor, (2) Heating chamber, (3) Water cooling jacket, (4) Reactor opening, (5) gas inlet nozzle (graphite), (6) Pressure sensor, (7) Pre-heater, (8, 9) MFC gas flow controller, (10) Liquid precursor pump, (11) AC/DC supply, (12) DAQ board, (13) Thermo-coil heating coil (inserted), (14) Reactor temperature control panel, (15) Computer CPU, (16) Computer monitor, (17) Computer key board, (18) Valve

Table 9.2 Selection of parameters and their levels for Taguchi method (Reprinted with permission from Ref. [105]. Copyright Bentham Science Publishers)

Parameter	Level 1	Level 2	Level 3
Temperature (°C)	1500	1300	1100
MTS flow rate (ml/min)	1.0	0.5	2.0
Hydrogen flow rate (lpm)	2.0	0.8	1.2
Total flow rate (Ar+H_2) (lpm)	2	4	6

ratio of silicon to carbon. Hydrogen along with argon was used as carrier gas to sweep the MTS vapors into the reactor.

It has been reported that the lower deposition rates and lower temperature of deposition for fcc crystals result in development of lowest surface energy plane, i.e., {111} plane Fig. 9.12 while higher temperature and higher deposition rates causes development of higher surface energy, i.e., the {110} plane [106]. So for the growth of SiC nanowires, the growth of SiC nuclei crystal should be in lowest surface energy plane. To achieve above objective, we have optimized the parameters for lowest deposition rate and growth of minimum diameter of SiC wires based on the concept of smaller the better function in Taguchi method. Various operating parameters which affect the growth morphology of SiC such as temperature of synthesis, MTS flow rate, hydrogen flow rate, and total carrier gas flow rate have been chosen and varied at three different levels (Table 9.2). So, this is a four factor three level system and L9 orthogonal array was used to design the experiments. Taguchi method allows us to

Table 9.3 Design of L-9 orthogonal array for experiments (Reprinted with permission from Ref. [105]. Copyright Bentham Science Publishers)

Exp. No.	Total flow (lpm)	Hydrogen flow (lpm)	MTS flow (ml/min)	Temperature (°C)	Experimental Results	
					Ave. Deposition rate (gm/min)[#]	Ave. Diameter (nm)*
1	2	2	1.0	1500	0.007	Dense (2000)
2	2	0.8	0.5	1300	0.184	265
3	2	1.2	2.0	1100	0.91	1000
4	4	2	0.5	1100	0.035	250
5	4	0.8	2.0	1500	0.29	Dense (2000)
6	4	1.2	1.0	1300	0.23	350
7	6	2	2.0	1300	0.35	497
8	6	0.8	1.0	1100	0.0193	559
9	6	1.2	0.5	1500	0.076	Dense (2000)

Since some of the average deposition values were too low, all deposition values have been multiplied by 10 to facilitate in calculations

* In case of dense deposition the average diameter was assumed to be 2000 nm to get the effects of different parameters on the growth diameter of SiC wires

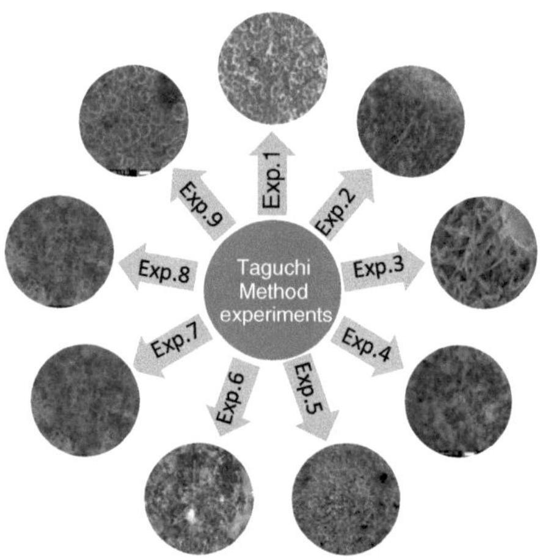

Fig. 9.9 SEM images of the SiC deposits developed by experiments designed by Taguchi method (Reprinted with permission from Ref. [105]. Copyright Bentham Science Publishers)

carry out only nine experiments to get optimum synthesis parameters for synthesizing pure nanodimensional SiC wires.

In the classical experimental design using full factorial experimentation the number of experiments to study the selected process space would be $3^4 = 81$. With the selection of L_9 (3^4) orthogonal array the number of required experiments can be drastically reduced to nine. The experimental plan and the results of average deposition rate (gm/min) and Average growth diameter of SiC wires (nm) are shown in the Table 9.3.

The XRD pattern showed the reflections from (111), (200), and (220) planes in all nine experiments results. It confirms the growth of β-SiC phase in all nine

Table 9.4 Analysis of variance (ANOVA) for average deposition rate

Source	Sum of squares	DOF	Variance	F-value	P-value	Contribution of factor (%)
Model	1470.86	8	183.86	–	–	100
Total flow rate (lpm)	21.228	2	10.613	–	–	1.43
Hydrogen flow rate (lpm)	341.95	2	170.97	–	–	23.42
MTS flow rate (ml/min)	832.75	2	416.375	–	–	56.65
Temperature (°C)	274.94	2	137.47	–	–	18.50
Error	0	0	–	–	–	–
Total	1470.86	8				100

Table 9.5 Results of the pooled ANOVA for average deposition rate

Source	Sum of squares	DOF	Variance	F-value	P-value	Pure sum	Contribution of factor (%)
Model	1470.86	6	246.46	23.22	0.0419	1385.96	93.726
Total flow rate (lpm)	(21.228)	(2)	–	pooled	pooled	–	–
Hydrogen flow rate (lpm)	341.95	2	167.44	16.108	0.0585	320.725	21.69
MTS flow rate (ml/min)	832.75	2	412.83	39.229	0.0249	811.523	54.88
Temperature (°C)	274.94	2	133.93	12.952	0.0717	253.715	17.156
Error	21.228	2	10.6139	–	–		6.274
Total	1470.86	8					

experiments. The morphology of SiC deposits obtained in all nine experiments is shown in Fig. 9.9. The SEM images showed that there is no growth of SiC wires (dense SiC deposition) at deposition temperature 1500 °C (Experiment no. 1, 5, and 9).

Statistical analysis of the results obtained from nine CVD experiments was performed to see whether the effect of process parameters are significant or not as well as to justify the adequacy of the models. The relative effect of different factors can be obtained by decomposition of total variation into its appropriate components, which is commonly called analysis of variance (ANOVA). The result of ANOVA analysis for minimum deposition rate and minimum growth diameter of SiC nanowires are presented in Tables 9.4, 9.5, 9.6, 9.7. For statistical analysis of the results a mathematical function named transformations may be needed to apply to the results to meet the assumptions that make the ANOVA valid. For example if the residuals are not normally distributed with a constant variance a suitable transformation must be applied. For this purpose average deposition rate and average growth diameter have

Table 9.6 Analysis of variance (ANOVA) for average growth diameter of SiC nanowire

Source	Sum of squares	DOF	Variance	*F*-value	*P*-value	Contribution of factor (%)
Model	457.02	8	57.127	–	–	100
Total flow rate (lpm)	22.0	2	11	–	–	4.81
Hydrogen flow rate (lpm)	15.36	2	7.68	–	–	3.361
MTS flow rate (ml/min)	51.73	2	25.865	–	–	11.329
Temperature (°C)	367.93	2	183.965	–	–	80.50
Error	0	0	–	–	–	–
Total	457.02	8				100

Table 9.7 Results of the pooled ANOVA for average growth diameter of SiC nanowire

Source	Sum of squares	DOF	Variance	*F*-value	*P*-value	Pure sum	Contribution of factor (%)
Model	457.02	4	114.255	12.232	0.0163	382.3	83.64
Total flow rate (lpm)	(22.0)	(2)	–	pooled	pooled	–	–
Hydrogen flow rate (lpm)	(15.36)	(2)	–	pooled	pooled	–	–
MTS flow rate (ml/min)	51.73	2	25.865	1.384	0.3493	33.05	7.23
Temperature (°C)	367.93	2	183.965	9.848	0.0285	349.25	76.41
Error	37.36	4	9.34	–	–		16.36
Total	457.02	8					–

been modified accordingly as shown in Table 9.3. The results of ANOVA analysis for minimum deposition rate are given in Table 9.4. According to percentage distribution of factors MTS flow rate has the maximum (56 %) and the total flow rate has the minimum effect (1.43 %) on the deposition rate of SiC produced by CVD technique. The temperature of deposition and hydrogen flow rate has 18.6 and 23 % effect respectively. It shows that the MTS flow rate, temperature of deposition and hydrogen flow rate are the most significant effective parameters for minimum deposition rate of SiC.

F-value is a criterion for comparing model (or factor) variance with residual (error) variance (Table 9.4). If the variances are close to the same, the ratio will be close to one and it is less likely that any of the factors have a significant effect on the response. *F*-value is calculated by model variance divided by error variance. According to Table 9.4 error degree of freedom (DOF) is zero and the error variance could not be calculated. Consequently it was not possible to calculate the *F*-value. In order to eliminate the zero DOF from the error term, a pooled ANOVA was applied. The values of the ANOVA analysis for the minimum deposition rate after pooling of the total flow rate from the model are given in Table 9.5. The model *F*-value 23.22

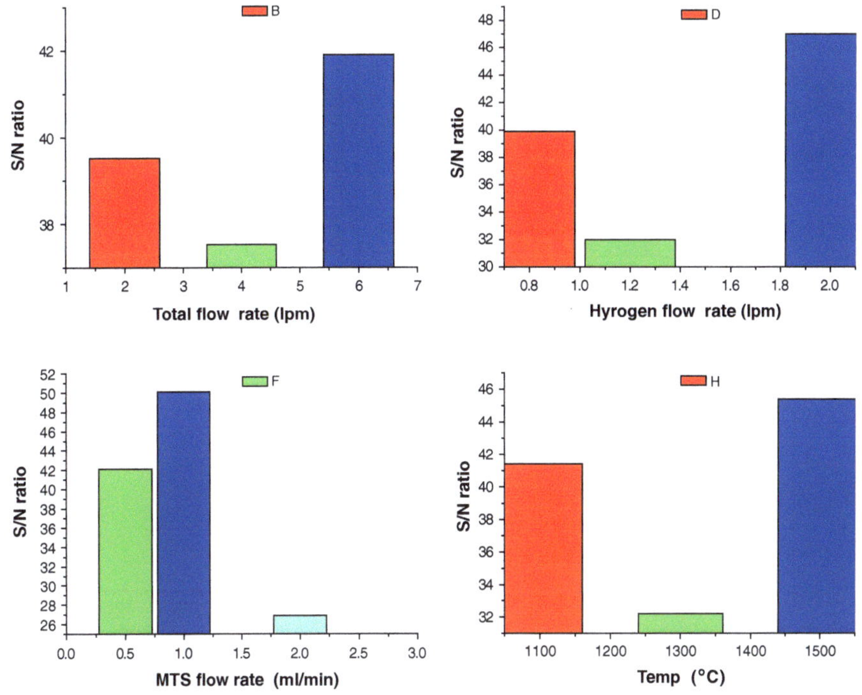

Fig. 9.10 Graphical representation of the average effect of different factors on the deposition rate of SiC (Reprinted with permission from Ref. [105]. Copyright Bentham Science Publishers)

implies the model was significant and there was only 4.19 % chance that a model F-value this large could occur due to noise. P-value less than 0.05 indicate that the effect of model factors is significant within the 95 % confidence interval. Therefore in the case of minimum deposition rate of SiC only three factors including MTS flow rate, temperature of deposition and hydrogen flow rate are significant model terms. The percentage distribution reveals that the total flow rate is not the statistically significant factor, at least in the range under study. Therefore minimum deposition rate could be achieved using an appropriate combination of the other operating parameters. Figure 9.10 shows the graphical representation of the average effect of the parameters on the desired result in terms of S/N ratio. It shows that the minimum deposition rate would be obtained at total flow rate 6 lpm, hydrogen flow rate 2 lpm, MTS flow rate 1 ml/min and 1500 °C deposition temperature.

The same procedure was followed for the second response (Average diameter of grown SiC wires) and the results are given in the Tables 9.6 and 9.7. In this case total flow rate and Hydrogen flow rate were pooled. The model F-value 12.23 implies that the model was significant and there was only 1.6 % chance that this F-value could occur due to noise. Base on the percentage contribution of factors the most significant process parameters on the minimum growth diameter are the temperature of deposition (80 %) and MTS flow rate (11 %). Moreover based on the P-value

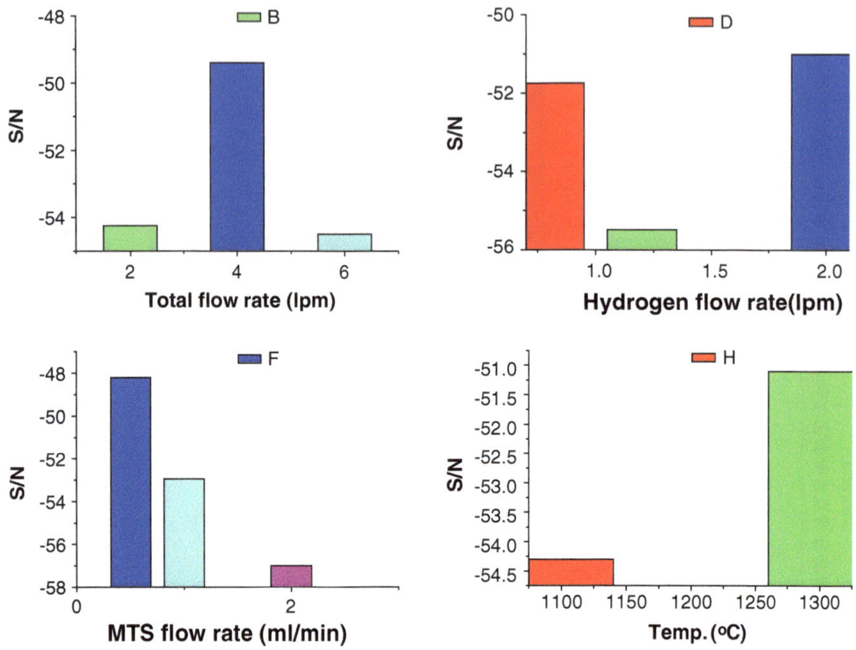

Fig. 9.11 Graphical representation of the average effect of different factors on the growth diameter of SiC (Reprinted with permission from Ref. [105]. Copyright Bentham Science Publishers)

Temperature of deposition is the factor with statistically significant effect on the growth diameter of the SiC wires. Figure 9.11 shows the graphical representation of the average effect of the parameters on the desired result in terms of S/N ratio. The minimum growth diameter of SiC wires would be obtained at total flow rate 4 lpm, hydrogen flow rate 2 lpm, MTS flow rate 0.5 ml/min and 1300 °C deposition temperature.

Meanwhile Tables 9.5 and 9.7 indicate that 93.7 and 83.6 % of the total variance in the deposition rate and growth diameter of SiC respectively were attributed to the studied experimental variables. Comparing ANOVA results (Tables 9.4 and 9.6) and the pooled ANOVA results (Tables 9.5 and 9.7) indicated that after pooling the factor with least variance, the percentage contributions of the remaining factors decreased slightly, but the ranking of the factor effects still remained the same.

From the statistical results obtained it was shown that the developed models are adequate to predict the desired responses in the range of variables under study.

From statistical analysis of both responses it has been concluded that Deposition temperature, MTS flow rate, and hydrogen flow rate are the most significant factors for nanodimensional SiC wire growth.

It has been observed that the optimum deposition temperature for minimum deposition rate was 1500 °C and for minimum growth diameter was 1300 °C but SEM images of experiments 1, 5 and 9 shows that there was grain growth of SiC deposits

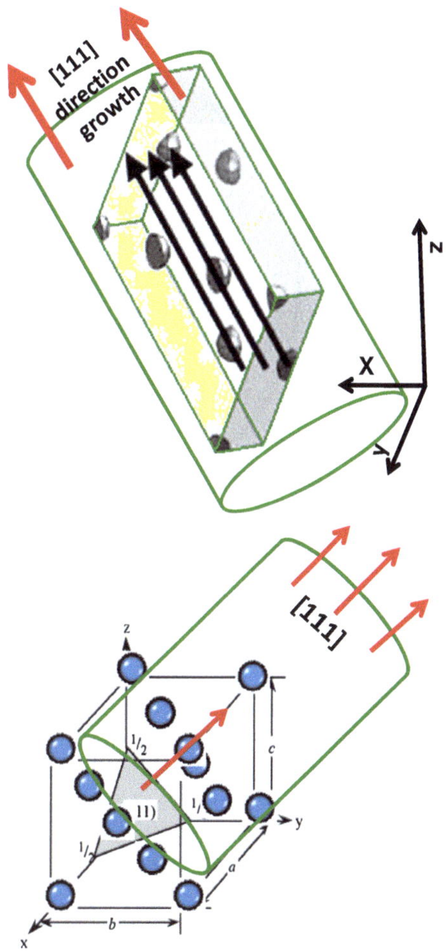

Fig. 9.12 Schematic representation of lowest surface energy growth direction in fcc crystal

at deposition temperature 1500 °C. It confirms that MTS flow rate and hydrogen flow rate are the two important factors which decide the final morphology and diameter of SiC nanowires below particular temperature.

The above observation was supported by Lee's [106] proposition that lower deposition rates and lower temperature for fcc crystals result in development of the lowest surface energy plane, i.e., the {111} plane (Fig. 9.12), while higher temperature and higher deposition rates causes development of higher surface energies, i.e., the {110} plane (Fig. 9.13). At higher temperatures and lower deposition rate the preferred orientation occurs in the most densely populated atom planes or the lowest surface energy crystal facets (i.e. {111} plane in SiC). At higher temperatures and high deposition rates the molecules attached to the substrate have the texture that places the less densely populated atomic plane or higher surface energy lattice plane (i.e. {110} plane in SiC) parallel to the substrate. The surface morphology of the SiC

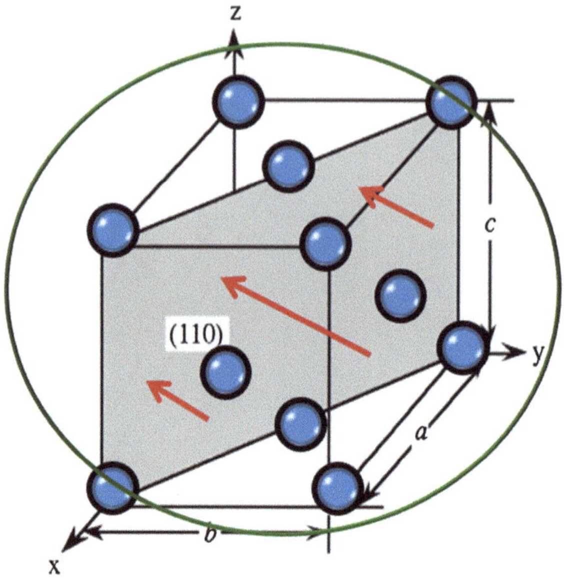

Fig. 9.13 Schematic representation of higher surface energy growth direction in fcc crystal

deposited films largely depend on the preferred orientation. The deposition temperature and the H_2/MTS ratio play an important role in determining the microstructure of deposit. The morphologies of as- deposited SiC crystals are affected primarily by the supersaturating degree of the gaseous reactants which are dependent on the deposition temperature and the concentration of MTS in the reactor. According to the stagnant gas layer model [107], growth rates are limited by surface reaction at lower temperature and by mass transfer of vapor at higher temperature. An increase in both temperature and MTS concentration can enhance the supersaturating degree of the gaseous reactants and the driving force of the chemical reaction. The reaction temperature and the supersaturating degree of gaseous reactants are the macroscales of reflecting the energy value which affect the morphology. The high reaction energy can promote SiC crystals to grow along several directions which results in the formation of grains.

It has been concluded from the above analysis that for nanodimensional diameter growth of SiC wire the temperature of deposition should be 1300 °C, total carrier gas flow rate 4 lpm, hydrogen flow rate 2 lpm, MTS flow rate 0.5–1 ml/min.

The hydrogen reduction of MTS controls the kinetics and chemistry of MTS decomposition. For better understanding of deposition kinetics of SiC by mass transfer mechanism, the knowledge of MTS decomposition mechanism and important reacting intermediates are required. Since in our present study SiC deposition using MTS decomposition was carried out mostly at higher temperature, we suppose that the SiC deposition kinetics was determined by mass transfer mechanism. MTS is a liquid at room temperature (bp = 67.3 C at 1 atm) which can be easily purified by distillation. It is abundant and relatively cheap. Conversely its decomposition on/in a heated substrate in a hydrogen atmosphere, occurs according to the following overall reaction equations.

$$CH_3SiCl_{3(g)} \longrightarrow SiC_{(s)} + 3HCl_{(g)} \tag{9.1}$$

This reaction yields significant amounts of corrosive HCl species. The actual reaction mechanism involved in the formation of SiC from MTS is much more complex and still not well established. It is generally accepted that the MTS decomposes into CH_3 and $SiCl_3$ radicals which in turn react in the gas phase to yield CH_4, $SiCl_2$, $SiCl_4$ and $SiHCl_3$ tentatively according to the following scheme [108].

$$CH_3SiCl_3{}^* \longrightarrow CH_3{}^{\cdot} + SiCl_3{}^{\cdot} \tag{9.2}$$
$$CH_3{}^{\cdot} + H_2 \longrightarrow CH_4 + H^{\cdot} \tag{9.3}$$
$$SiCl_3{}^{\cdot} + SiCl_3{}^{\cdot} \longrightarrow SiCl_2 + SiCl_4 \tag{9.4}$$
$$SiCl_3{}^{\cdot} + H^{\cdot} \longrightarrow SiHCl_3{}^* \tag{9.5}$$
$$SiCl_3{}^* + H^{\cdot} \longrightarrow SiCl_2 + HCl \tag{9.6}$$

where * and · stand for an activated molecule and free radical, respectively. Hence CH_4 and $SiCl_2$, which are very reactive species, could be the main carbon and silicon precursors during SiC deposition. Considering CH_4 and $SiCl_2$ as reactive intermediates the overall reaction can be depicted as follows:

$$CH_3SiCl_3 + H_2 \rightarrow CH_4 + SiCl_2 + HCl \tag{9.7}$$
$$CH_4 + SiCl_2 \rightarrow SiC + 2HCl + H_2. \tag{9.8}$$

First step of the coating process (equation 9.7) associated with the pyrolysis of MTS assisted by hydrogen and production of two major intermediate species namely CH_4 and $SiCl_2$. In second step, CH_4 and $SiCl_2$ react on the supporting substrate surface to form SiC. It has been reported in literature [109] that rate law for deposition of SiC (step 2) is based on the assumption that reaction between CH_4 and $SiCl_2$ is fast. Whereas, the pyrolysis reaction (step 1) to form CH_4 and $SiCl_2$ is slow (rate limiting) and taking place isothermally inside hot chamber. The gas stream (inert gas Ar and MTS + H_2) which enters from the bottom of the reactor isothermally goes up and MTS cracked and react with hydrogen to form CH_4 and $SiCl_2$ and finally arrived to the solid surface where heterogeneous deposition takes place. Pyrolysis reaction is slow may be due to low temperature of mixed gas steam compared to the substrate plate surface. Deposition reaction is fast may be due to the presence of activated intermediate species at the hot surface of the substrate plate and react immediately due to high temperature environment. The above reaction confirms that MTS and hydrogen concentration are very important to decide the concentratration of reactants in gas phase on the deposition surface. It means the hydrogen to MTS ratio affects the gas phase dynamics and adsorption behavior of reactants. As the hydrogen to MTS ratio is an important parameter which decides deposition kinetics, its effect on

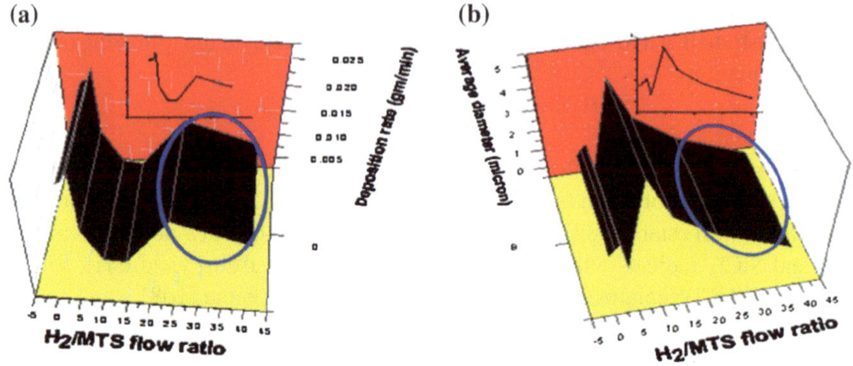

Fig. 9.14 Graphical representation of variation of **a** deposition rate and **b** average growth diameter with hydrogen to MTS ratio. (Reprinted with permission from Ref. [105]. Copyright Bentham Science Publishers)

growth morphology and diameter of SiC nanowires have been studied. The hydrogen to MTS ratio in total carrier gas composition were varied at optimized temperature, i.e., 0 1300 °C and corresponding morphology of SiC wires were studied. The total amount of MTS used for all experiment was kept constant.

The variation of deposition rate and average growth diameter with hydrogen to MTS ratio are shown in Fig. 9.14. It has been observed that there were three different regions depending on the variation of diameter of SiC and deposition rate with H_2/MTS ratio. The Ist region was below H_2/MTS ratio 4 and the second region was between H_2/MTS ratio 4–10. The third region was above H_2/MTS ratio 10 (incircled in the Fig. 9.14).

In the first region it has been observed that when H_2/MTS ratio decreases the diameter of grown SiC wires also decreases but there is a critical position of H_2/MTS ratio below which there is no formation of SiC wires. At this point there was multi dimensional growth of SiC crystals resulting in growth of SiC grains. It has been observed that when H_2/MTS ratio was zero there was no growth of SiC wires. In this region the deposition rate also decreases with decreasing H_2/MTS ratio. The average diameter of grown SiC wires in this region were in several microns. As the deposition rate decreases there was decrease in the diameter of SiC wires and formation of smoother surface with equal diameter distribution of SiC.

In the second region there was random variation in the diameter of the of SiC wires. The SiC wires were randomly grown with unequal deposition.

In the third region it was observed that the diameter of the grown SiC decreases from micron to nanometer range with increasing H_2/MTS ratio. With increasing the H_2/MTS ratio purity of SiC wires gradually increased and when H_2/MTS ratio is about 40, SiC wires without other SiC crystallites were deposited on the C/C composite. The SiC wires grown at this H_2/MTS ratio value were equally distributed on the surface with almost equal distribution of diameter.

These experiments have proved that for uniform diameter growth of SiC, smoothness of surface and homogeneous growth of SiC on the substrate, the hydrogen to

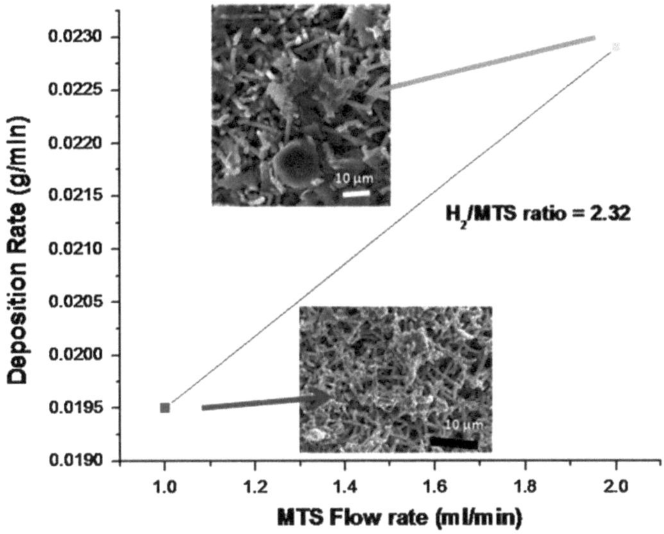

Fig. 9.15 Variation of deposition rate and morphology of SiC deposits with MTS flow rate at constant hydrogen to MTS ratio

MTS ratio must be high (for nanodimensional diameter growth H_2/MTS ratio should be greater than 20). At particular H_2/MTS ratio the MTS flow rate should not be too high. As it has been seen from the Fig. 9.15, that at particular H_2/MTS ratio as the MTS flow rate increases, the diameter of SiC wires increases. The results indicate that the reaction is very fast at higher concentration of MTS, i.e., the decomposition of MTS was fast. It means enough reaction intermediates were generated at high MTS flow. This high concentration of intermediates leads to the increase of diameter of SiC wires. In this situation still the intermediate concentration does not reach the saturation point. As discussed earlier at hydrogen to MTS ratio zero there was grain growth in the SiC deposits. It means at this point the MTS concentration was very high leading to drastic increase in the intermediate concentration. It means enough intermediates were generated which reached the saturation level and this intermediate concentration was enough for grain growth.

Further it was concluded from above observation that the temperature of deposition should be around 1000–1300 °C for crystalline SiC wires and total flow rate should be on moderate side for particular H_2/MTS ratio. It has been observed in Fig. 9.16, that at particular H_2/MTS ratio 14 as the total gas flow rate increases from 2 to 6 lpm the morphology of SiC changes from wire to grain. It can be attributed from above result that at lower total gas flow rate the intermediate reactant concentration was below saturation leading to wire morphology growth. According to the mass transfer mechanism there was less transport of reactive intermediates on the substrate surface. But when the total gas flow rate increases there was efficient transport of intermediates on the surface leading to the drastic increase in the intermediate

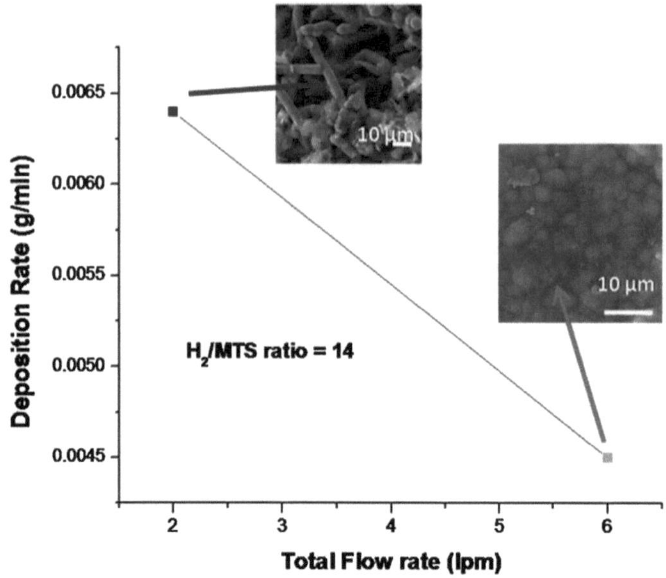

Fig. 9.16 Variation of deposition rate and morphology of SiC deposits with total flow rate at constant hydrogen to MTS ratio

concentration. This concentration reaches to the saturation level leading to the grain growth.

9.5 Growth Mechanism of SiC Nanowires

In the growth process of SiC micron and nanowires, the whole reaction involves the solid and vapor phase. So the SiC wires grown in the present process follow the Vapor Solid (VS) mechanism.

In a typical VS mechanism (Fig. 9.17) when the gaseous reactants come close in contact with the substrate, the absorption of gaseous reactants on the surface occurs. When the concentration of absorbed gaseous reactants reach to a particular value nucleation phenomena starts. In the present case when the concentration of silicon and carbon intermediates reach to a particular value on the surface nucleation of silicon carbide occur. With the continuous supply of feed intermediate reactants on the surface growth of silicon carbide crystals through these nuclei occur. The morphologies of as deposited SiC crystals are affected primary by the supersaturating degree of gaseous reactant, which was depended on the deposition temperature and the concentration of MTS in the process. As it has been seen in taguchi experiments, at higher deposition temperature, the SiC crystal dimensions increased and the morphologies of the as grown SiC crystal change from nanowires to grains.

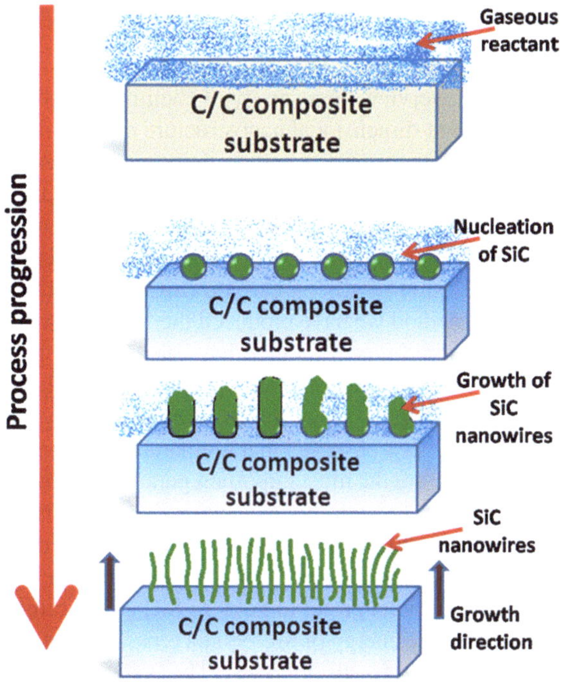

Fig. 9.17 Schematic representation growth mechanism of SiC nanowires

The concentration of MTS plays a vital role in deciding the morphology and nature of the grown SiC. The SiC crystal nuclei apt to grow along the least energy direction, i.e., along <111> direction, until the chemical reaction driving force are less than a certain value. As it has been seen in the first region (Fig. 9.14) that the SiC wires have diameter in micron range and with increasing the MTS concentration the wire shape changes to grain shape. It means in this region the supersaturating degree of the gaseous reactant and the driving force of chemical reaction crosses the limit after a certain concentration of MTS and results in the growth of SiC crystal in several directions.

It has been observed in the third region (Fig. 9.14) that the SiC wires have diameter in nanometer range and with decreasing the MTS concentration the wire diameter decreases. It has been also observed in this region that the deposition rate of SiC decrease with increase in H_2/MTS ratio. The decrease in the concentration of decomposition products at the growth site occur for higher H_2/MTS ratio. This low concentration of reactive intermediates restrains the multi directional growth of SiC. Then the radial directional growth rate of the SiC wires becomes slow and the SiC wires grow along a certain direction and the purity of SiC whiskers is relatively high.

9.6 Conclusions

This chapter provides an overview on a variety of chemical based methods that has been developed to grow one-dimensional nanostructure of silicon carbide. We have briefly discussed about these methods. Among these methods except vapor phase method all other methods have their own limitations for bulk production of morphological pure SiC nanowires. In the vapor phase method we have specially discussed about the chemical vapor deposition hydrogen reduction of methyltrichlorosilane (MTS) method for production of silicon carbide (SiC) nanowires with controlled morphology. Using this method we have produced the SiC nanowires of high purity and homogeneous diameter. Taguchi method has been used to design the experiments to get the optimum parameters for bulk production of silicon carbide wires of uniform diameter in nanometer range. From the analysis of variance (ANOVA) of experimental data it has been observed that temperature of deposition, hydrogen flow rate and MTS flow rate are the most important parameters which affect the final morphology of SiC deposit. The optimum deposition conditions have been obtained by analyzing Signal to Noise (S/N) ratio corresponding to lowest deposition rate and minimum growth diameter of SiC wires. At higher temperature ($\geq$1400 °C) it has been observed that the SiC nuclei prefer to grow as SiC grains rather than wires. At higher temperature the surface energy of SiC crystal crosses a particular value which leads to the multidimensional growth of crystal. The optimum deposition temperature for the growth of crystalline SiC wires should be 1100–1300 °C. A wide range of hydrogen to MTS ration was chosen to study its effect on the growth diameter and deposition rate of SiC nanowires. It has been observed that the hydrogen to MTS flow rate ratio value should be above 20 for the growth of SiC wires of nanometer diameter. At below hydrogen to MTS flow rate ratio value 20, the growth of SiC wires of micron size diameter was more prominent. Even at much lower hydrogen to MTS flow rate ratio value grain growth of SiC deposit occur. It has been observed that the total flow rate of carrier gas comprising of argon and hydrogen should be in moderate range for particular hydrogen to MTS ratio. At comparative higher total flow rate lead to the supersaturation of reactive intermediates on the substrate surface resulting in multidimensional growth of SiC nuclei occur.

A comprehensive study on synthesis of SiC nanowires of high purity and homogeneous diameter by hydrogen reduction of MTS by a simple atmospheric pressure CVD method without using a metallic catalyst has been discussed. Still there are many research opportunities to explore the application of these synthesed SiC nanowires. In coming years it is expected that many research laboratories across the world will take this synthesis method and application of these SiC nanowires to the higher level.

References

1. Wang, Z.L., et al.: Characterizing the structure and properties of individual wire-like nanoentities. Adv. Mater. **12**, 1295 (2000)

2. Hu, J., et al.: Chemistry and physics in one dimension: synthesis and properties of nanowires and nanotubes. Acc. Chem. Res. **32**, 435 (1999)
3. Levenson, M.D., et al.: Welcome to the duv revolution. Solid State Technol. **38**(9), 81–98 (1995)
4. Gibson, J.M., et al.: Reading and writing with electron beams. Phys. Today **50**, 56 (1997)
5. Hong, S.H., et al.: Multiple ink nanolithography: toward a multiple-pen nano-plotter. Science **286**, 523 (1999)
6. Xia, Y., et al.: Unconventional methods for fabricating and patterning nanostructures. Chem. Rev. **99**, 1823 (1999)
7. Yang, P., et al.: Inorganic semiconductor nanowires. Int. J. Nanosci. **1**, 1–39 (2002)
8. Xia, Y., et al.: One-dimensional nanostructures: synthesis, characterization, and applications. Adv Mater. **15**, 353–389 (2003)
9. Givargizov E.I., et al.: Highly anisotropic crystals. In: Senechal M., College S. (eds.) Reidel, Dordrecht (1987)
10. Noll, W., et al.: Synthesis in system $MgO/SiO_2/H_2O$. Z. Anorg. Chem. **261**, 1 (1950)
11. Cheetham, A.K., et al.: Solid State Chemistry (Compounds), p. 31. Clarendon Press, Oxford (1992)
12. Cheetham, A.K., et al.: Chemistry and Physics of One-Dimensional Metals. In: Keller, H.J. (ed.) Plenum Press, New York (1977)
13. Huiyu, C., et al.: Selenium nanowires and nanotubes synthesized via a facile template-free solution method. Mater. Res. Bull **45**, 699–704 (2010)
14. Gates, B., et al.: Synthesis and characterization of uniform nanowires of trigonal selenium. Adv. Funct. Mater. **12**, 219–227 (2002)
15. Mayers, B., et al.: One-dimensional nanostructures of trigonal tellurium with various morphologies can be synthesized using a solution-phase approach. J. Mater. Chem. **12**, 1875–1881 (2002)
16. Messer, B., et al.: Surfactant-induced mesoscopic assemblies of inorganic molecular chains. Adv. Mater. **12**, 1526–1528 (2000)
17. Song, J., et al.: MMo3Se3 (M = Li+, Na+, Rb+, Cs+, NMe4+) nanowire formation via cation exchange in organic solution. J. Am. Chem. Soc. **123**, 9714–9715 (2001)
18. Meyer, K.H., et al.: Propriété des polymères on solution VI. Energie libre et chaleur de solution. Système nitrocellulose-cyclo-hexanone. Helv. Chim. Acta. **61**, 783–790 (1937)
19. Stryer, L., et al.: Biochemistry, 3rd edn. W.H. Freeman and Company, New York (1988)
20. Jones, E.T.T., et al.: Preparation and characterization of molecule-based transistors with a 50-nanometer source-drain separation with use of shadow deposition techniques. Toward faster, more sensitive molecule-based devices. J. Am. Chem. Soc. **109**, 5526 (1987)
21. Jorritsma, J., et al.: General technique for fabricating large arrays of nanowires. Nanotechnology **7**, 263 (1996)
22. Penner, R.M., et al.: Mesoscopic metal particles and wires by electrodeposition. J. Phys. Chem. B **106**, 3339 (2002)
23. Gao, T., et al.: Template synthesis of Y-junction metal nanowires. Appl. Phys. A **74**, 403 (2002)
24. Masuda, H., et al.: Ordered metal nanohole arrays made by a two-step replication of honeycomb structures of anodic alumina. Science **268**, 1466 (1995)
25. Brumlik, C.J., et al.: Template synthesis of metal microtubule ensembles utilizing chemical, electrochemical, and vacuum deposition techniques. J. Mater. Res. **9**, 1174 (1994)
26. Cao, H., et al.: Sol–gel template synthesis of an array of single crystal CdS nanowires on a porous alumina template. Adv. Mater. **13**, 1393 (2001)
27. Hoyer, P., et al.: Formation of a titanium dioxide nanotube array. Langmuir **12**, 1411 (1996)
28. Li, Y., et al.: Ordered semiconductor ZnO nanowire arrays and their photoluminescence properties. Appl. Phys. Lett. **76**, 2011 (2000)
29. Kwan, S., et al.: Synthesis and assembly of $BaWO_4$ nanorods. Chem. Commun. **5**, 447 (2001)
30. Yu, Y.-Y., et al.: Gold nanorods: electrochemical synthesis and optical properties. J. Phys. Chem. B **101**, 6661 (1997)

31. Yin, Y., et al.: Silver nanowires can be directly coated with amorphous silica to generate well-controlled coaxial nanocables of silver/silica. Nano Lett. **2**, 427 (2002)
32. He, R.R., et al.: Functional bimorph composite nanotapes. Nano Lett. **2**, 1109 (2002)
33. Zhang, Y., et al.: Metal coating on suspended carbon nanotubes and its implication to metal-tube interaction. Chem. Phys. Lett. **331**, 35 (2000)
34. Trentler, T.J., et al.: Solution–liquid–solid growth of crystalline III–V semiconductors: an analogy to vapor–liquid–solid growth. Science **270**, 1791–1794 (1995)
35. Trentler, J.J., et al.: Solution-liquid-solid growth of indium phosphide fibers from organometallic precursors: elucidation of molecular and nonmolecular components of the pathway. J. Am. Chem. Soc. **119**, 2172–2182 (1997)
36. Markowitz, P.D., et al.: Phase separation in $Al_xGa_{1-x}As$ nanowhiskers grown by the solution-liquid-solid mechanism. J. Am. Chem. Soc. **123**, 4502–4511 (2001)
37. Lourie, O.R., et al.: CVD growth of boron nitride nanotubes. Chem. Mater. **12**, 1808–1810 (2000)
38. Yao, J., et al.: Solvothermal synthesis and characterization of CdS nanowires/PVA composite films. Mater. Lett. **59**, 3652–3655 (2005)
39. Cheng, H.M., et al.: Formation of branched ZnO nanowires from solvothermal method and dye-sensitized solar cells applications. J. Phys. Chem. C. **112**(42), 16359–16364 (2008)
40. Peng, X.G., et al.: Shape control of CdSe nanocrystals. Nature **404**, 59 (2000)
41. Manna, L., et al.: Synthesis of soluble and processable rod-, arrow-, teardrop-, and tetrapod-shaped CdSe nanocrystals. J. Am. Chem. Soc. **122**, 12700 (2000)
42. Sun, Y., et al.: Crystalline silver nanowires by soft solution processing. Nano Lett. **2**, 165 (2002)
43. Korgel, B.A., et al.: Self-assembly of silver nanocrystals into two-dimensional nanowire arrays. Adv. Mater. **10**, 661 (1998)
44. Wyrwa, D., et al.: One-dimensional arrangements of metal nanoclusters. Nano Lett. **2**, 419 (2002)
45. Taylor, G.F.: A method of drawing metallic filaments and a discussion of their properties and uses. Phys. Rev. **23**, 655 (1924)
46. Penner, R.M., et al.: Fabrication and use of nanometer-sized electrodes in electrochemistry. Science **250**, 1118 (1990)
47. Levitt, A.P.: Whisker Technology, Wiley-Interscience, New York (1970)
48. Volmer, M., et al.: The mechanism of molecular condensation on crystals. Z. Phys. **7**, 13 (1921)
49. Sears, G.W., et al.: A mechanism of whisker growth. Acta Metall. **3**, 367 (1955)
50. Zhang, Y.: A simple method to synthesize nanowires. Chem. Mater. **14**, 3564 (2002)
51. Dai, Z.R., et al.: Tin oxide nanowires, nanoribbons, and nanotubes. J. Phys. Chem. B **106**, 1274 (2002)
52. Hayashi, S., et al.: Growth of magnesia whiskers by vapor-phase reactions. J. Cryst. Growth **24/25**, 345 (1974)
53. Wolfe, E.G., et al.: Growth and morphology of magnesium oxide whiskers. J. Am. Ceram. Soc. **48**, 279 (1965)
54. Duan, X.F., et al.: Laser-assisted catalytic growth of single crystal GaN nanowires. J. Am. Chem. Soc. **122**, 188 (2000)
55. Zhang, Y.J., et al.: Synthesis of thin Si whiskers (nanowires) using $SiCl_4$. J. Cryst. Growth **226**, 185 (2001)
56. Chen, C.C., et al.: Catalytic growth and characterization of gallium nitride nanowires. J. Am. Chem. Soc. **123**, 2791 (2001)
57. Wang, Y.W., et al.: Catalytic growth and photoluminescence properties of semiconductor single-crystal ZnS nanowires. Chem. Phys. Lett. **357**, 314 (2002)
58. Huang, M.H., et al.: Catalytic growth of Zinc oxide nanowires by vapor transport. Adv. Mater. **13**, 113 (2000)
59. Iijima, S.: Helical microtubules of graphitic carbon. Nature (London) **354**, 56 (1991)
60. Chopra, N.G., et al.: Boron nitride nanotubes. Science **269**, 966 (1995)

61. Tenne, R., et al.: Polyhedral and cylindrical structures of tungsten disulphide. Nature (London) **360**, 444 (1992)
62. Feldman, Y., et al.: High-rate, gas-phase growth of MoS_2 nested inorganic fullerenes and nanotubes. Science **267**, 222 (1995)
63. Weng-Sieh, Z., et al.: Synthesis of BxCyNz nanotubules. Phys. Rev. B **51**(11), 229 (1995)
64. Han, W.Q., et al.: Continuous synthesis and characterization of silicon carbide nanorods. Chem. Phys. Lett. **265**, 374 (1997)
65. Yang, P.D., et al.: Nanorod-superconductor composites: a pathway to materials with high critical current densities. Science **273**, 1836 (1996)
66. Xu, X.L., et al.: Tem characterization of calcium-oxygen nanorods. Nanostruct. Mater. **8**(3), 373 (1997)
67. Han, W.Q., et al.: Synthesis of gallium nitride nanorods through a carbon nanotube-confined reaction. Science **277**, 1287 (1997)
68. Han, W.Q., et al.: Synthesis of silicon nitride nanorods using carbon nanotube as a template. Appl. Phys. Lett. **71**(16), 2271 (1997)
69. Ono, T., et al.: Si nanowire growth with ultrahigh vacuum scanning tunneling microscopy. Appl. Phys. Lett. **70**(14), 1852 (1997)
70. Wong, E.W., et al.: Nanobeam mechanics: elasticity, strength, and toughness of nanorods and nanotubes. Science **277**, 1971 (1997)
71. Chu, Yanhui, et al.: Influence of SiC nanowires on the properties of SiC coating for C/C composites between room temperature and 1500oC. Corros. Sci. **53**, 3048–3053 (2011)
72. Menga, Shuai: et al.: Tailoring and application of SiC nanowires in composites. Materials Science and Engineering A **527**, 5761–5765 (2010)
73. Chena, Jianjun, et al.: Field emission performance of SiC nanowires directly grown on graphite substrate. Mater. Chem. Phys. **126**, 655–659 (2011)
74. Karuppanan, S., et al.: Enhanced field emission from density-controlled SiC nanowires. Mater. Chem. Phys. **112**, 88–93 (2008)
75. Li, Ke-Zhi, et al.: Photoluminescence of hexagonal-shaped SiC nanowires prepared by sol–gel process. Mater. Sci. Eng. A **460–461**, 233–237 (2007)
76. Hao, J.Y., et al.: Photocatalytic hydrogen production over modified SiC nanowires under visible light irradiation. Inter. J. Hydrogen Ener **37**, 15038–15044 (2012)
77. Wanga, H.Y.: Capacitive humidity sensing properties of SiC nanowires grown on silicon nanoporous pillar array. Sens Actuators, B **166—-167**, 451–456 (2012)
78. Liu, H.: Porous SiC nanowire arrays as stable photocatalyst for water splitting under UV irradiation. Mater. Res. Bull. **47**, 917–920 (2012)
79. Chu, Y.: SiC coating toughened by SiC nanowires to protect C/C composites against oxidation. Ceram. Inter. **38**, 189–194 (2012)
80. Choi, H.-J.: Continuous synthesis of silicon carbide whiskers. J. Mater. Sci. **30**, 1982 (1995)
81. Meng, G.W., et al.: Preparation of β-SiC nanorods with and without amorphous SiO_2 wrapping layers. J. Mater. Res. **13**, 2533 (1998)
82. Yang, W.Y., et al.: Synthesis of silicon carbide nanorods by catalyst-assisted pyrolysis of polymeric precursor. Chem. Phys. Lett. **383**, 441 (2004)
83. Shen, G., et al.: Silicon carbide hollow anospheres, nanowiresand coaxial nanowires. Chem. Phys. Lett. **375**, 177–184 (2003)
84. Chiu, S.C., et al.: SiC nanowires in large quantities: synthesis, band gap characterization, and photoluminescence properties. J. Cryst. Growth **311**, 1036 (2009)
85. Zhu, J., et al.: Nanostructure of GaN and SiC nanowires based on carbon nanotubes. J. Mater. Res. **14**, 1175 (1999)
86. Park, B.T., et al.: Growth and characterization of silicon carbide nanowires. Surf. Rev. Lett. **11**, 373 (2004)
87. Ryu, Y., et al.: Direct growth of core-shell SiC–SiO_2 nanowires and field emission characteristics. Nanotechnology **16**, S370 (2005)
88. Liu, D.F., et al.: A simple large-scale synthesis of coaxial nanocables:silicon carbide sheathed with silicon oxide. Chem. Phys. Lett. **375**, 269–272 (2003)

89. Choi, H.-J., et al.: Growth and modulation of silicon carbide nanowires. J. Cryst. Growth **269**, 472 (2004)
90. Zhou, X.T., et al.: Thin β-SiC nanorods and their field emission properties. Chem. Phys. Lett. **318**, 58 (2000)
91. Kim, W.J., et al.: Growth of SiC nanowires within stacked SiC fiber fabrics by a noncatalytic chemical vapor infiltration technique. J. Cryst. Growth **300**, 503–508 (2007)
92. Yang, W., et al.: In situ growth of SiC nanowires on RS-SiC substrate(s). J. Cryst. Growth **264**, 278–283 (2004)
93. Takao, S., et al.: MOCVD growth of spherical aggregates of SiC nanocrystallites. Appl. Surf. Sci. **254**, 7630–7632 (2008)
94. Attolini, G., et al.: A new growth method for the synthesis of 3C-SiC nanowires. Mater. Lett. **63**, 2581–2583 (2009)
95. Ju, Z., et al.: High-yield synthesis of single-crystalline 3C-SiC nanowires by a facile autoclave route. Mater. Lett. **61**, 3913–3915 (2007)
96. Lespiaux, D., et al.: Chemisorption on β-SiC and amorphous SiO_2 during CVD of silicon carbide from the Si–C–H–Cl system. Correlations with the nucleation process. Thin Solid Films **265**, 40–51 (1995)
97. Fu, Q.G., et al.: Synthesis of silicon carbide nanowires by CVD without using a metallic catalyst. Mater. Chem. Phys. **100**, 108–111 (2006)
98. Wang, F.L., et al.: SiC nanowires synthesized by rapidly heating a mixture of SiO and arc-discharge plasma pretreated carbon black. Nanoscale Res. Lett. **4**, 153 (2009)
99. Chen, J., et al.: A simple catalyst-free route for large-scale synthesis of SiC nanowires. J. Alloys and Compd. **509**, 6844–6847 (2011)
100. Longkullabutra, H., et al.: Large-scale: synthesis, microstructure, and FT-IR property of SiC nanowires. Curr. Appl. Phys. **12**, S112–S115 (2012)
101. Zhou, W.M., et al.: Large-scale synthesis and characterization of SiC nanowires by high-frequency induction heating. Appl. Surf. Sci. **252**, 5143–5148 (2006)
102. Wei, J., et al.: Large-scale synthesis and photoluminescence properties of hexagonal-shaped SiC nanowires. J. Alloys Compd. **462**, 271–274 (2008)
103. Sharma, P., et al.: Process parameter selection for strontium ferrite sintered magnets using Taguchi L9 orthogonal design. J. Mate. Process. Technol. **168**, 147–151 (2005)
104. Khoeia, A.R., et al.: Design optimisation of aluminium recycling processes using Taguchi technique. J. Mater. Process. Technol. **27**, 96–106 (2002)
105. Prakash, J., et al.: Taguchi method optimization of parameters for growth of nano dimensional SiC wires by chemical vapor deposition technique. Curr. Nanosci. **8**, 161–169 (2012)
106. Lee, D.N.: A model for development of orientation of vapour deposits. J. Mater. Sci. **24**, 4375–4378 (1989)
107. Alwndrof, M.D., et al.: A model of silicon carbide chemical vapor deposition. J. Electrochem. Soc. **138**(3), 841–852 (1991)
108. Sotirchos, S.V., et al.: On the homogeneous chemistry of the thermal decompositionof methyl-trichlorosilane: thermodynamic analysis and kinetic modeling. J. Electrochem. Soc. **141**, 1599–1627 (1994)
109. Joseik, A., et al.: Residence-time dependent kinetics of CVD growth of SiC in the MTSH2 system. J. Cryst. Growth **160**, 253–260 (1996)
110. Yang, W., et al.: Fabrication in-situ SiC nanowires/SiC matrix composite by chemical vapour infiltration process. Mater. Lett. **58**, 3145–3148 (2004)
111. Wei, J., et al.: Fabrication of composite structure of carbon fibers and high density SiC nanowires. Physica E **41**, 1810–1813 (2009)
112. Sun, X.H., et al.: Formation of silicon carbide nanotubes and nanowires via reaction of silicon (from disproportionation of silicon monoxide) with carbon nanotubes. J. Am. Chem. Soc. **124**, 14464–14471 (2002)
113. Chen, J., et al.: Growth mechanism of twinned SiC nanowires synthesized by a simple thermal evaporation method. Physica E **42**, 2335–2340 (2010)

114. Chiew, Y.L., et al.: Growth of SiC nanowires using oil palm empty fruit bunch fibres infiltrated with tetraethyl orthosilicate. Physica E **44**, 2041–2049 (2012)
115. Jian, W., et al.: Growth and morphology of one-dimensional SiC nanostructures without catalyst assistant. Mater. Chem. Phys. **95**, 140–144 (2006)
116. Zhaoqian, L., et al.: Growth mechanism of silica nanowires without a metal catalyst via oxyacetylene torch ablation. Mater. Lett. **74**, 118–120 (2012)
117. Dhiman, R., et al.: Growth of SiC nanowhiskers from wooden precursors, separation, and characterization. Ceram. Inter. **37**, 3759–3764 (2011)
118. Lee, J.S., et al.: Improvement of porous silicon carbide filters by growth of silicon carbide nanowires using a modified carbothermal reduction process. J. Alloys Compd **467**, 543–549 (2009)
119. Lee, J.S., et al.: In situ growth of SiC nanowires by carbothermal reduction using a mixture of low-purity SiO_2 and carbon. J. Alloys Compd **456**, 257–263 (2008)
120. Huang, H., et al.: In situ growth of silicon carbide nanowires from anthracite surfaces. Ceram. Inter. **37**, 1063–1072 (2011)
121. Chu, Y., et al.: Microstructure and growth mechanism of SiC nanowires with periodically fluctuating hexagonal prisms by CVD. J. Alloys Compd **508**, L36–L39 (2010)
122. Xin, L., et al.: Morphological evolution of one-dimensional SiC nanomaterials controlled by sol–gel carbothermal reduction. Mater. Charact. **65**, 55–61 (2012)
123. Wei, J., et al.: Photoluminescence performance of SiC nanowires, whiskers and agglomerated nanoparticles synthesized from activated carbon. Physica E **41**, 1616–1620 (2009)
124. Taguchi, T., et al.: Preparation and characterization of single-phase SiC nanotubes and C–SiC coaxial nanotubes. Physica E **28**, 431–438 (2005)
125. Guo, J.Z., et al.: Preparation of SiC nanowires with fins by chemical vapor deposition. Physica E **39**, 262–266 (2007)
126. Li, X., et al.: Preparation of silicon carbide nanowires via a rapid heating process. Mater. Sci. Eng B **176**, 87–91 (2011)
127. Li, G., et al.: SiC nanowires grown on activated carbon in a polymer pyrolysis route. Mater. Sci. Eng. B **166**, 108–112 (2010)
128. Zhao, H., et al.: Silicon carbide nanowires synthesized with phenolic resin and silicon powders. Physica E **41**, 753–756 (2009)
129. Khongwong, W., et al.: Simple approach to fabricate SiC–SiO_2 composite nanowires and their oxidation resistance. Mater. Sci. Eng. B **173**, 117–121 (2010)
130. Li, B., et al.: Simultaneous growth of SiC nanowires, SiC nanotubes, and SiC/SiO_2 core–shell nanocables. J. Alloys Compd. **462**, 446–451 (2008)
131. Kang, P., et al.: Synthesis of SiO_2 covered SiC nanowires with milled Si,C nanopowders. Mater. Lett. **65**, 3461–3464 (2011)
132. Zhang, L.D., et al.: Synthesis and characterization of nanowires and nanocables. Mater. Sci. Eng. **A286**, 34–38 (2000)
133. Chen, J., et al.: Synthesis and photoluminescence of needle-shaped 3C-SiC nanowires on the substrate of PAN carbon fiber. J. Alloys Compd **456**, 320–323 (2008)
134. Zhang, E., et al.: Synthesis and photoluminescence property of silicon carbon nanowires synthesized by the thermal evaporation method. Physica E **41**, 655–659 (2009)
135. Wei, J., et al.: Synthesis of centimeter-scale ultra-long SiC nanowires by simple catalyst-free chemical vapor deposition. J. Cryst. Growth **335**, 160–164 (2011)
136. Niu, J.J., et al.: Synthesis of macroscopic SiC nanowires at the gram level and their electrochemical activity with Pt loadings. Acta Mater. **57**, 3084–3090 (2009)
137. Zhang, H.X., et al.: Synthesis of nanostructured SiC using the pulsed laser deposition technique. Materi. Res. Bull. **44**, 184–188 (2009)
138. Raman, V., et al.: Synthesis of silicon carbide nanofibers from pitch blended with sol–gel derived silica. Mater. Lett. **60**, 3906–3911 (2006)

Chapter 10
Adhesion and Indentation Fracture Behavior of Silicon Carbonitride Nanocomposite Coatings Deposited by Magnetron Sputtering

S. K. Mishra and A. S. Bhattacharyya

10.1 Introduction

The ternary nanocomposite material Si–C–N was first introduced to the scientific community as a high temperature oxidation resistant polymer-derived ceramic (PDCs) [1]. PDCs are a new class of ceramics that combine the functional properties of polymers with the mechanical and chemical durability of ceramics. Polymer-derived ceramics (PDCs) have a polymer-like nanostructure and ceramic-like properties, e.g., creep and oxidation resistance [2, 3].

Si–C–N has also been produced as nanocomposite thin films and are very promising material for the future. They have been proved ideal for applications as wear and oxidation resistance, optoelectronic material in hostile environments. Si–C–N coatings are desired for many industrial applications such as turbine engines, blades, and wear resistant coatings for the automotive industry to enhance the life as well as the performances of the components.They also find application in MEMS device fabrication in the form of field emission displays, catalyst support high temperature semiconducting devices, metal and polymer matrix composites, and high temperature applications, smooth surface morphology, stable amorphous structure, and good prospect electronic devices [4, 5]. Elastic and plastic deformation behavior of precursor-derived Si–C–N ceramics at room temperature under contact loading was investigated using nano- and microindentation experiments [6].

An alternative way of forming Si–C–N films is by *magnetron sputtering* using SiC targets in argon/nitrogen plasma. This method allows one to produce mainly Si–C–N amorphous films. The lower pressure in the chamber helps create a cleaner

S. K. Mishra (✉)
CSIR-National Metallurgical Laboratory, Jamshedpur , Jharkhand 831007, India
e-mail: suman@nmlindia.org

A. S. Bhattacharyya
Central University of Jharkhand, Ranchi 835205, India

H. Li et al. (eds.), *Silicon-based Nanomaterials*,
Springer Series in Materials Science 187, DOI: 10.1007/978-1-4614-8169-0_10,
© Springer Science+Business Media New York 2013

film. Target temperature is lower with magnetron sputtering enhancing the deposition of high quality films [5].

The X-ray diffraction (XRD) analysis of the SiC target used as the sputtering source showed the presence of $\beta-$SiC and C phases on the diffraction pattern. However, the XRD patterns of all the films deposited at different pressures showed amorphous nature. Even if a small amount of crystallization had taken place during deposition of these films, the XRD was not able to show any characteristic peak in the diffraction pattern and due to small volume percentage, only peak corresponding to the substrate was visible [7, 8]. This is a common phenomenon for the crystallization of Si–C–N materials, which has been observed by other groups [9].

For engineering applications, hardness must be complemented with high toughness, which is a property of equal importance as hardness. Absolute hardness and fracture toughness measurement of coatings are, however, complicated issues as there is a significant influence of the substrate. The hardness values obtained through microindentation are basically the measured hardness or composite hardness of the film-substrate systems. This composite hardness is an index of indentation load [10–12]. In case of coating-substrate system, a majority of the authors apply models separating the contribution of the substrate from the hardness measurement, considering various factors [13–21]. Depth sensing nanoindentation is, however, a more appropriate means for obtaining absolute mechanical properties of the coatings [22]. We have used both microindentation and nanoindentation to get SiCN coating hardness in our earlier publications [23–25].

For fracture toughness calculation corresponding to the half-penny (HP) crack and/or the Palmqvist crack, knowledge of the shape of the crack is required; Lube [26] has amply shown that the crack shape can get changed from a Palmqvist to an HP one, according to the applied load; and which is why, to avoid such a difficult experimental determination, Chicot et al. [27] have recently developed a methodology leading to a toughness parameter that was calculated by a relationship independent of the crack shape.

Evaluation of coating substrate adhesion strength is also very important, as it is often the primary failure mechanism of the coatings [17]. Adhesion measurements with the help of scratch tests are well-known techniques because of ease of operation and good reproducibility [28]. A primary study on the adhesion of Si–C–N coatings by micro-scratching was reported by us previously [29]. Measure and control of coating substrate adhesion strength is important, as it is often the primary failure mechanism of the coatings [30]. A stylus made of diamond is allowed to scratch the coating with an increasing load. In principle, the stylus induces a shear force in the nearby coating which is proportional to the applied load. The induced shear force is transmitted through the bulk of the composite sample. Scratch test is most effective for hard coatings on hard substrate as the substrate does not deform plastic to a great extent [31]. As the mechanical properties of the coating and the substrate are different, there is a discontinuity in the shear stress at the interface; which when sufficiently high, induces adhesive failure at the critical load [32].

The energy required to form the new surface generated by spallation of the coating must be balanced by release of stored elastic energy. Therefore, the calculated work of adhesion provides us with the values of interfacial energies assuming the condition that they are released when interfacial failure occurs. The energy released comprises three components; (1) stored energy around indentation, (2) stored energy due to internal stress present, and (3) stored energy due to frictional stresses. The interfacial toughness is also a measure of the nature of adhesion of the coating. Similar to the work of adhesion, it possesses three components related to indentation, internal stress, and friction. Failure of the film occurs when the coating starts to peel off from the substrate. The normal applied load at that point is called the critical load (Lc), which gives a qualitative estimation of the adhesion of the film. The change in the slope of the tractional force curve is the point at which the peeling starts to occur. The failure of coating in scratch tests depends on the hardness of both the coating and the substrate. The critical load depends on many factors like the mechanical strength (adhesion, cohesion) of coating substrate composite, loading rate scratch speed, and indenter tip radius indenter material. The coating substrate-specific parameters that are important for scratch adhesion are mainly substrate hardness and roughness, coating thickness, frictional coefficients between coating and indenter, and internal stresses in the coating. In this chapter, our investigation of the adhesion and deformation behavior of the magnetron sputtered silicon carbon nitride (SiCN) nanocomposite thin film on various substrates has been presented.

10.2 Deposition of SiCN Coating by Magnetron Sputtering and Measurements

The nanocomposite SiCN thin films were deposited using 50 mm disk SiC targets, sintered at 1950 °C in a graphite furnace. The SiCN film deposition was carried out at 1 Pa, at RF 400W, and at 400 °C substrate temperature by magnetron sputtering (HHV, Banglore, India). The films were deposited on Silicon (100), Borosilicate glass and SS 304 substrates. Microhardness tester (Leica, Germany) was used to determine the hardness and indentation fracture studies of the films by indenting at different loads. The nanoindentation tests were carried out for 500 nm depths in continuous stiffness mode (CSM) using XP nanoindenter (MTS) having a Berkovich indenter. The standard diamond Rockwell C indenter tip of 200 μm radius was used during scratch with increasing load from 0–50 N. The scratch rate was fixed as 0.2 mm/s. The scratch experiments were also carried out at different rate of loadings during the adhesion studies to understand the failure mechanism.

10.3 Indentation Fracture Behavior of the Nanocomposite SiCN Coatings

A typical AFM picture and TEM of nanocomposite film is shown in Fig. 10.1. The films had particle size in the range of 10–20 nm in amorphous matrix. The maximum hardness and modulus of the films were about 22 and 240 GPa, respectively, as measured by nanoindentation [23, 24].

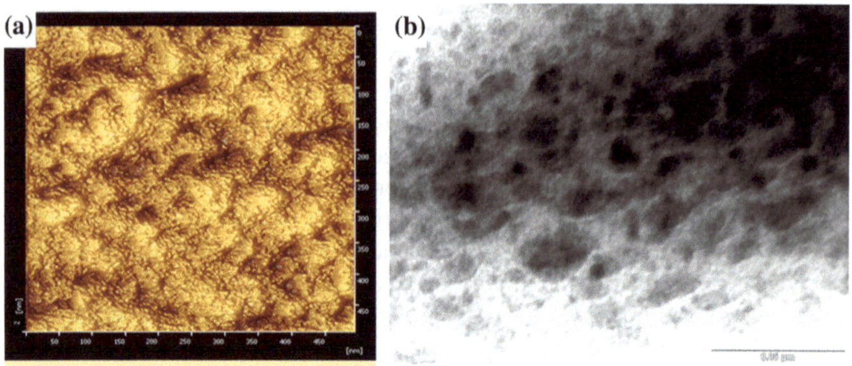

Fig. 10.1 **a** AFM and **b** TEM images of nanocompositeSiCN coatings

Failure in terms of development of cracks was observed by indenting at higher loads on coatings on all three substrates (304SS, Glass and Silicon). The failure behavior was found to be dependent on thickness, hardness, and nature of substrate. During indentation, the irreversible deformation within the contact zone does not allow full recovery of the initial stress-free state in the surrounding elastic matrix. This results in the building up of the residual stress field which provides the primary driving force for the radial crack formation. The initiation and propagation of these cracks are the two most important aspects of indentation fracture mechanics. Crack initiation begins from flaws which are either pre-existing or induced by the indentation process itself. The material-specific behavior under indentation is reflected in the central contact impression and surface traces of the crack systems. The major crack types are *conical, radial, and lateral.* The contact stress field developed beneath the indenter is of primary importance for failure studies. Indentations which are elastic in nature up to the point of fracture are called blunt and are usually observed in case of spherical indenter. The indentations produced by sharp pyramid indenters like Vickers in this case, are plastic in nature and are called *sharp.* However, due to the presence of an elastic matrix around the plastically deformed region, the contact is termed as *elastic-plastic* [33, 34]. The advantage of sharp indenters of the constant geometry of residual impression makes it more applicable for the fracture studies.

The elastic-plastic indentation can be divided into three zones: (1) an outwardly expanding core immediately below the indenter, (2) an ideally plastic region encasing the core, and (3) an elastic matrix beyond the plastic region (Fig. 10.2). The major crack types are *conical, radial, and lateral* [33, 34].

Figure 10.3a shows formation of conical cracks on coatings deposited on steel substrate using Rockwell C indenter besides a Vickers Indentation. Cone cracks are usually generated by the elastic loading of spherical indenters. However these cracks can also occur for sharp Vickers Indenter. At higher loads the crack propagates downwards which results in cone cracks. Upon unloading the cone crack tends to

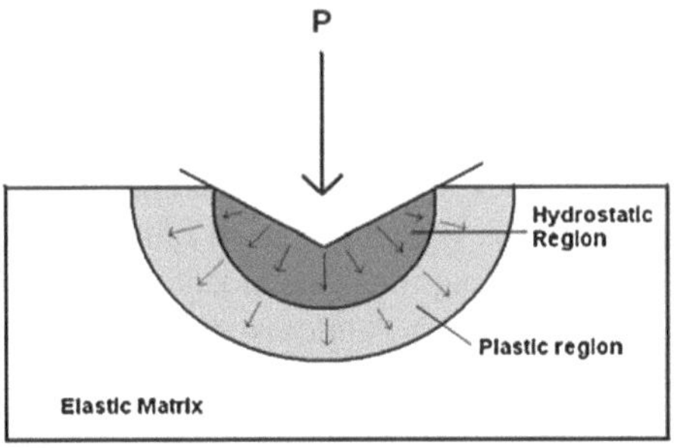

Fig. 10.2 The elastic-plastic zone during indent [33, 34]

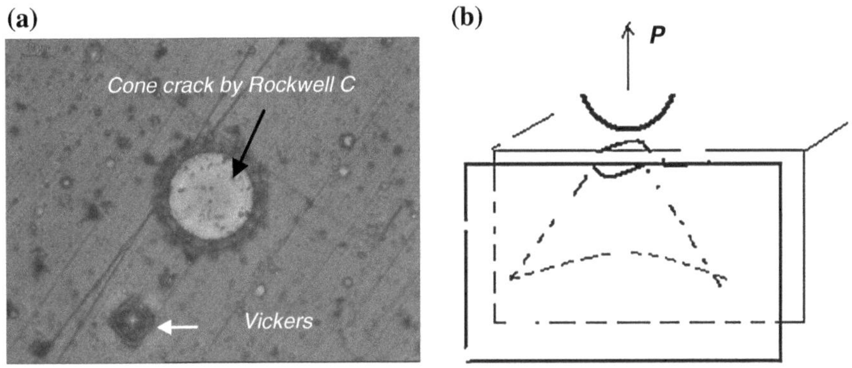

Fig. 10.3 **a** Optical micrograph and **b** schematic representation of cone crack by Rockwell C indenter on Si–C–N coatings deposited on 304 SS substrates

close and a surface circular crack becomes visible as observed in the above-mentioned figure and shown schematically in Fig. 10.3b.

The radial cracks are of two types: median and palmqvist. The median crack initiates below the contact point at the elastic-plastic interface where the stress concentration is maximum. These types of cracks are a combination of two mutually perpendicular penny-shaped cracks which lie in directions parallel to the indenter diagonal. With increase in loads the median cracks become unstable and propagate downwards until their maximum size is reached at the peak load. On unloading they close up below the surface but expand toward the indented surface forming the final half-penny shaped crack. Since the surface cracks are radial in nature, such crack systems are called the *median-radial*. When the applied loads are smaller than the critical load for median crack formation or when the specimen has rough surface, shallow radial cracks termed as Palmqvist cracks are formed. On unloading, nucleation of shallow radial cracks occur on the specimen surface at the edge of the plastic indentation impression. Propagation of these cracks is prevented until unloading. Just prior to complete removal of the indenter, they start to grow both in length and depth parallel to the load axis. The maximum depth is of the order of the indentation depth while the crack shape differs from elliptical to circular; such partially formed cracks tend to propagate downward and merge beneath the plastic zone into half-penny geometry. The distinction between radial-median and radial-palmqvist cracks has been made by using the crack length *(a)* to indentation diagonal *(d)* ratio *(a/d)*. For $a/d \leq 2.5$ the crack profile is Palmqvist. In other words, if the radial crack length is greater than or equal to four times the indentation diagonal length then only is it a median crack, otherwise it is a Palmqvist crack. The lateral cracks are generated below the surface regardless of whether the threshold for the median cracks initiation is exceeded. The maximum stress for lateral crack propagation occurs at the depth of the contact impression. Lateral cracks nucleate during unloading and grow outwards from the site of plastic deformation in a plane parallel to the surface. If the applied load is too high for the given specimen, then these lateral cracks tend to delaminate the coating by diverting upwards toward the surface and result into material removal from the surface called *chipping*. The radial cracks are associated with strength degradation, whereas the lateral cracks are linked to erosion and wear due to tendency of chipping.

10.4 Coatings on Silicon Substrates

Figure 10.4 shows the optical micrographs of indentations made at different loads on the SiCN coatings deposited on silicon substrates having thickness of 2.5 μm. A bright area surrounding the indentation impression on increasing the load was observed due to pileup (Fig. 10.4b). Radial cracks were first observed at 100 gf which were accompanied by lateral cracks containing concentric shear faults or Newton's rings owing to the interference of partially delaminated coating (Fig. 10.4c). The length of the cracks increased for higher loads (Fig. 10.4b, d). These lateral cracks eventually lead to chipping of the coating. The radial cracks are associated with strength degradation, while the lateral cracks are linked to erosion and wear. Details regarding the mechanism of initiation and propagation of these cracks are provided in the literature [33–35] and also explained above.

10.5 Evaluation of Fracture Toughness

The crack length of the radial crack was found to increase with increase in load. The variation of $\boldsymbol{a}^{3/2}$ with load is shown in Fig. 10.5. The slope of the curve was used in the Antis formula to determine the fracture toughness of the coating given in Equation 10.1

$$K_C = \beta \cdot \left(\frac{E}{H}\right)^{\frac{1}{2}} \cdot \left(\frac{P}{a^{3/2}}\right) \tag{10.1}$$

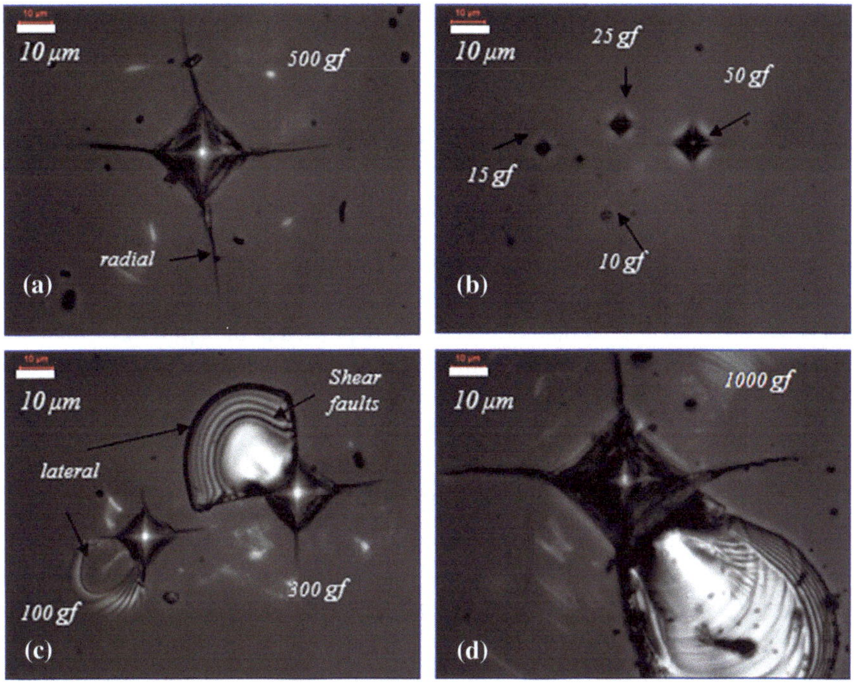

Fig. 10.4 Optical micrograph showing radial and lateral cracking at **a** 100 gf **b** 10, 15, 25, and 50 gf **c** 300, 500 gf, and **a** 1000 gf of SiCN coatings deposited on silicon substrates

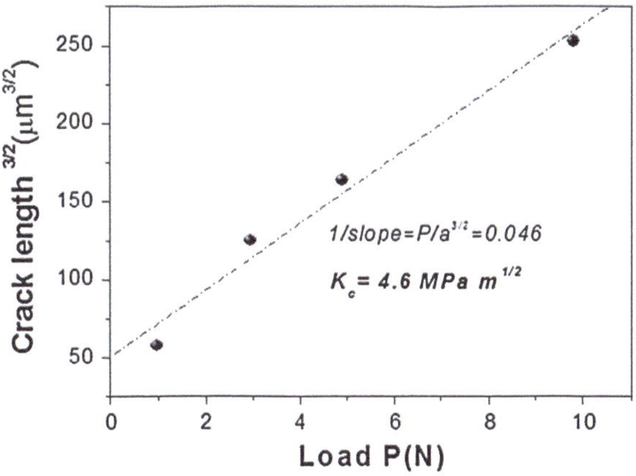

Fig. 10.5 The variation of crack length $a^{3/2}$ with load during indentation

where β is a parameter depending on the geometry of the target and is equal to 0.032 for Vickers indenters. P is the applied load and 'a' is the crack length. The ratio of E/H, where E is the bulk modulus and H is the hardness, is ~10 typical for hard coatings, as measured by nanoindentation. The fracture toughness of the coating came out to be 4.6 MPa$\sqrt{m}$, which is a good value of toughness along with hardness of 22 GPa.

One interesting phenomenon that was observed was the diversion of radial cracks, which occurred as the delamination process started at which higher loads led to *four blade table fan*-like deformation morphology on chipping (Fig. 10.6) The threshold load for radial crack formation was 300 gf for the coating which was 4.4 μm thick and showed the highest hardness of 4445 HV deposited at the optimized condition mentioned in the previous section. For loads up to 200 gf perfect indentation impression with no visible radial crack was observed (Fig. 10.6a) . On increasing the load to 300gf, radial cracks were seen to form at the edges of the indenter impression (Fig. 10.6b). As the length of the cracks was smaller than the indentation diagonal, palmqvist cracks beneath the surface must have occurred. At 500 gf, lateral crack initiation joining two adjacent radial cracks causing delamination (evident from the

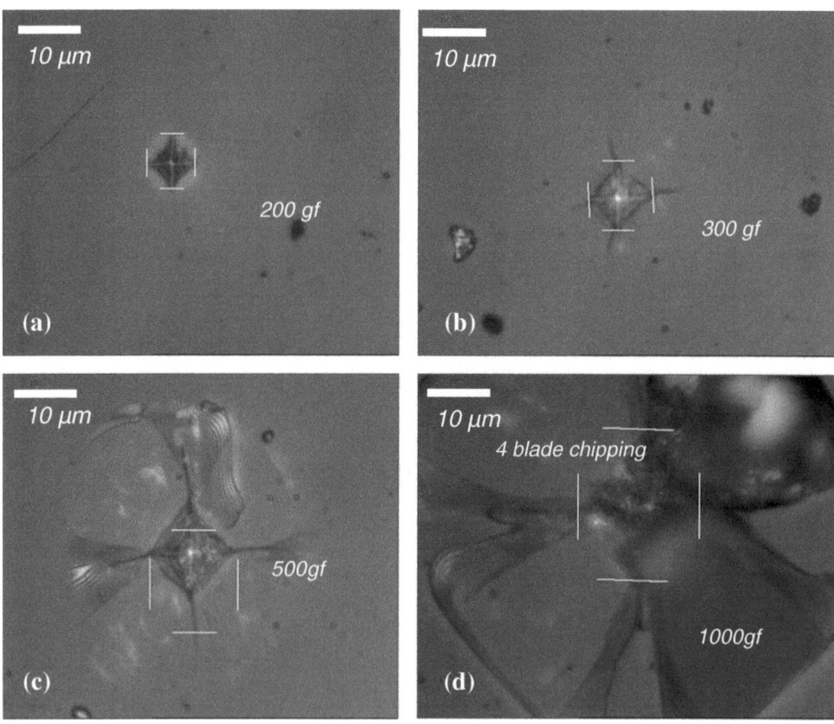

Fig. 10.6 Failure behavior of hardest Si–C–N coatings on silicon substrates by Vickers microindentation at **a** 200 gf, **b** 300 gf, **c** 500 gf, **d** 1000 gf

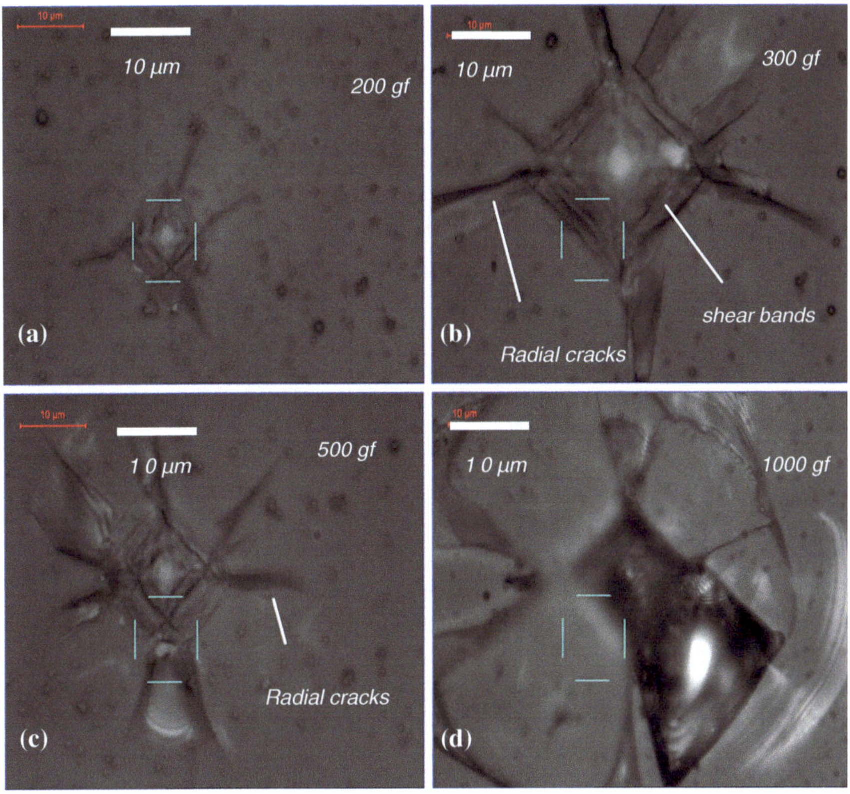

Fig. 10.7 Radial crack deviation followed by failure of Si–C–N coatings on silicon substrates by Vickers microindentation at **a** 200 gf, **b** 300 gf, **c** 500 gf, and **d** 1000 gf

bright area) were observed (Fig. 10.6c) which eventually led to chipping of the coating at 1000 gf (Fig. 10.6d).

The phenomenon was more severe for coatings with comparatively lower hardness (2788 HV) as shown in Fig. 10.7. In this case, formation of two radial cracks were observed to be arising from the same corner of the indentation impression. Coating delamination due to initiation of lateral cracking process was seen between the two radial cracks originating from the same point. At higher loads similar to the previous case *four blade* chipping was observed along with separate chipping phenomenon between two adjacent blades. The entire chipping phenomenon combined together gave rise to a circumferential chipping surrounding the indentation [33–35].

Median cracks were seen on Si–C–N coatings deposited at lower temperature as shown in Fig. 10.8. It can be observed that the length of the radial cracks formed is more than four times the indentation diagonal which makes it median below the surface. Lateral cracks followed by delamination were seen along with the radial cracks.

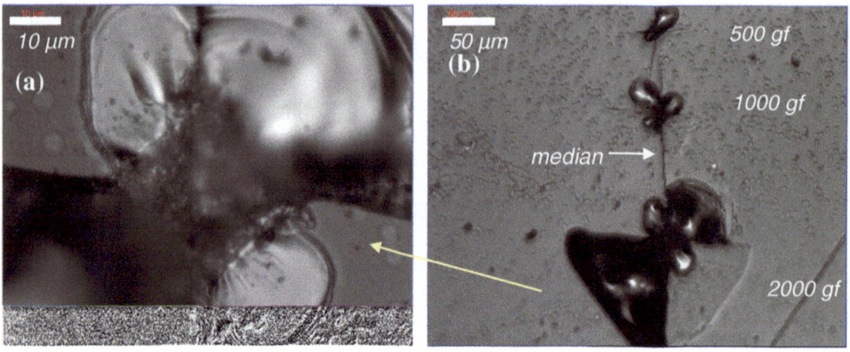

Fig. 10.8 **a** Median cracks and extensive delamination followed by chipping on Vickers microindentation at **b** 500 gf, 1000 gf, and 2000 gf

10.6 Mixed Mode Failure

The loading during indentation is of mixed mode type, which is a combination of all the three Irwin's modes—opening mode, in plane shear sliding mode, and out of plane tearing mode. In mixed mode condition, studies are carried out in finding crack extension direction, critical load, and stability of crack path. For constructive fracture-based design estimates, one needs to characterize the crack under mixed mode loading [36, 37]. Various models have been proposed to characterize the mixed mode cracks .The models are essentially based either on energy or stresses [37–39]. We have used the energy-based criterion proposed by Griffith. According to Griffith's criterion, the total energy release rate (G) for a crack in a coating subjected to mode I and mode II loading is given by

$$G = G_I + G_{II} \tag{10.2}$$

where

$$G_I = \alpha K_I^2/E \tag{10.3}$$

and

$$G_{II} = \alpha K_{II}^2/E \tag{10.4}$$

The crack extension will occur in the direction where total energy release rate is maximum and the extension will take place when the maximum energy release rate reaches a critical value G_C which is obtained from pure mode I loading
where

$$Gc = \alpha K_{IC}^2/E \tag{10.5}$$

where K_{IC} is the critical mode I stress intensity factor or the fracture toughness of the material. The elastic modulus (E) of the coating from nanoindentaion was around

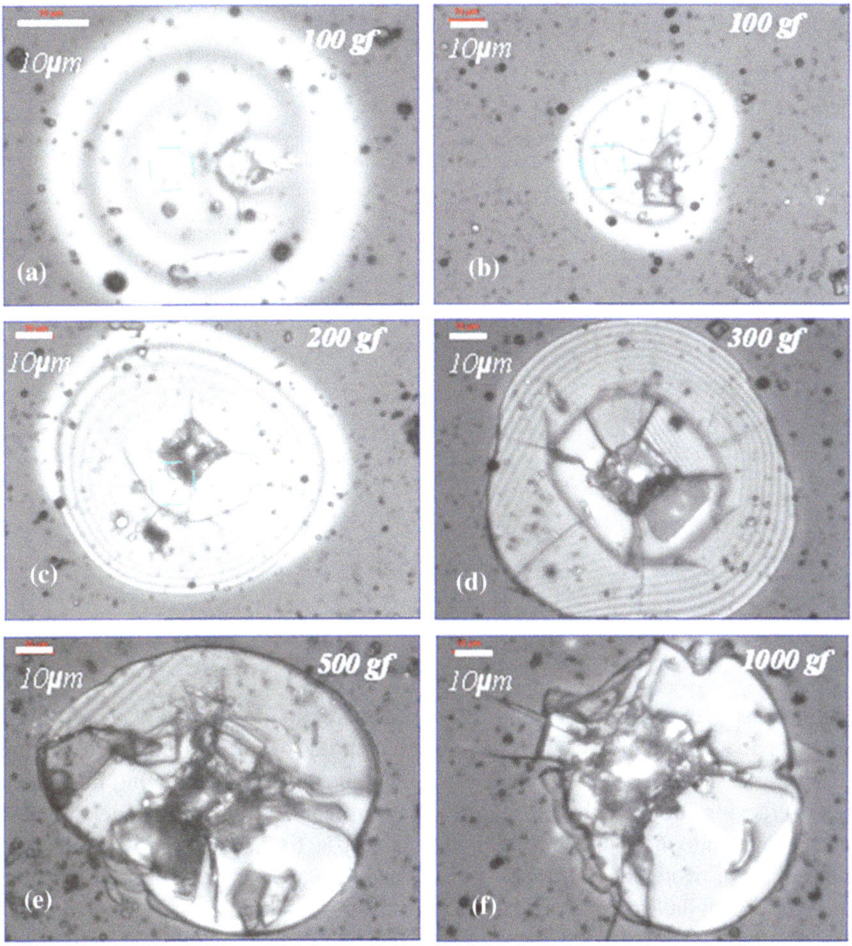

Fig. 10.9 Optical micrographs showing fracture mechanisms of SiCN coatings deposited on silicon substrates having lower hardness and adhesion at indentation loads of **a** 100 gf, **b**100 gf, **c** 200 gf, **d** 300 gf, **e** 500 gf, and **f** 1000 gf

240 GPa. Considering the condition for plane stress ($\alpha = 1$) the critical strain energy release rate or the critical mixed mode delamination energy came out to be 88 Jm^{-2}.

10.7 Coating Failure: Fracture Analysis

Figure 10.9 shows the deformation behavior of poorly adhered SiCN coating deposited on silicon substrates. Figure 10.10 gives the schematic representation of the different failure mechanisms. The circular flow of shear stress causing delamination of

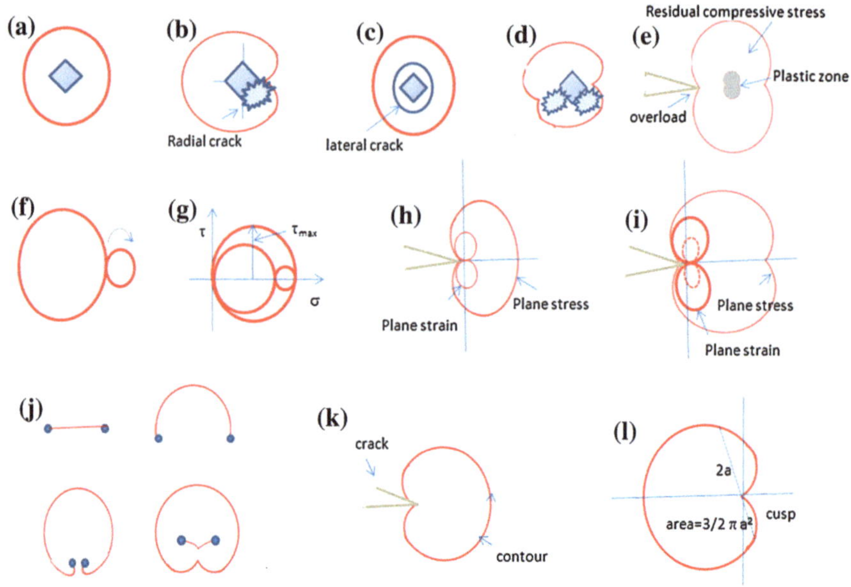

Fig. 10.10 The schematic representation of the different failure mechanisms

coating can be clearly observed in Fig. 10.9a where the indentation was performed at 100 gf load. Radial cracks of very low magnitude were also observed. Figure 10.9b was taken just after 5 s of the same indenter impression showing intensified radial cracking along with film failure. The circular circumferential delaminated area changed into the shape of a cardioid (Fig. 10.10a, b). This region corresponds to the enclave of residual compressive pulse which occurs as the plastic zone near the crack tip which has grown in size due to overload, is unable to regain its shape after unloading unlike the elastic zone similar to the case of fatigue failure (Fig. 10.10e). The shape of the cardioid is formed when a circle traverses a path along the circumference of another circle of lower area (Fig. 10.10f) which resembles the Mohr circle (Fig. 10.10g). The plastic zones are obtained based on Tresca and Mises yield criterions for an inelastic deformation at the crack tip (Fig. 10.10h, i) which are also similar in shape to a cardioid and compressive residual stress region mentioned in Fig. 10.10e [37].

As the plastic deformation is caused by nucleation and motions of dislocations, such type of cardioid shapes are also obtained in the operation of a Franck-Read source during multiplication of dislocations. The mechanism is associated with annihilation of two interaction points of dislocations at the cusp to form a new dislocation inside the cardioid (Fig. 10.10j) [36]. A reverse analogy could thus be found between the two cases, as the formation of facture corresponds to the formation of a new dislocation and the dislocation loop area corresponds to the delaminated area. The cardioid surrounding the indentation with its cusp at the crack tip can be used to define the J-integral (Fig. 10.10k) [36]. Further research keeping in consideration the above-mentioned phenomena may yield interesting new results.

From geometry, any chord passing through a cusp has a length of 2a and the area of the cardioid is $3/2\ \pi a^2$ (Fig. 10.10l) [40]. The chord length was 40 μm and hence the area of the plastically flown region was 1884 μm^2. The load applied was 0.98 N (100 gf) hence the delamination shear flow stress, which is nothing but load divided by area came out to be 520 MPa. Formations of Newton rings were also observed inside the delaminated area. At higher loads of 200gf (Fig. 10.9c) lateral cracks appeared which got intensified at 300 gf (Fig. 10.9d) leading to films failure though chipping at 500 gf (Fig. 10.9e) and 1000 gf (Fig. 10.9f). The circular plastic flow region surrounding the lateral crack changes to a shape on chipping to a shape which corresponds to a cardioid or combination of cardioids. If the chipping is only from one end of the four-sided square impression, then the region resembles only one cardioid. If the chipping is through two sides, then the region resembles two cardioids and so on and so forth (Fig. 10.10c, d). Coatings with lower thickness of 750 nm showed the delamination through lateral crack formation and at comparatively low loads as shown in Fig. 10.11.

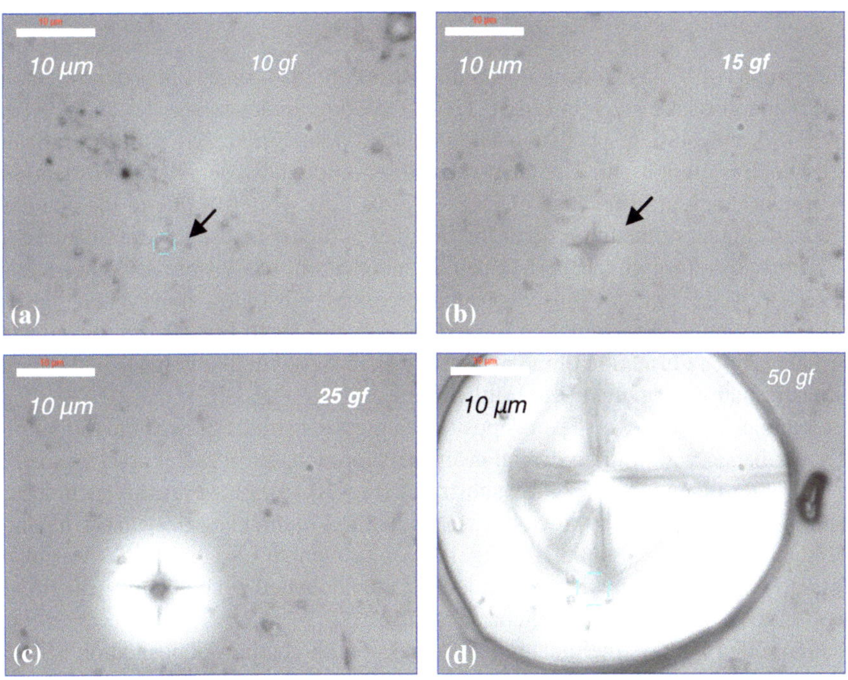

Fig. 10.11 Vickers microindentation of 750 nm thick Si–C–N coatings on Silicon substrates at **a** 10 gf, **b** 15 gf, **c** 25 gf , **d** 50 gf

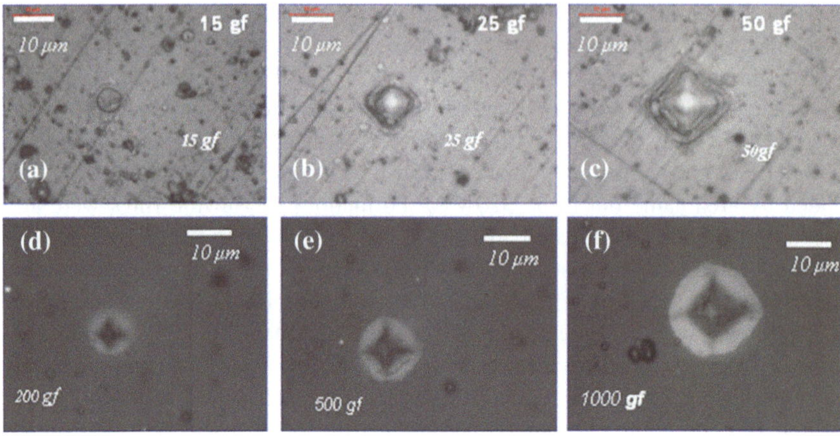

Fig. 10.12 Optical micrographs showing Vickers microindentations of SiCN coatings deposited on **a**, **b**, **c** 304SS and **d**, **e**, **f** glass substrates

10.8 Coatings on Steel and Glass Substrates

Micrographs of Vickers indentation at three different loads on coatings deposited on 304SS and glass are given in Fig. 10.12. The coating thickness was 1.5 and 3.9 μm on 304SS and glass respectively. The indentation size increased with increase in load which is quite usual. The interesting observation, however, was the formation of circular crack at lower load of 15 gf (Fig. 10.12a). This was due to the circular Hertzian contact of the blunt tip of the indenter at higher loads, however, the circular nature disappeared and perfect square indentation was visible at higher loads (Fig. 10.12b, c). For coatings deposited on glass substrates, however, a plastically deformed area surrounding the indentation similar to Fig. 10.4b were visible (Fig. 10.12d–f). Formation of radial cracks was observed at higher loads, indicating a transformation from ductile to brittle failure.

Coatings of thickness 0.5 μm deposited on glass substrates showed the formation of shear bands within the indentation impression along with radial cracks usually found while indenting bare glass substrates (Fig. 10.13). However, no such shear bands were observed in the previous case for coatings of thickness 3.9 μm and having the highest hardness of 4305 Hv (Fig. 10.13). The length of radial cracks was also less and the indentation was surrounded by a delaminated zone due to shear stress as observed earlier in case of silicon substrates.

Figure 10.14 shows the indentation impression on coatings deposited on SS305 substrates with increasing loads. At lower loads the Vickers contact is similar to a spherical contact, hence circular deformed area is observed. As the indentation load increased the square-shaped typical Vickers indentation impression was observed. At higher loads lateral cracking encircling the indented zone was observed. On increasing the load further shear bands similar to that in the case of glass substrates were seen.

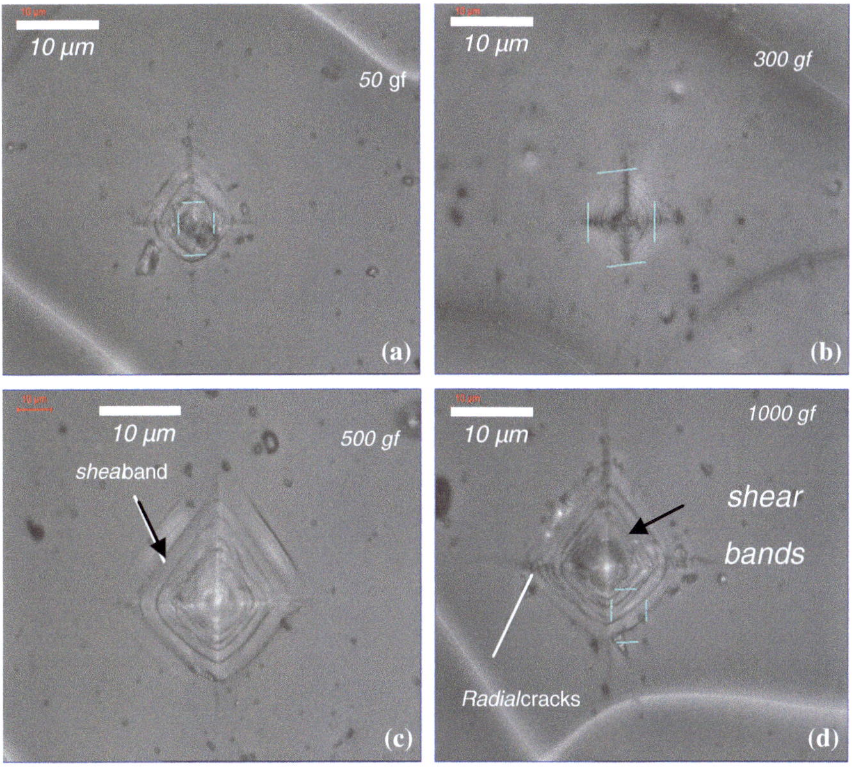

Fig. 10.13 Vickers microindentation of 0.5 μm thick Si–C–N coatings on glass substrates at **a** 50 gf, **b** 300 gf, **c** 500 gf, **d** 1000 gf

Indentations made on Si–C–N coatings having lower thickness (750 nm) deposited on 304 SS substrates showed mainly substrate effect as shown in Fig. 10.15 with the coating being chipped out in the form of debris seen besides the indentation impression.

10.9 Adhesion Studies of SiCN Nanocomposite Coatings on Different Substrates

The critical loads and the nature of failure, both of which indicate the adhesion property of the coatings which were found to be depending upon deposition conditions and the nature of substrate. The effect of loading rate also has an effect on the failure mechanism of the coating. A sliding speed of 0.2 mm/s was used and the loading rate was kept at 2, 5, 10, and 20 N/mm.

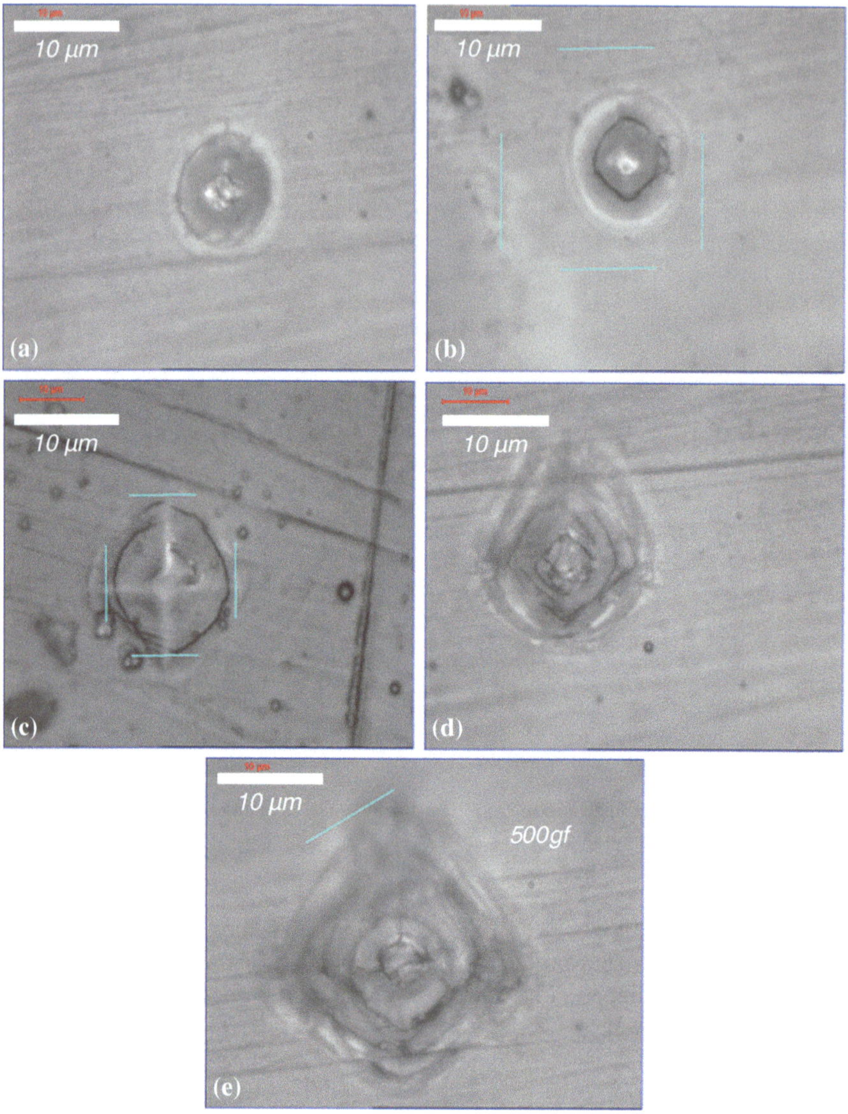

Fig. 10.14 Vickers microindentation of 2 μm thick Si–C–N coatings on 304 SS substrates showing circular cracks at **a** 25 gf, **b** 50 gf, **c** 100 gf, **d** 200 gf, **e** 500 gf

The scratching activity has three contributions: (1) ploughing, (2) friction, and (3) stress. The continuous increase in the coefficient of friction during scratching is attributed to increased ploughing of the sample by the tip with increasing normal load. The indentation stress and the flow stress on the coating are the major parameters of the plough which also increase with the increase in normal applied sliding load

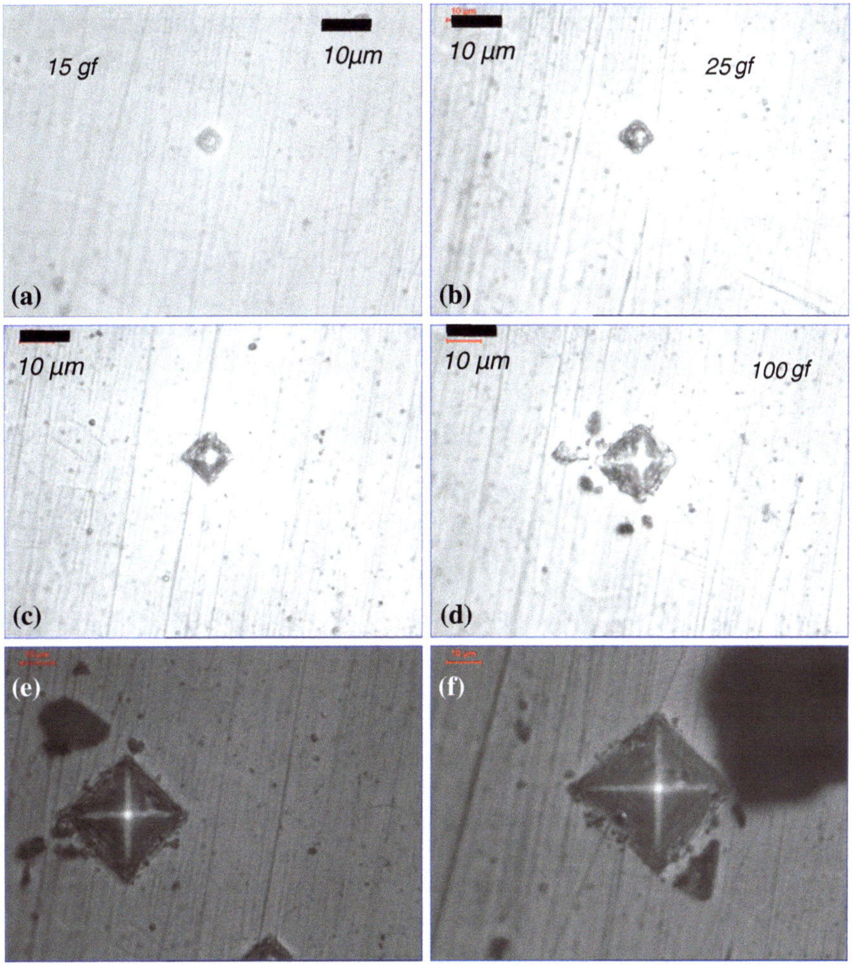

Fig. 10.15 Vickers microindentation of 750 nm thick Si–C–N coatings on 304 SS substrates **a** 15 gf, **b** 25 gf, **c** 50 gf, **d** 100 gf, **e** 200 gf, **f** 500 gf

[17, 28–32, 43–47]. After the coating failure, coating removal takes place as observed from the change in the nature of scratch track in the micrograph where piling ups at the edges of the scratch are observed. Highest critical load of 24 N was found for coatings deposited on steel substrates. The test was performed at a loading rate of 10 N/mm.

For 5 N/mm loading rate, the tractional force curve Fig. 10.16a shows a sudden and prominent slope change at 21 N (Lc_1) and 23 N (Lc) with μ_{eff} around 0.32 (Fig. 10.16b). The reason is the lower loading rate of 5 N/mm in which results allow the coating to offer resistance with less influence from the substrate. The point of

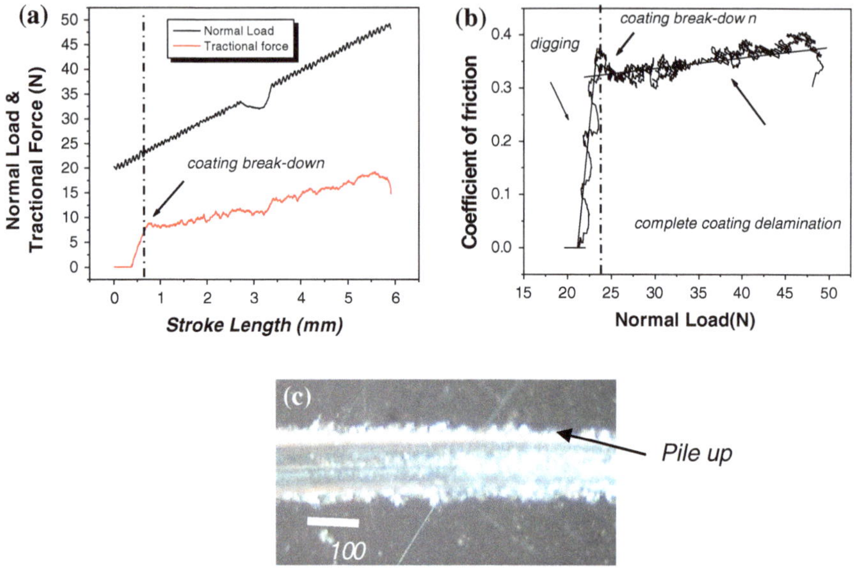

Fig. 10.16 Scratch tests with 5N/mm loading rate for steel substrates showing plots of **a** normal load and tractional force versus stroke length, **b** C.O.F versus normal load and **c** the optical micrograph of the scratch track

coating failure is indicated in the micrograph Fig. 10.16c. Spallation accompanied by chipping was seen at the edges indicating total delamination of the coating.

A loading rate of 20 N/mm showed higher substrate influence evident from the comparatively lower slopes of the tractional force curves during the scratching process Fig. 10.17. The lower critical load was 16 N while the upper critical load was 22 N. The μ_{eff} was also higher (0.44).

The adhesive failure occurs at a critical load is 21 N and 0.27 μ_{eff}. Complete removal of the coating occurs at a stroke length of 4 mm which is evident from the sudden rise in the tractional force curve which also shows the release of stored elastic energy. In the micrograph the removed coating can be pileups at the edges of the scratch track (Fig. 10.18). Total delamination by spallation occurred at higher loads [28–32, 43–47].

Better adhesion in terms of coating failure and corresponding second critical load has been observed in case of coatings on silicon substrates compared to that of steel substrates. The interfacial shear stress generated at the coating—substrate interface plays a major role in the adhesion properties. Figures 10.19, 10.20, 10.21 show the different scratch tests performed on Si–C–N coatings deposited on silicon substrates.

Tensile cracking was observed in some cases (Figs. 10.19, 10.20) for films deposited at higher deposition temperatures (~500°C) at a power of 400 W and pressure of 1 Pa in Ar and nitrogen atmosphere. It is indicative of strong adhesion to the substrate and consists of cracking within the scratch only. Semicircular trajectories of cracks are seen parallel to the trailing edge of the indenter. These cracks arise

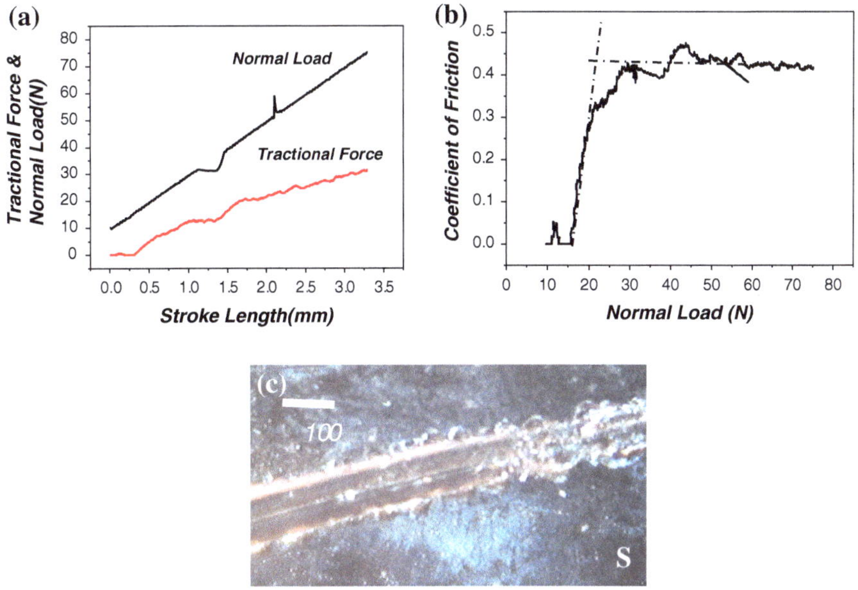

Fig. 10.17 Scratch tests with 20 N/mm loading rate for Silicon substrates showing plots of **a** normal load and tractional force versus stroke length, **b** C.O.F versus normal load and **c** the optical micrograph of the scratch track

due to tensile frictional force present behind the stylus and balance the compressive frictional force ahead [28–32, 43–47]. Coating removal by buckling of the film at higher loads occurred which were evident from the circular flakes coming out from the side of the scratch trajectory which is also an indication of plastic deformation of the film and the substrate.

Cohesive spallation leading to brittle failure was observed in coatings deposited on silicon substrates. Not much variation is observed in the tractional force curve apart from a little deviation at higher loads due to failure. Circular Hertzian tensile cracks were also seen in some scratch tracks which arise due to union of two symmetrical tracks at the edge of the scratch line (Fig. 10.19).

Different scratch tests performed on coatings deposited on glass substrates are given in Figs. 10.21, 10.22. In Fig. 10.20a, the failure of the film starts at 0.5 mm at a load of 7 N (lower critical load) indicating cohesive failure. The coating breakdown becomes more and more severe with increase in normal load evident from the rise in amplitude of vibration of the tractional force curve. Severe brittle failure accompanied by coating removal started at the upper critical load of 10 N evident from the high and low amplitudes of the curve.

The phenomenon of indenter penetration (digging), coating breakdown, and removal were prominent as shown in Fig. 10.22 and comparatively high adhesive or upper critical load of 15 N was obtained indicating strong adhesion. Film failure by *wedge spallation* is seen (Fig. 10.23) and the adhesive critical load was about

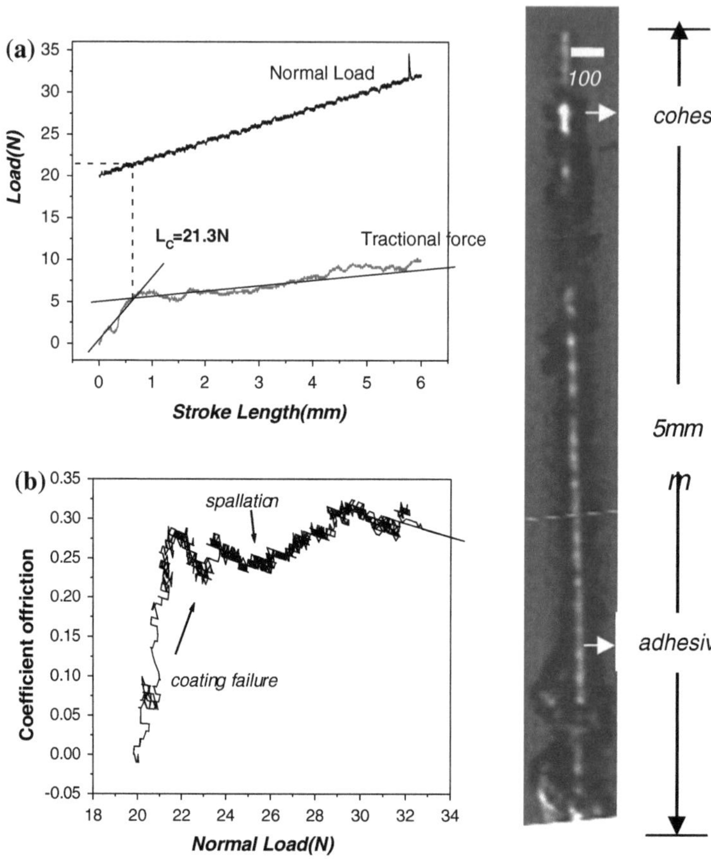

Fig. 10.18 Scratch tests with 2 N/mm loading rate for steel substrates showing plots of **a** normal load and tractional force versus stroke length, **b** C.O.F versus normal load and **c** the optical micrograph of the scratch track

16.5 N. The maximum critical load obtained for steel and silicon substrate was 24 and 18 N respectively. The μ_{eff} was very low for silicon and glass of the order of 0.01–0.001, whereas for steel substrate it was around 0.2–0.4.

The films with higher thickness usually show higher critical loads as more load is required to detach the coating from the substrate. With increase in substrate temperature the films are generally likely to be better crystallized and are better adherent showing a rise in critical load at higher temperature but the amount of thermal stresses generated due to CTE mismatch between the coating and the substrate is also more at higher temperature which causes a detrimental effect in the film adhesion [29].

The nanoscratch studies of the coatings showing highest hardness were done with a Berkovich indenter scratching with a speed of 50 μm/s to a maximum load of 50 mN on a 300 μm scratch track for all the three substrates (*silicon, glass, and steel*). The three curves indicate before, during, and after scratch displacements

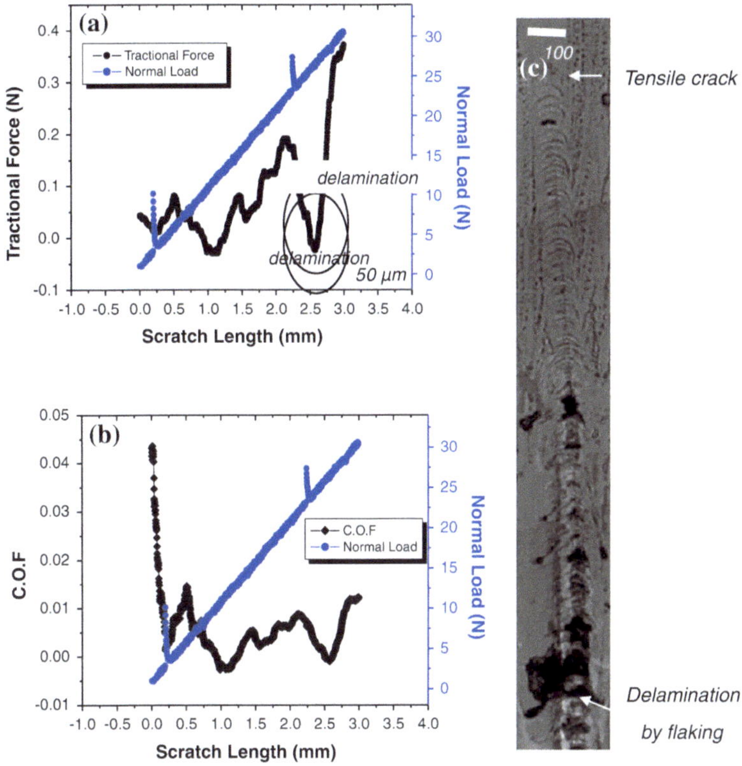

Fig. 10.19 Tensile cracking and flaking during scratching with 10 N/mm loading rate for Si–C–N coated silicon substrates showing plots of **a** normal load and tractional force versus stroke length, **b** C.O.F and normal load versus scratch length and **c** the optical micrograph of the scratch track

(Fig. 10.24a, 10.25a, 10.26a). The indenter depth increased linearly with the ramping load without abrupt change on the curve due to progressive peeling of the coating indicating good adhesion. The curves after scratching and before scratching are separately observable indicating permanent plastic residual deformation in the coating. The penetration depths before scratch, after scratch, and during scratch are shown in Table 10.1.

The final displacement of the indenter before scratching and simple scanning of the surface was 26 , 66, and 1827 nm for SS304, glass, and silicon substrate respectively .The final depth at 100 mN during scratching changed to 831, 714, and 1342 nm whereas the residual depth obtained after unloading were 287, 330, and 1676 nm for SS304, glass, and silicon substrate respectively.

The ratio of difference between the "after scratch depth" and "before scratch depth" to the difference between the "during scratch" and "before scratch" depth gave the measure of the plasticity. The coatings showed 70 % elastic recovery on all the three substrates. Although the initiation of coating failure was not indicative in

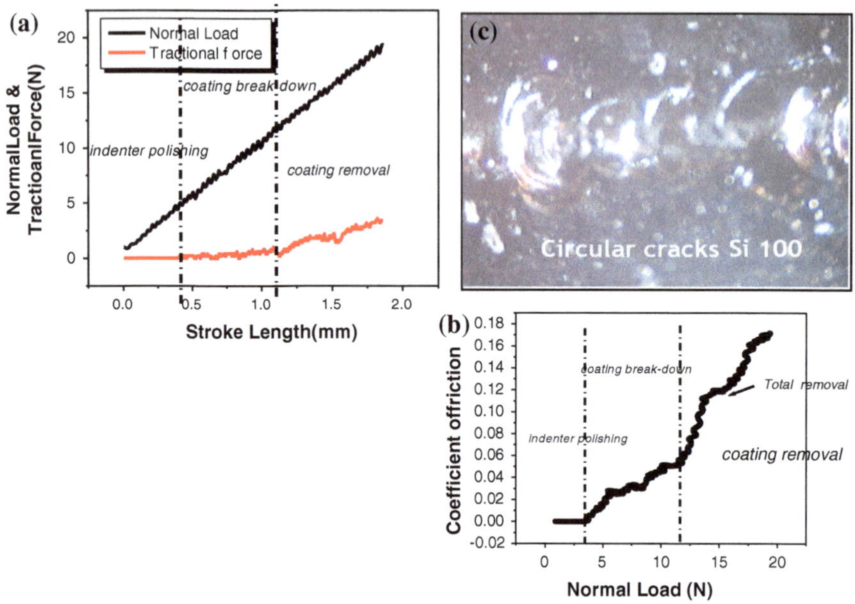

Fig. 10.20 Scratch test with 10 N/mm loading rate for Si Substrates showing plots of **a** normal load and tractional force versus stroke length, **b** C.O.F versus normal load and **c** the optical micrograph of the scratch track showing failure by circular cracking. The films were deposited at 100°C in RF mode

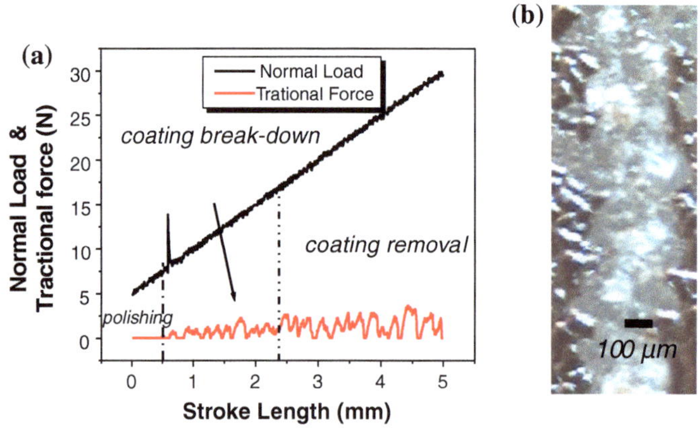

Fig. 10.21 Scratch tests with 5 N/mm loading rate for glass substrates showing plots of **a** normal load and tractional force versus stroke length, and **b** the optical micrograph of the scratch track

the plots presented, the micrograph of the scratch tracks given however showed the point of coating failure (Figs. 10.24, 10.25, 10.26b).

A continuous increase in the scratch track width leading to spallation of the coating at higher loads was observed for coatings on silicon substrate and on glass. Ductile

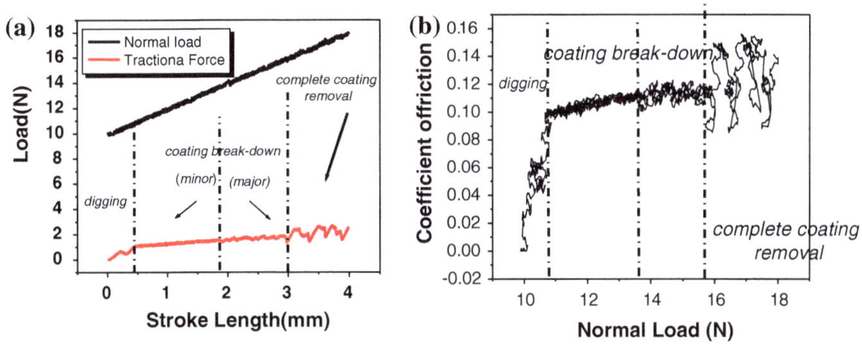

Fig. 10.22 Scratch tests with 2 N/mm loading rate for glass substrates showing plots of **a** normal load and tractional force versus stroke length and **b** C.O.F versus normal load

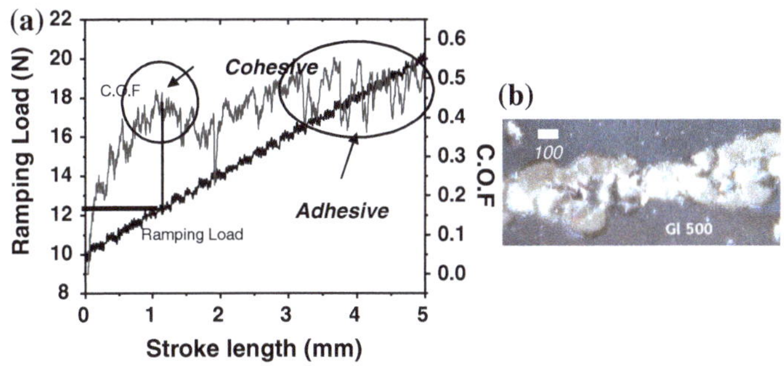

Fig. 10.23 Scratch with 2 N/mm loading rate for glass substrates showing plots of **a** normal load and C.O.F versus stroke length and **b** the optical micrograph of the scratch track showing failure at 11 N

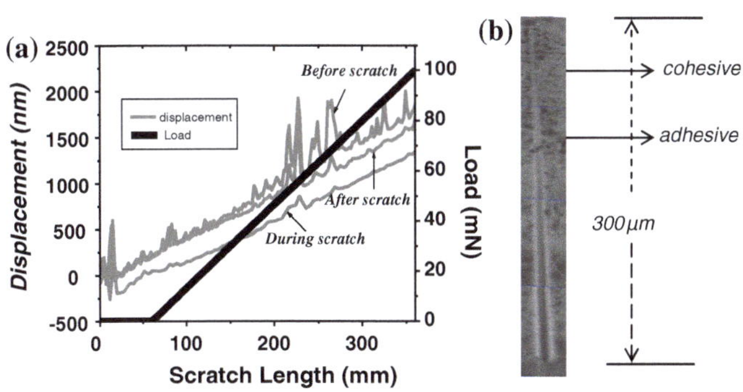

Fig. 10.24 Nanoscratch performed on Si–C–N coatings deposited on Si substrate showing plot of **a** displacement versus scratch length and **b** the optical micrograph of the scratch track

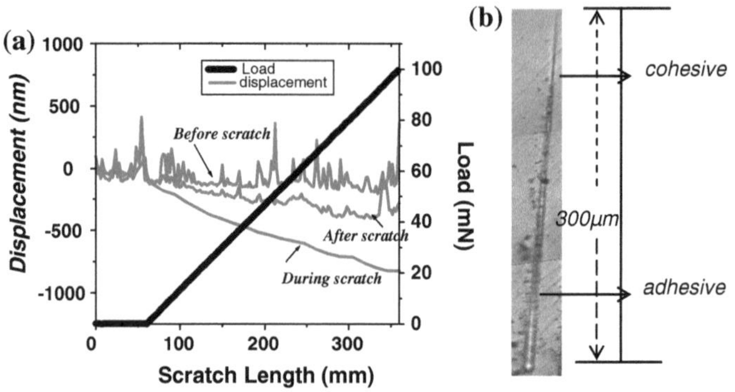

Fig. 10.25 Nanoscratch performed on Si–C–N coatings deposited on 304 SS substrate showing plot of **a** displacement versus scratch length and **b** the optical micrograph of the scratch track

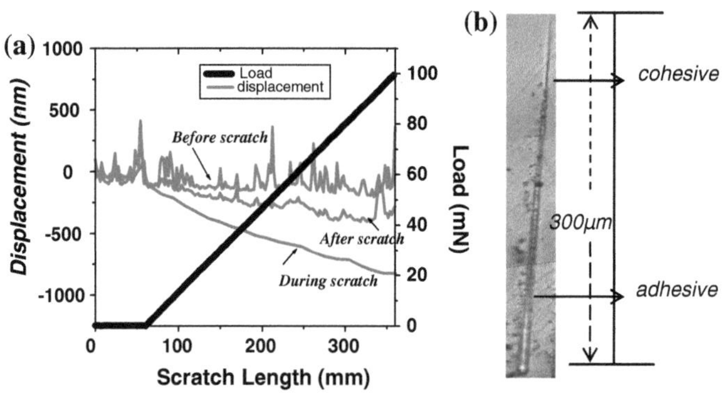

Fig. 10.26 Nanoscratch performed on Si–C–N coatings deposited on Glass substrates showing plot of **a** displacement versus scratch length and **b** the optical micrograph of the scratch track

Table 10.1 The depths obtained during different stages of nanoscratch tests

S. No	Substrate	Before scratch (nm)	During scratch (nm)	After scratch (nm)	Elastic (%)	Plastic (%)
1.	SS304	26	831	287	68	32
2.	Glass	166	714	330	70	30
3.	Silicon	1827	1343	1676	69	31

failure accompanying pileups at the two sides of the scratch track was observed for 304 SS substrates, whereas brittle failure with chipping was observed in case of coating on glass substrates [48–50]. These observations indicate that even at Nanoscratch there is significant contribution from the substrates.

Summary

Investigation of deformation under indentation and Scratch adhesion behavior of magnetron sputtered Si–C–N film deposited on silicon, glass, and steel substrates at various deposition parameters have been investigated. Both microindentation and nanoindentations were used to evaluate the coating hardness, modulus, failure behavior of the films. Adhesion studies using scratch tests showed a strong adherence between the coating and the substrate. The highest critical load of 24 N was obtained for coatings deposited on 304SS substrates. The failure morphology was found to be dependent on scratch velocity, loading rate, and the nature of the substrate. High elastic recovery of 70 % was obtained through nanoscratch tests for coatings deposited on all the three substrates. Ductile failure accompanying pileups at the two sides of the scratch track and by indentation were observed for 304 SS substrates, whereas brittle failure with chipping was observed in case of coating on silicon and glass substrates.

References

1. Riedal, R., Kleebe. H.J., Schoenfelder, H., Aldinger, F.: A covalent micro/nano composite resistant to high temperature oxidation. Nature **374**, 526–528 (1995)
2. Bill, J., Seitz, J., Thurn, G., Dürr, J., Canel, J., Janos, B.Z., Jalomecki, A., Santez, D., Schempp, S., Lamparter, H.P., Mayer, J., Aldinger, F.: Structure analysis and properties of Si–C–N ceramics derived from polysilazanes, Phys. Status Solidif A **166**, 269 (1998)
3. Raj, R., Riedel, R., Soraru, G.D.: Introduction to the special topical issue on ultrahigh-temperature polymer-derived ceramics. J. Am. Ceram. Soc. **84**(10), 2158–2159 (2001)
4. Ferreira, I., Fortunato, E., Vilarinho, P., Viana, A.S., Ramos, A.R., Alves, E., Martins, R.: Hydrogenated silicon carbon nitride films obtained by HWCVD PA-HWCVD and PECVD techniques. J. Non-Crystal. Solids **352**, 1361–1366 (2006)
5. Berlind, T., Hellgren, N., Johansson, M.P., Hultman, L.: Microstructure, mechanical properties, and wetting behavior of Si–C–N thin films grown by reactive magnetron sputtering Surf. Coat. Technol. **141**, 145–155 (2001)
6. Janakiraman, N., Aldinger, F.: Indentation analysis of elastic and plastic deformation of precursor-derived Si–C–N ceramics. J. Eur. Ceram. Soc. **30**, 775–785 (2010)
7. Mishra, S.K., Gaur, H., Rupa, P.K.P., Pathak, L.C.: Deposition of nanostructured Si–C–N superhard coatings by rf magnetron sputtering. J. Vac. Sci. Technol. B **24**, 1311 (2006)
8. Mishra, S.K., Shekhar, C., Rupa, P.K.P., Pathak, L.C.: Effect of pressure and substrate temperature on the deposition of nano-structured silicon–carbon–nitride superhard coatings by magnetron sputtering. Thin Solid Films **515**, 4738 (2007)
9. Kroke, E., Li-Li, Y., Konetschney, C., Lecomte, E., Fasel, C., Riedel, R.: Mater. Sci. Eng. **26**, 97 (2000)
10. Meyer, E.: Untersuchungen uber Harteprugung und Harte. Phys. Z. **9**, 66–74 (1980)
11. Thomas, A.: Microhardness measurement as a quality control technique for thin, hard coatings. Surf. Eng. **3**, 117–22 (1987)
12. Tabor, D.: The Hardness of Metals. Clarendon Press, Oxford (1951)
13. Jönsson, B., Hogmark, S.: Hardness measurements of thin films. Thin Solid Films **114**, 257–69 (1984)
14. Sargent, P.M.: Ph.D. Thesis, University of Cambridge, UK (1979)

15. Burnett, P.J., Rickerby, D.S.: The mechanical properties of wear-resistant coatings: I: modelling of hardness behaviour. Thin Solid Films **148**, 41–50 (1987)
16. Burnett, P.J., Rickerby, D.S.: The mechanical properties of wear-resistant coatings: II: experimental studies and interpretation of hardness. Thin Solid Films **148**, 51–65 (1987)
17. Bull, S.J., Rickerby, D.S.: New developments in the modeling of the hardness and scratch adhesion of thin films. Surf. Coat. Technol. **42**, 149–64 (1990)
18. Chicot, D., Lesage, J.: Absolute hardness of films and coating. Thin Solid Films **254**, 123–130 (1995)
19. Ichimura, H., Rodriguez, F.M., Rodrigo, A.: The composite and film hardness of TiN coatings prepared by cathodic arc evaporation. Surf. Coat. Technol. **127**, 138–143 (2000)
20. Korsunsky, A.M., McGurk, M.R., Bull, S.J., Page, T.F.: On the hardness of coated systems. Surf. Coat. Technol. **99**, 171–183 (1998)
21. Puchi-Cabrera, E.S.: A new model for the computation of the composite hardness of coated systems. Surf. Coat. Technol. **160**, 177–186 (2002)
22. Bull, S.J.: Nanoindentation of coatings. J. Phys. D: Appl. Phys. **38**, 393–413 (2005)
23. Bhattacharyya, A.S., Das, G.C., Mukherjee, S., Mishra, S.K.: Effect of radio frequency and direct current modes of deposition on protective metallurgical hard silicon carbon nitride coatings by magnetron sputtering. Vacuum **83**, 1464–1469 (2009)
24. Bhattacharyya, A.S., Mishra, S.K., Das, G.C.: Mukherjee shot properties. Eur. Coat. J. **3**, 108–114 (2009)
25. Bhattacharyya, A.S., Mishra, S.K., Mukherjee, S., Das, G.C.: A comparative study of Si–C–N films on different substrates grown by RFmagnetron sputtering. J. Alloys Comp. **478**, 474–478 (2009)
26. Lube, T.: Indentation crack profiles in silicon nitride. J. Eur. Ceram. Soc. **21**, 211–218 (2001)
27. Chicot, D., Duarte, G., Tricoteaux, A., Jorgowski, B., Leriche, A., Lesage, J.: Vickers indentation fracture (VIF) modeling to analyze multi-cracking toughness of titania, alumina and zirconia plasma sprayed coatings. Mater. Sci. Eng. A **527**, 65–76 (2009)
28. Burnett, P.J., Rickery, D.S.: The scratch adhesion test an elastic-plastic indentation analysis. Thin Solid Films **157**, 233–254 (1988)
29. Mishra, S.K., Bhattacharyya, A.S.: Effect of substrate temperature on the adhesion properties of magnetron sputtered nano-composite Si–C–N hard thin films. Mater. Lett. **62**, 398–402 (2008)
30. Gonczy, S.T., Randall, N.: An ASTM standard for quantitative scratch adhesion testing of thin, hard ceramic coatings. In: 29th International Conference on Advance Ceramic and Composites, Cocoa beach, FL, 23–28 Jan 2005
31. Bull, S.J., Berasetegui, E.G.: An overview of the potential of quantitative coating adhesion measurement by scratch testing. Tribol. Int. **39**, 99–114 (2006)
32. Ashrafizadeh, F.: Adhesion evaluation of PVD coatings to aluminium substrate. Surf. Coat. Technol. **130**, 186–194 (2000)
33. Burghard, Z.: PhD-Thesis, Max-Planck Institute for Metals Research & University of Stuttgart, Germany (2004)
34. Cook, R.F., Pharr, G.M.: Lateral cracks and microstructural effects in the indentation fracture of yttria. J. Am. Ceram. Soc. **73**, 787–817 (1990)
35. Lawn, B.R.: Fracture of Brittle Solids, 2nd edn. Cambridge University Press, Cambridge (1993) (1993).
36. Dieter, G.E.: Mechanical Metallurgy. McGraw Hill Book Company (UK) limited, New York (1988)
37. Kumar, P.: Elements of Fracture Mechanics, 1st edn. A.H. Wheeler & Co. Ltd., Allahabad (1999)
38. Krueger, R., O'Brien, T.K.: A shell/3D modeling technique for the analysis of delaminated composite laminates. Comp. Part A: Appl. Sci. Manuf. **32**, 25–44 (2001)
39. Davies, P.: Round Robin analysis of G_{Ic} interlaminar fracture test. Appl. Comp. Mater. **3**, 135–140 (1996)
40. Weisstein, E.W.: "Cardioid." from mathworld-a Wolfram Web resource. http://mathworld.wolfram.com/Cardioid.html

41. Zhou, Y., Probst, D., Thissen, A., Kroke, E., Riedel, R., Hauser, R., Hoche, H., Broszeit, E., Kroll, P., Stafast, H.: Hard silicon carbonitride films obtained by RF-plasma enhanced chemical vapour deposition using the single source precursor bis(trimethylsilyl)-carbodiimide. J. Eur. Ceram. Soc. **26**, 1325–1335 (2006)
42. Probst, D., Hoche, H., Zhou, Y.Y., Hausier, R., Stelzner, T., Scheerer, H., Brotzeit, E., Berger, C., Riedel, R., Stafast, H., Koke, E.: Development of PECVD Si/C/N:H films for tribological and corrosive complex load conditions. Surf. Coat. Technol. **200**, 355–359 (2005)
43. Bushan, B., Gupta, B.K.: Handbook of Tribology (Materials Coating and Surface Treatments). pp. 15.45–15.58. ISBN: 0-07-005249-2
44. Mehrotra, P.K., Quinto, D.T.: Techniques for evaluating mechanical 236 properties of hard coatings. J. Vac. Sci. Technol. A **3**(6), 237, 2401–2405 (1985)
45. Perry, A.J.: The adhesion of chemically vapour deposited hard coatings to steel—the scratch test. Thin Solid Films **78**, 77–93 (1981)
46. Steinmann, P.A., Hintermann, H.E.: Adhesion of TiC and TiCN coatings on steel. J Vac. Sci Technol. A **3**, 2394–400 (1985)
47. Valli, J.: A review of adhesion test methods for thin hard coatings. J. Vac. Sci. Technol. A **4**, 3001–3014 (1986)
48. Mittal, K.L.: Adhesion Measurement of Films and Coatings, pp. 1–13. The Netherlands: VSP, Utrecht (1995)
49. Bhushan, B.: Handbook of Micro/Nanotribology, 2nd edn. CRC Press, Boca Raton (1999)
50. Huang, L.Y., Zhao, J.W., Xu, K.W., Lu, J.: Analysis of nano-scratchbehavior of diamond-like carbon films. Surf. Coat. Tech. **154**, 232–236 (2002)

Chapter 11
Impact of Defects and Doping on Electron Transport in SiCNTs

Sudhanshu Choudhary and S. Qureshi

Abstract The effects of vacancy defects and boron-nitrogen dopants on electron transport properties of SiCNT are investigated. The results of geometry optimization of vacancy defects show that single vacancies and di-vacancies in SiCNTs have different reconstructions. A single vacancy when optimized, reconstructs into a 5-1DB configuration in both zigzag and armchair SiCNTs, and a di-vacancy reconstructs into a 5-8-5 configuration in zigzag and into a 5-2DB configuration in armchair SiCNTs. Analysis of frontier molecular orbitals (FMO) and transmission spectrum show that the introduction of vacancy reduces the bandgap of (5,5) semiconducting SiCNTs and increases the bandgap of (4,0) conducting SiCNTs (converts them to semi-metallic nanotubes). Bias voltage-dependent current characteristics show reduction in overall current in metallic SiCNT and an increase in overall current in semiconducting SiCNT which is due to introduction of new electronic states around the Fermi level followed by conduction through the defect sites. The results of the study on the effect of BN co-doping in SiCNTs suggests that co-doping BN impurities suppresses the important negative differential resistance (NDR) property. NDR suppression is attributed to the introduction of new electronic states near the Fermi level followed by weak orbital localization. BN co-doping results in exponential current-voltage (I–V) characteristics which is in contrast to linear I–V characteristics for individual boron and nitrogen doped SiCNTs. HOMO has no contribution from B impurity, whereas LUMO has contribution from N impurity at low and high bias.

S. Choudhary (✉) · S. Qureshi
Department of Electrical Engineering, I. I. T. Kanpur, Kanpur 208016, India
e-mail: hellosudhanshubit@gmail.com

S. Qureshi
e-mail: qureshi@iitk.ac.in

H. Li et al. (eds.), *Silicon-based Nanomaterials*,
Springer Series in Materials Science 187, DOI: 10.1007/978-1-4614-8169-0_11,
© Springer Science+Business Media New York 2013

11.1 Introduction

To address the issue of scaling related to silicon ICs, there have been several approaches that could be classified as long-term approaches and near-future approaches [1]. Long-term approach is based on completely novel devices through novel concepts such as spintronic devices, single-molecule devices, and quantum computing which are expected to be widely implemented in the far future. Whereas the devices that have current-voltage characteristics, similar to current devices with novel materials and nanoscale operating principles, are explored for the near–far realization [2]. These devices include graphene field effect transistors (FETs), silicon nanowire FETs, and carbon nanotube field effect transistors (CNTFETs). Among these devices, CNTFET draws more attention due to its superior electronic transport properties. Furthermore, CNTs behave as either conductor or semiconductor depending on the direction in which a graphene sheet is rolled which is also advantageous from the IC viewpoint since the whole IC can be implemented using semiconducting CNTs as channel in CNTFETs and metallic CNTs as interconnects [3–7]. Recently, the use of silicon carbide nanotubes (SiCNTs) as a possible alternative to CNTs for field emitting applications has been suggested. SiCNTs are analogous to CNTs and can offer better device operation at high temperatures, high power, and high frequency and in harsh environments [8].

The objective of this chapter is to present the results of investigation carried out on the electronics properties of SiCNTs in the presence of vacancy defects and impurity doping. The investigation was carried out computationally using the ab-intio method-based simulator. Before advancing on this study, it is necessary to understand the electronic properties of CNTs that are believed to offer superior performance when used as interconnects in ICs and in CNTFET structures. These properties are described in the next section.

11.2 Electron Transport Properties of Carbon Nanotubes

Carbon is the sixth element of the periodic table with atomic number 6. Each carbon atom has six electrons which occupy $1s^2$, $2s^2$ and $2p^2$ atomic orbital. It can hybridize in sp (e.g. C_2H_2), sp^2 (e.g. graphite) or sp^3 (e.g. CH_4) forms. Discoveries of very stable nanometer size sp^2 carbon bonded materials such as fullerenes [9], carbon nanotubes [10], and graphene [11] have boosted the research interest in this field. Since carbon nanotubes derive most of the physical properties from graphene, it is therefore useful to understand this material first [12].

Graphene is a flat single layer of carbon atoms tightly packed into a two-dimensional (2D) honeycomb lattice. It is the basic building block for graphitic materials of all other dimensionalities [13]. These graphene sheets could be wrapped into 0-D fullerenes, rolled into 1-D nanotubes or stacked into 3-D graphite [12].

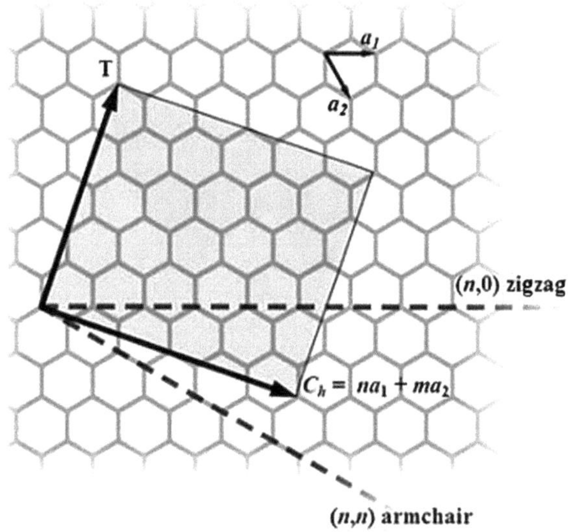

Fig. 11.1 Graphene sheet [14]

Due to the presence of sp2 hybridized carbon atoms in one plane, graphene is two-dimensional.

Figure 11.1 shows a graphene sheet and Fig. 11.2 shows its band structure; a_1 and a_2 are the unit vectors in real space. In a hexagonal lattice like this, the real space unit vectors can be written as:

$$\vec{a}_1 = \left(\frac{\sqrt{3}}{2}a, \frac{a}{2}\right), \quad \vec{a}_2 = \left(\frac{\sqrt{3}}{2}a, \frac{-a}{2}\right) \tag{11.1}$$

where 'a' is the lattice constant given by $\sqrt{3}a_{C-C}$ is the carbon–carbon bond length (0.142 nm). In the Brillouin zone there are three high symmetry points at the center, corners, and center edge. The electrical properties of graphene can be described by a conventional tight-binding model; in this model the energy of the electrons with wave vector k is [15, 16]:

$$E = \pm\sqrt{\gamma_0^2\left(1 + 4\cos^2\frac{K_y a}{2} + 4\cos\frac{K_y a}{2}.\cos\frac{k_x\sqrt{3}a}{2}\right)} \tag{11.2}$$

with the nearest-neighbor hopping energy $\gamma 0 \approx 2.8$ eV and the lattice constant $a \approx 2.46$ Å. Conduction and valence band, respectively, correspond to the different signs in the above dispersion relation; they touch at the six k-points of the first Brillouin zone.

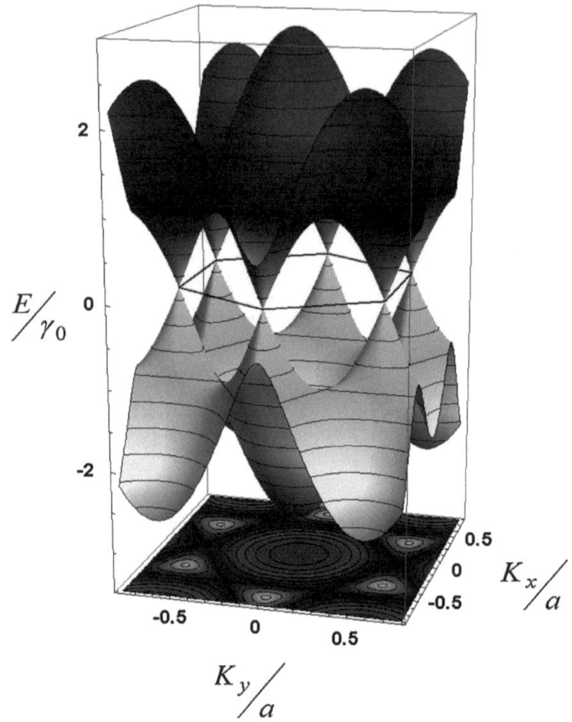

Fig. 11.2 Band structure of graphene [17]

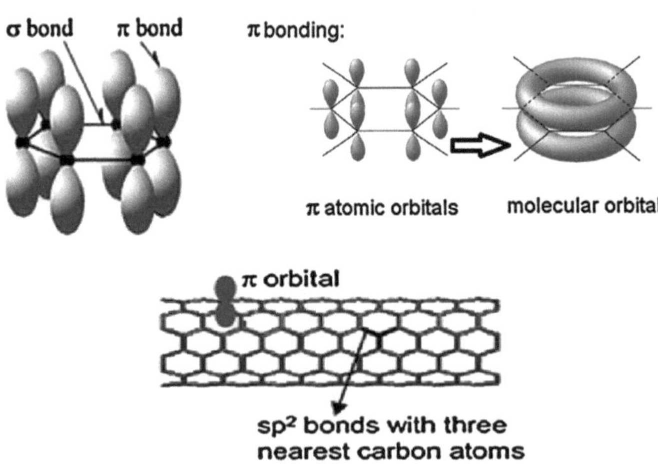

Fig. 11.3 Bonds and π-orbitals in a CNT [20, 21]

Carbon has four valence electrons of which three are strongly bound to neighbor atoms giving graphene its very high in-plane rigidity. The fourth electron is delocalized and shared by all the atoms (see Fig. 11.3), thus allowing for electronic current transport. In such sp^2 hybridize system only π electrons contribute to electronic transport; hence the above energy dispersion (see Fig. 11.2) describes the electronic band structure of graphene plotted in the Brillouin zone using Equation 11.2. The upper half of the energy curve describes anti-bonding (π*) orbital, while the lower part describes the bonding (π) orbital. Furthermore, π and π* bands are degenerate at the *k*-points in the Brillouin zone, through which Fermi energy passes. Since the valence band and conduction band meet at these six points, graphene is considered semi-metallic or zero-gap semiconductor. The form of the *E*–*k* function given in Equation 11.2 is in the form of the *E*–*k* function of photons, the graphene dispersion becomes linear (conical) near the *k*-points, thus the electrons in graphene-based structures are expected to behave like massless particles, which makes their velocity faster than conventional Si or GaAs structures [18]. The ballistic transport and nearly massless behavior are also verified by experiments [19].

Carbon nanotubes are allotropes of carbon with a cylindrical nanostructure. Single-wall carbon naotubes (SWCNT) are long, hollow structures with the cylindrical walls formed by one-atom-thick sheets of graphene. These graphene sheets are rolled in specific and discrete (chiral) angles, and the combination of rolling angle and tube radius determines whether the nanotube is metallic or semiconducting. The chiral vector (C_h), which has the length *L*, circumference of the nanotube is given in terms of two integers (n, m) in the directions of two basis vectors of the honeycomb lattice a_1 and a_2 (see Fig. 11.1). Depending on the (n, m) values, three types of tubules are defined: armchair when $n = m$, zigzag when $m = 0$, and chiral nanotube when $0 < m < n$ (see Fig. 11.4). Multi-walled nanotubes (MWNT) consist of multiple rolled layers (concentric tubes) of graphite.

The electronic structure of carbon nanotubes can be obtained from the dispersion of graphene through applying periodic boundary conditions along chiral vector C_h. This concept can be used to obtain the one-dimensional dispersion for a nanotube

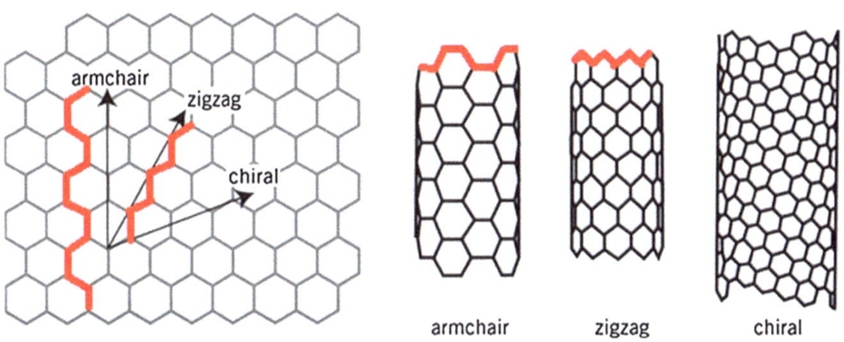

Fig. 11.4 Graphene sheet rolled into carbon nanotubes

through taking out periodic slices from the graphene dispersion relation, a gap will be opened or not at the k-point. Since each unit cell contains 2N carbon atoms, the final band structure of the nanotube possesses N bonding (π) and N anti-bonding (π*) atomic orbitals. Using this approach, the dispersion relationships of CNTs can be obtained from Equation 11.2 by the quantization of the energy levels depending on the type of the CNT. Dispersion relations of zigzag and armchair carbon nanotubes are given respectively as in Equations 11.3 and 11.4 [18].

$$E_z(k) = \pm \sqrt{\gamma_0^2 \left(1 + 4\cos^2 \frac{\pi c}{n} + 4\cos \frac{\pi c}{n} . \cos \frac{k_x \sqrt{3}a}{2} \right)} \tag{11.3}$$

$$E_a(k) = \pm \sqrt{\gamma_0^2 \left(1 + 4\cos^2 \frac{K_y a}{2} + 4\cos \frac{\pi d}{n} . \cos \frac{K_y a}{2} \right)} \tag{11.4}$$

where c and d are nonzero integers.

In carbon nanotubes, the density of states generally diverges at the band edges. These singularities of the density of states are called van-Hove singularities and are typical for 1D nano structures. However, in the particular case of metallic nanotubes in an energy interval centered at the Fermi level the density of states is nonzero and constant.

11.3 Silicon Carbide Nanotubes (SiCNTs)

Recently, scientists have shown interest in understanding the properties of silicon carbide-related materials. Since bulk silicon carbide materials possess unique physical properties, such as high hardness, high thermal conductivity, and superior radiation resistance, which are very suitable for applications in electronic and optical devices, scientists were interested in understanding the properties of silicon carbide related materials. Silicon carbide nanotubes (SiCNTs) were successfully synthesized by a reaction of silicon with a multi-walled carbon nanotube at different temperatures with Si to C at different ratios [22, 23]. The SiCNTs obtained are analogous to CNTs [8].

Theoretical studies on single-wall SiCNTs (SWSiCNT) have shown that their energetically favorable structure consists of Si and C atoms alternating as their nearest neighbors [24–26], as shown in Fig. 11.5; C and Si atoms are represented by different colors. As the fraction of Si over C atoms increases, SiCNTs start losing their stability [27]. After relaxation, SiCNTs show the radial buckling behavior, where Si atoms move toward the tube axis and C atoms move in the opposite direction. Thus, the radial geometry of the tubular structures for the small-sized SiCNTs seems like two concentric cylindrical tubes [24, 25]. Unlike CNTs, SiCNTs are always semiconductors with direct bandgaps for zigzag tubes and indirect gaps for

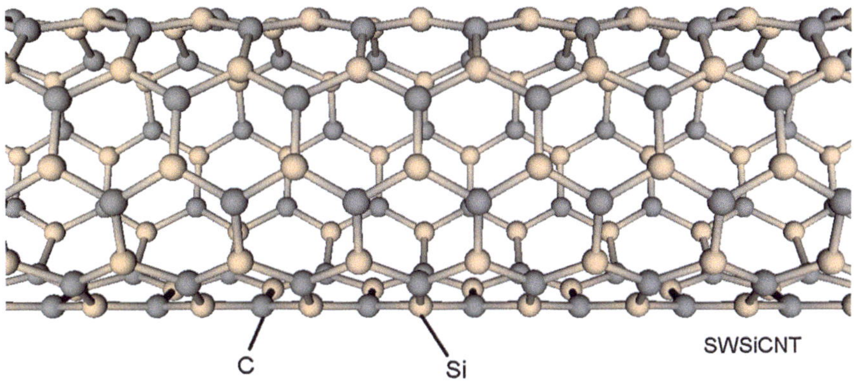

Fig. 11.5 Single-wall SiCNT

armchair and chiral tubes [26, 28–31], which are intrinsically suitable for gas sensor application [32, 33]. However, as the diameter is smaller than 10 Å for SWSiCNTs in zigzag configurations, bandgap energy, and Young's modulus strongly depend on tube diameter [34]. The electronic conductivity of ultrathin SWSiCNTs, such as (3,0) and (4,0) nanotubes, exhibit metallic behavior [26, 28–31], also confirmed by our study [35, 36].

11.4 Defects in Nanotubes

11.4.1 Vacancy Defects

Defects such as 5–7 rings, kinks, junctions, vacancies, and impurities [37–41] may be present in nanotubes. These defects can significantly change the electronic, chemical, and mechanical properties of the CNTs. Vacancies are the most common defects in CNTs which may inevitably occur during the growth process of CNT or by mechanical deformation, they have drawn much research attention in recent years [37, 40] [41]. An ideal single vacancy containing three hanging bonds is unstable [42]; two of the three dangling bonds form a new C–C bond and the structure becomes a pentagonal ring coupled with one dangling bond, which forms a so-called 5-1DB defect [42–46]. The reconstruction of di-vacancy in CNTs results into an octagonal ring coupled to two pentagonal rings [42, 45, 46], the new 5-8-5 structure does not have a hanging bond. Another study [39] suggests that an introduction of vacancy on single wall carbon nanotubes (SWCNTs) can have a large impact on electronic properties, as bandgap of semiconducting CNTs reduces and conducting CNTs increases when a vacancy is introduced.

The above works mainly concentrated on vacancy defects in CNTs. There have been very few studies performed on the vacancy defects in SiCNTs, and most of them

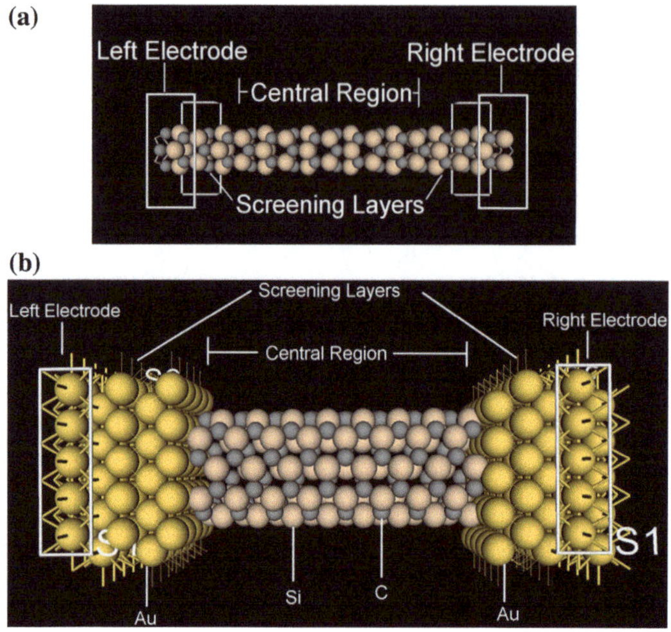

Fig. 11.6 Two-probe system made of left electrode, right electrode, and central nanotube region: **a** (4,0) zig-zag SiCNT and **b** (5,5) armchair SiCNT. [Reprinted with permission from Modern Physics Letters B, Vol. 25, No. 28 (2011) 2159–2170.]

lack a detailed explanation of their effects on transport properties. In our previous study [35], vacancy reconstruction in SiCNTs and its effect on electron transport properties was reported. A perfect (no defects) (4,0), and (5,5) SiCNTs is shown in Fig. 11.6, SiCNT structures with single and di-vacancy defects are shown in Fig. 11.7. A two-probe geometry [47] that consists of a left electrode, a right electrode, and a central region (scattering region) was constructed. Si-C bond length of 1.78 Å [37, 48] and 1:1 Si-C ratio was considered.

Since a (4,0) SiCNT is metallic by nature, we used it for constructing electrodes (see Fig. 11.6a), so as to minimize the contact dependency. A (5,5) SiCNT is semiconducting by nature hence it cannot be used as electrodes, as semiconducting leads will give no current. Unless the electrode is conducting, there will be no injection and hence no current. We used gold (Au) electrodes for building two-probe geometry with (5,5) SiCNT in the central region (see Fig. 11.6b), gold electrodes have been used previously in studying the electronic properties of CNTs and SiCNTs by [49] [50] and also by [51] for doped SiCNTs, linear current characteristics were reported in these works. Length dependency is neglected by considering a minimal length segment of nanotube in the central region.

For studying vacancy reconstruction, the whole setup was optimized (electrodes fixed) using steepest descent method of geometry optimization. Perfect (4,0) and (5,5) SiCNTs were studied for comparison with defected SiCNTs. In the case of

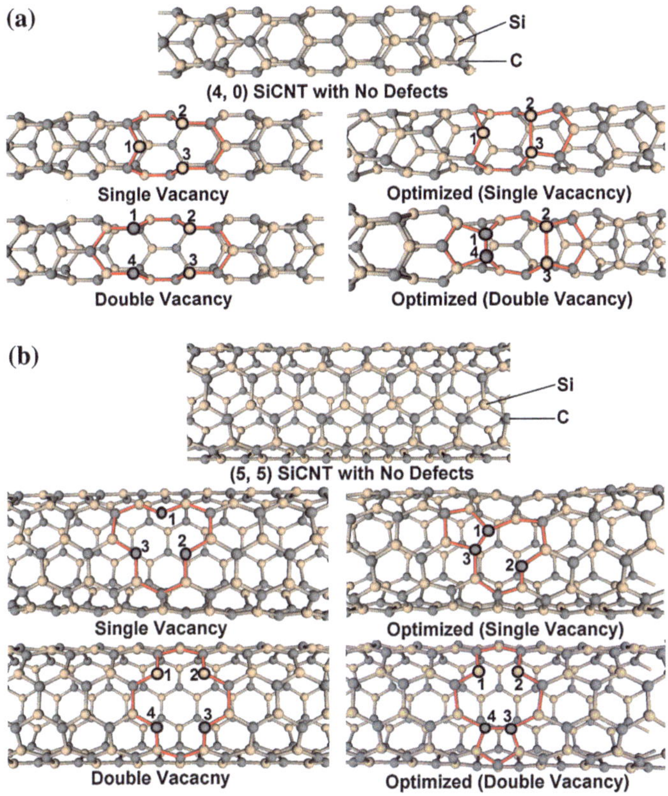

Fig. 11.7 **a** Single and double vacancies and their optimized structures in (4,0) SiCNT. **b** Single and double vacancies and their optimized structures in (5,5) SiCNT. [Reprinted with permission from Modern Physics Letters B, Vol. 25, No. 28 (2011) 2159–2170]

(4,0) zigzag SiCNT, a single vacancy reconstructs into 5-1DB defect similar to that in CNTs [43] and SiCNTs [29]; here two of the three dangling bonds (marked as 2 and 3 in Fig. 11.7a) form a new Si–Si bond (2.36 Å bond length) and the structure becomes a pentagonal ring coupled with one dangling bond. The reconstruction of di-vacancy in (4,0) SiCNT is also similar to that in CNTs [42, 45]; here the reconstruction results into an octagonal ring coupled to two pentagonal rings (see Fig. 11.7a), the new 5-8-5 structure has a C–C and Si–Si bond, it does not have a hanging bond [35].

In the case of (5,5) armchair SiCNT, a single vacancy reconstructs into 5-1DB defect similar to that in (4,0) SiCNT and CNTs [43]. However, the reconstruction of a di-vacancy forms a new 5-2DB defect in (5,5) armchair SiCNT; two of the four dangling bonds (marked as 3 and 4 in Fig. 11.7b) form a new C–C bond and the structure becomes a pentagonal ring coupled with two dangling bonds (marked as 1 and 2 in Fig. 11.7b). The separation width between these two Si atoms associated with hanging bonds is 2.55 Å [35].

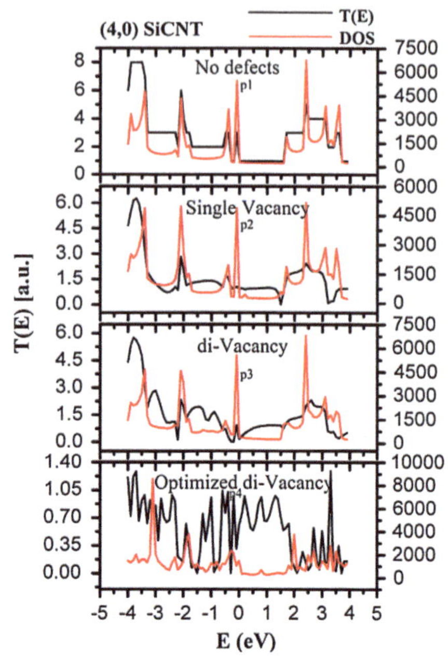

Fig. 11.8 Zero bias transmission spectra T(E) and density of states DOS profile of (4,0) SiCNT. DOS is number of electrons/energy, i.e., unit 1/eV. [Note: a.u. is arbitrary units and E is energy in electron volts]. [Reprinted with permission from Modern Physics Letters B, Vol. 25, No. 28 (2011) 2159–2170.]

The transmission spectrum T(E) is plotted in Fig. 11.8 by dark (black) color curve, here the Fermi energy is set as 0 eV [55], *DOS* is shown by light (red) color curve. The equilibrium transport properties (no bias voltage applied) of the SiCNTs are studied first. A perfect (4,0) SiCNT is metallic by nature [41] and a (5,5) SiCNT is semiconducting, which is confirmed by the transmission gap (transmission coefficient is zero) around Fermi level for semiconducting tube and by no gap in transmission for the case of metallic SiCNT in Fig. 11.8.

The eigenstates of molecular projected self-consistent Hamiltonian (MPSH) are closely related to the poles of Green's function, which roughly corresponds to the peaks (p1, p2, p3, p4, p1′, p2′, p3′ and p4′) of the transmission spectrum [52] in Figs. 11.8 and 11.9. Frontier molecular orbitals of the MPSH are calculated and shown in Fig. 11.10a, b to study the physical origin of these peaks, here red (light) and blue (dark) color indicate the positive and negative wave phases in the lobe. The peaks p1, p2, p3 originate from orbitals O1, O2, O3 whose localization is weak around the defect site. Peaks p4, p1′, p2′, p3′, and p4′ originate from orbitals O4, O1′, O2′, O3′, and O4′, which are localized near the defect site (see Fig. 11.10a, b).

There is a close relationship between I–V characteristics of Fig. 11.11 and peaks in the transmission spectrum of Figs. 11.8, 11.9. For a (4,0) SiCNT, the peaks in T(E) near the Fermi level get smaller with the introduction of vacancy (Fig. 11.8), indicating a reduction in electronic states which in turn affects the current (Fig. 11.11a). However, new peaks appear in T(E) near the Fermi level in a (5, 5) SiCNT with the introduction of vacancy (Fig. 11.9), which means more electronic states appear near the Fermi level that improves conducting properties resulting in an increase in

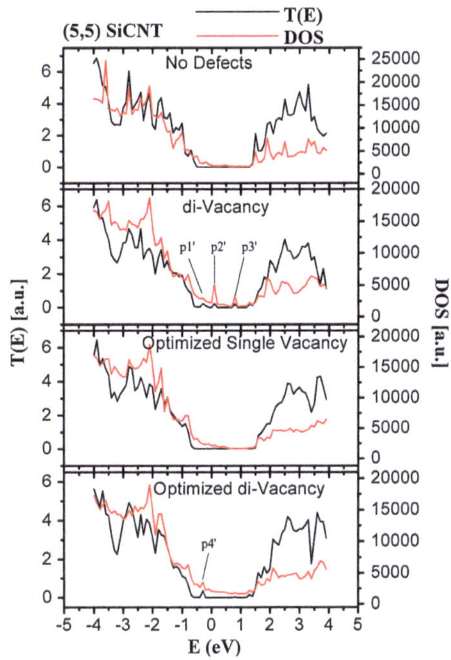

Fig. 11.9 Zero bias transmission spectra T(E) and density of states DOS profile of (5,5) SiCNT. DOS is number of electrons/energy, i.e., unit 1/eV. [Note: a.u. is arbitrary units and E is energy in electron volts]. [Reprinted with permission from Modern Physics Letters B, Vol. 25, No. 28 (2011) 2159–2170.]

current in (5,5) SiCNT (Fig. 11.11b). Further, an increase in current with increase in bias voltage implies that more molecular orbitals contribute to the formation of the current [51].

11.5 Boron-Nitrogen Co-doped SiCNTs

The electronic properties of CNTs are strongly related to the delocalized electron system and any (chemical) modification of the CNTs will influence these properties. Therefore, by the proper choice of the type of modification the electronic properties of a CNT can be tuned. The tuning of the electronic properties of the CNT can be achieved by so-called impurity doping. In this chapter we review the field of CNT and SiCNT doping and present the investigation results of our study on Boron and Nitrogen co-doped SiCNTs.

There are a wide variety of possible approaches to dope CNTs and thereby change their physical and electronic properties [53]. One of the traditional semiconductor approaches is to replace carbon atoms with impurities such as boron and nitrogen. We cover the transport properties of boron and nitrogen doped CNTs in the next section, which is based on [54].

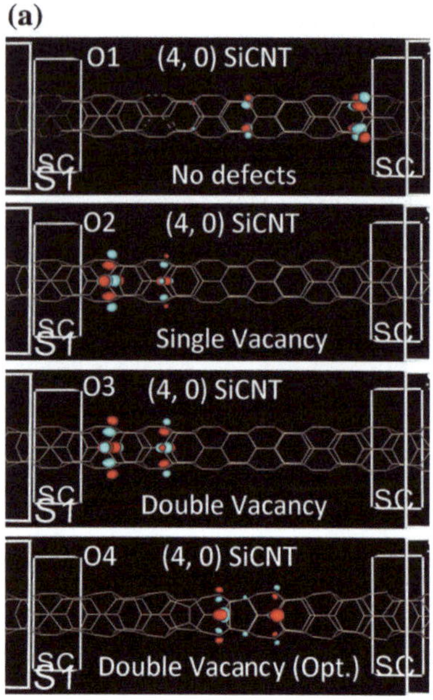

Fig. 11.10 Orbitals corresponding to the peaks around the Fermi level in the transmission spectrum of **a** (4,0) SiCNT; Orbitals O1, O2, O3, O4 corresponding to peaks p1, p2, p3, p4 of T(E) in (**a**). **b** (5,5) SiCNT; Orbitals O1′, O2′, O3′, O4′ corresponding to peaks p1′, p2′, p3′, p4′ of T(E) in (**b**). Optimized structures have more localized states at the defect site. [Reprinted with permission from Modern Physics Letters B, Vol. 25, No. 28 (2011) 2159–2170.]

11.5.1 Substitution Impurity Doped Nanotubes

The substitution of carbon atoms in the honeycomb lattice of a CNT by impurity atoms with a different number of valence atoms introduces additional states in the density of states (DOS) of the CNT. These additional states in the DOS could be electron-donor states, electron-acceptor states, or neither of these two, depending on the local bonding arrangements of the heteroatoms. In this section we will focus on the substitution of boron and nitrogen atoms in SWCNT and SWSiCNT.

11.5.1.1 Nitrogen Doped Carbon Nanotubes

We consider first the substitution of a carbon by a nitrogen atom into the CNT lattice of a metallic SWCNT. Since nitrogen has 5 valence electrons and carbon only 4, the nitrogen state will be located above the Fermi-energy of the undoped SWCNT. For an SWCNT with a diameter of about 1 nm, the nitrogen states will be found approximately a few hundred meV below the first van-Hove singularity of the π^*-band [55, 56]. The exact position of the nitrogen states depends on diameter, helicity, and number of substitution nitrogen atoms.

When the incorporated nitrogen ionizes, that is, liberates its excess electron to the π-electron system of the SWCNT host, the Fermi-energy will shift upwards in energy relative to its initial position.

Fig. 11.10 (continued)

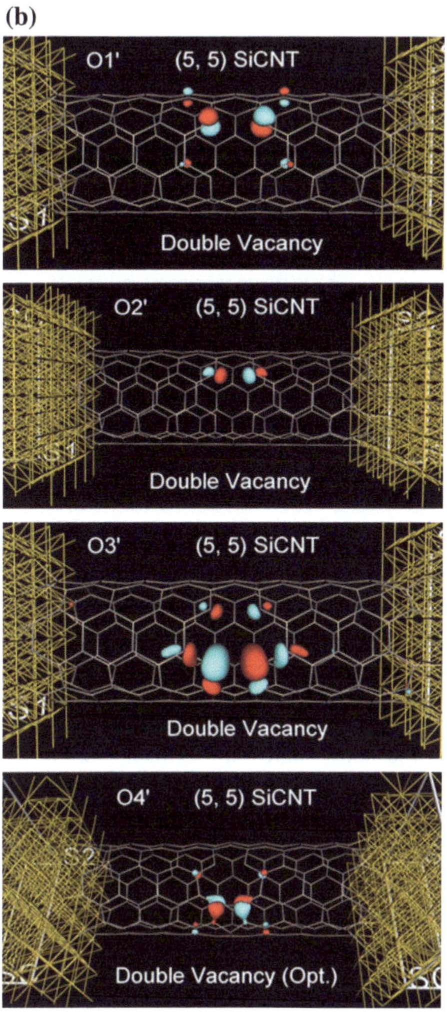

In case of a semiconducting SWCNT with an incorporated nitrogen impurity, similar to the metallic SWCNT, the nitrogen states are found above the initial position (in the centre of the energy gap) of the Fermi-energy. The nitrogen states lies about 150 to 200 meV below the conduction band of a nanotube with a diameter of one nm, depending on the tube helicity and the number of nitrogen atoms.

As the nitrogen states ionize, the liberated electrons will occupy empty states at the edge of the conduction-band of the tube and an n-type semiconducting SWNT will be created. Charge transport occurs solely through the π^*-band channel. The Fermi energy of the nanotube will shift (relatively) upwards in energy and position itself at the nitrogen states in the DOS.

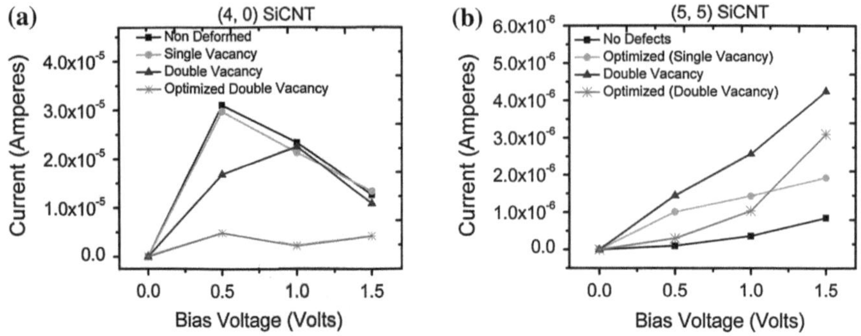

Fig. 11.11 Bias voltage-dependent current characteristic (I–V characteristic) of **a** (4,0) and **b** (5, 5) SiCNT two-probe system. Current is seen to be different in perfect and vacancy defected SiCNTs. [Reprinted with permission from Modern Physics Letters B, Vol. 25, No. 28 (2011) 2159–2170]

11.5.1.2 Boron Doped Carbon Nanotubes

The case of boron-doping by substitution of carbon atoms in a CNT is principally the inverse of the nitrogen-doping case. Boron has only three valence electrons, the boron states will be located below the Fermi-energy of the undoped SWCNT. Due to their reduced valence compared to carbon, the boron atoms can act as electron traps for the delocalized π-electrons of undoped SWCNT. If these traps are occupied, that is, the boron states are ionized, the Fermi-energy will shift downwards in energy toward the first van-Hove singularity of the π-band. Similar to nitrogen doping, the shift of the Fermi-energy depends on diameter, helicity, and number of boron atoms. Also, the boron states for a 1 nm diameter SWCNT lie some few hundred meV above the first van-Hove singularity of the π-band, similar to nitrogen doped SWCNTs.

In semiconducting SWCNTs, doping by boron substitution creates non-occupied states in the energy gap between the initial position of the Fermi energy and the valence-band edge. In SWCNTs of about 1 nm in diameter these states lie at $\sim$100 meV above the valence band. If these empty boron-states are occupied, holes at the edge of the valence band are created and a p-type semiconducting SWCNT is created. The Fermi energy will then shift down toward the valence band and boron states.

11.5.1.3 Nitrogen Doped SiCNTs

In the case of nitrogen doped SiCNTs, the unpaired electron goes to a delocalized state split from the conduction band edge in the (8,0) SiC nanotube. LDA bandgap of (8,0) SiCNT is 1.2 eV, the occupied shallow donor level is at about 0.33 eV below the conduction band edge [30]. Similar to the bulk SiC, doped nitrogen acts as a shallow effective mass-like donor in the (8,0) SiC nanotube. For larger bandgap SiCNTs like a (6,6) SiCNT which has an LDA bandgap of 2.0 eV, nitrogen acts as a deep donor in the SiCNT.

The nitrogen cannot act as an n-type dopant in wide bandgap SiC nanotubes unlike the case of bulk SiC, it is rather a compensating center both for the holes and for the electrons. Nevertheless, the nitrogen can be used in the small bandgap SiC nanotubes as an n-type dopant [30].

11.5.1.4 Boron Doped SiCNTs

Boron forms acceptor states in SiCNTs, it can form shallow and relatively deep acceptor states depending on the stoichiometry of SiC nanotubes. If the concentration and the site selection of boron can be controlled, then the Fermi level can be close to the valence band top. The manipulation of the Fermi level in SiC nanotubes can be an important issue in biological applications and in chemical sensors [30].

11.5.2 Boron and Nitrogen Co-doped CNTs

The substitution of a BN pair in the carbon network does not modify significantly the quantum transport of the nanotube at low gate voltage, whereas impurity states reduce the conductance by half when B and N atoms are far apart. For larger BN nanodomains and rings, the conductance plateau decreases quasilinearly as a function of the density of BN pairs and localization phenomenon could be observed or not depending on the geometry of the domain [57].

11.5.3 Boron and Nitrogen Co-doped SiCNTs

The electron transport properties for individual boron and nitrogen doped SiCNTs could be found in previous research works [51, 58–60]. In our previous study [61], we reported the transport properties in the presence of both boron and nitrogen impurities in SiCNT. For this purpose, in Fig. 11.12, a silicon atom toward the left electrode is replaced by a boron atom forming a p-type (acceptor) region and a silicon atom toward the right electrode is replaced by a nitrogen atom making an n-type region. The whole geometry in Fig. 11.12 could be visualized as a p-n junction formed by a doped (8,0) SiCNT placed between the gold (Au) electrodes.

The transmission spectrum T(E) for various bias voltages is plotted in Fig. 11.13, here the Fermi energy is set as 0 eV. The equilibrium transport properties (no bias voltage applied, Fig. 11.13a) of the BN doped SiCNTs are studied first. A (8, 0) SiCNT is semiconducting by nature, a BN doped (8,0) SiCNT is also semiconducting which is confirmed by the transmission gap (transmission coefficient is zero) around Fermi level in Fig. 11.13a. The eigenstates of molecular projected self-consistent Hamiltonian (MPSH) are closely related to the poles of Green's function, which roughly corresponds to the peaks (p1, p2, p3, p4, p5 and p6) of the transmission spectrum in Fig. 11.13a [52]. The frontier molecular orbitals of the MPSH are calculated and shown in Fig. 11.14 to study the physical origin of these peaks.

The highest occupied molecular orbital (HOMO) lies at −0.42 eV and has no contribution from BN atoms. The lowest unoccupied molecular orbital (LUMO) lies at 0.23 eV and is mainly contributed by N atom. The difference in HOMO-LUMO levels gives the bandgap of BN co-doped SiCNT as ∼0.65 eV, which is comparable with the bandgap of individual B-doped SiCNT (∼0.7 eV) and N-doped SiCNT (∼ 0.5 eV) reported in [30]. The transmission peaks (p1, p2, p3) in Fig. 5.2a with lower values than Fermi level originates mainly from orbital O1, O2 and O3 in Fig. 11.14, whose localization is high. The transmission peaks (p4, p5, p6) with higher values than Fermi level primarily come from O4, O5 and O6, which exhibit weaker localization.

Negative differential resistance (NDR) is a useful property of molecular devices and switches. It is meaningful to find whether BN co-doped (8,0) SiCNT has the property of NDR or not. A BN co-doped SiCNT junction formed as in Fig. 11.12 is examined for NDR region and also for rectification, if any. To do so, we examine the bias voltage-dependent current characteristics of this device. We neglect SiCNT length dependency on the results by considering a minimum length central region. Contact dependency is resolved by using Au electrodes which have been used in the study on the electronic transport properties of CNTs and SiCNTs by [49, 62], linear current characteristics with NDR region were reported in these works and also in [51] for boron doped SiCNT with Au electrodes.

I–V characteristics of BN co-doped SiCNT in Fig. 11.15 is different from that of the individual boron and nitrogen doped SiCNT reported in the literature [51, 58, 59]. Most importantly, both currents under positive and negative bias voltages show an exponential growth and decay, which is in contrast to the linear current growth and decay for positive and negative bias voltages reported in previous research works [51, 58, 59]. Rectification observed is negligible as the current is seen to increase exponentially after a forward and reverse cut-in (threshold) voltage of ∼0.5 volts in Fig. 11.15. Also, the negative differential resistance (NDR) region seems to be suppressed in BN co-doped SiCNTs in the bias range −2.1 V to +2.1 V. However, NDR was observed for SiCNTs and individual boron and nitrogen doped SiCNTs in previous research works [51, 54, 59, 60]. Further, NDR was also seen in our previous research works on deformed and vacancy defected SiCNTs [35, 36], but no NDR is seen here for BN co-doped SiCNT.

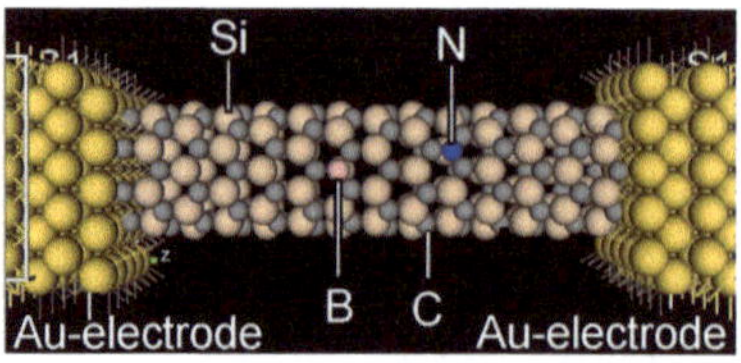

Fig. 11.12 Two-probe geometry of a BN doped (8,0) SiCNT with Au electrodes. [Reprinted with permission from Physics Letters A 375 (2011 3382-3385).]

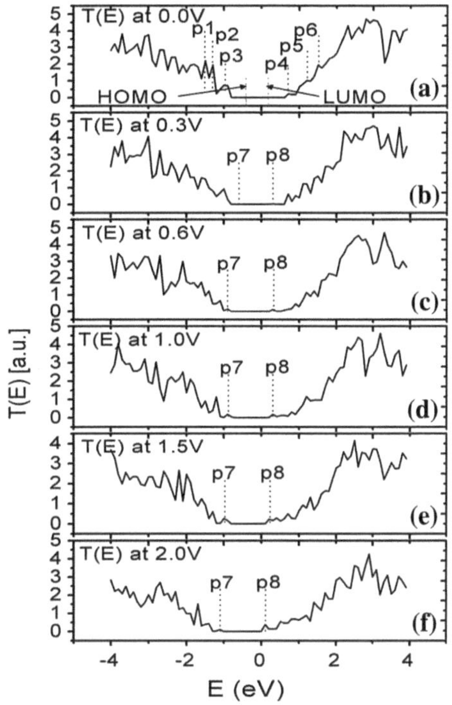

Fig. 11.13 Transmission spectrum T(E) in the bias range 0 to +2.0 V. New peaks (p7 and p8) appear near the Fermi level at high bias voltages. [Reprinted with permission from Physics Letters A 375 (2011 3382–3385).]

It is possible to investigate the suppression of NDR effect by comparing the transmission spectrum under the bias range 0 V to +2.0 V. Under both negative bias and positive bias, the current increases exponentially with the increase in bias voltage. Let us consider the positive bias range 0.3–2.0 V, the magnitude of current increases (see Fig. 11.15), which is determined by integral of the transmission spectrum in a particular energy range (called bias window). An increase in bias window implies that more molecular orbitals contribute to the formation of the current [58]. Comparing the transmission spectrum under this bias range in Fig. 11.13b–f, it can be seen that new transmission peaks p7 and p8 develop around the Fermi level and these peaks get enhanced with an increase in bias voltage. These peaks (p7 and p8) are the new electronic states developed around the Fermi level, which play an important role in the formation of frontier orbitals (HOMO and LUMO).

There is a close relationship between the peaks in the transmission spectrum and the localization of molecular orbitals, which is evident from the equilibrium transport properties. The localization of frontier orbitals is lowered with increase in bias voltage, which could be observed by comparing frontier orbitals (FO) at small bias with the orbitals O7 and O8 (at high bias of 1.5 and 2.0 volts) in Fig. 11.16 corresponding to peaks p7 and p8 of Fig. 11.13e, f. From the analysis of Fig. 11.16, it is observed that FOs at low bias exhibit strong localization. However, FOs exhibit reduced localization at high bias, where LUMO shares a contribution from nitrogen

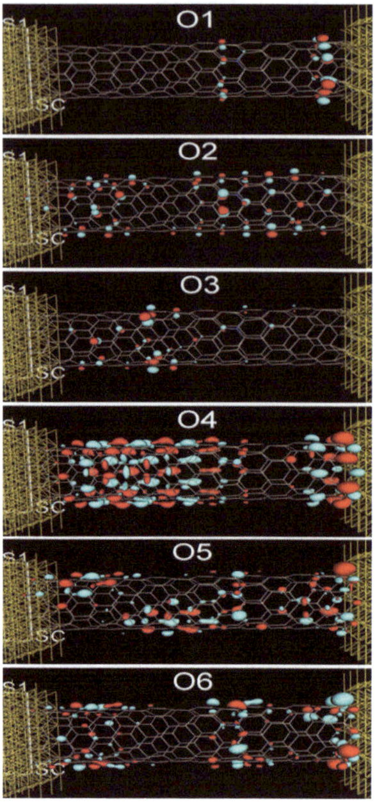

Fig. 11.14 Frontier orbitals of the MPSH for BN doped SiCNT at zero bias; O1, O2, O3, O4, O5, O6 are the orbitals corresponding to the peaks p1, p2, p3, p4, p5, p6 of the Transmission spectrum in Fig. 11.13. [Reprinted with permission from Physics Letters A 375 (2011 3382-3385).]

and its nearby atoms (see Fig. 11.16). The introduction of new electronic states around the Fermi level (see Fig. 11.13c–f) followed by lowering in orbital localization (see Fig. 11.16) facilitates tunneling from left electrode to right electrode, which is the main reason for the suppression of NDR region.

11.6 Conclusions

It is observed that the reconstruction of di-vacancy in armchair SiCNT forms a 5-2DB defect, which is different from the reconstruction of CNTs and zigzag SiCNT. The introduction of vacancy reduces the bandgap of (5,5) semiconducting SiCNTs and increases the bandgap of (4,0) conducting SiCNTs (converts them to semi-metallic nanotubes). A large increase in current in (5,5) semiconducting SiCNTs is observed, which is due to the introduction of new electronic states around the Fermi level followed by conduction through the defect sites. Transport properties of BN co-doped SiCNTs are different from that of individual boron and nitrogen doped SiCNTs.

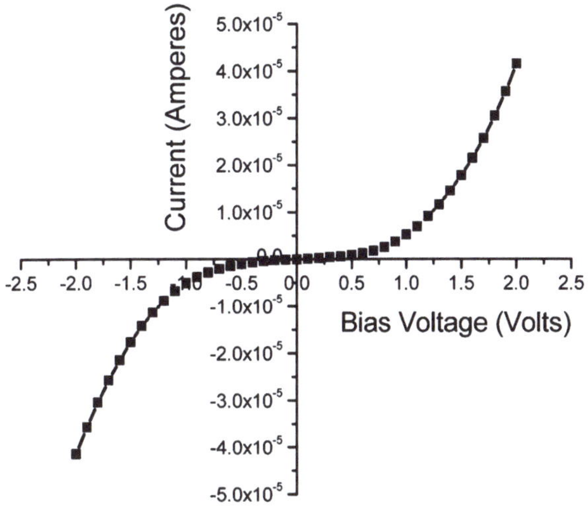

Fig. 11.15 I–V characteristics of BN doped SiCNTs. BN co-doping suppresses NDR; rectification is negligibly small. (Reprinted with permission from Physics Letters A 375 (2011 3382–3385).)

Fig. 11.16 Frontier orbitals of the MPSH. [Reprinted with permission from Physics Letters A 375 (2011) 3382–3385]

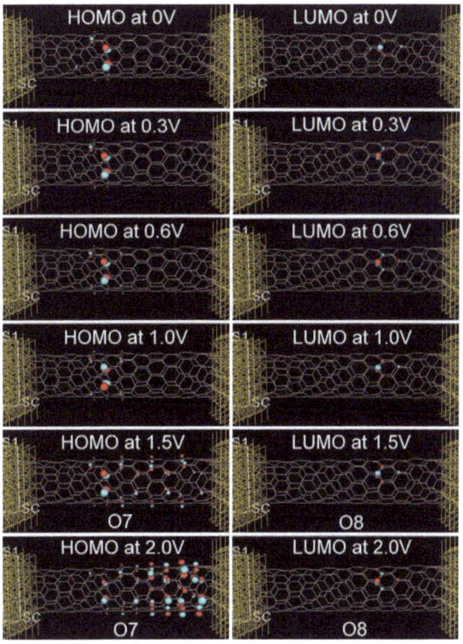

At higher bias voltages, NDR effect is suppressed due to introduction of new electronic states around Fermi level followed by reduction in orbital localization which makes tunneling easier from left electrode to right electrode. I–V characteristics show an exponential growth and reduction which is in contrast to the linear current characteristics reported in previous research works for individual boron and nitrogen doped SiCNTs.

References

1. Yamacli, S.: Development of ab initio based models for carbon nanotube technology. Ph.D. Thesis submitted at Cukurova University (2011)
2. Avouris, P., Radoslavjevic, M., Wind, S.: Carbon Nanotube Electronics and Optoelectronics. In: Carbon Nanotubes. Springer-Verlag, Berlin (2004)
3. Liu, Z., Bajwa, N., Ci, L., Kar, S., Ajayan, P.M., Lu, J.Q.: Densification of carbon nanotube bundles for interconnect application. In: Proceedings of IEEE Interconnect Conference, pp. 201–203 (2007)
4. Li, J., Ye, Q., Cassell, A., NG, H.T., Stevens, R., Han, J., Meyappan, M.: Bottom-up approach for carbon nanotube interconnects. Appl. Phys. Lett. **82**, 2491–2493 (2003)
5. Sun, Y., Zhu, L., Jiang, H., Lu, J., Wang, W., Wong, C.P.: A paradigm of carbon nanotube interconnects in microelectronic packaging. J. Electron. Mater. **37**, 1691–1697 (2008)
6. Pesetski, A.A., Baumgardner, J.E., Krishnaswamy, S.V., Zhang, H., Adam, J.D., Kocabas, C., Banks, T., Rogers, J.A.: A 500MHz carbon nanotube transistor oscillator. Appl. Phys. Lett. **93**, 123506 (2008)
7. Harris, P.: Carbon Nanotube Science: Synthesis, Properties and Applications. Cambridge University Press, UK (2011)
8. Chen, H.-L., Ju, S.-P., Lin, J.-S., Jijun, Z., Chen, H.-T., Chang, J.-G., Weng, M.H., Lee, S.-C., Lee, W.-J.: Electronic properties of a silicon carbide nanotube under uniaxial tensile strain: a density function theory study. J. Nanopart. Res. **12**(8), 2919–2928 (2010)
9. Dresselhaus, M.S., Eklund, P.C.: Phonons in carbon nanotubes. Adv. Phys. **49**(6), 705–814 (2000)
10. Biel, B., Garcı'a-Vidal, F.J., Rubio, A., Flores, F.: Anderson localization in carbon nanotubes: defect density and temperature effects. Phys. Rev. Lett. **95**(26), 266801 (2005)
11. Park, J.Y.: Electrically tunable defects in metallic single-walled carbon nanotubes. Appl. Phys. Lett. **90**, 23112 (2007)
12. Chandra, B.: Synthesis and electronic transport in known chirality single wall carbon nanotubes. Ph.D. Thesis submitted at Columbia University (2009)
13. Saraswat, K., Cho, H., Kapur, P., Koo, K.H.: Performance comparison between copper, carbon nanotube and optical interconnects. In: Proceedings of IEEE International Symposium on Circuits and Systems, pp. 2781–2784 (2008)
14. http://en.wikipedia.org/wiki/File:CNTnames.png
15. Geim, A.K., Novoselov, K.S.: The rise of graphene. Nat. Mater. **6**(3), 183–191 (2007)
16. Wallace, P.R.: The band theory of graphite. Phys. Rev. **71**(9), 622–634 (1947)
17. http://en.wikipedia.org/wiki/File:GrapheneE2.png
18. Semenof, G.W.: Condensed-matter simulation of a three-dimensional anomaly. Phys. Rev. Lett. **53**(26), 2449–24452 (1984)
19. Hanson, G.W.: Fundamentals of Nanoelectronics. Prentice-Hall Inc., New Jersey (2008)
20. Anantram, M.P., Leonard, F.: Physics of carbon nanotube electronic devices. Rep. Prog. Phys **69**, 507–561 (2006)
21. Loiseau, A., Launois, P., Petit, P., Roche, S., Salvetat, J.-P.: Understanding carbon nanotubes from basics to applications. Lect. Notes Phys. **677**, XVI–552 (2006)
22. Sun, X.H., Li, C.P., Wong, W.K., Wong, N.B., Lee, C.S., Lee, S.T., Teo, B.K.: Formation of silicon carbide nanotubes and nanowires via reaction of silicon (from disproportionation of silicon monoxide) with carbon nanotubes. J. Am. Chem. Soc. **124**(48), 14464–14471 (2002)

23. Pham-Huu, C., Keller, N., Ehret, G., Ledoux, M.J.: The first preparation of silicon carbide nanotubes by shape memory synthesis and their catalytic potential. J. Catal. **200**(2), 400–410 (2001)
24. Alam, K.M., Ray, A.K.: A hybrid density functional study of zigzag sic nanotubes. Nanotechnology **18**(49), 495706 (2007)
25. Alam, K.M., Ray, A.K.: Hybrid density functional study of armchair sic nanotubes. Phys. Rev. B **77**(3), 035436 (2008)
26. Menon, M., Richter, E., Mavrandonakis, A., Froudakis, G., Andriotis, A.N.: Structure and stability of SiC nanotubes. Phys. Rev. B **69**(11), 115322 (2004)
27. Mavrandonakis, A., Froudakis, G.E., Schnell, M., Muhlhauser, M.: From pure carbon to silicon-carbon nanotubes: an ab initio study. Nano Lett. **3**(11), 1481–1484 (2003)
28. Zhao, M.W., Xia, Y.Y., Li, F., Zhang, R.Q., Lee, S.T.: Strain energy and electronic structures of silicon carbide nanotubes: density functional calculations. Phys. Rev. B **71**(8), 085312 (2005)
29. Baierle, R.J., Piquini, P., Neves, L.P., Miwa, R.H.: Ab initio study of native defects in sic nanotubes. Phys. Rev. B **74**(15), 155425 (2006)
30. Gali, A.: Ab initio study of nitrogen and boron substitutional impurities in single-wall sic nanotubes. Phys. Rev. B **73**(24), 245415 (2006)
31. Gali, A.: Ab initio theoretical study of hydrogen and its interaction with boron acceptors and nitrogen donors in single-wall silicon carbide nanotubes. Phys. Rev. B **75**(8), 085416 (2007)
32. Wu, R.Q., Yang, M., Lu, Y.H., Feng, Y.P., Huang, Z.G., Wu, Q.Y.: Silicon carbide nanotubes as potential gas sensors for CO and HCN detection. J. Phys. Chem. C **112**(41), 15985–15988 (2008)
33. Zhao, J.X., Ding, Y.H.: Can silicon carbide nanotubes sense carbon dioxide. J. Chem. Theor. Comput. **5**(4), 1099–1105 (2009)
34. Baumeier, B., Kruger, P., Pollmann, J.: Structural, elastic, and electronic properties of SiC, BN, and BeO nanotubes. Phys. Rev. B **76**(8), 085407 (2007)
35. Choudhary, S., Qureshi, S.: Theoretical study on the effect of vacancy defect reconstruction on electron transport in Si-C nanotubes. Mod. Phys. Lett. B **25**(28), 2159-2170 (2011)
36. Choudhary, S., Qureshi, S.: Vacancy defect reconstruction and its effect on electron transport in Si-C nanotubes. J. Nano Electron. Phys **3**(1), 1035–1040 (2011)
37. Tombler, T.W., Zhou, C., Alexseyev, L., Kong, J., Dai, H., Liu, L., Jayanthi, C.S., Tang, M., Wu, S.: Reversible electromechanical characteristics of carbon nanotubes under local-probe manipulation. Nature **405**, 769–772 (2000)
38. Miyamoto, Y., Yu, B.D.: Computational designing of graphitic silicon carbide and its tubular forms. Appl. Phys. Lett. **80**(4), 586–588 (2002)
39. Vincent Liu, L., Tian, W.Q., Wang, Y.A.: Ab initio studies of vacancy-defected fullerenes and single-walled carbon nanotubes. Int. J. Quantum Chem. **109**(14), 3441–3456 (2009)
40. Chibotaru, L.F., Compernolle, S., Ceulemans, A.: Electron transmission through atom-contacted carbon nanotubes. Phys. Rev. B **68**(12), 125412–125424 (2003)
41. Cai, Y., Zhang, A., Feng, Y.P., Zhang, C., Teoh, H.F., Ho, G.W.: Strain effects on work functions of pristine and potassium-decorated carbon nanotubes. J. Chem. Phys. **131**(22), 224701–224705 (2009)
42. Yuan, J., Liew, K.M.: Effects of vacancy defect reconstruction on the elastic properties of carbon nanotubes. Carbon **47**(6), 1526–1533 (2009)
43. Lu, A.J., Pan, B.C.: Nature of single vacancy in a chiral carbon nanotubes. Phys. Rev. Lett. **92**(10), 105504–105507 (2004)
44. Rossato, J., Baierle, R.J., Fazzio, A., Mota, R.: Vacancy formation process in carbon nanotubes: first-principles approach. Nano Lett. **5**(1), 197–200 (2005)
45. Ajayan, P.M., Ravikumar, V., Charlier, J.C.: Surface reconstructions and dimensional changes in single-walled carbon nanotubes. Phys. Rev. Lett. **81**(7), 1437–1440 (1998)
46. He, Y., Zhang, C., Cao, C., Cheng, H.P.: Effects of strain and defects on the electron conductance of metallic carbon nanotubes. Phys. Rev. B **75**(23), 235429–235434 (2007)
47. Yamacli, S., Avci, M.: Simple and accurate model for voltage- dependent resistance of metallic carbon nanotube interconnects: an ab initio study. Phys. Lett. A **374**(2), 297–304 (2009)

48. Wang, Z., Zu, X., Xiao, H., Gao, F., Weber, W.J.: Tuning the band structures of single walled silicon carbide nanotubes with uniaxial strain: a first principles study. Appl. Phys. Lett. **92**(18), 183116–183118 (2008)
49. Li, X.F., Chen, K.Q., Wang, L., Long, M.Q., Zoul, B.S., Shuai, Z.: Effect of length and size of heterojunction on the transport properties of carbon-nanotube devices. Appl. Phys. Lett. **91**(13), 133511–133513 (2007)
50. Yang, Y.T., Song, J.X., Liu, X.H., Chai, C.C.: Negative differential resistance in single-walled SiC nanotubes. Chin. Sci. Bull. **53**(23), 3770–3772 (2008)
51. Yang, Y.T., Ding, R.X., Song, J.X.: Transport properties of boron-doped single-walled silicon carbide nanotubes. Physica B **406**(2), 216–219 (2010)
52. Li, Z., Kosov, D.S.: Dithiocarbamate anchoring in molecular wire junctions: a first principles study. J. Phys. Chem. B **110**(20), 9893–9898 (2006)
53. Mintmire, J.W., White, C.T.: Properties: theoretical predictions. In: Ebbesen, T.W. (ed.) Carbon Nanotubes – Preparation and Properties, pp. 191–209, CRC Press, Boca Raton (1997)
54. Wei C.: Doped Nanomaterials and Nanodevices (2007)
55. Nevidomskyy, A.H., Csanyi, G., Payne, M.C.: Chemically active substitutional nitrogen impurity in carbon nanotubes. Phys. Rev. Lett. **91**(10), 105502 (2003)
56. Kang, H.S., Jeong, S.: Nitrogen doping and chirality of carbon nanotubes. Phys. Rev. B **70**(23), 233411–233414 (2004)
57. Khalfoun, H., Hermet, P., Henrard, L.: B and N codoping effect on electronic transport in carbon nanotubes. Phys. Rev. B **81**(19), 193411–193414 (2010)
58. Taguchi, T., Igawa, N., Yamamoto, H., Jitsukawa, S.: Synthesis of silicon carbide nanotubes. J. Am. Ceram. Soc. **88**(2), 459–461 (2005)
59. Song, J., Yang, Y., Liu, H.: Electronic structures and optical properties of the nitrogen-doped SiC nanotube. In: Proceedings of IEEE EDSSC, pp. 509–512 (2009)
60. Guo, W., Hu, Y.B., Zhang, Y.Y., Du, S.X., Gao, H.J.: Transport properties of boron nanotubes investigated by ab initio calculation. Chin. Phys. B **18**(6), 2502 (2009)
61. Choudhary, S., Qureshi, S.: Theoretical study on transport properties of a BN co-doped SiC nanotube. Phys. Lett. A **375**(38), 3382–3385 (2011)
62. Yang, Y.T., Song, J.X., Liu, X.H., Chai, C.C.: Negative differential resistance in single-walled SiC nanotubes. Chin. Sci. Bull. **53**(23), 3770–3772 (2008)

Chapter 12
Synthesis, Properties, and Applications of One-Dimensional Transition Metal Silicide Nanostructures

Guangwei She, Hailong Liu, Lixuan Mu and Wensheng Shi

Abstract One-dimensional (1D) nanostructures of transition metal silicide (TMS) have attracted more attention due to their unique properties and potential applications in microelectronics. A variety of synthetic approaches were developed to fabricate 1D TMS nanostructures. Chemical vapor deposition (CVD) is the most widely used method to synthesize 1D TMS nanostructures. Various precursors and growth mechanisms are involved in the CVD processes. Other methods such as chemical vapor transport (CVT) method, silicidation method, reactive epitaxial method, hydrothermal method, were also successfully employed for the formation of 1D TMS nanostructures. Electrical transport measurements reveal that many TMS nanowires exhibit metallic behavior with extremely high conductivity. At the same time, semiconducting behaviors were observed in some situations. Some silicide nanowires (silicides of Fe, Co, Ni, Cr, Mn, etc.) have ferromagnetic properties, even at room temperature. Unique properties such as field-emission, optical, thermoelectric, mechanical were also investigated in the 1D TMS nanostructures. Based on these properties, 1D TMS nanostructures were utilized in microelectronic devices, lithium ion batteries, memory device and capacitor, etc.

W. Shi (✉) · H. Liu · L. Mu · G. She
Key Laboratory of Photochemical Conversion and Optoelectronic Materials, Technical Institute of Physics and Chemistry, Chinese Academy of Sciences,
100190 Beijing, People's Republic of China
e-mail: shiws@mail.ipc.ac.cn

H. Liu
e-mail: liuhailong@mail.ipc.ac.cn

L. Mu
e-mail: mulixuan@mail.ipc.ac.cn

G. She
e-mail: shegw@mail.ipc.ac.cn

H. Li et al. (eds.), *Silicon-based Nanomaterials*,
Springer Series in Materials Science 187, DOI: 10.1007/978-1-4614-8169-0_12,

© Springer Science+Business Media New York 2013

12.1 Introduction

Transition metal silicides (TMS), a class of intermetallic compounds between metals and silicon, have been intensively studied for over 70 years. These studies revealed some unique properties such as electrical conductivity, semiconducting behavior, ferromagnetism, thermoelectricity, and high-temperature stability. Metallic, intermetallic, and semiconducting behaviors were observed in the electrical transport measurements. Most TMS are metallic with resistivity from ~10 to 100s of $\mu\Omega\,cm$. The high conductivity promises applications in integrated circuits as ohmic contacts, interconnectors, and gate materials. Several TMS are semiconductors, including β-$FeSi_2$, ε-FeSi, Cr_2Si, etc. Among them, β-$FeSi_2$ is a good semiconducting material with direct band gap of about 0.85 eV. Furthermore, it has large optical absorption coefficient and good physical–chemical stability at high temperature [1, 2]. Cr_3Si has a melting point as high as 2043 K, promising applications in high temperature environments. Ferromagnetic properties were observed in some TMS, which is useful for spintronic devices. The higher manganese silicides are good thermoelectric materials with figures of merit (ZT) up to 0.7–0.8. Besides the above-mentioned interesting properties and merits, the good compatibility with silicon makes it easy for TMS to integrate into conventional silicon-based electronics.

In addition to exploring the intrinsical properties of TMS, reducing the size of the HMS to nanoscale would bring novel properties and interesting applications. With the rapid development of nanoscience and nanotechnology in the last two decades, TMS nanostructures, especially one-dimensional (1D) structures have attracted more attention. Research groups, all over the world, such as Jin's group, Wang's group, and Kim's group have devoted many efforts to developing new synthetic strategies and exploring novel phenomena and applications. In this chapter, we will introduce the present synthesis method for 1D TMS nanostructures, as well as their electric, magnetic, optical properties, and relative applications.

12.2 Synthesis of 1D TMS Nanostructures

Many methods have been successfully employed for the synthesis of 1D nanostructures of many materials. But it is rather difficult to synthesize 1D nanostructures of TMS, due to their numerous stoichiometry and complicated crystallographic phase. For example, there are five known iron silicide compound phase (Fe_2Si, Fe_5Si_3, FeSi, α-$FeSi_2$, β-$FeSi_2$). Fortunately, with consistent efforts of researchers in this field, some general growth methods have been developed for the 1D TMS nanostructures fabrication.

12.2.1 Chemical Vapor Deposition

Chemical vapor deposition (CVD) is almost the most successful and widely used method to synthesize 1D TMS nanostructures. A variety of metal and Si precursors (sources) were used in the CVD process. Different precursors determine the growth conditions, growth behavior, and growth mechanism of the 1D nanostructures.

Si + Metal Compounds

Metal compounds can directly react with the Si substrate or Si powder to form metal silicides. Generally, the metal sources simultaneously act as the catalyst for the 1D nanostructures growth. ε-FeSi nanowires were synthesized by using $FeCl_3$ powder as the metal source to react with the silicon substrate at 1100 °C [3]. As shown in Fig. 12.1, the nanowires are single crystalline with uniform diameter.

Two plausible reaction paths were suggested for the formation of FeSi:

$2FeCl_3(s) \leftrightarrow Fe_2Cl_6(g)$

$2Fe_2Cl_6(g) + 7Si(s) \rightarrow 4FeSi(s) + 3SiCl_4(g)$

or

$Fe_2Cl_6(g) \leftrightarrow 2FeCl_2(s) + Cl_2(g)$

$FeCl_2(s) + Cl_2(g) + 2Si(s) \rightarrow FeSi(s) + SiCl_4$

It is suggested that the FeSi nanowires were produced by a vapor-solid (VS) mechanism under a relatively low supersaturation ratio. This growth mechanism was supported by several experimental observations. First, no metal catalyst particles were found at the ends of the FeSi nanowires, thus the vapor-liquid-solid (VLS) mechanism can be ruled out. Second, vapor pressure of the iron precursor was critically important in nanowire growth. Third, branched nanowires were formed in the

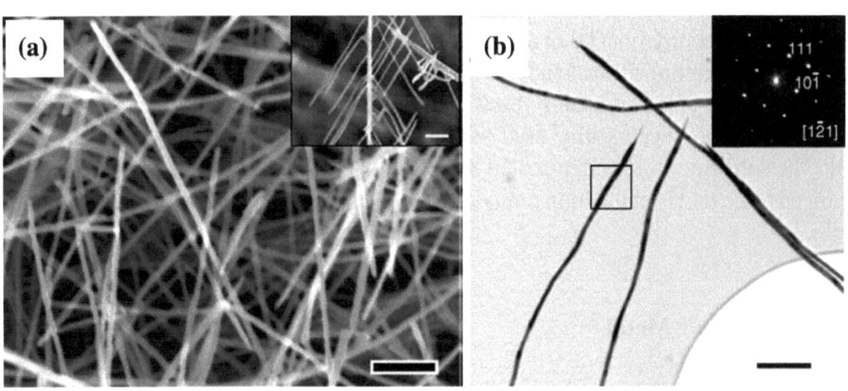

Fig. 12.1 **a** SEM and **b** TEM images of ε-FeSi nanowires. Reprinted with permission from Ref. [3]. Copyright (2006) WILEY-VCH Verlag GmbH & Co. KGaA, Weinheim

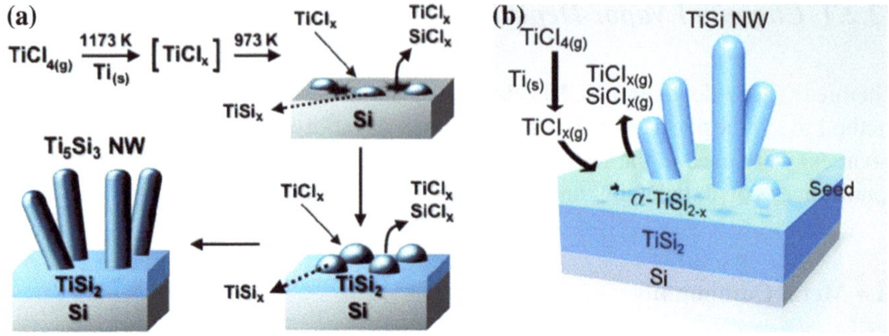

Fig. 12.2 Proposed growth pathways of Ta_5Si_3 nanowires (**a**) and TiSi nanowires (**b**). Reprinted with permission from Refs. [4] and [5]. Copyright (2008, 2009) American Chemical Society

experiments, being consistent with previous study that high supersaturation ratios can lead to nanowires with branched morphologies.

Ti_5Si_3 nanowires were prepared on Si substrate by using $TiCl_4$ as the Ti precursor [4]. Interestingly, Ti powder was simultaneously loaded in the reaction system, which is important for successful growth of the nanowires. The growth steps of Ta_5Si_3 nanowires are illustrated in Fig. 12.2a. $TiCl_4$ first reacted with Ti in the high temperature zone to form $TiCl_x$. Then the $TiCl_x$ either decomposed to $TiCl_4$ and Ti on the substrate, or direct reacted with Si to form $TiSi_x$ clusters. The limited diffusion of Si from the substrate led to the 1D growth of Ti_5Si_3, which has the highest melting point and lowest heat energy of formation. The authors further prepared TiSi nanowires using the same process by adjusting the reaction parameters [5]. As shown in Fig. 12.2b, an amorphous titanium silicide interlayer was found between the TiSi nanowires and the $TiSi_2$ film, which played key role for TiSi nanowire growth.

In et al. synthesized single crystalline V_5Si_3 nanotubes and nanowires using VCl_3 and Si powder as the V and Si precursors, respectively [6]. The morphology of the 1D V_5Si_3 nanostructures can be tuned by controlling the reaction temperature, duration, and precursor composition. MnSi nanowires were synthesized by the CVD of $MnCl_2$ on Si substrate [7]. The usually used metal precursors are metal halides. However, Fe_2O_3 was also successfully used as the Fe precursor to prepare FeSi nanowires [8]. The substrate is Si wafer coated with 5 nm thick gold thin film. Fe_2O_3 was first reduced to Fe by H_2 at the high temperature. Then the Fe reacted with silicon to form FeSi.

Si Compounds + Metal

TMS can also be obtained by the reaction between Si compounds and metal. For the Ni silicides, silane (SiH_4) and metal Ni were often used as the Si source and Ni source, respectively. Decker et al. first report the synthesis of nickel silicide nanowires by thermal decomposition of silane on a Ni surface [9]. The substrates, Ni foil, or Ni films on SiO_2-coated silicon wafers were placed on a Mo holder, whose temperature was monitored with a thermalcouple. Silane (10 % silane, balance He) was flowed through the reactor, decomposed and reacted at the surface of the hot Ni substrates. The temperature, pressure, silane flow rate and substrates could influence the diameter, length and density of the product nanowires. Depending on growth conditions, single crystalline NiSi, Ni_2Si and Ni_3Si_2 nanowires sheathed with thin oxide layers were prepared. The authors proposed three sequential processes for the growth of the nickel silicide nanowires. Silane was first decomposed into Si on Ni surface. Then Ni diffused into Si and nickel silicide nanowires were formed. Generally, thermal decomposition of silane will take place at temperature above 600 °C. But on clean single-crystal Ni (111) surface, silane can be catalytic decomposed at temperature below 0 °C [10]. At the reaction temperature of 370–420 °C, silane will be decomposed to form an amorphous film on Ni substrate. Since Ni diffuses faster than Si, the Ni atoms will diffuse into this amorphous Si film and form the nickel silicides. Yan et al. synthesized branched Ni_2Si nanowires using CVD of silane on Ni plates at 500 °C [11]. Because no catalytic particles were observed at the ends of the nanowires, which is the typical characteristic of the VLS mechanism, growth of the branched Ni_2Si nanowires should act on other mechanisms. The authors suggested that silane was first decomposed to Si[$SiH_4(g) \rightarrow Si(s) + 2H_2(g)$] and nucleated at the Ni substrate. Ni atoms diffused outward from the nickel substrate and reacted with Si atoms to form the Ni_2Si nanowires. Lee et al. employed a similar process with a Ni-loaded Si wafer as the substrate [12]. At a higher temperature of 800 °C, the authors obtained NiSi nanowires coated with carbon, which probably came from the graphite susceptor. They guessed that the Ni metal either acted as a catalyst or participated into the NiSi nanowire growth reaction as an element of the product.

Although nickel silicide nanowires were successfully prepared by CVD of silane on Ni substrate, the exact growth mechanism of these nanowires is unavailable. In this connection, the Jo's group did some experiment aiming to explore the growth mechanisms [13]. Specially, they investigated the influence of reaction temperature and pressure on the morphology of the products. At lower temperature of 250 °C, the decomposition of silane and Ni out-diffusion is not so sufficient, leading to a very thin NiSi film. At higher temperature of 600 °C, the decomposition of silane and Ni out-diffusion is excessive, resulting in thick films of various phases. Only at the moderate temperature of 400 °C, nickel silicide nanowires could be obtained. They found that optimum conditions for nickel silicide nanowire growth are at a constant temperature of 400 °C with silane pressure of 50 torr. Increasing pressure to 75 and 100 torr, the nanowires are not separated from each other. Since the pressure is a measure of supersaturation of gaseous Si during the reaction, it can be concluded that lower supersaturation favors the nanowire formation. The authors further studied the composition distribution in a single nanowire, as shown in Fig. 12.3. EDX analysis indicates a composition gradient of Ni element exists in the nanowire. Bottom end (locations a, b) is Ni-rich and top end (locations d, e) is Si-rich. Whereas in the middle regions (location c), the ratio of Ni to Si is about 1:1. This result suggests that Ni is supplied by 1D diffusion through the nanowire from the Ni substrate and reacted with Si vapor to form the nickel silicides.

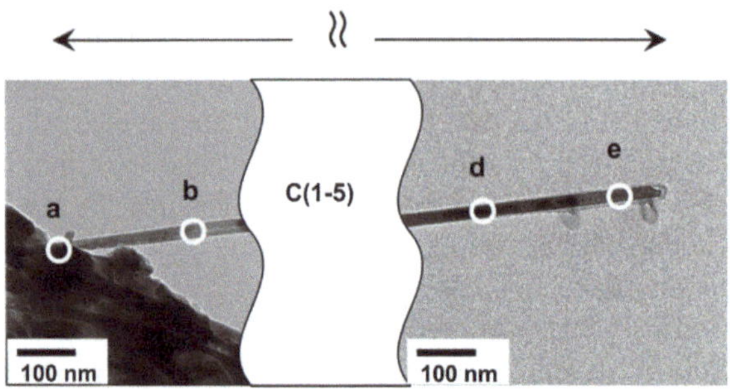

Fig. 12.3 Composition gradient of a single nickel silicide nanowire. Reprinted with permission from Ref. [13]. Copyright (2007) WILEY-VCH Verlag GmbH & Co. KGaA, Weinheim

It was further found that oxide layer over Ni is quite important for the growth of $NiSi_x$ nanowires [14]. Before CVD of silane, the Ni substrate was annealed in different atmospheres, including H_2, N_2 and O_2. High density of $NiSi_x$ nanowires can only be observed in the O_2-annealed sample, as shown in Fig. 12.4. It seems that NiO_x layer formed in the annealing process is critical for the $NiSi_x$ nanowires growth. At the same time, no $NiSi_x$ nanowires were observed on thicker NiO_x layer on Ni and stand-alone NiO_x. Thus, the spontaneous growth of $NiSi_x$ nanowires originated from the surface reaction of the out-diffused Ni from the seed Ni layer and the silane vapor. In the growth process, NiO_x overlayer acted as a diffusion barrier to kinetically and spatially control the out-diffusing Ni supply from Ni films underneath at a given thickness. Limited nucleation controlled by the NiO_x led to the 1D growth of $NiSi_x$

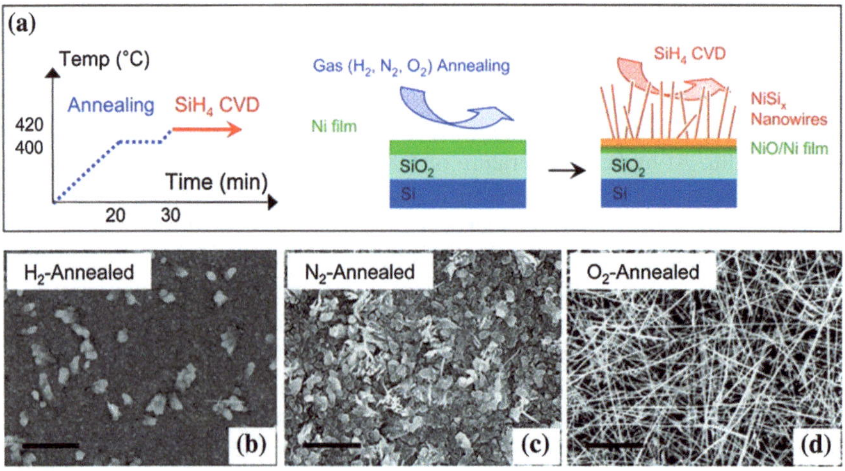

Fig. 12.4 **a** Schematic diagram of the $NiSi_x$ nanowire growth. **b–d** The representative SEM images of products on H_2-, N_2-, and O_2-annealed samples. Reprinted with permission from Ref. [14]. Copyright (2008) American Chemical Society

at an optimally low supersaturation of Ni for the vapor-solid reaction. Zhang et al. synthesized NiSi nanobelts and nanosheets by chemical reaction of Ni substrate with $SiHCl_3$ [15].

Other than silane, silicides were also used as Si source to synthesize other silicide nanowires. For example, $TaSi_2$ nanowires were synthesized by the reaction of Ta vapor and $FeSi_2$ or $NiSi_2$ thin film [16–18]. The growth of $TaSi_2$ nanowires is consistent with the VS mechanism. Si atoms were segregated from $FeSi_2$ and reacted with Ta to form $TaSi_2$ nanodots with a semispherical cap shape on Si. The lateral growth was restricted due to the edge energy and Si base out of $FeSi_2$ film. Continuous reaction of Ta and Si led to 1D growth along the long axis since (−2110) plane has the lowest surface energy.

Si+Metal

The direct reaction between Si and metal can also form 1D nanostructures of TMS silicides in proper conditions. Tang et al. reported the growth of $NiSi_2$ nanowires filling in the BN nanotubes [19]. They used a single-crystal silicon wafer with approximately 300 nm thickness nickel film as the substrate to deposit BN nanotubes. Interestingly, they found that the as-synthesized BN nanotubes were filled with nanowires, which were later determined to be $NiSi_2$ nanowires. The filled $NiSi_2$ nanowires grow along the 110 direction and the direction of the epitaxial crystallization on the BN shell is 112. The authors believed that the $NiSi_2$ nanowires were formed simultaneously with the growth of BN nanotubes. At the reaction temperature of 1300 °C, the liquid Ni–Si alloys acted as catalyst for the BN nanotubes growth via the VLS mechanism. At the same time, Ni–Si alloys might be absorbed into the BN nanotubes and formed $NiSi_2$ nanowires. The Yu's group reported the synthesis of $TiSi_2$ nanowires using Ti powder as the source and Si wafer as the substrate [20, 21]. The author defined the growth process as physical vapor deposition (PVD). Actually, it is a CVD process, because chemical reactions took place during the growth. A self-catalytic VS mechanism was proposed for the growth of $TiSi_2$ nanowires. Quai-liquid Ti–Si alloy nanoislands were formed at the reaction temperature and led to 1D growth of $TiSi_2$ crystals. α-$FeSi_2$ nanowires were synthesized by reaction of Fe and Si powder at 1350 °C for 2 h [22]. Enough Fe supply and high temperature are the two important conditions for synthesis of the α-$FeSi_2$ nanowire. A VLS growth process was suggested for the α-$FeSi_2$ nanowire formation. α-$FeSi_2$ heads were first formed and attached to the silicon substrate, then α-$FeSi_2$ nanowires grew from the α-$FeSi_2$ heads, which acted like metal catalysts in VLS process for 1D nanostructures growth.

Joshi et al. used Pd nanoparticles-coated Si wafer as the only source to prepare Pd_2Si nanowires by microwave plasma enhanced CVD (MPECVD) [23]. The Pd nanoparticles were electrochemically deposited on the Si substrate. Ultrapure H_2 was introduced to the reaction system. Reacting at 500 for 5 min, vertical aligned Pd_2Si nanowires were obtained on the Si substrate. Figure 12.5a, b shows the typical SEM and TEM images of the as-synthesized Pd_2Si nanowires. TEM analysis indicates that body of the nanowire is polycrystalline hexagonal Pd_2Si, and neck of the nanowire is

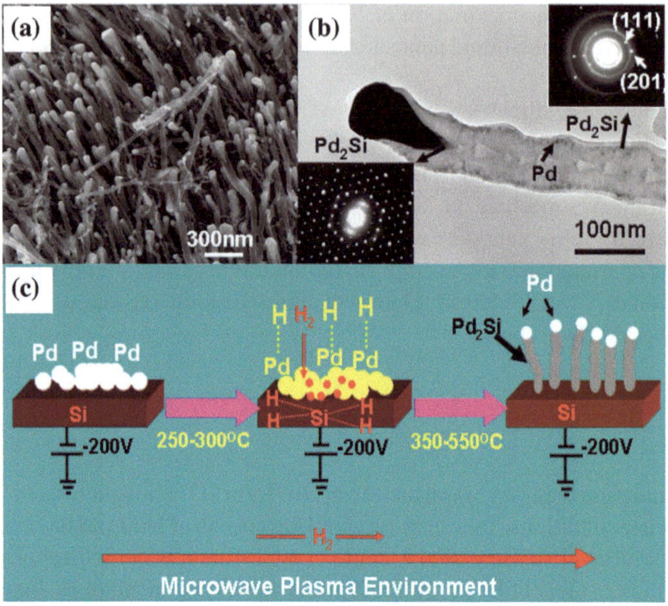

Fig. 12.5 **a** SEM and **b** TEM images of Pd_2Si nanowires, **c** Schematic diagram of the Pd_2Si nanowire growth process. Reprinted with permission from Ref. [23]. Copyright (2008) American Chemical Society

fully crystalline Pd_2Si. A Pd particle was found at the tip of the nanowire, implying the VLS growth mechanism of the Pd_2Si nanowires. In the growth process, Pd acted as catalyst as well as Pd source for the Pd_2Si nanowire formation. Figure 12.5c illustrates the Pd_2Si nanowire growth process. H_2 was introduced to the reaction system to accelerate palladium silicide formation by altering the diffusion barrier at Pd/Si interface. In addition, H_2 is also a necessary catalytic gaseous environment for the formation of Pd_2Si nanowires. When H_2 was replaced by Ar and N_2, no nanowires were observed.

Chang et al. prepared Cr_5Si_3 nanopillars embedded in Si nanorods by codeposition of Si and Cr sources on gold-coated silicon substrates [24]. Figure 12.6 illustrates the formation of the core-shell nanorods. Silicon nanorods were first formed through a VLS process. Cr atoms continuously impinged onto the Au catalyst, agglomerated around the center region of the growing nanorods and reacted with Si to form chromium silicide nanorod.

Using Mn powder and Si as the precursors, Higgins et al. prepared Mn-rich silicide nanowires, including α-Mn_5Si_3, β-Mn_5Si_3, and β-Mn_3Si [25]. The Mn-rich silicides originated from the high Mn vapor concentration and limited gaseous Si species. Chang et al. synthesized Ti_5Si_4 nanobats using Ti foil and Si powder as the precursors via a self-catalytic VS growth process [26]. Si powder first reacted with the Ti foil and to form Si-Ti alloys. Continuous feeding of Si to the quasiliquid Si-Ti alloys led to 1D growth of Ti_5Si_4 crystals.

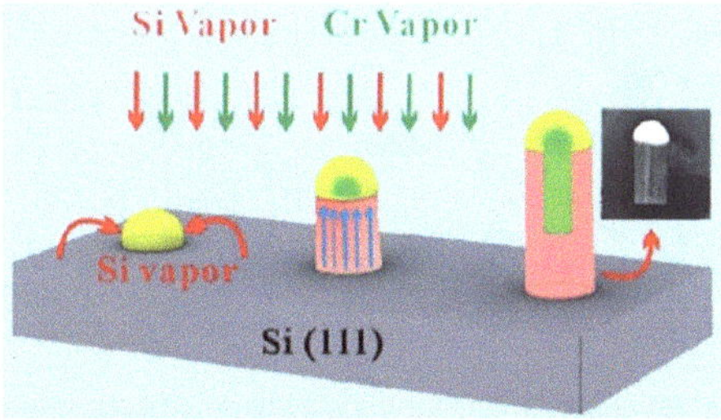

Fig. 12.6 Growth model of the Cr_5Si_3/Si nanorods. Reprinted with permission from Ref. [24]. Copyright (2009) American Chemical Society

Besides single TMS silicide, heterostructures of different TMS can also be obtained by using multiple sources. For example, Chen et al. prepared $Ni_2Si/TiSi_2$ core/shell nanowires by using source materials containing Ni, Si, and Ti [27]. First, a 30 nm thick Ni thin film was deposited on the Si substrate. Then the substrate was annealed at 600 °C for 30 min to form a $NiSi_2$ nanodot thin film. The $NiSi_2$ thin film coated Si substrate reacted with Ti vapor at high temperature and formed the coaxial $Ni_2Si/TiSi_2$ nanowires, as shown in Fig. 12.7. The formation of the coaxial nanowires was ascribed to a solid-liquid-solid (SLS) mechanism.

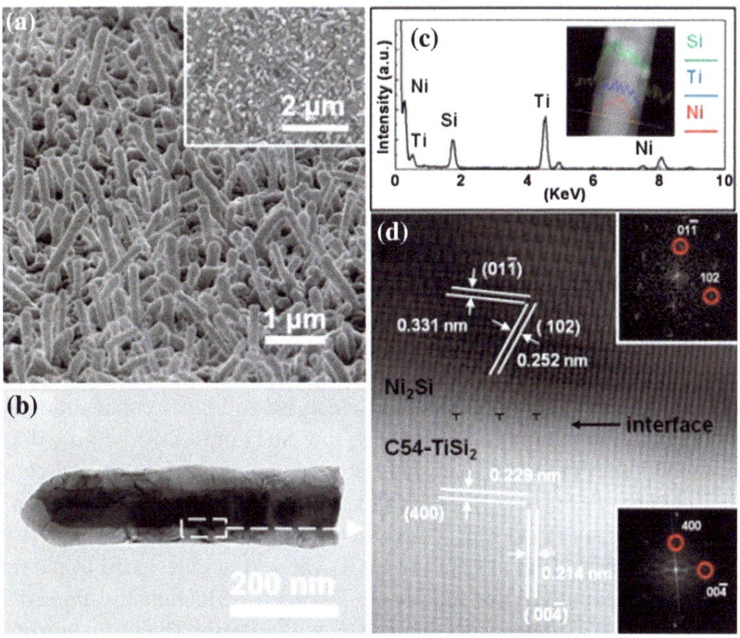

Fig. 12.7 **a** SEM and **b** TEM images of the $Ni_2Si/TiSi_2$ core/shell nanowires. **c** Chemical composition analysis of a single nanowire. **d** HRTEM image of the core-shell nanowire. Reprinted with permission from Ref. [27]. Copyright (2012) American Chemical Society

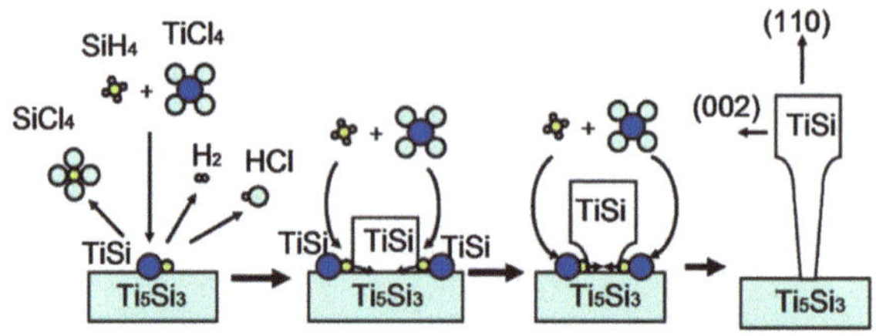

Fig. 12.8 Schematic diagram of the TiSi nanopins growth process. Reprinted with permission from Ref. [28]. Copyright (2007) American Chemical Society

Si Precursor + Metal Precursor

Using SiH_4 and $TiCl_4$ as silicon and metal precursors, TiSi nanopins were prepared by an atmospheric pressure CVD process [28, 29]. It seems that the ratio of SiH_4 to $TiCl_4$ is quite important to the composition of the products. When this ratio is 1.5, TiSi nanopins were obtained. But when SiH_4 and $TiCl_4$ were in the same molar mass, Ti_5Si_3 thin films were prepared. In these two conditions, the following reactions took place correspondingly.

$5TiCl_4(g) + 5SiH_4(g)\ Ti5Si_3(s) + 2SiCl_4(g) + 12HCl(g) + 4H_2(g)$

$2TiCl_4(g) + 3SiH_4(g)\ 2TiSi(s) + SiCl_4(g) + 4HCl(g) + 4H_2(g)$

The schematic diagram shown in Fig. 12.8 illustrates the growth process of the TiSi nanopins. At high temperature, TiSi islands were first formed on the Ti_5Si_3 substrate. Continuous feeding of Ti and Si led to 1D growth of TiSi crystals. The TiSi islands were pushed up and finally TiSi nanopins were formed. Using this method, TiSi nanowires were also fabricated [30–33].

$TiSi_2$ nanowires, two-dimensional (2D) nanonets and three-dimensional (3D) complexes composed of $TiSi_2$ belts have been synthesized by using SiH_4 and $TiCl_4$ as the Si and Ti precursors, respectively [34–36]. Figure 12.9 shows the typical SEM and TEM images of the as-synthesized products. The 1D nanowires are C54 $TiSi_2$ while the 2D nanonets and 3D nanocomplexes are C49 $TiSi_2$. The authors conjectured that the 1D growth following the VS mechanism. The surface energy differences in different crystal planes caused the unique morphologies. Through varying important parameters in the CVD process, they found that higher pressures, lower temperatures, more SiH_4 and less $TiCl_4$ feeding, which led to higher concentrations of Si species, favored C49 nanonets. Conversely, low Si:Ti ratio favored growth of C54 $TiSi_2$ nanowires due to surface energy differences among various crystal planes.

Radial aligned Ti_5Si_3 nanowire architectures were fabricated on carbon paper substrate [37]. The substrate was successively deposited with 1 μm thick Ti film as Ti source and 5 nm Au film as catalyst. $SiCl_4$ vapor was introduced to the reaction system as the Si source. After reacted at 750 °C for 5 h, radially aligned Ti_5Si_3 nanowires on the microfibers were obtained (Fig. 12.10a–d). The formation of Ti_5Si_3 nanowires was ascribed to a catalytic growth process following the VLS mechanism. As shown in Fig. 12.10e, $SiCl_4$ decomposed to Si atoms and formed droplet of the

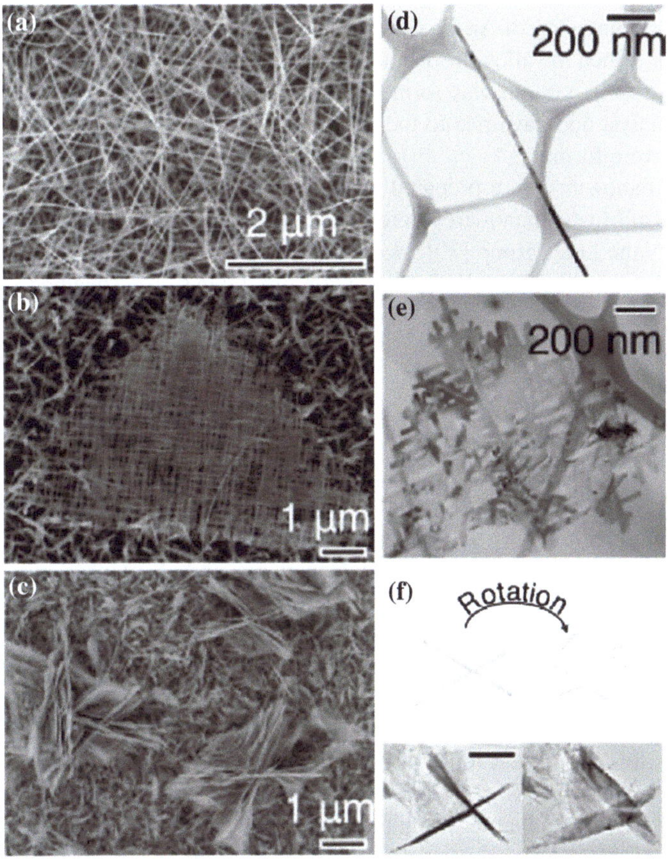

Fig. 12.9 **a, b, c** SEM and **d, e, f** corresponding TEM images of the 1D nanowires, 2D nanonets and 3D complexes of $TiSi_2$. Reprinted with permission from Ref. [35]. Copyright (2009) American Chemical Society

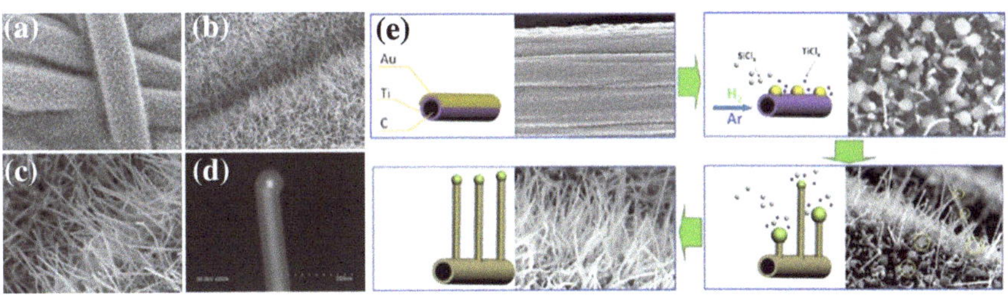

Fig. 12.10 **a–d** SEM images of the Ti_5Si_3 nanowires at different magnifications. **e** Growth process of the Ti_5Si_3 nanowires. Reprinted with permission from Ref. [37]. Copyright (2011) American Chemical Society

low eutectic point alloy of Au-Si. Simultaneously, $SiCl_4$ reacted with Ti and formed $TiCl_x$ vapor. When the alloy droplet absorbed became supersaturated with absorbed element, Au-Si-Ti system was formed and Ti_5Si_3 nuclei were generated. The Au acted as catalyst and gave rise to the 1D growth of Ti_5Si_3 crystals. Finally, Ti_5Si_3 nanowires were formed.

Mn_4Si_7 nanowires were prepared using CVD of SiO and $MnCl_2$ powders [38]. The growth of Mn_4Si_7 nanowires were explained by the oxide-assisted growth model proposed by the Lee's group [39]. According to this mechanism, SiO vapor reacted with $MnCl_2$ vapor resulting in the formation of silicides core sheathed with oxide shell.

Organometallic Precursor

All the above-mentioned methods need two sources (silicon source and metal source) for TMS growth. The Jin's group developed a CVD process using organometallic single-source precursors (SSP) to synthesize TMS nanowires, where both silicon and metal were delivered by the SSP. Nanowires of FeSi, CoSi, and $Mn_{19}Si_{33}$ were successfully synthesized by using $Fe(SiCl_3)_2(CO)_4$, $Co(SiCl_3)(CO)_4$ and $Mn(CO)_5SiCl_3$ as SSP, respectively [40–42]. The metal silicides are formed by the decomposition of precursors according to the following equation, where M represents the transition metal.

$$M(SiCl_3)_x(CO)_y \rightarrow MSi_x + y\,CO + 3x/2Cl_2$$

In 2006, this group first reported the synthesis of free-standing FeSi nanowires through the CVD of trans-$Fe(SiCl_3)_2(CO)_4$, which was synthesized from $Fe_3(CO)_{12}$ and $SiHCl_3$ [42]. The advantages of this method over conventional multisource metal organic CVD (MOCVD) are elimination of highly hazardous precursors, easier and precise control over stoichemistry and higher quality products. The authors suggested that the growth of FeSi nanowires might be a self-seeded VS process, or a VLS process that involves nonmetal catalysts such as silicon or silicon oxides. The same procedure was also utilized to synthesize CoSi nanowires [41]. Figure 12.11 shows representative SEM images of FeSi and CoSi nanowires. As an extension of these works, ternary alloy $Fe_{1-x}Co_xSi$ nanowires was prepared by using two-component SPP composed of $Fe(SiCl_3)_2(CO)_4$ and $Co(SiCl_3)(CO)_4$ [43]. Compositions of the products can be tuned by controlling the ratio of Co to Fe in the precursor mixture. Ferromagnetic semiconductor $Fe_{0.94}Co_{0.06}Si$ nanowires and helical magnet $Fe_{0.56}Co_{0.44}Si$ nanowires were obtained from precursors with Co to Fe ratio of 5:4 and 5:2, respectively. During the synthesis of FeSi, CoSi, and $Fe_{1-x}Co_xSi$ nanowires, the authors found that silicon substrate covered with 1–2 nm silicon oxide film is the key to the nanowire growth. When silicon substrates without oxide layer, covered with metal thin film (Fe, Ni, Cu) or colloids (Cu, Au, Ag), covered with thick thermal SiO_2 or Al_2O_3, were used, no nanowires were produced. Based on this substrate-selective deposition, patterned growth of FeSi nanowires was also realized [44].

Interestingly, unlike the synthesis of FeSi, CoSi, and FeCoSi nanowires, no thin film of oxide is required to synthesize the high manganese silicides (HMS) $Mn_{19}Si_{33}$

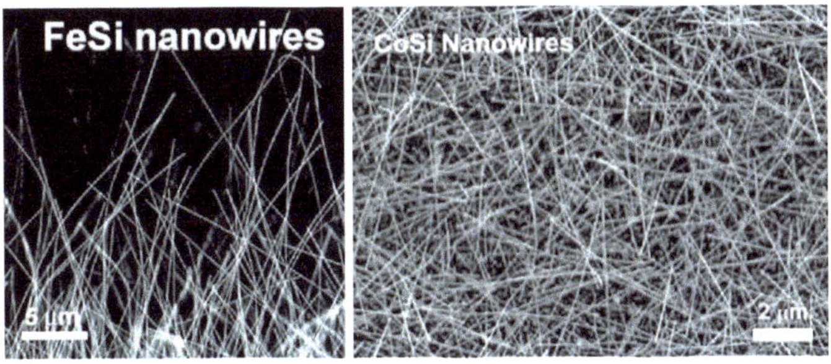

Fig. 12.11 SEM images of FeSi and CoSi Nanowires. Reprinted with permission from Refs. [41, 42]. Copyright (2006) American Chemical Society

nanowires. H-terminated silicon substrates or substrates with thick oxides were successfully used for the growth of $Mn_{19}Si_{33}$ nanowires. It was found that the nanowires always nucleated from HMS "island" on the substrate surface. Thus, the authors suggested that the $Mn_{19}Si_{33}$ nanowires might grow via the vapor-solid-solid (VSS) process with catalyst particles remaining on the substrate surface. Another possible mechanism is that Si nanowires are being synthesized and silicided during the growth process.

Although various TMS nanowires have been successfully synthesized by CVD of SSP, a clear understanding of how the nanowires growth has remained elusive. To better understand the growth mechanism, the Jin's group did some experiments to explore the roles substrate films and precursor decomposition play in the formation and doping of TMS silicide nanowires [45]. Different substrates including silicon, germanium, $CoSi_2$ films on Si, and CoSi films on Si were used to study CVD of $Fe(SiCl_3)_2(CO)_4$. A thin film of FeSi was first formed on the substrate. This is an important prerequisite for the FeSi nanowires. The diffusion of Si and Ge from the substrate to the produced nanowires was observed. Thus, the Si elements in the nanowire products came from both the vapor phase precursor and substrate. In contrast, vapor phase delivery of metal source is far more quicker than diffusion from the substrate.

12.2.2 Chemical Vapor Transport

Chemical vapor transport (CVT) is an established solid-state technique often used for growing single crystals. CVT process generally utilizes the thermodynamically reversible reaction between the transport reagent (often halogens) and the products. Ni_2Si nanowires were synthesized via the CVT method using I_2 as transport reagent and Ni_2Si powder as source material [46]. As illustrated in Fig. 12.12a, I_2 was transported by the Ar carrier gas to the high temperature zone T1 (1100 °C) and reacted with Ni_2Si powder to produce gaseous NiI_2 and SiI_4. The gaseous products were

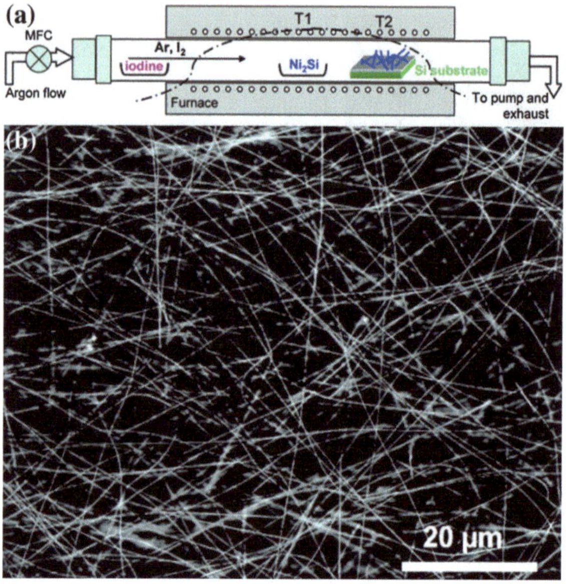

Fig. 12.12 Schematic setup for the synthesis of Ni_2Si nanowires using chemical vapor transport (CVT). **b** SEM image of the as-synthesized Ni_2Si nanowires. Reprinted with permission from Ref. [46]. Copyright (2007) American Chemical Society

carried downstream by the Ar flow to the low temperature zone T2 (850 °C), where the reverse reaction took place. The thermodynamically reversible reaction in this process is described as follows.

$Ni_2Si(s) + 4I_2(g) \rightarrow 2NiI_2(g) + SiI_4(g)$

Minute amount of $Ni(NO_3)_2$ was loaded on the SiO_2/Si substrate. The presence of $Ni(NO_3)_2$ is crucial for the large amount of Ni_2Si nanowires formation. The growth of Ni_2Si nanowires might be through the VLS or related mechanism. Trace amount of Ni decomposed from $Ni(NO_3)_2$ provided nucleation sites for the nanowire growth and participated in the growth reaction latter. Therefore, no catalytic caps were found at the ends of the nanowires. $CrSi_2$ nanowires were synthesized by the same method using $CrSi_2$ as the Cr source [47], where $Ni(NO_3)_2$ was also loaded on the Si substrate. Interestingly, using $TiSi_2$ powder and I_2 as the transport reagent, the Jin's group synthesized Ni_3Si nanowires instead of titanium silicides [48]. In this process, $TiSi_2$ only acted as the Si source and the Ni source was provided by $Ni(NO_3)_2$ on the silicon substrate. The Ti must be flushed out or reacted with trace amount oxygen along the way. In the above examples, $Ni(NO_3)_2$ was used and found to be crucial for the nanowire growth, while the roles of $Ni(NO_3)_2$ are different. In another report, however, FeSi nanowires can also be synthesized without $Ni(NO_3)_2$ [49].

Using the vapor transport method, the Kim's group synthesized single crystalline CoSi, Cr_2Si, MnSi, and iron silicide (Fe_5Si_3 and FeSi) nanowires [50–53]. Typical SEM images of these nanowires are shown in Fig. 12.13. Metal halides, i.e., $CoCl_2$,

Fig. 12.13 SEM images of CoSi, Cr_2Si, FeSi and Fe_5Si_3 nanowires. Reprinted with permission from Refs. [50–52]. Copyright (2007) American Chemical Society

$CrCl_2$ and FeI_2 powders, were used as the metal precursors for the metal silicdes growth. In a typical process, the metal halides were heated to vapor phase in the upstream zone, and then carried downstream by the inert gas to react with the Si substrate and form the metal silicide nanowires.

When Si wafer alone was used as the substrate, single phase metal silicide (CoSi or Cr_2Si) nanowires were obtained. However, in the case of Fe silicides, the composition of the products (Fe_5Si_3 or FeSi) can be tuned when a sapphire substrate was placed on the Si wafer. FeI_2 precursor was evaporated at the upstream zone (low temperature) and transported to the downstream zone (high temperature) by the carrier gas. Sapphire substrate loaded on a silicon wafer was placed at the downstream zone, where the iron silicide nanowires were formed through the following plausible reaction pathways.

$$2FeI_2(g) + 3Si(s) \leftrightarrow 2FeSi(s) + SiI_4(g)$$
$$5FeI_2(g) + 3SiI_4(g) \leftrightarrow Fe_5Si_3(s) + 11I_2(g)$$
$$FeI_2(g) + SiI_4(g) \leftrightarrow FeSi(s) + 3I_2(g)$$

FeI_2 vapor reacted with the Si wafer to form the FeSi and SiI_4. The gaseous SiI_4 can further reacted with FeI_2 to form Fe_5Si_3 and FeSi. The composition of the product nanowires (Fe_5Si_3 or FeSi) can be controlled by tuning the concentration ratio of $FeI_2(g)$ and $SiI_4(g)$, which can be achieved by controlling the FeI_2 vapor pressure. A relatively higher upstream zone temperature and carrier gas flow rate favored the formation of Fe_5Si_3 nanowires. On the contrary, FeSi nanowires were formed. But when Si wafer alone was used as the substrate, only FeSi nanowires can be obtained.

Similar phase controllable synthesis was also realized in the case of Co silicides [54]. As illustrated in Fig. 12.14, $CoCl_2$ precursor was placed at the upstream zone,

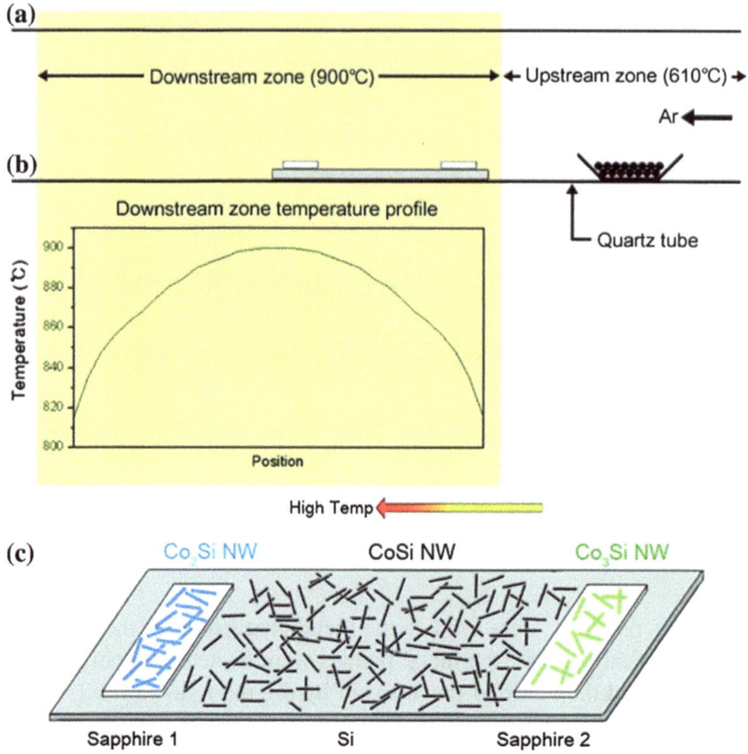

Fig. 12.14 Illustrations of the experimental setup (**a**), temperature profile (**b**) and Co silicides with different phases at different locations of the substrate. Reprinted with permission from Refs. [54]. Copyright (2009) American Chemical Society

Si wafer with two sapphire substrates at each end was placed at the downstream zone. The $CoCl_2$ vapor was transported downstream by Ar, reacted with Si, and formed the diverse Co silicides. The reaction pathways of different phase Co silicides are described as the following equations.

$$2CoCl_2(g) + 3Si(s) \leftrightarrow 2CoSi(s) + SiCl_4(g) \quad \text{[on a Si substrate]} \tag{12.1}$$

$$6CoCl_2(g) + 3SiCl_4(g) \leftrightarrow 3Co_2Si + 12Cl_2(g) \quad \text{[on the sapphire 1]} \tag{12.2}$$

$$6CoCl_2(g) + 2SiCl_4(g) \leftrightarrow 2Co_3Si + 10Cl_2(g) \quad \text{[on the sapphire 2]} \tag{12.3}$$

Identical results were obtained when the sapphire substrate was replaced by quartz substrate. This indicates that sapphire is not the reason for the phase variation. The different Si to Co ratio led to the diverse phases of the products. On the Si substrate, CoSi was formed through reaction (12.1), where Si came from the Si substrate. In reactions (12.2) and (12.3), the compositions of the products were determined by the concentration of Si source $SiCl_4$, which is dependent on the reaction temperature.

The temperature at location of sapphire 1 is higher than that of sapphire 2. Higher temperature led to high concentration of $SiCl_2$ by the more Si supply in reaction (12.1). Thus, products on sapphire 1 (Co_2Si) is Si-richer than that on sapphire 2 (Co_3Si).

Zhang et al. synthesized Cu_3Si nanowires by a vapor transport process using CuO and graphite mixture as the Cu precursor [55]. Au nanoparticles were pre-coated on the Si substrate before deposition. The Au nanoparticles played important roles for the nanowire growth, ether in absorbing Cu vapor and penetrating the oxide barrier. Lee et al. prepared free-standing $NiSi_2$ nanowires on Ni substrate [56]. Ni foil and Si powder were used as the Ni and Si sources, respectively. The phases of nickel silicides depend on the ratio of Ni and Si sources. Because Ni is the dominant diffusion species in the Ni-Si reactions, the Ni-rich phase nickel silicides are usually obtained when using Si powder as the sole source to grow nickel silicides on Ni substrate. Therefore, $NiCl_2$ powders were placed in front of the Ni foil to form the Si-rich $NiSi_2$ phase. The possible reaction pathways are as follows.

$$2NiCl_2(g) + Ni(s) + 7Si(g) \leftrightarrow 3NiSi_2(s) + SiCl_4(g)$$
$$NiCl_2(g) + 2SiCl_4(g) \leftrightarrow NiSi_2(s) + 5Cl_2(g)$$
$$Ni(s) + Cl_2(g) \leftrightarrow NiCl_2(g)$$

The $NiCl_2$ played two important roles for the $NiSi_2$ nanowires formation. First, it can react with Si to form the highly reactive $SiCl_4$ vapor and enhance the Si condensation. Second, the surplus Cl_2 can react with Ni to form $NiCl_2$. The enhanced condensation of Si and diminished supply of Ni contributed to the formation of Si-rich $NiSi_2$ phase.

TMS nanostructures with strange morphologies were also prepared by the CVT method. For example, $CrSi_2$ nanostructures with a hexagonal spider-web-like morphology were reported by Wang et al. [57], as shown in Fig. 12.15. The nanowebs composed of nanowires with thickness of 10–30 nm. This interesting nanostructure was obtained by abruptly ramping up the flow rate of Ar during the growth process. The special crystal structure and minimization of electrostatic energy associated with charged sidewalls were suggested as the driven force for the formation of the web-like structure.

Except the above-mentioned binary materials, ternary TMS silicides can also be prepared by the CVT method. Using the FeI_2 and CoI_2 powders as the precursor, $Fe_{1-x}Co_xSi$ nanowires were prepared by this method [58].

12.2.3 Silicidation of Silicon Nanowires

Si Nanowires + Metal

The Lee's group first reported silicidation of Si nanowires by metal vapor vaccum arc (MEVVA) implantation [59]. $NiSi_2$ and $CoSi_2$ were formed on the surface of Si nanowires by Ni^+ and Co^+ implantation followed by thermal annealing. In 2004, Lieber's group realized fabrication of NiSi nanowires by in situ silicidation of Si

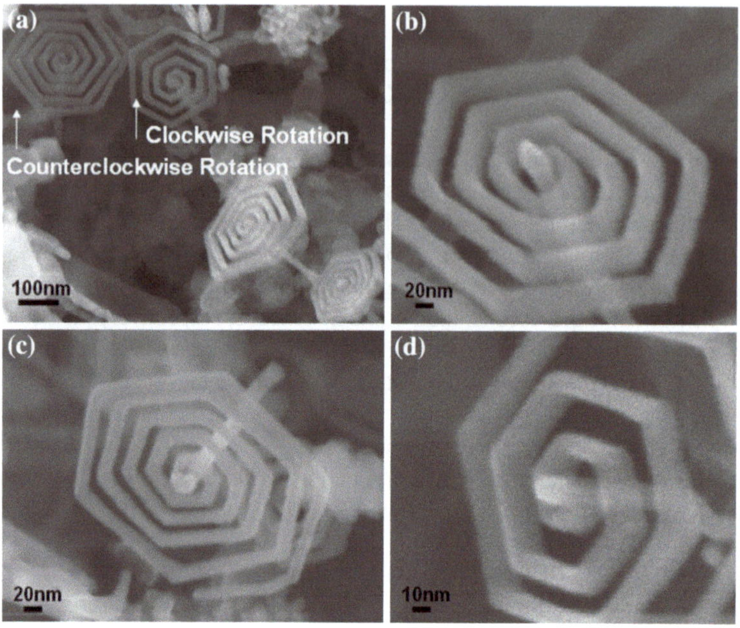

Fig. 12.15 SEM images of the $CrSi_2$ nanowebs. Reprinted with permission from Ref. [57]. Copyright (2010) American Chemical Society

nanowires [60]. The fabrication process is illustrated in Fig. 12.16. Si nanowires were synthesized by a CVD method using Au nanoclusters as catalyst. Ni metal with a thickness comparable to the nanowire diameter was deposited on the surfaces of the Si nanowires. Then the Ni-metal-coated Si nanowires were annealed at 500 °C and NiSi nanowires were formed. The residual Ni metal on the nanowires

Fig. 12.16 Fabrication process of the NiSi nanowire. Reprinted by permission from Ref. [60], copyright (2004) Macmillan Publishers Ltd: (Nature)

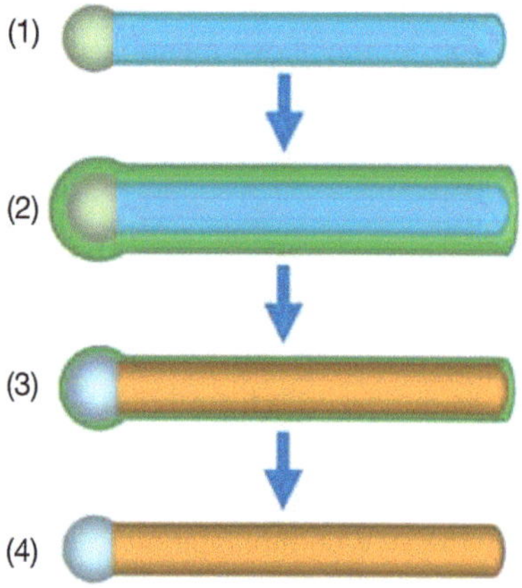

could be removed by chemical etching. NiSi nanowires prepared by this solid-state reaction are single crystalline. Combining with the lithography technique, NiSi/Si heterostructures were also prepared. In the silicidation process, Si nanowires act as Si sources for the silicides, as well as template for the formation of the 1D nanostructures. Therefore, compared with the CVD and CVT processes, it is easier to fabricate 1D TMS nanostructures through the silicidation process in not so rigorous conditions.

PtSi nanowires were prepared through the silicidation of Si nanowire by Pt [61]. The Si nanowires were deposited with Pt films of 20–50 nm, and then annealed at temperatures of 250–700 °C for 5–60 min. Morphologies of the product nanowires were greatly influenced by the annealing temperature, platinum film thickness, silicon nanowire diameter, and level of unintentional oxygen contamination in the annealing furnace. Annealing Si nanowires with appropriate diameter and Pt thickness at temperature above 400 °C in pure N_2 can achieve the conversion of uniform PtSi nanowires. Oxygen contamination in the reaction system will greatly affect the silicidation process.

For the silicidation of Si nanowires, generally, a thin film of metals was pre-coated on the surface of nanowires. Lu et al. developed an entirely different method, i.e. using point contact reaction to silicidation Si nanowires [62, 63]. The point contacted polycrystalline Ni nanowires and single crystalline Si nanowires were in situ annealed by the high resolution transmission electron microscopy (HRTEM). Si and Ni nanowires were dispersed on Si_3N_4 membrane with point contacts (Fig. 12.17a). Then the nanowires were annealed at 700 °C, leading to the reaction between Si and Ni and formation of NiSi. The NiSi growth process was in situ observed by TEM (Fig. 12.17b–f). Figures 12.17g, h illustrate the growth mechanism of the NiSi nanowires in Si and NiSi/Si/NiSi heterostructures. Ni nanowires as the Ni source were point contacted with Si nanowires. Ni atoms were dissolved in Si nanowires and interstitial diffused from the point of contact to the both ends of Si nanowires. Growth of NiSi started from the tips rather than from the point contact due to two reasons. First, supersaturation at point contact is impeded by the limit contact area and oxide layer over the nanowires. Second, Ni atoms diffuse easier through Si than through silicides. The growth of NiSi is linear with time.

Chou et al. performed Si nanowires silicidation via point contact reaction using Co nanodots as the metal source [64]. At different temperatures of 700 and 800 °C, CoSi and $CoSi_2$ were formed, respectively. The silicidation process was investigated by an ultrahigh vaccum HRTEM. It was found that the $CoSi_2$ undergoes a stepwise epitaxial growth in Si nanowires. The stepwise growth occurred repeatedly in the form of repeating nucleation and atomic sweeping across the $CoSi_2$/Si interface. Point contact samples of Si and Co nanowires were heated and Co nanowires were disintegrated into Co nanodots on the Si nanowire surface. In situ TEM image sequences of Fig. 12.18a, b show CoSi grows while Co particle disappears at 700 °C. The dark region is Co silicide. As time goes on, some Co nanodots disappeared and Co silicide grew promptly. The growth rate is about 0.88 nm/s.

The supply of Ni can also realized by depositing large amounts of Ni as Ni reservoirs on the ends of Si nanowires [65]. As shown in Fig. 12.19, upon annealing,

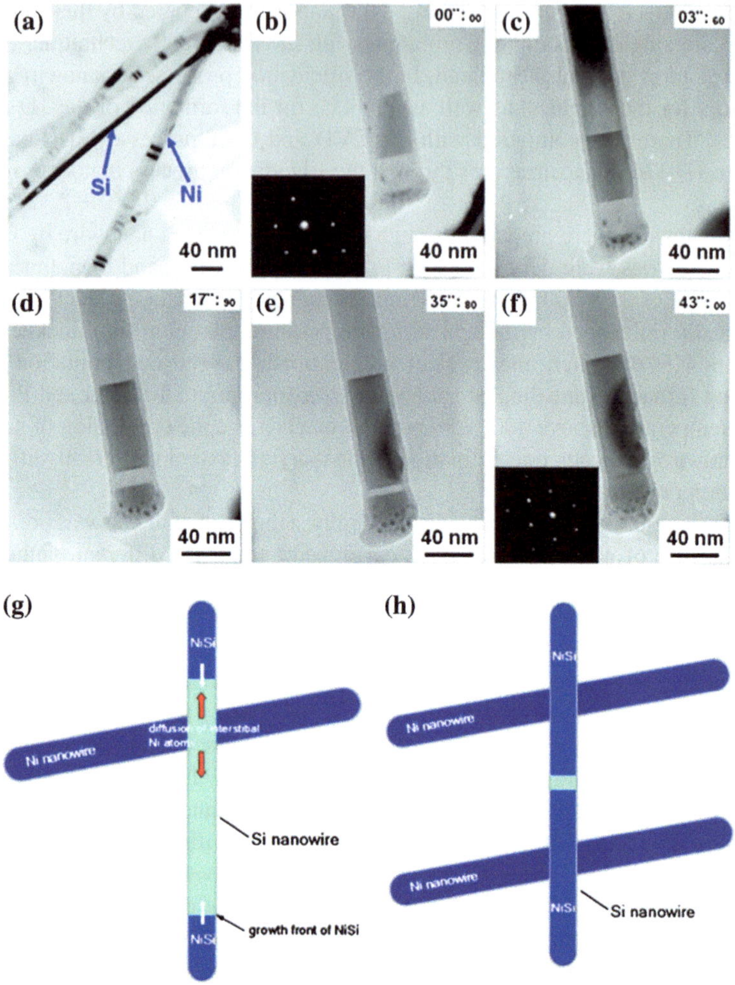

Fig. 12.17 Overview of NiSi formation within Si nanowires by point contact reaction. **a** A TEM image of Si and Ni nanowires dispersed on a Si_3N_4 membrane. **b–f** Sequence of in situ TEM images depicting the growth of a bamboo-type grain of NiSi within a Si nanowire at. **g**, **h** Schematic illustrations of NiSi growth within Si nanowires. Reprinted with permission from Ref. [63]. Copyright (2007) American Chemical Society

Ni diffused axially into Si nanowires and $NiSi_2$ was formed at the diffusion path. Unlike the point contact reactions, the silicidation starts from the contact point owing to the sufficient supply of Ni and removing of the oxide layer.

Lin et al. prepared PtSi nanowires through controlled reactions between lithographically defined Pt pads and Si nanowires [66]. The schematic diagram and SEM images of the silicidation process are shown in Fig. 12.20. After annealing at 520 °C

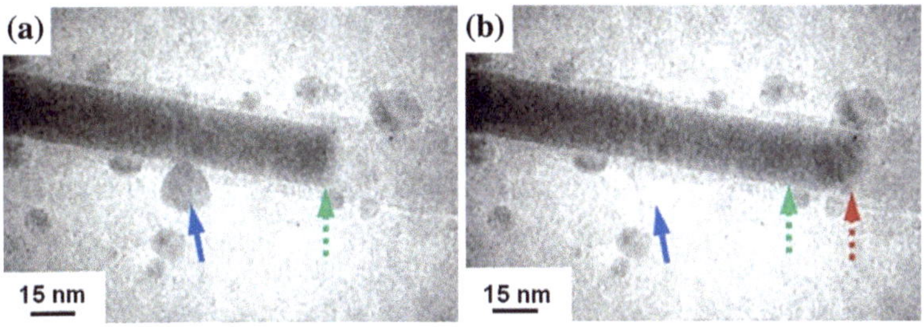

Fig. 12.18 In-situ TEM image at 0 s (**a**) and 0.07 s (**b**). Reprinted with permission from Ref. [64]. Copyright (2008) American Chemical Society

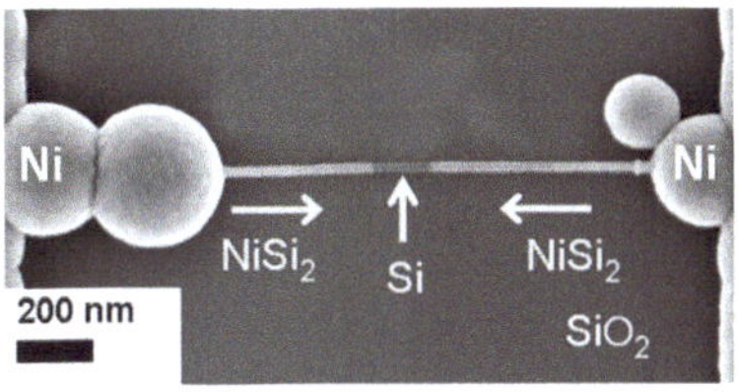

Fig. 12.19 SEM image shows the $NiSi_2$ growth path. Reprinted with permission from Ref. [65]. Copyright (2007) WILEY-VCH Verlag GmbH & Co. KGaA, Weinheim

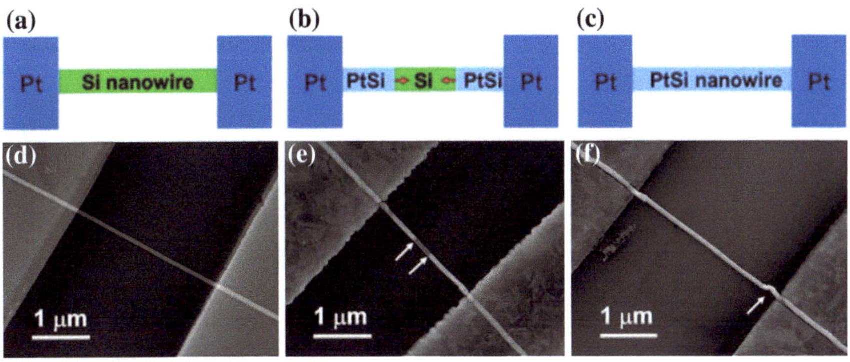

Fig. 12.20 Formation of PtSi nanowire and PtSi/Si/PtSi nanoheterostructures. **a, b, c** Schematic illustrations of the PtSi/Si/PtSi heterostructure and PtSi nanowire growth process; **d,e,f** SEM images corresponding to the structures illustrated in a,b,c, respectively. Reprinted with permission from reference [66]. Copyright (2008) American Chemical Society

for 30 s, PtSi/Si/PtSi heterostructure was formed. Increase the annealing duration to 60 s led to the fully silicidation of Si nanowires to PtSi nanowires.

From the HRTEM observation shown in Fig. 12.21, it was found that the PtSi/Si heterostructure has an atomically sharp interface with epitaxial relationships of Si[1-10]//PtSi[010] and Si(111)//PtSi(101).

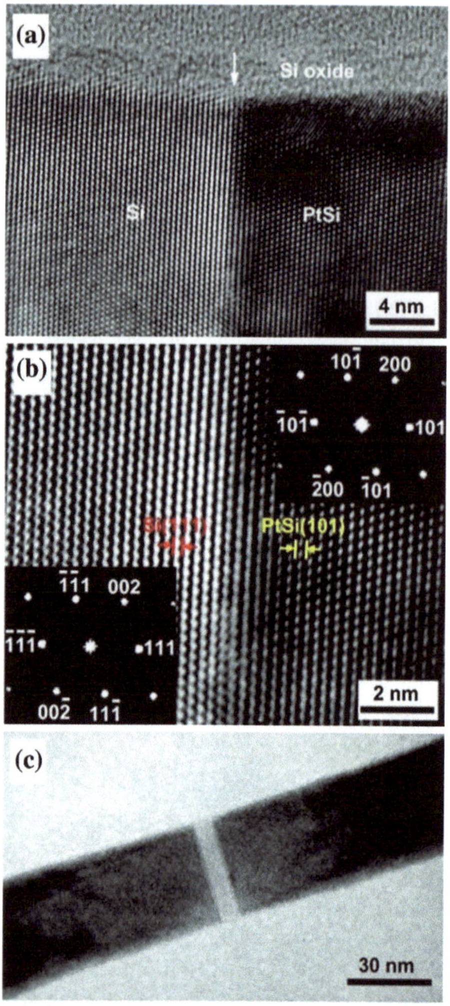

Fig. 12.21 Epitaxial relationship of PtSi formation within a Si nanowire. **a, b** HRTEM images show the epitaxial relationship between the PtSi and Si; **c** AvTEM image of the PtSi/Si/PtSi heterostructure. Reprinted with permission from Ref. [66]. Copyright (2008) American Chemical Society

$NiSi_x$ nanowires and $NiSi_x/Si$ heterostructures were fabricated by the same process of PtSi [67]. In situ TEM investigation observed that Ni silicide grows from the Ni pad and continuously into Si nanowires laterally. TEM images in Fig. 12.22 shows the Ni silicide growth process. After annealing at 550 °C for 5 s, $NiSi_2$ was formed near the Ni pad. With increasing of the annealing time, another phase $Ni_{31}Si_{12}$ with volume larger than initial Si nanowire appeared. The observation suggests that $NiSi_2$ forms as the initial phase and continuous grew into Si nanowires. Then $NiSi_2$ region near the Ni pad transform into Ni-rich phase $Ni_{31}Si_{12}$ with continued Ni supply. Ni silicide growth in Si nanowires with 5–50 nm oxide shell was retarded dramatically. The retarding effect increased with increase of the oxide shell thickness.

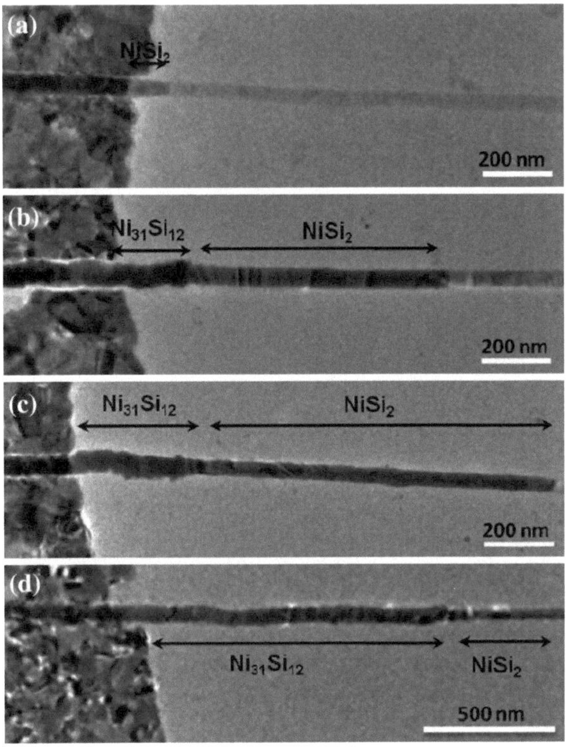

Fig. 12.22 Silicide growth in various duration of annealing time at 550 °C. **a** 5 s. **b** 60 s. **c** 2 min. **d** 8 min. Reprinted with permission from Ref. [67]. Copyright (2010) American Chemical Society

Si Nanowires + Metal Compounds

The Shi's group developed a method to prepare TMS silicides by silicidation of Si nanowires using metal compounds as the metal source. Employing this method, they prepared nanowire arrays of Mn silicides, Ni silicides, and Mn silicides/Si, Ni silicides/Si heterostructures [68, 69]. Si nanowire arrays prepared by the chemical etching method were coated with $MnCl_2$ on the surface and then annealed at 750 °C for 5 h. Structure characterization indicated that the Si nanowires were transformed to Si/$Mn_{27}Si_{47}$ nanowires, as shown in Fig. 12.23. The shell thickness can be controlled by the concentration of $MnCl_2$ solutions. The Si nanowires can be entirely conversed to $Mn_{27}Si_{47}$ nanowires when the Mn source supply is high enough.

When $MnCl_2$ was replaced by $NiCl_2$, Ni silicide nanowires can be formed by the same process [69]. The phases of Ni silicide obtained dependent on the reaction temperature, as illustrated in Fig. 12.24. At lower temperature of 700 °C, Si/NiSi core/shell nanowires or NiSi nanowires were obtained, depending on the concentration of $NiCl_2$ solutions. When the temperature increased to 800 and 900 °C, other phases including Ni_2Si, $Ni_{31}Si_{12}$ and $NiSi_2$ were formed. The dependence of

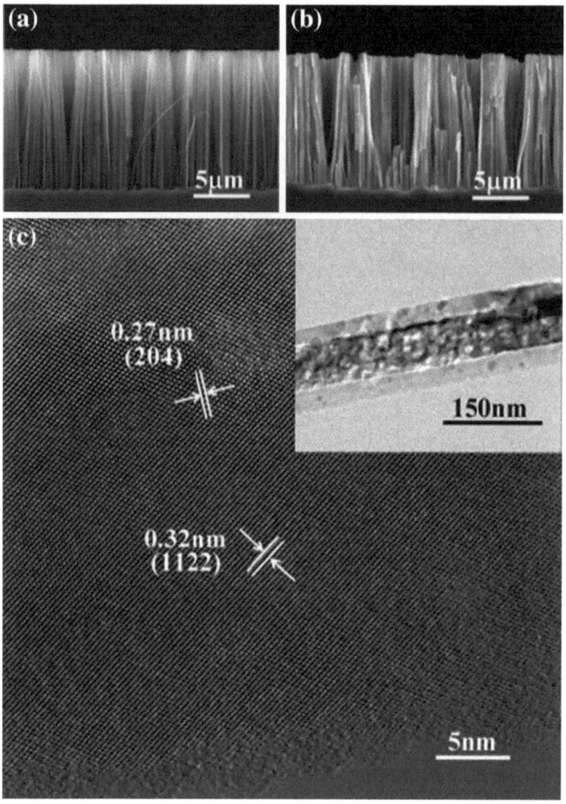

Fig. 12.23 **a** SEM image of Si nanowire arrays. **b** SEM image of Si/$Mn_{27}Si_{47}$ nanowire arrays. **c** HRTEM and TEM *(inset)* image of the Si/$Mn_{27}Si_{47}$ nanowire. Reprinted with permission from Ref. [68]. Copyright (2011) American Institute of Physics

diffusion rate on the temperature was suggested to be responsible for the phase variations. At lower temperature, Ni is the dominant diffusing species in the Ni-Si solid state reaction. Ni diffuses into and reacts with Si to form NiSi. With increase of temperature, the Ni diffusion is accelerated and Ni-rich phase Ni_2Si is formed. But the Ni_2Si will retard Ni diffusion, leading to the formation of Si/NiSi/Ni_2Si heterostructures.

12.2.4 Reactive Epitaxial Growth

Several TMS have small crystal mismatch with silicon in certain crystal directions. This makes epitaxial of TMS on single-crystalline Si substrate possible. More importantly, TMS nanowires can be grown epitaxially on Si substrate by the reaction of

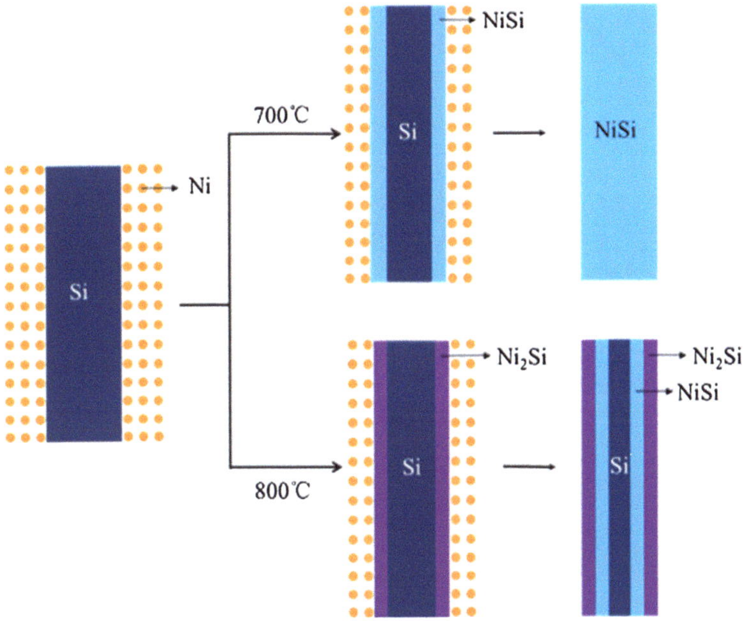

Fig. 12.24 Schematic diagram for Ni silicides growth at different temperatures. Reprinted with permission from Ref. [69]. Copyright (2012) Elsevier

metal sources and Si. The formation of 1D nanowires originates from different mechanisms.

Lattice Mismatch Mechanism

Flat 2D layer can be grown on a single crystalline substrate with small lattice mismatch. Increased lattice mismatch will limit the epitaxial growth in some directions. Thus, we can use this limitation to fabricate 1D nanostructures through 1D epitaxial growth. Chen et al. realized the preparation of metal silicide nanowires by this method [70, 71]. Nanowires of TMS $ScSi_2$ and rare-earth metal silicides $ErSi_2$, $DySi_2$, $GdSi_2$ were synthesized on Si (001) substrate. As for $ScSi_2$, it has lattice mismatches of $-4.6\,\%$ along one of the Si $\langle 110 \rangle$ directions and $0.8\,\%$ along the perpendicular Si$\langle 110 \rangle$ direction. The lattice mismatches satisfy the proposed conditions for nanowire growth. The annealing temperature, annealing duration, and amount of deposited metal greatly influence the products morphology. Figure 12.25 shows the STM images of metal silicide nanowires on Si substrates by deposition metals at 600 °C for 2 min and then annealed at 600 °C for 2 min. The difference in lattice mismatches leads to different width and height of the nanowires.

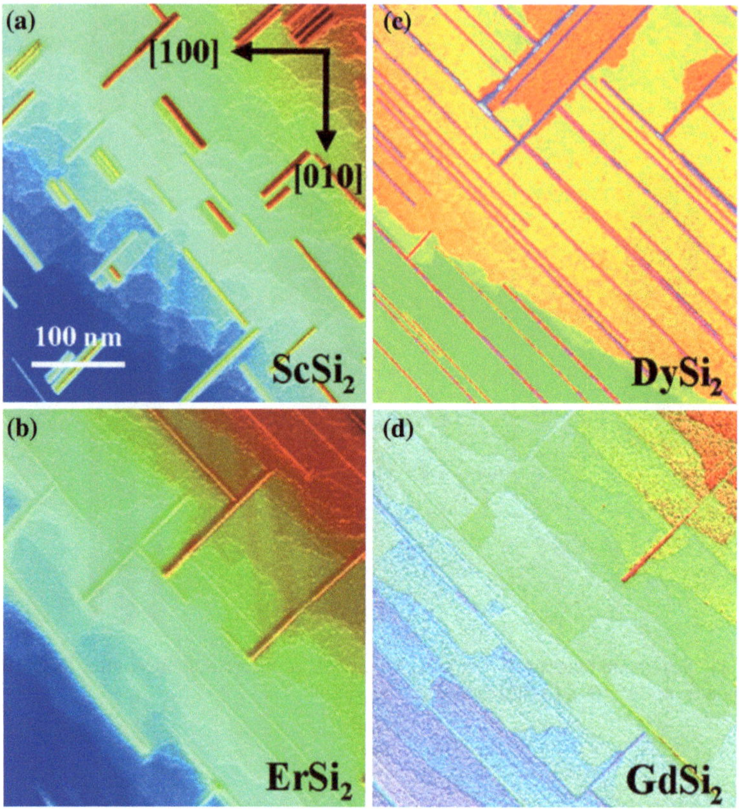

Fig. 12.25 STM images of **a** $ScSi_2$, **b** $ErSi_2$, **c** $DySi_2$ and **d** $GdSi_2$ nanowires Si substrates. Reprinted with permission from Ref. [71]. Copyright (2002) American Institute of Physics

$MnSi_{1.7}$ nanowires were also successfully fabricated on Si⟨111⟩ substrate by this method [72–74]. The d-spacing of $MnSi_{1.7}$ (110) plane is 0.391 nm, corresponding to a +1.8 % mismatch with that of Si (-110), (10-1), (0-11) planes. $MnSi_{1.7}$ (011) plane has a −17 % mismatch with Si (11-2), (1-21), (-211) planes. Thus, the growth is preferential in the $MnSi_{1.7}$ 110 direction and restrained in the perpendicular direction, leading to the formation of nanowires. Except nanowires, islands with other shapes were also formed. It was found that high growth temperature and low Mn deposition rate are favorable for growth of long nanowires with high aspect ratio. Although there are some successful examples, we still cannot exactly predict the formation of 1D nanostructures only by the lattice mismatch of bulk materials, because nanostructures have low-coordinated atoms at the boundaries and can support large strains [75].

Endotaxial Mechanism

TMS nanowires prepared by the lattice mismatch mechanism usually lie on the surface of the substrate. The Bennett's group explored a method that can fabricate TMS nanowires in an "endotaxial" configuration. Nanowires prepared by this method are partially embedded in the Si substrate. They first fabricated TMS nanowires on Si substrate via a mechanism that does not require anisotropic lattice mismatch. This mechanism involves growing into the substrate along inclined Si⟨−111⟩ planes, which breaks the symmetry of the surface and leads to 1D growth. They prepared Ti silicide and Co silicide on Si substrate by this endotaxial mechanism [76–79]. The Ti silicide nanowires were directly formed on the substrate after depositing 1–2 monolayer (ML) Ti on Si⟨111⟩ at 850 °C [76]. Three different morphologies including nanowires, rectangular plate, and isolated islands were observed, as shown in Fig. 12.26. The predominant structure is nanowire. Two different types of Ti silicides were obtained by the TEM characterizations. One type product is C49 $TiSi_2$ nanowires oriented with long axis along Si⟨2 − 20⟩. Another type product is B27 TiSi nanowires with long axis along Si⟨224⟩. In the case of growing Co silicide nanowires, different morphologies were obtained by tuning the substrate orientation, temperature and growth duration. It was found that on the Si (110) substrate only Co silicide nanowires occurred. This endotaxial growth mechanism is expected to apply to other system that meets two requirements. First, the material should have reasonable match on some incline planes. Second, significant ingrowth must occur, which can be predicted from bulk solubility.

According to the endotaxial mechanism, $NiSi_2$ and α-$FeSi_2$ nanowires were fabricated by Chen et al [80–82]. It was found that coating a Si_3N_4 layer on the Si substrate can greatly improve the aspect ratio of the nanowires. Take α-$FeSi_2$ for example, large amount of equiaxed islands were formed on blank Si substrate. When Si substrate was capped with a layer of Si_3N_4, nanowires with large aspect ratio were successfully fabricated. The Si_3N_4 layer can diminish the flux of Ni atoms to Si surface, so that the rearrangement of Ni adatoms could be completed in the stage

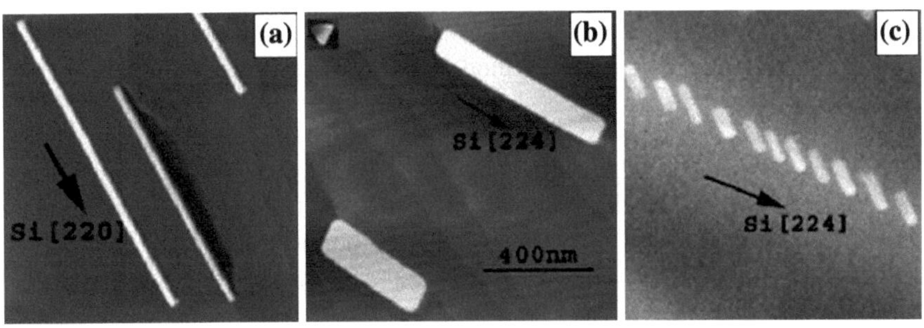

Fig. 12.26 AFM images of different $TiSi_2$ nanostructures on Si⟨111⟩ substrate. **a** Nanowires; **b** rectangular plate; **c** island-chain. Reprinted with permission from Ref. [76]. Copyright (2003) Elsevier

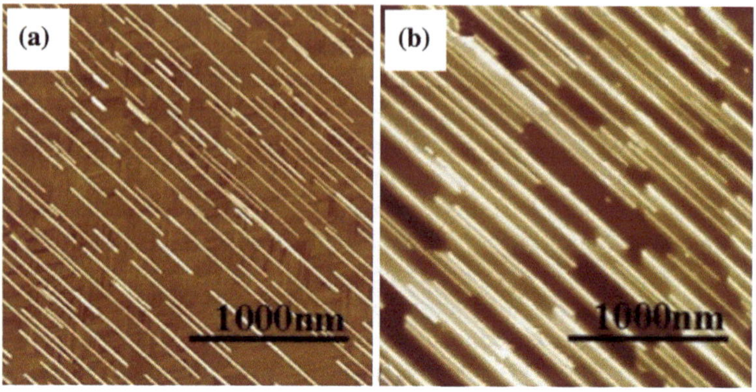

Fig. 12.27 AFM images of **a** s-$FeSi_2$ nanowires and **b** β-$FeSi_2$ nanowires produced by annealing s-$FeSi_2$ nanowires at 800 °C for 1 h. Reprinted with permission from Ref. [83]. Copyright (2006) Elsevier

of shape transition during nanowire growth. S-$FeSi_2$ nanowires were fabricated on Si⟨110⟩ substrate and transformed to β-$FeSi_2$ nanowires after annealed at 800 °C for 1 h, as shown in Fig. 12.27 [83]. The phase transformation temperature is higher than that observed in thin films (200–500 °C) due to the small thickness and large interface areas. Peng et al. reported the fabrication of Fe, Co, and Ni silicide nanowires on various Si substrates [84].

12.2.5 Other Methods

Besides the above-mentioned methods, there have been some other methods developed to prepare TMS nanowires. For example, Qian's group developed a co-reduction method to synthesize Cr_3Si nanorods. [85]. $CrCl_3$, $SiCl_4$ and metallic Mg were put in a sealed autoclave and reacted at 600 °C for 10 h. After this process, Cr_3Si nanorods with diameters of 40–80 nm and length of about 600 nm were obtained. Kim et al. reported a metal-induced growth method to prepare NiSi nanowires [86, 87]. Si nanoparticles were sputtered on Ni layer on SiO_2 covered Si substrate at 575 °C. The NiSi formation undergoes Ni grooving and agglomeration, clustering, fibering, and nanowire growth and segregation processes. Tanaka et al. fabricated Fe silicide nanorods on Si substrate by electron-beam induced deposition at elevated temperature in an ultrahigh vaccum TEM [88]. The vicinity of the substrate surface is supposed to give driving force for the formation of nanorod . Zhang et al. combined a sidewall transfer lithography technology and a self-aligned silicide process to fabricate PtSi and Pt_2Si nanowires [89].

A DC arc-discharge approach was exploited to prepare NiSi nanowires [90]. Si and Ni powders were put into a graphite crucible as the anode. NiSi nanowires were obtained by applying a direct current of 40 A with a voltage of 40 V between the

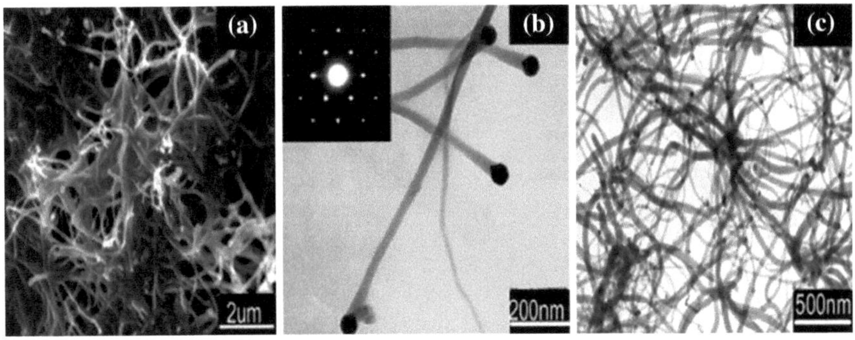

Fig. 12.28 SEM (**a**) and TEM (**b**, **c**) images of NiSi nanowires. Reprinted with permission from Ref. [90]. Copyright (2008) Elsevier

electrodes for 15 min. Sufficient Ni supply and high temperature were considered as two important conditions for the nanowire formation. From the TEM images shown in Fig. 12.28, catalyst heads can be clearly seen at each nanowire, suggesting a VLS growth mechanism for the NiSi nanowires.

α-$FeSi_2$ nanorods were fabricated at Si (100) surface via a low temperature implantation followed by a post-annealing process [91]. The anisotropy in interfacial energy is the main cause for the formation of nanorod. CuSi nanorod arrays were prepared by an oblique angle codeposition (OACD) technique [92]. The OACD setup used for CuSi nanorods fabrication is illustrated in Fig. 12.29. Formation of the nanorod arrays is mainly based on self-shadowing effect. Composition of the products can be tailored by precisely controlling the deposition rates of the source materials. Different CuSi composite nanorods with 0–100 at. % Cu could be fabricated by this method [93].

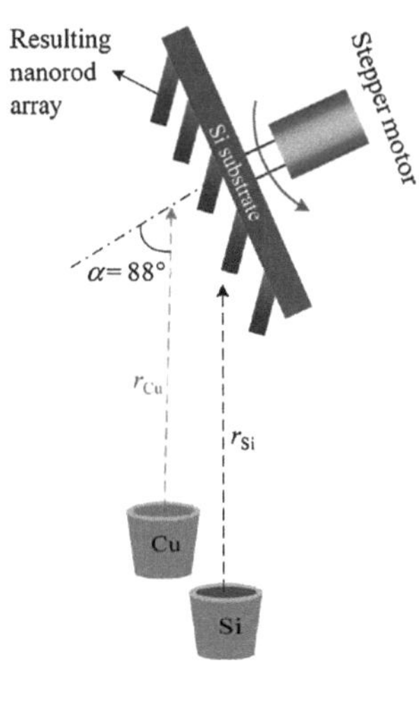

Fig. 12.29 A sketch of OACD setup for CuSi nanorods fabrication. Reprinted with permission from Ref. [92]. Copyright (2010) American Chemical Society

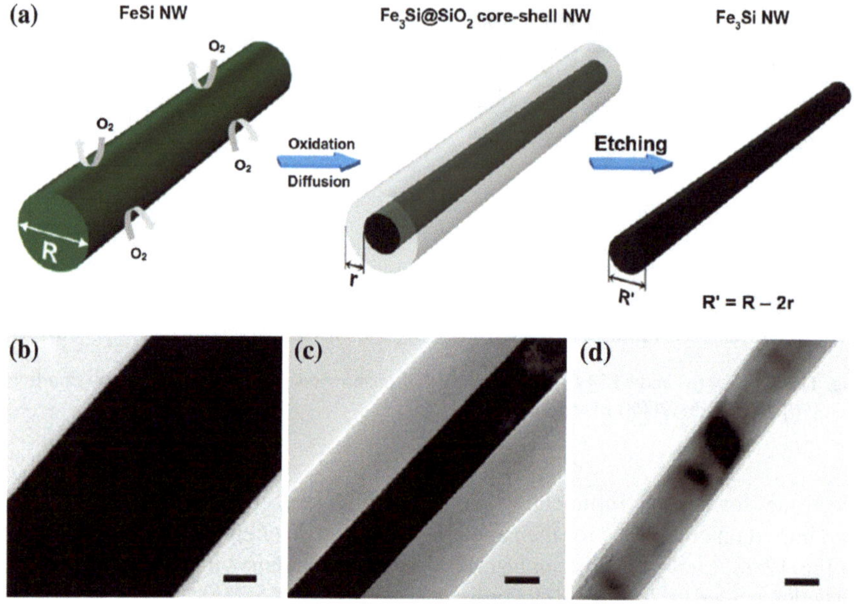

Fig. 12.30 **a** Schematic illustration of NW crystal transformation process. TEM images of FeSi (**b**), $Fe_3Si/SiO2$ core-shell (**c**), and Fe_3Si nanowire (**d**). Scale bars are 20 nm. Reprinted with permission from Ref. [94]. Copyright (2010) American Chemical Society

A diffusion-driven crystal structure transformation method was used to prepare Fe_3Si nanowires [94]. The transformation process is illustrated in Fig. 12.30. FeSi nanowires synthesized by CVT method was annealed at 900 °C in diluted O_2 condition and subsequent wet etched by 5 % HF. After these processes, FeSi nanowires were converted to Fe_3Si nanowires. The crystal transformation is thought to be controlled by free energy change and thermally driven reaction kinetics. The formation of Fe_3Si nanowires was explained by the thermodynamic stability and diffusion-controlled growth of Fe_3Si phase. Free energy calculation shows that Fe_3Si is the most stable in Fe-rich silicide phases and would form preferentially. The reaction kinetics is mostly governed by diffusion of Fe and Si. It is expected that $Fe_xSi(x \geq 1)$ nanowires with other phases could be prepared by more precisely controlling the oxidation and diffusion processes.

Cu_3Si nanowires were synthesized by annealing Cu/Ge bilayer film on a SiO_2/Si substrate [95]. Polycrystalline Cu and amorphous Ge were deposited on the substrate by magnetron sputtering method. Then the samples were annealed at 650 °C. Ge acts as catalyst to enhance Cu diffusion and penetrate the SiO_2 layer. Upon annealing, the Cu/Ge layer melts and form eutectic alloy droplets on the substrate. Cu diffuses into SiO_2 and forms Cu_3Si. Continuous supply of Cu from the eutectic liquid leads to further growth of Cu_3Si crystals and formation of the Cu_3Si nanowires. Figure 12.31 illustrates the growth mechanism and SEM images of the Cu_3Si nanowires.

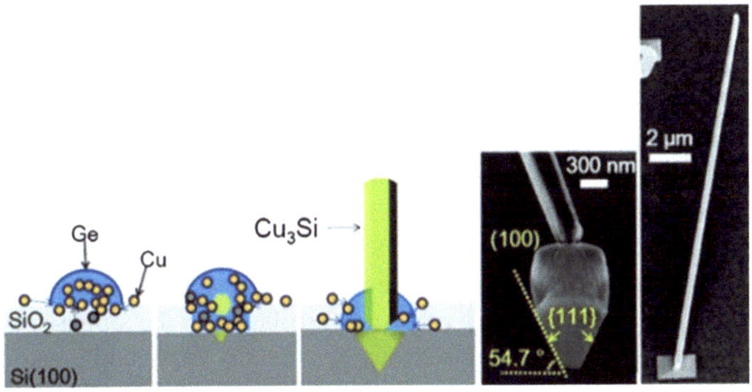

Fig. 12.31 Schematic illustration of the Cu_3Si nanowires growth mechanism and SEM images of Cu_3Si nanowires. Reprinted with permission from Ref. [95]. Copyright (2012) American Chemical Society

12.3 Properties of 1D TMS Nanostructures

12.3.1 Electrical Transport Properties

Many TMS nanowires exhibited a metallic transport behavior. NiSi is a typical one. Fig. 12.32 shows the electron transport data recorded from single NiSi nanowire with

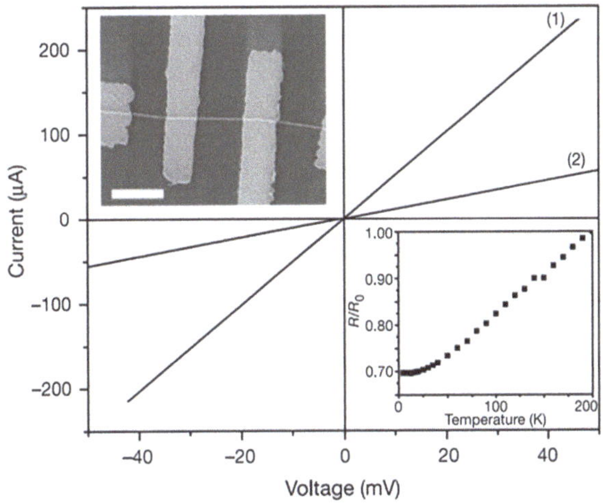

Fig. 12.32 Current versus voltage curves recorded from individual NiSi nanowire. Line (*1*) and line (*2*) corresponds to two- and four-terminal measurements, respectively. *Lower inset* is temperature-dependent normalized resistance obtained from a four-terminal device; the resistance, R, is normalized by the value at 200 K, R_0. Reprinted by permission from Macmillan Publishers Ltd: (Nature) reference [60], copyright (2004)

diameter of 29 nm prepared by silicidation of Si nanowires [60]. The I–V curve shows a linear with two- and four-terminal resistance of 886 and 184Ω, respectively. Resistivity based on the four-terminal data is calculated to be 9.5μΩcm, close to the value of NiSi single crystals and thin films (10 μΩ cm) [96]. Temperature-dependent resistance curve in inset of Fig. 12.32 shows that the resistance monotonically decreases with decrease of temperature, and saturates at about 30 K, indicating a diffuse transport. A maximum current density (J_{max}) of 3×10^8 A cm^{-2} is observed before the device breakdown due to melting. The value is reproducible and independent of nanowire diameter. The high J_{max} value was attributed to the single-crystal structures. The J_{max} value is comparable to the best results from single-walled carbon nanotubes, and about two orders of magnitude larger than that observed in lithography defined noble metal lines. There have been other reports about electron transport measurement of NiSi nanowires [13]. Although metallic behavior is observed in most measurements, the conductance of NiSi nanowires is no better than the above results. The difference in resistivity might be originated from the difference in microstructure due to growth method, device fabrication process, and other unknown reasons. Like NiSi phase, Ni silicides with other phases also exhibit metallic electron transport behavior. For example, Ni-rich phase β_3-Ni_3Si single-crystal nanowires prepared by CVT method have a resistivity of 72 μΩ cm and J_{max} of 1.7×10^7 A cm^{-2} [48]. Ni_3Si_2 nanowires prepared by CVD method have a resistivity of 208 μΩ cm and J_{max} of 3.36×10^7 A cm^{-2} [97]. Si-rich phase $NiSi_2$ nanowires exhibit a resistivity of 30 μΩ cm and J_{max} of 3.0×10^8 A cm^{-2} [56].

Electronic properties of FeSi nanowires were investigated using the X-ray absorption spectroscopy (XAS) and X-ray emission spectroscopy (XES) [42]. The XAS and XES represent the density of states of empty Fe 3*d* states and occupied Fe 3*d* states. The Fermi level (E_f) can be defined by the crossover from occupied to empty DOS, as designating by a dotted line in Fig. 12.33. Metallic behavior was observed from the XAS and XES, in agreement with the general expectation that the Kondo insulator FeSi behaves like a metal at room temperature. The authors also used

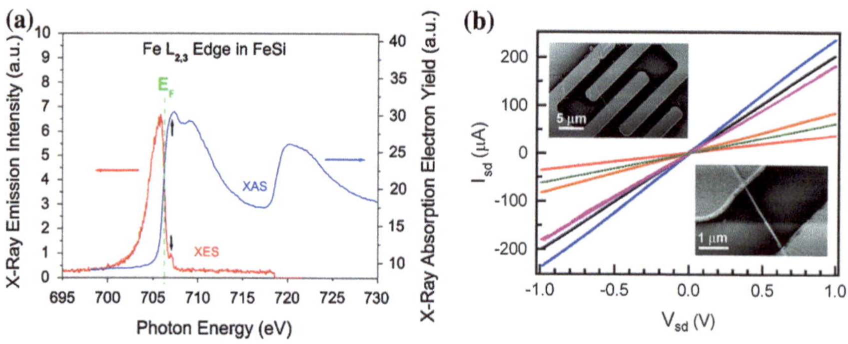

Fig. 12.33 **a** XAS and XES of FeSi nanowires, **b** Room-temperature transports properties of FeSi nanowires. Reprinted with permission from Ref. [42]. Copyright (2006) American Chemical Society

two-terminal devices to study the transport properties of the FeSi nanowires. The current (I_{sd}) versus voltage (V_{sd}) curve shows a linear behavior at room temperature, in consistent with the XAS and XES results and theoretical predictions [98, 99]. Resistance variation can be observed from the I–V curves, which might be originated from the diversity in nanowire diameter, distribution of effective channel device length and multiwire devices.

Electrical transport properties of other metallic TMS nanowires were evaluated by I–V curves. CoSi nanowires synthesized by the CVD method have resistivity of $510\,\mu\Omega\,cm$ and J_{max} of $5 \times 10^6\,A\,cm^{-2}$ [41]. The resistivity of Co_2Si nanowires was found to be $200\,\mu\Omega cm$ and monotonically decreased with decreasing temperature from room temperature to 2 K [54]. $TaSi_2$ nanowires have resistivity of $102\,\mu\Omega\,cm$ and failure current density of $3 \times 10^8\,A\,cm^{-2}$ [18]. PtSi nanowires have a low resistivity of $72\,\mu\Omega cm$ and breakdown current density $>1 \times 10^8\,A\,cm^{-2}$ [66]. Cu_3Si nanowires were reported to have resistivity lower than $30\,\mu\Omega cm$ [95]. The lowest resistivity was observed from $TiSi_2$ nanowires, which value is about $5\,\mu\Omega\,cm$. At the same time, the J_{max} is extremely high up to $10^{11}\,A\,cm^{-2}$ [21].

Some of the TMS are semiconductors, such as $CrSi_2$, β-$FeSi_2$ and NCL phase $MnSi_{1.7}$. $CrSi_2$ is an indirect narrow-gap semiconductor with Eg of 0.35 eV. Two-terminal electrical transport measurements indicate that $CrSi_2$ nanowires have an average resistivity of $4.4 \times 10^{-3}\,\Omega\,cm$ and behave like a degenerate semiconductor [47]. I–V curve in Fig. 12.34a shows a linear behavior for the $CrSi_2$ nanowire. The rather lower resistivity suggests that the $CrSi_2$ nanowires are heavily doped. Temperature-dependent measurements exhibit two different behaviors. For some nanowires, resistance increases with decreasing temperature (Fig. 12.34b), as expected for semiconductors. However, sometimes the resistance decreases with decrease of temperature (Fig. 12.34c), which is the typical behavior for a metal. The conflicting behavior was attributed to the fact that the temperature ranges from low T to room temperature is in the extrinsic region of the $CrSi_2$ semiconductor. The varying levels of acceptor-doping concentration in nanowires from different batches of synthesis might be another reason for this phenomenon.

Iron silicide is the only TMS occurs both in semiconducting and metallic phase. Another semiconducting TMS, β-$FeSi_2$ nanowires show resistivity of $1.23 \times 10^{-3}\,\Omega\,cm$, which is one order lower than that of thin films. The resistivity was found to decrease with decrease of nanowire's diameter due to enhanced surface conductivity in thinner nanowires [100]. Linear I–V curves were observed from $Mn_{19}Si_{33}$ nanowires, indicating the nanowires are degenerately doped semiconductor [40]. The average resistivity was calculated to be $17.4 \times 10^{-3}\,\Omega\,cm$. It was found that the resistivity monotonically increasing with increase of temperature from 5 to 300 K, which is a typical metallic behavior. The metallic character is in agreement with a semiconductor operating in extrinsic regime at room temperature. A nonlinear I–V behavior was reported in the NCL phase $Mn_{15}Si_{26}$ nanowires [101]. P-type semiconducting behavior was deduced from the very small variation in current with applied gate voltage. The $Mn_{15}Si_{26}$ nanowires exhibited high resistivity, which was attributed to the anisotropic resistivities observed in $MnSi_{1.75}$ systems.

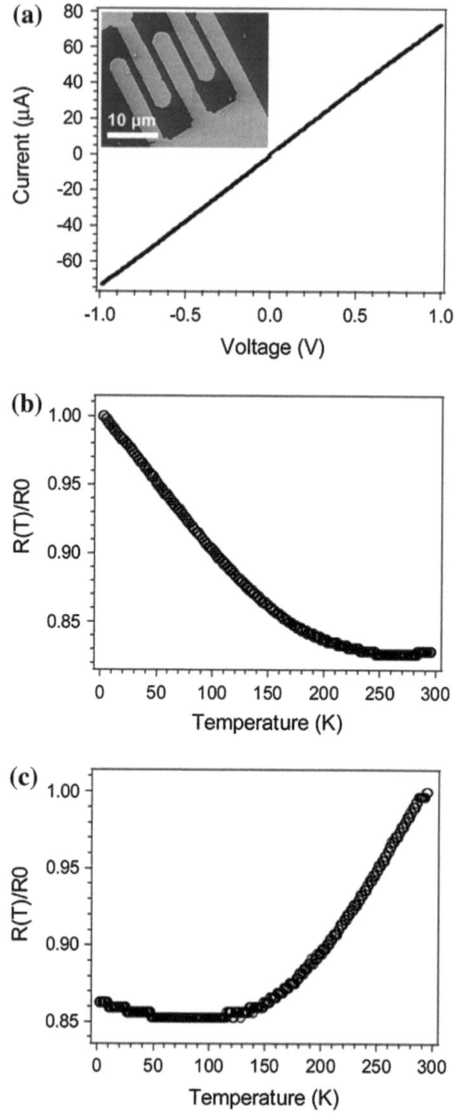

Fig. 12.34 **a** I–V curve recorded from the Cr_2Si nanowire. **b**, **c** Temperature-dependent resistance variation of the Cr_2Si nanowire. Reprinted with permission from Ref. [47]. Copyright (2007) American Chemical Society

ε-FeSi is a hybridization-semiconductor or Kondo insulator with a narrow band gap. Electrical transport properties of individual ε-FeSi nanowire were measured as a function of temperature (T) [3]. Two different types of transport behavior were found in the single-stem nanowires and branched nanowires, as shown in Fig. 12.35. In the curve recorded from single-stem nanowire without branches (Fig. 12.35a), two drops in resistivity can be observed. The drop below 20 K is attributed to the excitation of states within the band gap, while the drop above 50 K reflects the

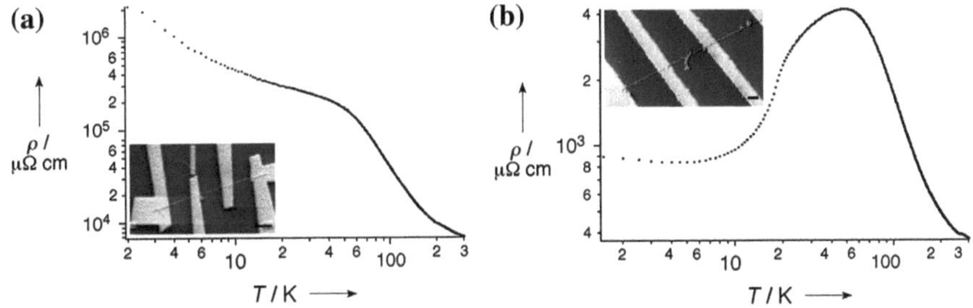

Fig. 12.35 Plots of resistivity (ρ) as a function of temperature (T) recorded from the single-stem (**a**) and branched FeSi nanowire (**b**). Reprinted with permission from reference [3]. Copyright (2006) WILEY-VCH Verlag GmbH & Co. KGaA, Weinheim

increase in charge-carrier concentration due to carrier excitation over the band gap. In the curve of the branched FeSi nanowires (Fig. 12.35b), it can be seen that the resistivity first rises and then falls. The rise of resistivity below 50 K is typical of a metal, while the fall at higher temperature is inconsistent with thermal-activation behavior of a semiconductor. This phenomenon is caused by transport competition between the carrier scattering and thermal excitation of carriers over the band gap. The metallic behavior at low temperatures suggests that the junctions between the stem and branches may act as a source of charge carriers, either by deviation from the ideal stoichiometry or by the introduction of lattice defects.

12.3.2 Field Emission

The good electrical conductivity and stability endow TMS with good field emission properties. Titanium silicide is one of promising cold cathode materials for field emission due to the thermal and chemical stability. Xiang et al. measured field emission property of $TiSi_2$ nanowires with diameter of about 40 nm and lengths of 150–250 μm [20]. The turn-on voltage is about 8V/μm, corresponding to a current density of about 0.1 $\mu A/cm^2$, much lower than the previous reports. Although the results are not so good, it is worth mentioning that no obvious degradation was observed in the life stability experiment period for over 40 h. Lin et al. reported a better performance of field emission from Ti_5Si_3 nanowires [4], where the turn-on voltage is only 5.4V/μm corresponding to a current density of 10 $\mu A/cm^2$. A much better field emission properties were observed from Ti_5Si_4 nanobats with current density of 1 $\mu A/cm^2$ at the turn-on voltage of 1.47 V/μm, as shown in Fig. 12.36 [26]. The field amplification factor was calculated to be 1350, which is much higher or comparable to those of other 1D nanostructures. The remarkable field emission properties were attributed to the intrinsic properties of Ti_5Si_4 nanowires as well as the good conductivity of the Ti substrate.

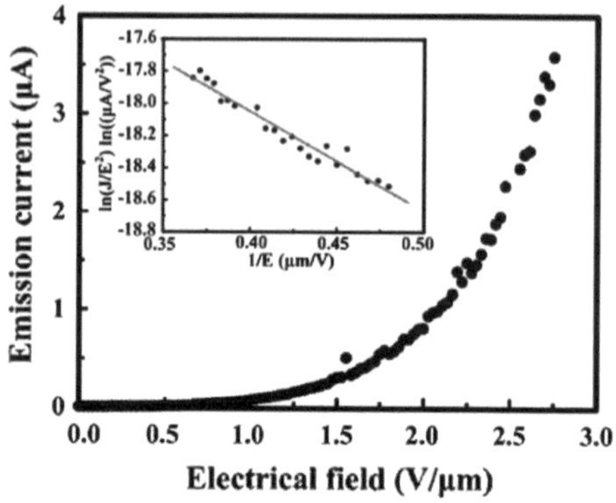

Fig. 12.36 I–V curves of field emission of the Ti_5Si_4 nanobats grown on Ti foil. The inset is the FN plot of the emission of the nanobats. Reprinted with permission from Ref. [26]. Copyright (2009) American Chemical Society

Field-emission measurements of $TaSi_2$ nanowires show that the turn-on field is low at 4–4.5 V/μm and the threshold field is down to 6 V/μm with the field enhancement factor as high as 1800 [17]. In the stability experiments, current density dropped sharply within 20 s when a fixed voltage of 900 V was applied. This degradation might be originated from melting, breakdown or detachment from the substrate of the nanowires. When the applied voltage was decreased to 700 V, the degradation in current density was found to be below 5 % over 3 h. Field emission in a single $TaSi_2$ nanowire was directed observed by means of electron holography [102].

Kim et al. measured the field emission property of NiSi nanowires [13]. As shown in Fig. 12.37, the turn-on voltage is below 4 V/μm, with turn on current density of about 10 μA/cm^2. From the linear fit of $J/E^2 - 1/E$ relation, the field enhancement factor (β) is extracted to be 2200. Ni silicide (NiSi and Ni_3Si) nanowires grown on different substrates (tungsten plate and SiO_2-coated Si substrate) exhibited different field emission performance [103]. The onset electric field and field enhancement factor was observed to be 3.23 V/μm and 3180, 3.47 V/μm and 3002 for nanowires on tungsten and Si substrate, respectively.

12.3.3 Magnetic Property

The magnetic properties of ε-FeSi nanowires were investigated by measuring the sample magnetization as a function of temperature and applied magnetic field [3]. In both the field-cooled (FC) and zero-field-cooled (ZFC) cases, the magnetization decreases as temperature increases to 90 K and then rises as temperature increases from 90 to 300 K. This behavior compares well with previous reports on polycrystalline and single-crystal bulk FeSi. Liang et al. measured magnetic properties of γ-$FeSi_2$ nanowires epitaxial grown on Si substrate [104]. A hysteresis at 2 K with a saturation magnetization of 1.2×10^{-6} emu and a coercive of 30 Oe were observed, corresponding to a magnetic moment of 0.3 Bohr magneton per Fe atom. Magnetoresistance (MR) measurements were done to epitaxial $FeSi_2$ nanowires at four

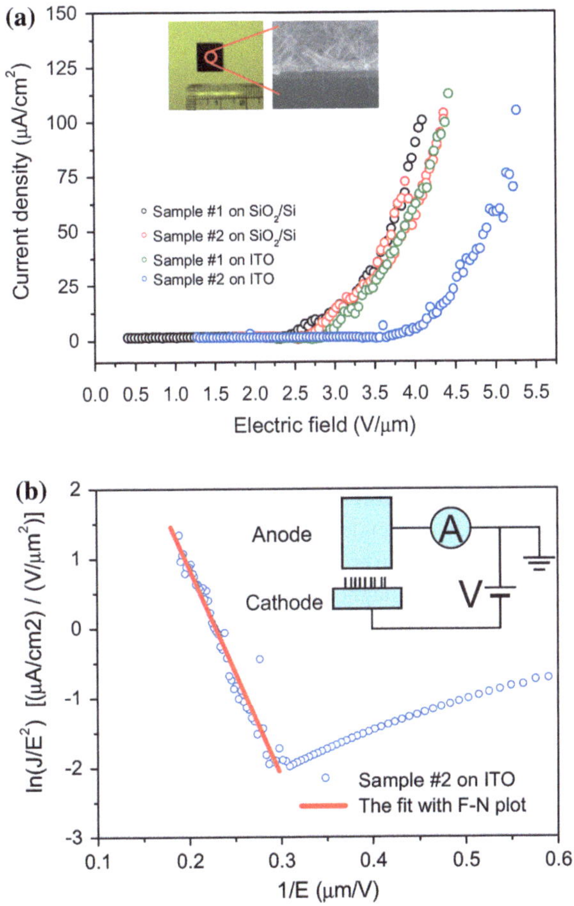

Fig. 12.37 **a** The field emission current versus the applied electrical field from the NiSi nanowire sample. **b** Linear fit to the $J/E^2 - 1/E$ relation. Reprinted with permission from Ref. [13]. Copyright (2007) WILEY-VCH Verlag GmbH & Co. KGaA, Weinheim

temperatures, which results are shown in Fig. 12.38 [105]. Clearly hysteresis was observed in MR, along with a pair of resistance peaks at ±0.46 T (indicated with arrows in the curves). The peaks arise when the nanowire magnetization rotates from out of plane to orient along the nanowire, corresponding to the anisotropic MR. Saturation of magnetization was achieved when $|B| > 5$ T. Small conductance fluctuations also appear in the whole field range. Furthermore, negative MR is clearly observed in the 1.5, 3.5, and 4.2 K data in Fig. 12.38, indicating ferromagnetic characteristic of the $FeSi_2$ nanowires. At temperature of 10 K, the MR has become virtually flat. This does not mean that the ferromagnetism is quenched, but that the coherence length of the carriers becomes too small to allow observation of weal localization effects. Kim et al. reported that the paramagnetic FeSi nanowires were transformed to

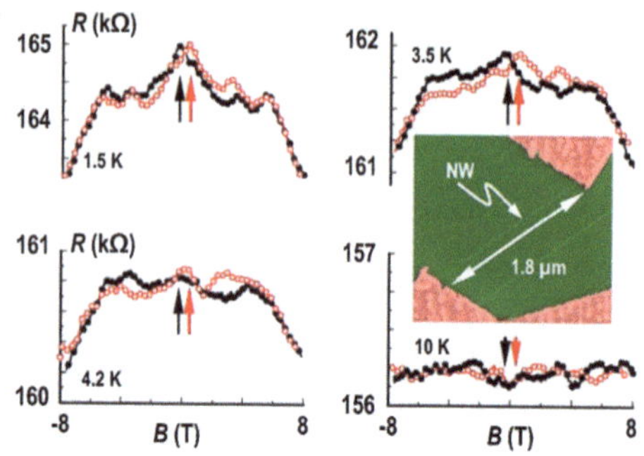

Fig. 12.38 MR of a 1.8-μm-long segment of $FeSi_2$ nanowire at four temperatures. Reprinted with permission from Ref. [105]. Copyright (2010) American Institute of Physics

ferromagnetic Fe_3Si nanowires with Currie temperature much high than 370 K [94]. In the FeSi crystal with cubic B20 structure, each Fe atom is surrounded by seven Si atoms. When transformed to fcc Fe_3Si crystal, Fe atoms replace Si atoms, resulting ferromagnetism at high temperatures.

CoSi is a diamagnetic semimetal. Magnetic properties of single crystalline CoSi nanowires with B20 structure were measured by the superconducting quantum interference device (SQUID) magnetometer [50]. The magnetization (M) versus applied magnetic field (H) curves at 2 and 300 K are nonlinear with nonzero remnant magnetization and coercivity, showing well pronounced ferromagnetic property of the nanowires. The ferromagnetism originates from undercoordinated Co atoms on the surface of the nanowires covering a diamagnetic core. Seo et al. measured the magnetic properties of Co_2Si nanowires, which results are shown in Fig. 12.39 [54]. Figure 12.39a is the temperature dependence magnetization curve measured at 100 Oe. At the high temperature region, the curve is linear and well fitted by the Currie-Weiss law. In the region of temperature lower than Tc (115 K), the inverse magnetization starts to deviate from linearity, which is attributable from ferromagnetic ordering. M-H curves of Fig. 12.39b show a rapid and nonlinear increase of M at lower fields and linear increase at higher fields when the temperatures are lower than Tc, indicating ferromagnetic properties of the Co_2Si nanowires. When the temperatures are higher than Tc, the M-H curves are almost linear response to the magnetic field, which is a typical paramagnetic behavior. Figure 12.39d shows the MR curves of individual Co_2Si nanowire recorded at different temperatures. A pair of MR peaks at $\pm$2T were observed, corresponding to the saturation field of magnetism. The monotonic decrease of MR with increasing field is attributed to the reduction of magnetic scattering by magnetic ordering.

Interesting ferromagnetic behaviors were observed in $Ni_2Si/TiSi_2$ coaxial nanowires [27]. Since $TiSi_2$ is nonmagnetic, the ferromagnetic property originates from Ni_2Si. The VSM hysteresis loop curves in Fig. 12.40a reveals that coaxial $Ni_2Si/TiSi_2$ nanowires exhibit ferromagnetism with a coercive field of about 50 Oe and saturation magnetization of 4000 Oe. However, extremely little hysteresis was found in the M-H curves of the substrate (Fig. 12.40b). In the SQUID curves shown in Fig. 12.40c, coercive fields of 200 and 100 Oe were observed from applying magnetic

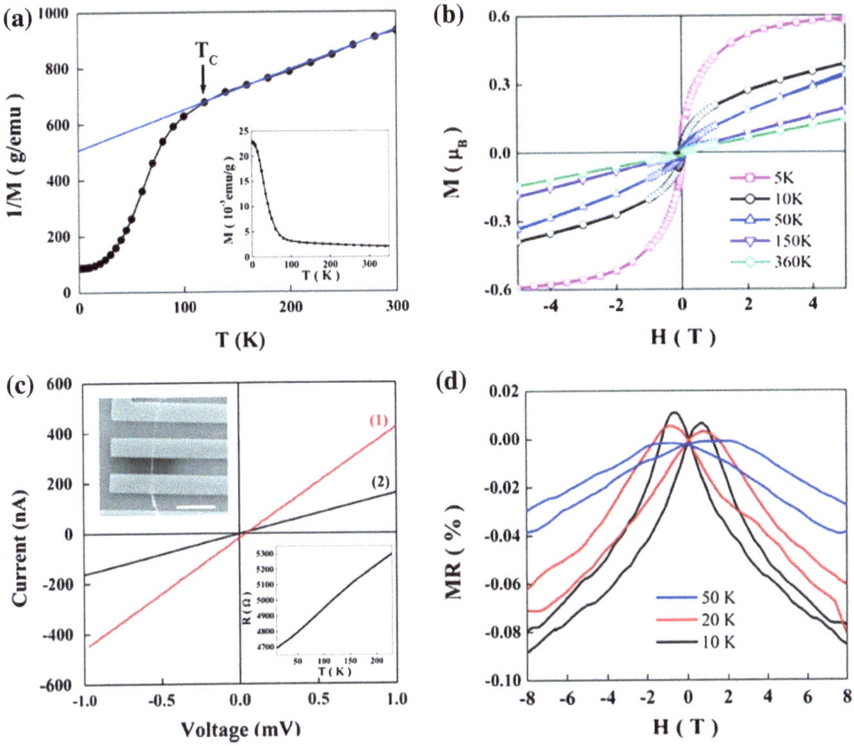

Fig. 12.39 Electrical and magnetic properties of Co_2Si nanowires. **a** Inverse magnetization versus T curve. *Inset shows* a plot of M versus T obtained from the NW ensemble at an applied field of 100 Oe. **b** M–H curves obtained from the nanowires at 5–360 K. **c** I–V curves recorded on a Co_2Si nanowire device with four- (line 1) and two- (line 2) probe measurements, respectively. **d** MR (%) curves of a single Co_2Si nanowire at various temperatures. Reprinted with permission from Ref. [54]. Copyright (2009) American Chemical Society

field out-of-plane and in-plane to the substrate, respectively. The hard remanences and coercive fields originate from the shape anisotropy owing to the vertical orientation of the nanowires on the substrate, which forces the magnetic moments to align mainly along the orientation of the nanowires [27]. Figure 12.40d shows the FC and ZFC curves from 354 to 4 K. As increasing of temperatures, magnetization decreased in FC curve, while first increased and then decreased in ZFC curve. A blocking temperature (T_b) of about 314 K could be revealed from these curves.

Ham et al. measured magnetic properties of Mn_4Si_7 nanowires using the SQUID magnetometer in the temperature range of 5–300 K [38]. As shown in Fig. 12.41, the Mn_4Si_7 nanowires exhibited a ferromagnetic property. The maximum magnetic moment was estimated to be 0.17 μB/Mn at 5 K, which is much higher than that of single crystalline bulk Mn_4Si_7. The T_c was found to be about 120 K, which is also much higher than that of bulk sample (T_c = 40K) in previous report [106]. The enhanced magnetic property might originate from the oxidation, clustering and

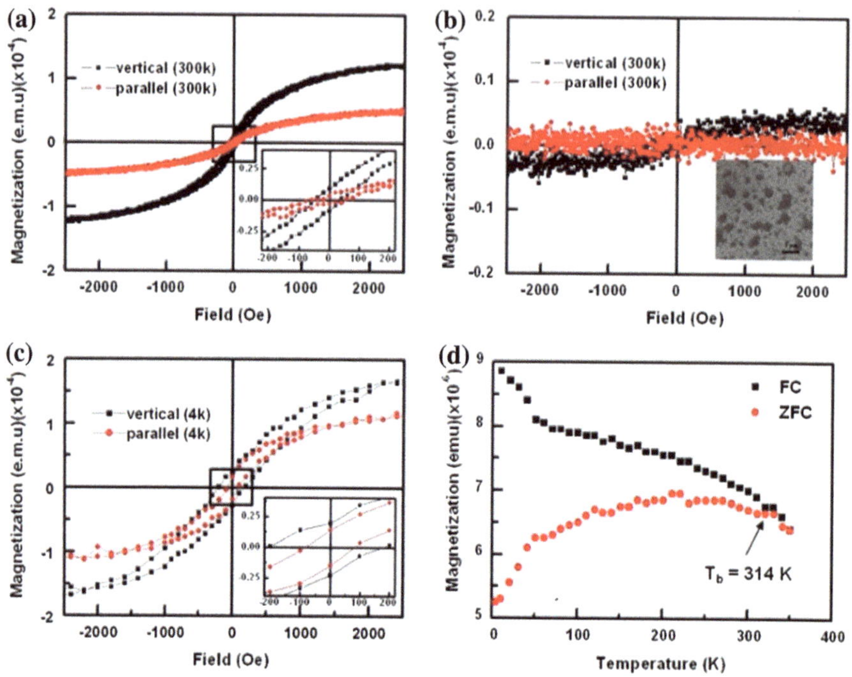

Fig. 12.40 The VSM hysteresis loop curves of the core/shell $Ni_2Si/TiSi_2$ nanowires (**a**) and the initial substrate (**b**). **c** The hysteresis loop curves of core/shell $Ni_2Si/TiSi_2$ nanowires in two different orientations measured by SQUID magnetometry at 4 K. **d** FC and ZFC curves of the core/shell $Ni_2Si/TiSi_2$ nanowires. Reprinted with permission from Ref. [27]. Copyright (2012) American Chemical Society

reduced coordination of Mn atoms at the nanowire surface. Seo et al. reported that two T_c (T_{c1} and T_{c2}) of about 30 k and over 300 k were observed in MnSi nanowires [53]. The T_{c1} of 30 K is consistent with that of bulk MnSi crystal, but the T_{c2} might be induced by the $MnSi_{1.7}$ film formed during the synthesis process. Jin's group revealed the helimagnetism in MnSi nanowires [7]. $Si/Mn_{27}Si_{47}$ core/shell nanowire arrays with a high T_c of about 150 K was reported by the Shi's group [68]. The enhanced ferromagnetism was ascribed to the reduced coordination of Mn atoms at the surface of the shells of the nanowires.

Both Cr_2Si and SiO_2 are diamagnetic in bulk. However, room-temperature ferromagnetism was observed in Cr_2Si/SiO_2 nanocables [107]. Nonlinear M-H curves with nonzero remnant magnetization and coercivity indicates ferromagnetic characteristic of the Cr_2Si/SiO_2 nanocables. Magnetization is much higher when applied field is perpendicular to the substrate than parallel to the substrate. This is because that the long axis of nanowires is almost perpendicular to the substrate, which is the easy direction of magnetization due to a lower demagnetization field as compared to that of the short axis. Figure 12.42 shows the simulated magnetization of Cr atoms vs. the distance to interface. It can be seen that the magnetization of Cr atoms at surface is

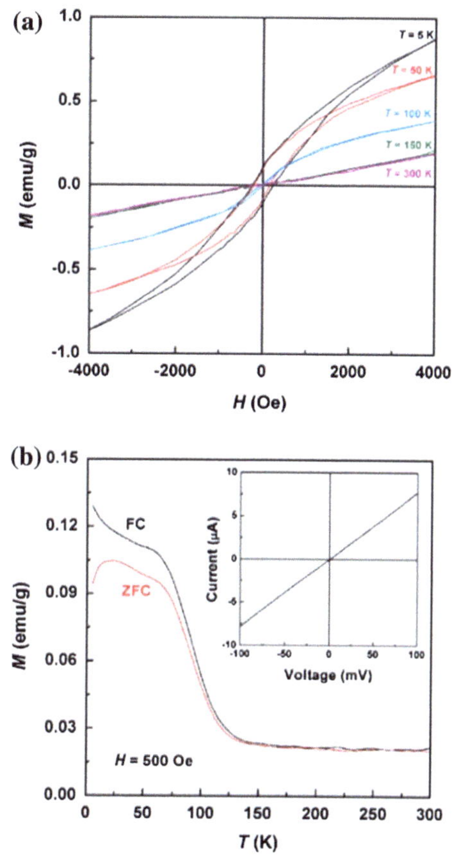

Fig. 12.41 Hysteresis loops for the Mn_4Si_7 nanowires at various temperatures (5–300 K). **b** Temperature dependence of FC and ZFC magnetizations with a magnetic field of 500 Oe. Reprinted with permission from Ref. [38]. Copyright (2009) American Chemical Society

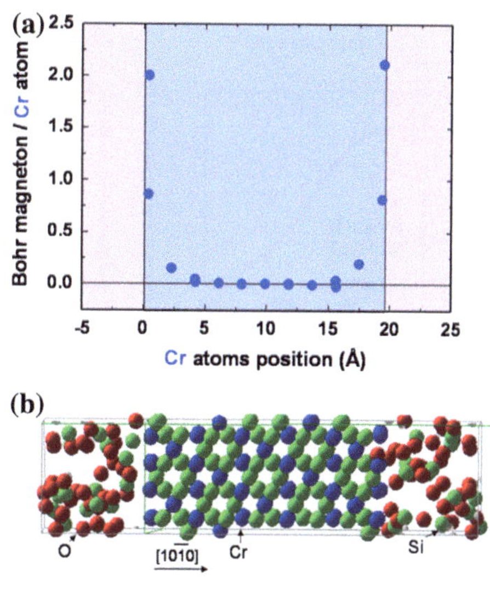

Fig. 12.42 **a** In-plane and perpendicular hysteresis loops measured at 5 and 300 K. **b** Simulated magnetization of Cr (*blue*) atoms versus the distance to interface as corresponding to the atomic arrangements around Si(*Green*)O(*Red*) $_2$/$CrSi_2$/SiO_2 interface shown below. Reprinted with permission from Ref. [107]. Copyright (2011) American Institute of Physics

as high as about 2 μ_B, and decreases to zero at 0.5 nm depth. The distorted/detangled bonds and surrounded oxygen atoms cause the magnetism. Oxygen plays important role for the magnetism, which is proved by the energy filtered TEM imaging that O distribution especially around the surface. The interface-induced magnetization was estimated to be 1.55×10^{-4} emu.

Ternary TMS $Fe_{1\text{-}c}Co_xSi$ nanowires were reported by Kim's group and Jin's group to have ferromagnetic properties [43, 58, 108]. $Fe_{1\text{-}c}Co_xSi$ nanowires synthesized by CVT method exhibit ferromagnetic properties with a sudden change in magnetization at applied magnetic field near zero [58]. This M-H behavior was attributed to the presence of helimagnetic order in the nanowires. T_c of the $Fe_{1\text{-}c}Co_xSi$ nanowires is about 38 K. Magnetoresistance measurements on a single nanowire device reveal a positive magnetoresistance without saturation up to the highest applied field. Magnetic properties of $Fe_{1\text{-}c}Co_xSi$ nanowires are found to be dependent on Co concentrations [43]. Magnetoresistance measurements were done to $Fe_{1\text{-}c}Co_xSi$ nanowires with $x = 0.06$ (sample A) and 0.44 (sample B), respectively. In the R-T curves shown in Fig. 12.43a, d, transition from semiconducting to metallic behavior was observed, indicating $Fe_{1\text{-}c}Co_xSi$ is a narrow band gap semiconductor. The MRs are positive in both samples, as shown in Fig. 12.43b, e. But the onset of large MR in sample A is at temperature (15 K) much lower that of sample B (50 K), which is due to higher T_c of samples with higher Co concentrations. M-H curves shown in Fig. 12.43c, f also

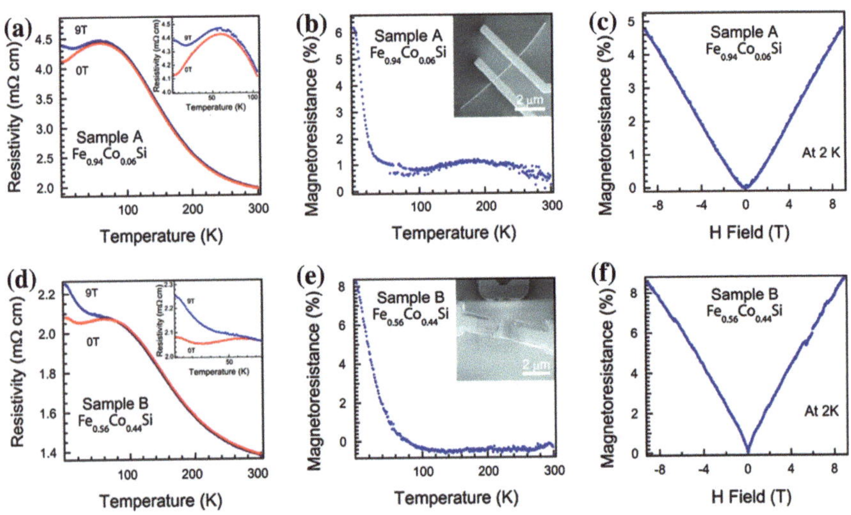

Fig. 12.43 Magnetotransport measurements of representative $Fe_{1\text{-}x}Co_xSi$ alloy nanowire devices for sample A ($x = 0.06$) and sample B ($x = 0.44$). **a, d** Resistivity as a function of temperature with (*blue curve*) and without (*red curve*) applied magnetic field of 9 T for samples A and B, respectively. Insets show the region of interest where the traces split. **b, e** MR calculated from the traces in a and d for samples A and B, respectively. *Insets* show the SEM images of the single nanowire devices measured. **c, f** Magnetic field dependence of the magnetoresistance for samples A and B at 2 K. Reprinted with permission from Ref. [109]. Copyright (2009) American Chemical Society

exhibit different shapes in sample A and sample B, indicating fundamental difference between the physical behaviors of these samples that stems from the different levels of Co incorporation.

12.3.4 Other Properties

Optical properties of CuSi nanorod arrays with different Cu concentrations (C_{Cu}) were measured using linear polarized incident light normal to the substrate [93]. In the polarized optical transmission spectra shown in Fig. 12.44a, b, when the photon energy is lower than E_g of Si, i.e., $\lambda > 1117$ nm, we can see that transmittance decreases with increase of C_{Cu} in both p- and s-polarization cases. However, the situation in the region of $\lambda > 1117$ nm is little complicated. The curves of transmittance as a function of C_{Cu} at two typical wavelength of 500 and 1500 nm are shown in Fig. 12.44c. In the case of 500 nm, T_{500} almost decreases with increase of C_{cu} with some exceptions due to the defect effect. While in the case of 1500 nm, T_{1500} monotonically decreased with increase of C_{Cu}. Furthermore, the transmittance in p-polarization is always bigger than that in s-polarization. This is caused by the different project width resulted from the sizes and alignment of the nanorods. Figure 12.44d, e show the angular-dependent polarized optical transmissions at 500 and 1500 nm from p-polarization direction. Optical anisotropy can be seen from the two-fold symmetry of the curves. A parameter called eccentricity (e) was introduced to characterize discrepancy in the degree of anisotropy. Where e is defined as $e = \sqrt{1 - \left(T_\lambda^p / T_\lambda^s\right)^2}$, $e = 0$ means optical isotropy. The curves of e as a function of C_{Cu} with $\lambda = 500$ and 1500 nm are shown in Fig. 12.44f. Different optical anisotropic behaviors were observed in 500 and 1500 nm conditions. In the case of 500 nm, the e values are always near 1. However, in the case of 1500 nm, the e values increase dramatically with increase of C_{Cu}, indicating the significant effect of Cu on the anisotropic response to the NIR light.

$CrSi_2$ is a semiconductor with an indirect band gap of 0.35 eV and a hexagonal C40 crystal structure. The special structure leads to anisotropic thermoelectric properties along different directions. Thermoelectric properties of individual single-crystal $CrSi_2$ nanowire were characterized using a suspended microdevice [109]. Figure 12.45a shows the measured Seebeck coefficient (S) from $CrSi_2$ nanowires with different diameters and previously reported S values of bulk $CrSi_2$ crystals. Along different crystal directions, the values of S are different. The measured S of $CrSi_2$ nanowires is comparable to that of bulk $CrSi_2$ crystal along the $\langle 0001 \rangle$ direction, which is in accordance with the TEM observation. Fermi energy (E_F) was calculated from S using both single band model and two band model. The curves of calculated E_F as a function of T (Fig. 12.45b) reveal that the $CrSi_2$ nanowires are degenerate p-type semiconductors. Hole concentration (p) in the semiconductor can be further calculated from the E_F. Fig. 12.45(c) shows the curves of p normalized

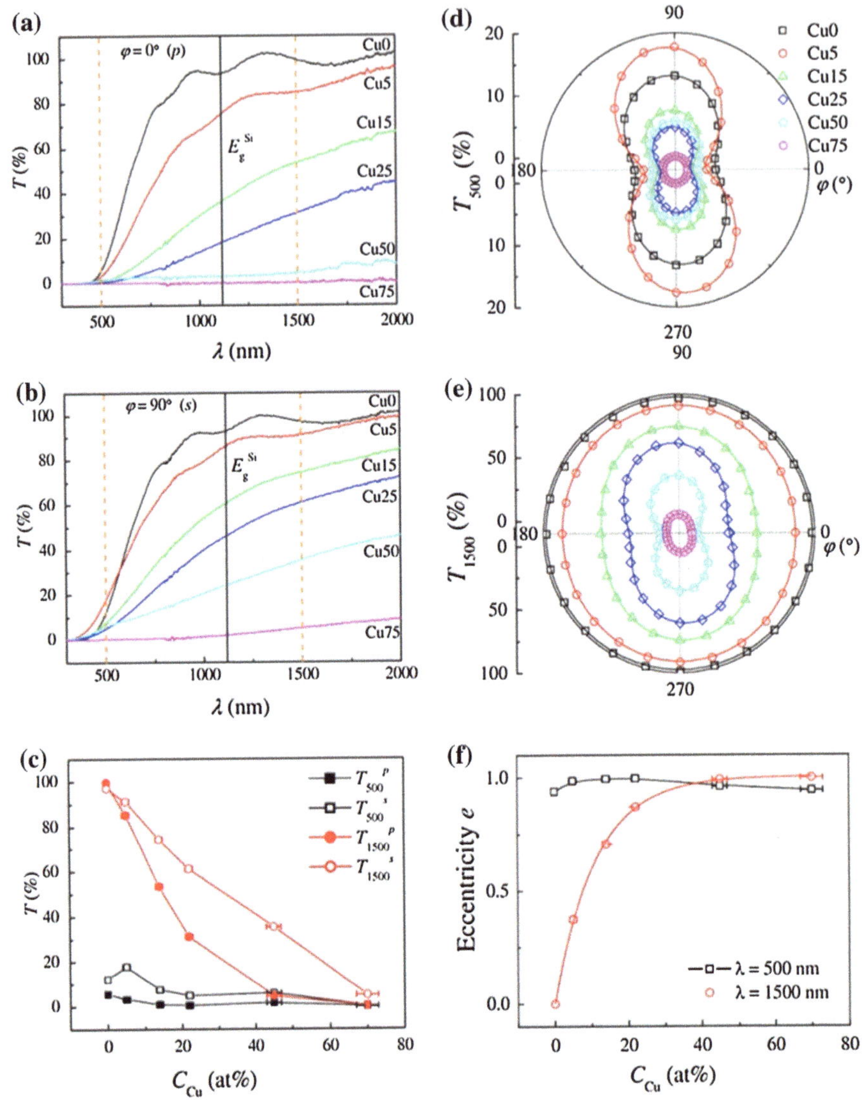

Fig. 12.44 Typical polarized optical transmission spectra for p-polarization at $\varphi = 0°$ (**a**) and s-polarization at $\varphi = 90°$ (**b**); **c** the p- and s-polarized optical transmission as functions of C_{Cu}; the incident light polarization angle dependent optical transmission T at $\lambda = 500$ nm, T_{500}—φ, in **d**, and at $\lambda = 1500$ nm, T_{1500}—φ, in (**e**); **f** The obtained eccentricity *e* versus C_{Cu} for both $\lambda = 500$ and 1500 nm

by *p* at 300 K as a function of 1000/T. The approximately linear curves indicate a thermally activated conduction process with an estimated thermal activation energy of 0.013 eV.

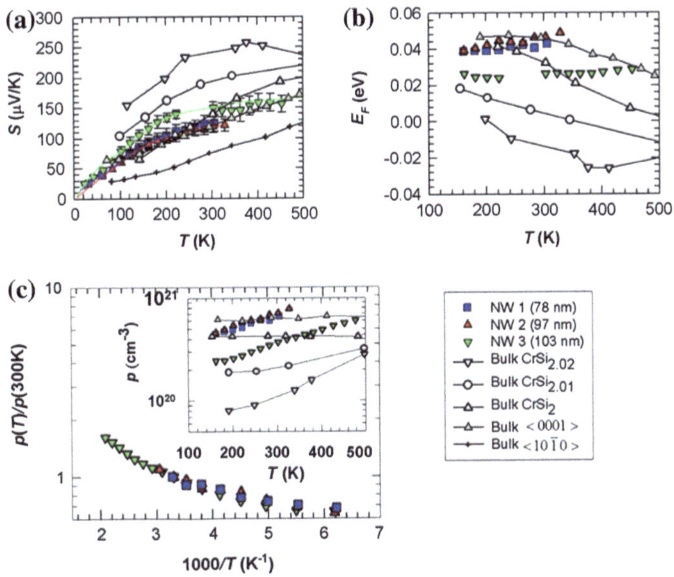

Fig. 12.45 **a** Seebeck coefficient (*S*) and **b** Fermi energy (E_F) as a function of temperature (T). **c** Hole concentration (*p*) normalized by *p* at 300 K as a function of 1000/T. *Inset*: p as a function of T. Reprinted with permission from Ref. [109]. Copyright (2007) American Chemical Society

Mechanical properties of $TiSi_2$ nanowires were investigated based on AFM bending test [110]. The average Young modulus of single $TiSi_2$ nanowire is calculated to be 194.5 GPa. Fracture experiments were done to nanowires with different diameters (D) and lengths (L). The fracture strengths of 9.55 GPa (D = 92 nm, L = 1444 nm), 8.81 GPa (D = 95 nm, L = 1467 nm) and 8.28 GPa (D = 100 nm, L = 1535 nm) were obtained. The fracture points locate either at the midpoints or fixed ends of the nanowires. No plastic deformation is observed from the cross section images of fractured nanowires, indicating the fracture is brittle. In addition, it was found that the nanowire fracture is easier to occur by using the lateral push method.

12.4 Applications of 1D TMS Nanostructures

12.4.1 Electrical Contact

Due to their excellent electrical conductivity, high breakdown current density and good compatibility with Si, TMS have been widely used in electronic devices as electrical contact material. 1D TMS nanostructures were also used as electrical contact in nanodevices. A NiSi/p-Si/NiSi heterojunction field effect transistor (FET) was fabricated by the silicidation process introduced above [60]. NiSi nanowire with

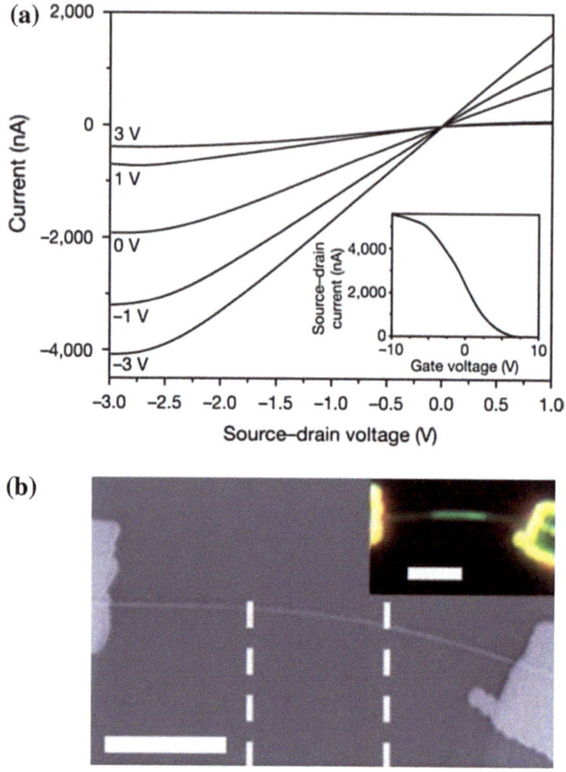

Fig. 12.46 **a** I versus V_{sd} curves recorded from the NiSi/p-Si/NiSi heterostructure FET. **b** SEM image of the FET device. *Inset* is the dark-field optical image of the device, where the *bright green* segment corresponds to silicon and the *dark segments* to NiSi. Reprinted by permission from Ref. [60]. Copyright (2004) Macmillan Publishers Ltd: (Nature)

low resistivity acted as source and drain contacting to Si nanowire. Figure 12.46a shows the current (I) as a function of source-drain voltage (V_{sd}) recorded from the NiSi/p-Si/NiSi heterojunction FET device shown in Fig. 12.46b. The I–V_{sd} curves are linear when $|V_{sd}| \leq 1V$, suggesting the ohmic contact at room temperature. I–V_{sd} data recorded at different back gate voltages (V_g) exhibit the behavior expected of a depletion mode p-FET with a high hole mobility of 325 cm^2 V_s.

Similar FET devices were reported by Weber et al. [65, 111]. In a Schottky barrier (SB) FET with undoped Si nanowire as the channel, nickel silicide ($NiSi_x$) was exploited as the source and drain (S/D) contacts [111]. Figure 12.47 shows the SEM image of the SBFET device. $NiSi_x$ was formed by silicidation of Si nanowires as introduced above. In an SBFET device with gate length $L_g = 1$ μm and Si nanowire diameter of 25 nm, a high on-current (I_{on}) of 0.7 μA at 1 V drain-source bias (V_{ds}) was obtained. The low off-current (I_{off}) allows a high I_{on}/I_{off} ratio up to 10^7. All the undoped Si nanowire based devices exhibit clear p-type behavior, which can be

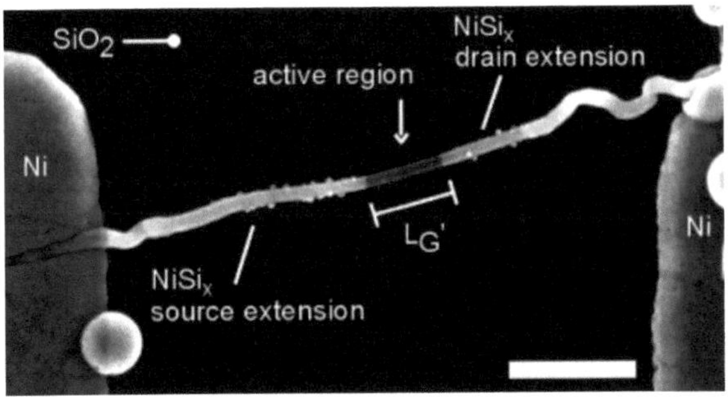

Fig. 12.47 SEM image of the SBFET device. The scalebar is 200 nm. Reprinted with permission from Ref. [111]. Copyright (2006) American Chemical Society

explained by the different transmission of electrons and holes through the Schottky barriers. In another FET device with Ni_2Si nanowire contact, a much higher I_{on} of 4.5 μA at 1V V_{ds} was achieved [65]. Hysteresis was observed in the I–V_g curve. The hysteresis was attributed to the charge traps located at nanowire surface or at the interface between the gate oxide and the nanowires. Using the same silicidation process, a PtSi/Si/PtSi heterostructure FET device was fabricated [66]. The device shows nearly perfect p-channel enhancement mode transistor behavior with a normalized transconductance of 0.3 mS/μm, field-effect hole mobility of 168 $cm^2/V \cdot s$, and on/off ratio $>10^7$.

Wang et al. used NiSi nanowires to build on-chip interconnects and repeaters [112]. It was found that NiSi nanowires offer better interconnect performances than commonly used Cu and another potential interconnect candidate, single-walled carbon nanotubes (CNT). At the same time, NiSi nanowires can integrate both active devices and high performance interconnect in a single nanoscale building blocks. Furthermore, NiSi interconnect does not suffer from the problem of electron-migration that plagues copper-based wires. In the comparison experiments, the NiSi nanowire interconnect/repeater can provide higher speed with similar power consumption than the structures based on Cu and CNTs. Although high performance can be achieved by using TMS as electrical contact, the degradation due to metal-dopant compound formation should be taken into account [113].

12.4.2 Lithium Ion Battery

1D TMS nanostructures were used in lithium ion battery aiming to achieve high performance. Some TMS can react with Li^+, thus can be used as anode materials. While most TMS are inactive for lithiation, they can be used as structural supporter and electron collector due to their unique morphology and good electrical conductivity.

Anode Materials

Silicide has attracted special interest as anode materials for high energy density lithium ion battery (LIB) due to the high capacity and good cyclability. Ni_3Si nanobelts and Ni silicides (Ni_3Si and $Ni_{31}Si_{12}$) nanosheets were first used as working electrode in a standard Li-ion half-cell [15]. Both nanobelts and nanosheets exhibit good cyclability. A high capacity of about $540\,mA\,h\,g^{-1}$ was obtained in the nanosheets, which value is better than the current used graphite anode ($330\,mA\,h\,g^{-1}$). The good performance was attributed to the large surface areas, small thickness and proper phase composition. Ni_2Si nanowire arrays were used as anode in LIB [114]. Although good cyclability was achieved in the anodes, the capacity ($362\,mA\,h\,g^{-1}$) is lower that of previous results in Ni silicides nanobelts and nanosheets.

Bulk Ti silicides are inactive when used as electrode materials for LIB. However, $TiSi_2$ nanonet was reported to have considerable activities in the reversible lithiation and delithiation processes [115]. EPS data in Fig. 12.48a clearly show lithiation of

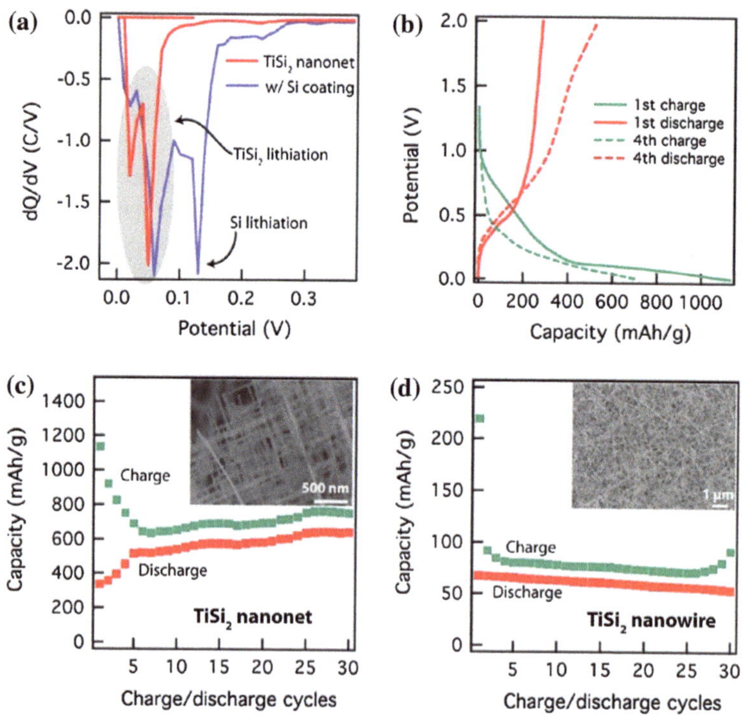

Fig. 12.48 **a** EPS data from $TiSi_2$ nanonets and $TiSi_2/Si$ heteronanostructures. **b** The charge/discharge behaviors of $TiSi_2$ nanonets. **c** Cyclability of $TiSi_2$ nanonets. **d** Cyclability of $TiSi_2$ nanowires have similar sizes to but have different crystal structures from $TiSi_2$ nanonets. Reprinted with permission from Ref. [115]. Copyright (2010) American Chemical Society

$TiSi_2$ with reaction peaks between 0 and 0.09 V. The activity of $TiSi_2$ nanonets was further proved by the high charge and discharge capacity, as shown in Fig. 12.48b, c. This unexpected phenomenon was ascribed to the unique crystal structure of $TiSi_2$ nanonet. In the C49 $TiSi_2$ crystal, a Si-only layer covers on the surface. This active Si layer serves as host for the Li^+ insertion. The explanation of the small size effect could be ruled out by a comparison experiment on $TiSi_2$ nanowires with similar size but different crystal structures to $TiSi_2$ nanonets, where the charge capacity is lower than 100 mA h g^{-1} (Fig. 12.48d). The $TiSi_2$ nanonet lithiation mechanism was revealed by the electrochemical impedance spectrum (EIS) method as well as TEM observation of the nanonet's surface morphology during the lithiation and delithiation processes. When the applied potential decreases to lithiation point ($<$0.09 V), reaction between Li^+ and $TiSi_2$ leads to an Si-rich amorphous layer on the surface. This amorphous layer will remain after the delithiation process.

The above example of $TiSi_2$ nanonet aroused two issues. First, the degradation of $TiSi_2$ structure by the formation of amorphous layer on the surface. Second, whether Li^+ was incorporated into the body of $TiSi_2$ or only on the surface? To, address these issues, a thin SiO_2 layer was introduced to the surface of $TiSi_2$ nanonet [116]. The charge-discharge experiments reveals that the oxide coated $TiSi_2$ nanonets have high specific capacities close to 800 mA h g^{-1} and excellent cyclability. More 80 % of the original capacity was remained after 500 cycles. The oxide layer was found to be critical for the high performance. HRTEM investigations reveal that the lattice spacing of $TiSi_2$ {020} planes increase from 0.667 to 0.675 nm, corresponding to a 1.2 % change. No measurable differences were observed in other crystal planes. As the crystal structure of C49 $TiSi_2$ shown in Fig. 12.49a, b, the Si-only layers reside in the {020} planes. Since Ti is inactive to Li^+, the change in lattice spacing should be originated from the Li^+ inserting into the $TiSi_2$ crystal and reacting with the Si-only layers, as shown in Fig. 12.49c.

Functional Components

Actually, most silicides are inactive for lithiation and cannot be directly used as electrode materials. In addition to acting as anode materials, 1D TMS nanostructures also played important roles in LIB with special functions. For example, a high power LIB with $NiSi_x$/a-Si core/shell nanowire anodes was reported by Kang et al. [117]. The metallic $NiSi_x$ core acts as a mechanical supporter and a kinetically unlimited charge supplier. A highly reversible capacitance of over 3000 mA h g^{-1} with stable cyclic retention was achieved even at a 2C rate. The electrochemical roles of $NiSi_x$ core were investigated by comparing the differential capacity plots of $NiSi_x$/a-Si nanowires with that of c-Si/a-Si nanowires. In the charge process, both nanowires exhibit two reduction peaks at 260 and 80 mV, which are associated with the formation of intermetallic compounds Li_xSi. The discharge processes are quite different for these two kinds of nanowires. In the case of c-Si/a-Si nanowires, a dominated peak located at 450 mV was observed. In contrast, there are two dissimilar peaks at 250 and 500 mV for the $NiSi_x$/a-Si nanowires. From these results, it can be deduced

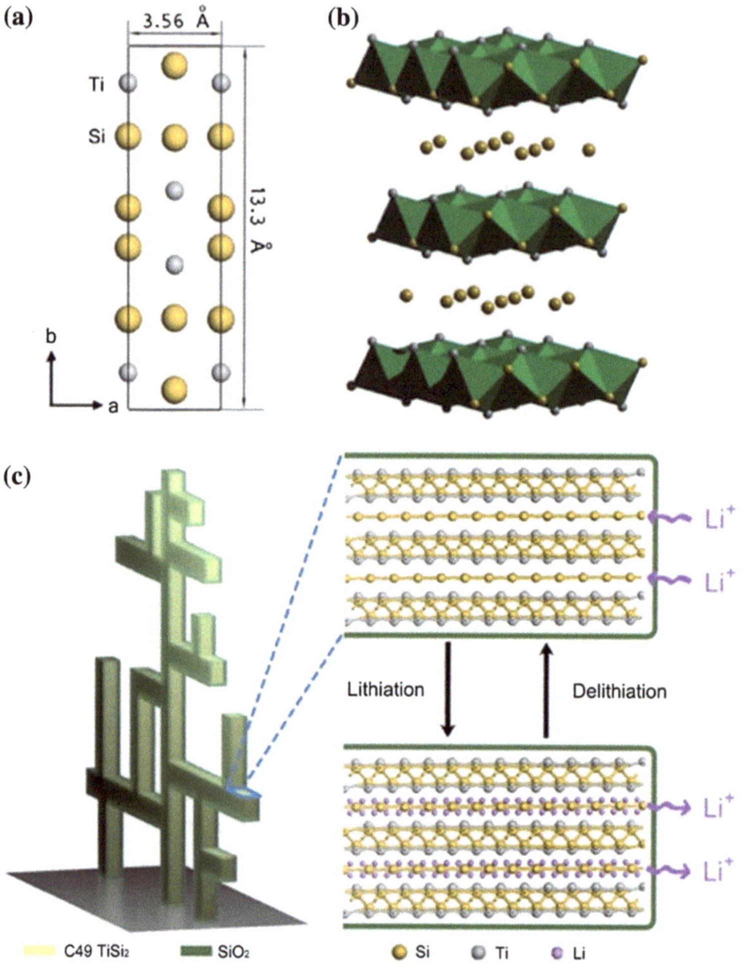

Fig. 12.49 Schematic illustration of the C49 $TiSi_2$ crystal structure. **a** *Side view* of the unit cell along the [001] direction. **b** Perspective view of a polyhedron representation of the structure. **c** Oxide-coated C49 $TiSi_2$ is stabilized by the nanonet morphology. The cartoons on the *right* illustrate the lithiation and delithiation processes. Reprinted with permission from Ref. [116]. Copyright (2012) American Chemical Society

that full lithiation into a-$Li_{x+y}Si$ in $NiSi_x$/a-Si nanowires, instead of c-Li_zSi in the c-Si/a-Si nanowires. Thus, the stable reversibility in the cyclic cell performance due to alleviated physical pulverization can be achieved. Furthermore, the metallic $NiSi_x$ cores can provide uniformly distributed electrochemical potential in the nanowires, leading to the reversible reaction route mediated by fully lithiated amorphous lithium silicides. $NiSi_x$ was also used as structural supports and current collector in a LIB anode made of $NiSi_x$/CoO core/shell nanowire arrays [118]. Higher cycle life and enhanced power performance were achieved comparing to planar CoO electrode.

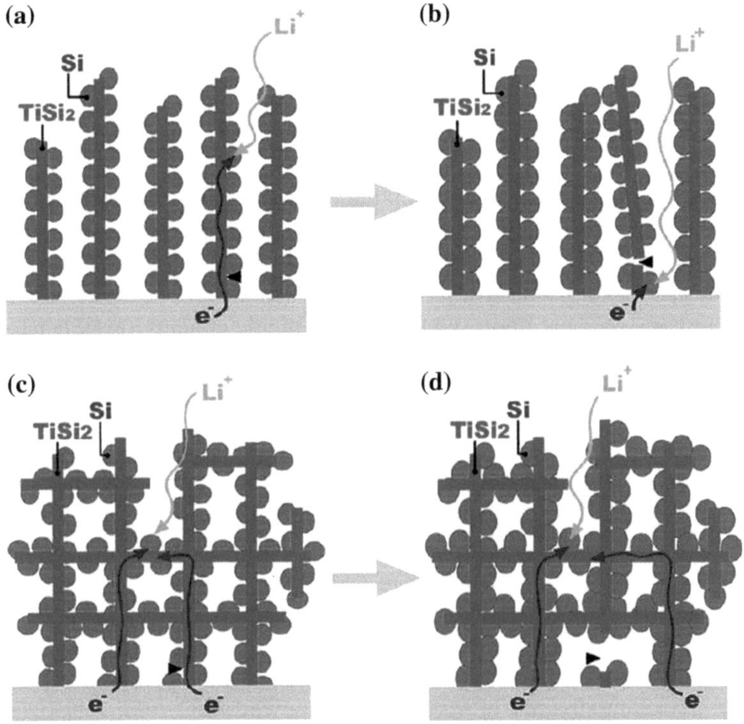

Fig. 12.50 Schematic comparison of the nanowires- and nanonets-based systems. Reprinted with permission from Ref. [119]. Copyright (2011) American Chemical Society

The improvement in performance was ascribed to the high surface-to-volume ratio and improved electronic/ionic conductivity of the nanostructured electrode.

Owing to their large surface areas, good electrical conductivity and superb mechanical strength, $TiSi_2$ nanowires and nanonets were used as structural supporter and charge transporter in LIB [119, 120]. The performances of 1D nanowires and 2D nanonets with the same composition ($TiSi_2$) and similar sizes were compared [119]. Both $TiSi_2$ nanostructures were coated with a layer of Si, which can react with Li^+. Figure 12.50 illustrates the schematic diagram of electrodes made of the 1D and 2D $TiSi_2$ nanostructures. The blue lines show the electron transport paths and the red lines show the Li^+ pathways. It can be seen that the 2D nanonets take advantages in electron transport, especially in the case of partially structure breaking (indicated with black arrows in Fig. 12.50), which is inevitable after some lithiation and delithiation processes. Therefore, the electrode based on 2D $TiSi_2$ nanonets exhibited better stability than that of the 1D nanowires.

12.4.3 Other Applications

The functions of structural supporter and charge transporter of $TiSi_2$ nanonets were also utilized in photoelectrochemical water splitting [121]. The highly conductive $TiSi_2$ nanonets were coated with photoactive TiO_2. In contact with electrolyte upon illumination, charge separate happens at the interface. Holes are transferred to the electrolyte for chemical reactions and electrons are quickly transported to the external circuit through the $TiSi_2$ nanonet pathways. Introducing $TiSi_2$ nanonets to the photoelectrode can maximize the TiO_2/electrolyte junction area. At the same time, the charge transport is greatly improved through the highly conductive $TiSi_2$ nanonets. Apart from the high conductivity, $TiSi_2$ was found to have good visible light activity for water splitting reaction [122]. A $TiSi_2$nanorod/TiO_2 nanotube core/shell structure was fabricated by impregnating $TiSi_2$ nanoparticles into TiO_2 nanotubes. As shown in Fig. 12.51, the $TiSi_2$–TiO_2 heterostructural photoanode exhibited much better photoelectrochemical performance than the pure TiO_2 nanotube photoanode. At the potential of 0.2 $V_{Ag/AgCl}$, a high photocurrent density of 3.49 mA/cm^2 was observed at the $TiSi_2$–TiO_2 electrode, which value is only 0.9 mA/cm^2 at the TiO_2 nanotube electrode. The enhancement in performance should be related with the enhanced absorption in the visible region as well as photoactivity of $TiSi_2$, which were proved by the absorption spectra and potentiodynamic measurements.

A nonvolatile memory device was fabricated using FeSi nanowires [123]. The device has a stacking structure of Al/SiO_2/FeSi nanowires/SiO_2/p-Si-sub/Al, as illustrated in Fig. 12.52a. Obvious hysteresis behavior can be observed in the capacitance as a function of voltage (C-V) curves (Fig. 12.52b). The hysteresis behavior indicates charge and discharge behavior of the $Fesi_2$ nanowires. The double C-V sweeps

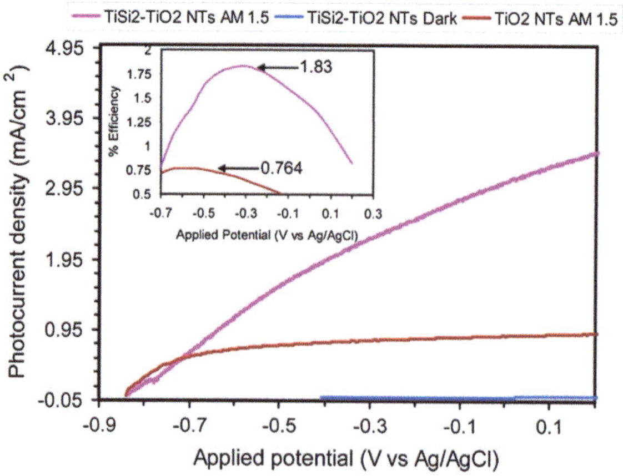

Fig. 12.51 I–V curves recorded from the $TiSi_2$–TiO_2 heterostructures and TiO_2 nanotubes. Inset is the efficiency curves. Reprinted with permission from Ref. [122]. Copyright (2011) American Chemical Society

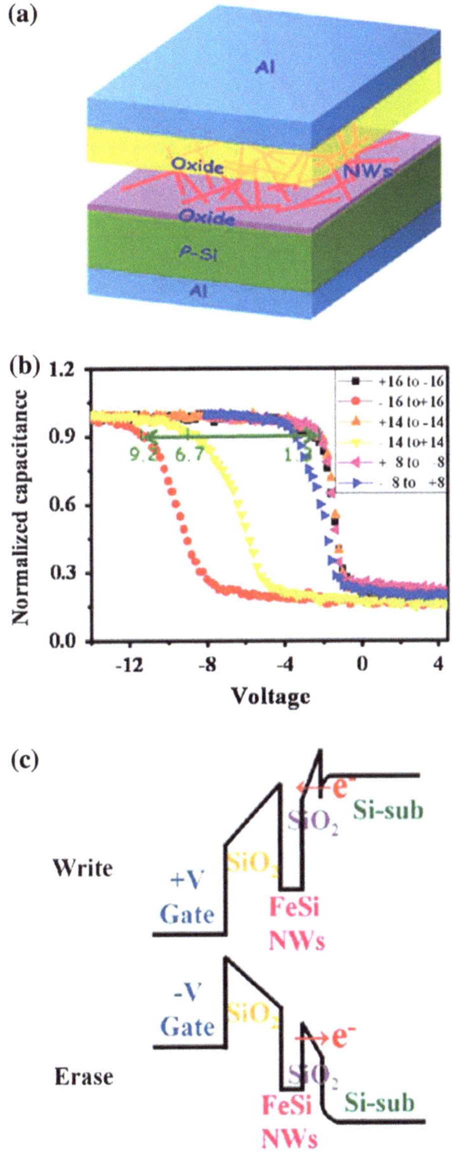

Fig. 12.52 **a** Schematic diagram of the stacking memory structure. **b** The C–V hysteresis curves of FeSi nanowire memory devices. **c** The band diagram of write and erase operations of the memory device. Reprinted with permission from Ref. [123]. Copyright (2011) American Chemical Society

exhibit a strong electron charging effect. A high memory window of 1.3 V was obtained when voltage swept from 8 to −8 V and back to 8 V. More significant C-V shifts of 6.7 and 9.2 V were observed when the swept gate voltages were increased to 14 and 16 V. The strong memory effect was attributed to the defects or dangling bonds on the surface of FeSi nanowires, which would enhance the trapping density in the devices. Figure 12.52c illustrates the band diagram of write and erase operations

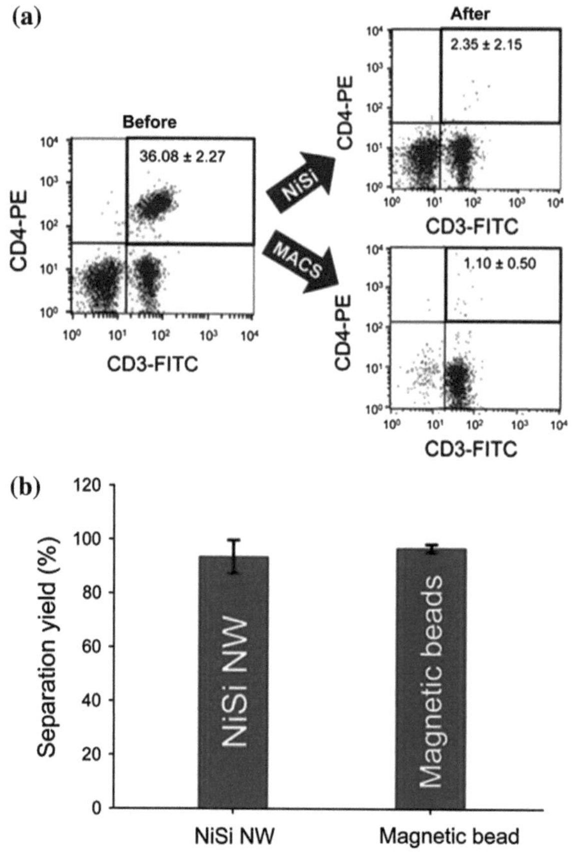

Samples	Separation efficiency (%)
NiSi NW	93.5 ± 6.2
Magnetic bead	96.8 ± 1.5

Fig. 12.53 **a** Flow cytometric analysis of CD4$^+$T lymphocytes using cell suspensions after binding to STR-conjugated NiSi nanowires (*top right panel*) and superparamagnetic beads (*bottom right panel*) before and after magnetic cell separation. **b** Average cell separation efficiency NiSi nanowires and commercial superparamagnetic beads. Reprinted with permission from Ref. [124]. Copyright (2012) American Institute of Physics

in the device. In the write operation, electrons tunneled through the oxide layer and trapped by FeSi. In the erase process, the trapped electron may tunnel back to the accumulation layer of the Si substrate.

The ferromagnetic NiSi nanowires were successfully used in cell separation with high separation efficiency [124]. The polycrystalline NiSi nanowires were prepared by silicidation of the chemical etched Si nanowire arrays. The NiSi nanowires were modified with a monoclonal anti-mouse CD4-antibody using a streptavidin (STR)-biotin immobilization method. The STR-conjugated NiSi nanowires were then placed in a suspension containing biotin-conjugated CD4 anti-mouse antibodies and mouse splenocytes. Cell separation was performed by placing the suspension under a magnetic field for 3–10 min. The separation result of NiSi nanowires was compared with that of superparamagnetic beads, as shown in Fig. 12.53. It was observed that only 2.35%$CD3^+/CD4^+$ remains after separation by NiSi nanowires. The average capture efficiency of $CD4^+$T cells was calculated to be about 93.5 ± 6.2 %, a little higher than that of the commercial superparamagnetic beads (96.8 ± 1%). Moreover, the STR-functionalized NiSi nanowires exhibited a high cell viability.

Furthermore, the Ti_5Si_3 nanowires exhibited electrochemical capacitance behavior [37]. The oxide layer on surface of Ti_5Si_3 nanowires was found to be disadvantageous for the capacitance properties. When the oxide layer was removed by etching in HF solution for 180 s, a rectangular shape was achieved in the voltammetric curve, implying an ideal capacitance behavior at a high scan rate of 50 mV s^{-1}.

12.5 Conclusions

Considering their unique properties and potential applications, various methods have developed to fabricate 1D nanostructures of TMS, including CVD, CVT, reactive epitaxial method, hydrothermal method, solid-state silicidation, etc. The as-prepared 1D TMS nanostructures exhibited fascinating physical properties, such as good electrical conductivity, semiconducting behavior, room temperature ferromagnetism, and thermoelectricity. Employing these properties, 1D TMS nanostructures have realized applications in microelectronic devices as ohmic contacts and gate materials, LIB as anode materials or structural supporter and electron collector, nonvolatile memory devices, capacitor and cell separation. Although many TMS nanowires have been successfully prepared, it is necessary to further synthesize more 1D TMS nanostructures. Growth mechanisms of the 1D nanostructures are not fully understood, which greatly limits the controllability of the nanostructures. Therefore, in-depth investigation of the growth mechanism of the 1D nanostructures is another important issue. Compared with the progress in materials preparation, the study in properties and applications is not enough. The electrical, magnetic, optical, thermoelectric properties and applications in nanoelectronics, spintronics, energy fields (solar cell, LIB, capacitor, water splitting, etc.) remain to be further exploited.

Acknowledgments The authors would like to acknowledge financial support from the Chinese Academy of Sciences, NSFC (grant nos. 21103211, 50902134, 51272258, 51272302, and 61025003), National Basic Research Program of China (973 Program) (grant nos. 2010CB934103 and 2012CB932400) and Beijing Natural Science Foundation (grant no. 2102043).

References

1. Dimitriadis, C.A., Werner, J.H., Logothetidis, S., Stutzmann, M., Weber, J., Nesper, R.: Electronic properties of semiconducting $FeSi_2$ films. J. Appl. Phys. **68**, 1726–1734 (1990)
2. Starke, U., Weiss, W., Kutschera, M., Bandorf, R., Heinz, K.: High quality iron silicide films by simultaneous deposition of iron and silicon on Si(111). J. Appl. Phys. **91**, 6154–6161 (2002)
3. Ouyang, L., Thrall, E.S., Deshmukh, M.M., Park, H.: vapor-phase synthesis and characterization of ε-FeSi nanowires. Adv. Mater. **18**, 1437–1440 (2006)
4. Lin, H.K., Tzeng, Y.F., Wang, C.H., Tai, N.H., Lin, I.N., Lee, C.Y., Chiu, H.T.: Ti_5Si_3 nanowire and its field emission property. Chem. Mater. **20**, 2429–2431 (2008)
5. Lin, H.K., Cheng, H.A., Lee, C.Y., Chiu, H.T.: Chemical vapor deposition of TiSi nanowires on C54 TiSi2 thin film: an amorphous titanium silicide interlayer assisted nanowire growth. Chem. Mater. **21**, 5388–5396 (2009)
6. In, J., Seo, K., Lee, S., Yoon, H., Park, J., Lee, G., Kim, B.: Morphology-tuned synthesis of single-crystalline V_5Si_3 nanotubes and nanowires. J. Phys. Chem. C **113**, 12996–13001 (2009)
7. Higgins, J.M., Ding, R., DeGrave, J.P., Jin, S.: Signature of helimagnetic ordering in single-crystal MnSi nanowires. Nano Lett. **10**, 1605–1610 (2010)
8. Liang, S., Fang, X., Xia, T.L., Qing, Y., Guo, Z.X.: Self-assembled magnetic nanohead-FeSi nanowire epitaxial heterojunctions by chemical vapor deposition. J. Phys. Chem. C **114**, 16187–16190 (2010)
9. Decker, C.A., Solanki, R., Freeouf, J.L., Carruthers, J.R., Evans, D.R.: Directed growth of nickel silicide nanowires. Appl. Phys. Lett. **84**, 1389–1391 (2004)
10. Dubois, L.H., Nuzzo, R.G.: The decomposition of silane and germane on Ni(111): implications for heterogeneous catalysis. Surf. Sci. **149**, 133–145 (1985)
11. Yan, X.Q., Yuan, H.J., Wang, J.X., Liu, D.F., Zhou, Z.P., Gao, Y., Song, L., Liu, L.F., Zhou, W.Y., Wang, G., Xie, S.S.: Synthesis and characterization of a large amount of branched N_2Si nanowires. Appl. Phys. A Mater. Sci. Process. **79**, 1853–1856 (2004)
12. Lee, K.S., Mo, Y.H., Nahm, K.S., Shim, H.W., Suh, E.K., Kim, J.R., Kim, J.J.: Anomalous growth and characterization of carbon-coated nickel silicide nanowires. Chem. Phys. Lett. **384**, 215–218 (2004)
13. Kim, C.J., Kang, K., Woo, Y.S., Ryu, K.G., Moon, H., Kim, J.M., Zang, D.S., Jo, M.H.: Spontaneous chemical vapor growth of NiSi nanowires and their metallic properties. Adv. Mater. **19**, 3637–3642 (2007)
14. Kang, K., Kim, S.K., Kim, C.J., Jo, M.H.: The role of NiO_x overlayers on spontaneous growth of $NiSi_x$ nanowires from Ni seed layers. Nano Lett. **8**, 431–436 (2008)
15. Zhang, H.L., Li, F., Liu, C., Chen, H.M.: The facile synthesis of nickel silicide nanobelts and nanosheets and their application in electrochemical energy storage. Nanotechnology **19**, 165606 (2008)
16. Chueh, Y.L., Chou, L.J., Cheng, S.L., Chen, L.J., Tsai, C.J., Hsu, C.M., Kung, S.C.: Synthesis and characterization of metallic $TaSi_2$ nanowires. Appl. Phys. Lett. **87**, 223113–223113 (2005)
17. Chueh, Y.L., Ko, M.T., Chou, L.J., Chen, L.J., Wu, C.S., Chen, C.D.: $TaSi_2$ nanowires: a potential field emitter and interconnect. Nano Lett. **6**, 1637–1644 (2006)
18. Chou, L.J., Chueh, Y.L., Ko, M.T.: Interconnect and contact for nanoelectronics: metallic $TaSi_2$ nanowires. Thin Solid Films **515**, 8109–8112 (2007)
19. Tang, C., Bando, Y., Golberg, D., Ding, X., Qi, S.: Boron nitride nanotubes filled with Ni and NiSi2 nanowires in situ. J. Phys. Chem. B **107**, 6539–6543 (2003)
20. Xiang, B., Wang, Q.X., Wang, Z., Zhang, X.Z., Liu, L.Q., Xu, J., Yu, D.P.: Synthesis and field emission properties of $TiSi_2$ nanowires. Appl. Phys. Lett. **86**, 243103–243103 (2005)
21. Zou, C., Zhang, X., Jing, G., Zhang, J., Liao, Z., Yu, D.: Synthesis and electrical properties of $TiSi_2$ nanocables. Appl. Phys. Lett. **92**, 253102–253103 (2008)
22. Zhang, Z.Y., Wu, X.L., Yang, L.W., Huang, G.S., Siu, G.G., Chu, P.K.: Catalytic growth of α-$FeSi_2$ and silicon nanowires. J. Cryst. Growth **280**, 286–291 (2005)

23. Joshi, R.K., Yoshimura, M., Tanaka, K., Ueda, K., Kumar, A., Ramgir, N.: Synthesis of vertically aligned Pd_2Si nanowires in microwave plasma enhanced chemical vapor deposition system. J. Phys. Chem. C **112**, 13901–13904 (2008)
24. Chang, M.T., Chen, C.Y., Chou, L.J., Chen, L.J.: Core-shell chromium silicide-silicon nanopillars: a contact material for future nanosystems. ACS Nano **3**, 3776–3780 (2009)
25. Higgins, J.M., Ding, R., Jin, S.: Synthesis and characterization of manganese-rich silicide (α-Mn_5Si_3, β-Mn_5Si_3, and β-Mn_3Si) nanowires. Chem. Mater. **23**, 3848–3853 (2011)
26. Chang, C.M., Chang, Y.C., Lee, C.Y., Yeh, P.H., Lee, W.F., Chen, L.J.: Ti_5Si_4 nanobats with excellent field emission properties. J. Phys. Chem. C **113**, 9153–9156 (2009)
27. Chen, C.Y., Lin, Y.K., Hsu, C.W., Wang, C.Y., Chueh, Y.L., Chen, L.J., Lo, S.C., Chou, L.J.: Coaxial metal-silicide Ni_2Si/C54-$TiSi_2$ nanowires. Nano Lett. **12**, 2254–2259 (2012)
28. Du, J., Du, P., Hao, P., Huang, Y., Ren, Z., Han, G., Weng, W., Zhao, G.: Growth mechanism of TiSi nanopins on Ti_5Si_3 by atmospheric pressure chemical vapor deposition. J. Phys. Chem. C **111**, 10814–10817 (2007)
29. Jun, D., Piyi, D., Peng, H., Yanfei, H., Zhaodi, R., Wenjian, W., Gaorong, H., Gaoling, Z.: Self-induced preparation of TiSi nanopins by chemical vapor deposition. Nanotechnology **18**, 345605 (2007)
30. Du, J., Ren, Z., Tao, K., Hu, A., Hao, P., Huang, Y., Zhao, G., Weng, W., Han, G., Du, P.: Self-induced preparation of assembled shrubbery TiSi nanowires by chemical vapor deposition. Cryst. Growth Des. **8**, 3543–3548 (2008)
31. Ren, Z., Hu, A., Tao, K., Du, J., Weng, W., Ma, N., Du, P.: Preparation of TiSi nanowires by APCVD used for improving dielectric properties of PST thin film. Thin Solid Films **517**, 5014–5017 (2009)
32. Ren, Z., Hao, P., Du, J., Han, G., Weng, W., Ma, N., Du, P.: Self-assembly of TiSi nanowires on $TiSi_2$ thin films by APCVD. J. Alloy. Compd. **509**, 7519–7524 (2011)
33. Ren, Z., Shen, M., Han, G., Weng, W., Ma, N., Du, P.: Influence of substrates on the formation of the TiSi nanowire by atmosphere pressure chemical vapor deposition. J. Nanosci. Nanotechnol. **11**, 11151–11155 (2011)
34. Zhou, S., Liu, X., Lin, Y., Wang, D.: Spontaneous growth of highly conductive two-dimensional single-crystalline TiSi2 nanonets. Angew. Chem. Int. Ed. **47**, 7681–7684 (2008)
35. Zhou, S., Liu, X., Lin, Y., Wang, D.: Rational synthesis and structural characterizations of complex $TiSi_2$ nanostructures. Chem. Mater. **21**, 1023–1027 (2009)
36. Zhou, S., Xie, J., Wang, D.: Understanding the growth mechanism of titanium disilicide nanonets. ACS Nano **5**, 4205–4210 (2011)
37. Zhang, Y., Geng, D., Liu, H., Banis, M.N., Ionescu, M.I., Li, R., Cai, M., Sun, X.: Designed growth and characterization of radially aligned Ti_5Si_3 nanowire architectures. J. Phys. Chem. C **115**, 15885–15889 (2011)
38. Ham, M.H., Lee, J.W., Moon, K.J., Choi, J.H., Myoung, J.M.: Single-crystalline ferromagnetic Mn4Si7 nanowires. J. Phys. Chem. C **113**, 8143–8146 (2009)
39. Lee, S.T., Zhang, Y.F., Wang, N., Tang, Y.H., Bello, I., Lee, C.S., Chung, Y.W.: Semiconductor nanowires from oxides. J. Mater. Res. **14**, 4503–4507 (1999)
40. Higgins, J.M., Schmitt, A.L., Guzei, I.A., Jin, S.: Higher manganese silicide nanowires of nowotny chimney ladder phase. J. Am. Chem. Soc. **130**, 16086–16094 (2008)
41. Schmitt, A.L., Zhu, L., D. Schmeiβer, F.J. Himpsel, Jin. S.: Metallic single-crystal CoSi nanowires via chemical vapor deposition of single-source precursor. J. Phys. Chem. B **110**, 18142–18146 (2006)
42. Schmitt, A.L., Bierman, M.J., Schmeisser, D., Himpsel, F.J., Jin, S.: Synthesis and properties of single-crystal FeSi nanowires. Nano Lett. **6**, 1617–1621 (2006)
43. Schmitt, A.L., Higgins, J.M., Jin, S.: Chemical synthesis and magnetotransport of magnetic semiconducting $Fe_{1-x}Co_xSi$ alloy nanowires. Nano Lett. **8**, 810–815 (2008)
44. Schmitt, A.L., Jin, S.: Selective patterned growth of silicide nanowires without the use of metal catalysts. Chem. Mater. **19**, 126–128 (2006)
45. Higgins, J.M., Carmichael, P., Schmitt, A.L., Lee, S., Degrave, J.P., Jin, S.: Mechanistic investigation of the growth of $Fe_{1-x}Co_xSi$ ($0 \leq x \leq 1$) and $Fe_5(Si_{1-y}Ge_y)_3$ ($0 \leq y \leq 0.33$) ternary alloy nanowires. ACS Nano **5**, 3268–3277 (2011)

46. Song, Y., Schmitt, A.L., Jin, S.: Ultralong single-crystal metallic Ni2Si nanowires with low resistivity. Nano Lett. **7**, 965–969 (2007)
47. Szczech, J.R., Schmitt, A.L., Bierman, M.J., Jin, S.: Single-crystal semiconducting chromium disilicide nanowires synthesized via chemical vapor transport. Chem. Mater. **19**, 3238–3243 (2007)
48. Song, Y., Jin, S.: Synthesis and properties of single-crystal β_3-N_3Si nanowires. Appl. Phys. Lett. **90**, 173122–173123 (2007)
49. Szczech, J.R., Jin, S.: Epitaxially-hyperbranched FeSi nanowires exhibiting merohedral twinning. J. Mater. Chem. **20**, 1375–1382 (2010)
50. Seo, K., Varadwaj, K.S.K., Mohanty, P., Lee, S., Jo, Y., Jung, M.H., Kim, J., Kim, B.: Magnetic properties of single-crystalline CoSi nanowires. Nano Lett. **7**, 1240–1245 (2007)
51. Seo, K., Varadwaj, K.S.K., Cha, D., In, J., Kim, J., Park, J., Kim, B.: Synthesis and electrical properties of single crystalline $CrSi_2$ nanowires. J. Phys. Chem. C **111**, 9072–9076 (2007)
52. Varadwaj, K.S.K., Seo, K., In, J., Mohanty, P., Park, J., Kim, B.: Phase-controlled growth of metastable Fe_5Si_3 nanowires by a vapor transport method. J. Am. Chem. Soc. **129**, 8594–8599 (2007)
53. Seo, K., Yoon, H., Ryu, S.W., Lee, S., Jo, Y., Jung, M.H., Kim, J., Choi, Y.K., Kim, B.: Itinerant helimagnetic single-crystalline MnSi nanowires. ACS Nano **4**, 2569–2576 (2010)
54. Seo, K., Lee, S., Yoon, H., In, J., Varadwaj, K.S.K., Jo, Y., Jung, M.-H., Kim, J., Kim, B.: Composition-tuned Co_nSi nanowires: location-selective simultaneous growth along temperature gradient. ACS Nano **3**, 1145–1150 (2009)
55. Zhang, Z., Wong, L.M., Ong, H.G., Wang, X.J., Wang, J.L., Wang, S.J., Chen, H., Wu, T.: Self-assembled shape- and orientation-controlled synthesis of nanoscale Cu_3Si triangles, squares, and wires. Nano Lett. **8**, 3205–3210 (2008)
56. Lee, C.Y., Lu, M.P., Liao, K.F., Lee, W.F., Huang, C.T., Chen, S.Y., Chen, L.J.: Free-standing single-crystal $NiSi_2$ nanowires with excellent electrical transport and field emission properties. J. Phys. Chem. C **113**, 2286–2289 (2009)
57. Wang, H., Wu, J.C., Shen, Y., Li, G., Zhang, Z., Xing, G., Guo, D., Wang, D., Dong, Z., Wu, T.: $CrSi_2$ hexagonal nanowebs. J. Am. Chem. Soc. **132**, 15875–15877 (2010)
58. In, J., Varadwaj, K.S.K., Seo, K., Lee, S., Jo, Y., Jung, M.H., Kim, J., Kim, B.: Single-crystalline ferromagnetic $Fe_{1-x}Co_xSi$ nanowires. J. Phys. Chem. C **112**, 4748–4752 (2008)
59. Li, C.P., Wang, N., Wong, S.P., Lee, C.S., Lee, S.T.: Metal silicide/silicon nanowires from metal vapor vacuum arc implantation. Adv. Mater. **14**, 218–221 (2002)
60. Wu, Y., Xiang, J., Yang, C., Lu, W., Lieber, C.M.: Single-crystal metallic nanowires and metal/semiconductor nanowire heterostructures. Nature **430**, 61–65 (2004)
61. Liu, B., Wang, Y., Dilts, S., Mayer, T.S., Mohney, S.E.: Silicidation of silicon nanowires by platinum. Nano Lett. **7**, 818–824 (2007)
62. Lu, K.C., Tu, K.N., Wu, W.W., Chen, L.J., Yoo, B.Y., Myung, N.V.: Point contact reactions between Ni and Si nanowires and reactive epitaxial growth of axial nano-NiSi/Si. Appl. Phys. Lett. **90**, 253111–253113 (2007)
63. Lu, K.C., Wu, W.-W., Wu, H.-W., Tanner, C.M., Chang, J.P., Chen, L.J., Tu, K.N.: In situ Control of atomic-scale si layer with huge strain in the nanoheterostructure NiSi/Si/NiSi through point contact reaction. Nano Lett. **7**, 2389–2394 (2007)
64. Chou, Y.C., Wu, W.W., Cheng, S.L., Yoo, B.Y., Myung, N., Chen, L.J., Tu, K.N.: In-situ TEM observation of repeating events of nucleation in epitaxial growth of nano $CoSi_2$ in nanowires of Si. Nano Lett. **8**, 2194–2199 (2008)
65. Weber, W.M., Geelhaar, L., Unger, E., Chèze, C., Kreupl, F., Riechert, H., Lugli, P.: Silicon to nickel-silicide axial nanowire heterostructures for high performance electronics. Phys. Status Solidi (b) **244** (2007) 4170–4175
66. Lin, Y.C., Lu, K.C., Wu, W.W., Bai, J., Chen, L.J., Tu, K.N., Huang, Y.: Single crystalline PtSi nanowires, PtSi/Si/PtSi nanowire heterostructures, and nanodevices. Nano Lett. **8**, 913–918 (2008)
67. Lin, Y.C., Chen, Y., Xu, D., Huang, Y.: Growth of nickel silicides in Si and Si/SiO_x core/shell nanowires. Nano Lett. **10**, 4721–4726 (2010)

68. Liu, H.L., She, G.W., Ling, S.T., Mu, L.X., Shi, W.S.: Ferromagnetic Si/$Mn_{27}Si_{47}$core/shell nanowire arrays. J. Appl. Phys. **109**, 4 (2011)
69. Liu, H., She, G., Mu, L., Shi, W.: Temperature-dependent structure and phase variation of nickel silicide nanowire arrays prepared by in situ silicidation. Mater. Res. Bull. **47**, 3991–3994 (2012)
70. Chen, Y., Ohlberg, D.A.A., Medeiros-Ribeiro, G., Chang, Y.A., Williams, R.S.: Self-assembled growth of epitaxial erbium disilicide nanowires on silicon (001). Appl. Phys. Lett. **76**, 4004–4006 (2000)
71. Chen, Y., Ohlberg, D.A.A., Williams, R.S.: Nanowires of four epitaxial hexagonal silicides grown on Si(001). J. Appl. Phys. **91**, 3213–3218 (2002)
72. Zou, Z.Q., Wang, H., Wang, D., Wang, Q.K., Mao, J.J., Kong, X.Y.: Epitaxial growth of manganese silicide nanowires on Si(111)-7 x 7 surfaces. Appl. Phys. Lett. **90**, 133111–133113 (2007)
73. Dan, W., Zhi-Qiang, Z.: Formation of manganese silicide nanowires on Si(111) surfaces by the reactive epitaxy method. Nanotechnology **20**, 275607 (2009)
74. Zou, Z.Q., Li, W.C., Liang, J.M., Wang, D.: Self-organized growth of higher manganese silicide nanowires on Si(111), (110) and (001) surfaces. Acta Mater. **59**, 7473–7479 (2011)
75. Bennett, P.A., He, Z., Smith, D.J., Ross, F.M.: Endotaxial silicide nanowires: a review. Thin Solid Films **519**, 8434–8440 (2011)
76. He, Z., Stevens, M., Smith, D.J., Bennett, P.A.: Epitaxial titanium silicide islands and nanowires. Surf. Sci. **524**, 148–156 (2003)
77. Stevens, M., He, Z., Smith, D.J., Bennett, P.A.: Structure and orientation of epitaxial titanium silicide nanowires determined by electron microdiffraction. J. Appl. Phys. **93**, 5670–5674 (2003)
78. He, Z., Smith, D.J., Bennett, P.A.: Endotaxial silicide nanowires. Phys. Rev. Lett. **93**, 256102 (2004)
79. Bennett, P.A., Ashcroft, B., He, Z., Tromp, R.M.: Growth dynamics of titanium silicide nanowires observed with low-energy electron microscopy, in AVS, Albuquerque, New Mexico, USA, pp. 2500–2504 (2002)
80. Chen, S.Y., Chen, L.J.: Nitride-mediated epitaxy of self-assembled $NiSi_2$ nanowires on (001)Si. Appl. Phys. Lett. **87**, 253111–253113 (2005)
81. Chen, S.Y., Chen, H.C., Chen, L.J.: Self-assembled endotaxial alpha-$FeSi_2$ nanowires with length tunability mediated by a thin nitride layer on (001)Si. Appl. Phys. Lett. **88**, 193114–193113 (2006)
82. Chen, S.Y., Chen, L.J.: Self-assembled epitaxial $NiSi_2$ nanowires on Si(001) by reactive deposition epitaxy. Thin Solid Films **508**, 222–225 (2006)
83. Liang, S., Islam, R., Smith, D.J., Bennett, P.A.: Phase transformation in FeSi2 nanowires. J. Cryst. Growth **295**, 166–171 (2006)
84. Peng, Z.L., Liang, S., Deng, L.G.: Transition metal silicide nanowires growth and electrical characterization. Chin. Phys. Lett. **26**, 127301 (2009)
85. Ma, J., Gu, Y., Shi, L., Chen, L., Yang, Z., Qian, Y.: Synthesis and oxidation behavior of chromium silicide (Cr_3Si) nanorods. J. Alloy. Compd. **375**, 249–252 (2004)
86. Kim, J., Anderson, W.A.: Spontaneous nickel monosilicide nanowire formation by metal induced growth. Thin Solid Films **483**, 60–65 (2005)
87. Kim, J., Anderson, W.A., Song, Y.J., Kim, G.B.: Self-assembled nanobridge formation and spontaneous growth of metal-induced nanowires. Appl. Phys. Lett. **86**, 253101–253103 (2005)
88. Tanaka, M., Chu, F., Shimojo, M., Takeguchi, M., Mitsuishi, K., Furuya, K.: Position- and size-controlled fabrication of iron silicide nanorods by electron-beam-induced deposition using an ultrahigh-vacuum transmission electron microscope. Appl. Phys. Lett. **86**, 183104–183103 (2005)
89. Zhang, Z., Hellström, P.-E., Lu, J., Östling, M., Zhang, S.-L.: A novel self-aligned process for platinum silicide nanowires. Microelectron. Eng. **83**, 2107–2111 (2006)
90. Geng, Z.R., Lu, Q.H., Yan, P.X., Yan, D., Yue, G.H.: Efficient preparation of NiSi nanowires by DC arc-discharge. Phys. E **41**, 185–188 (2008)

91. Gao, Y., Shao, G., Chen, R.S., Chong, Y.T., Li, Q.: TEM study of self-assembled $FeSi_2$ nanostructures by ion beam implantation. Solid State Commun. **149**, 97–100 (2009)
92. He, Y., Fan, J., Zhao, Y.: Engineering a well-aligned composition-graded CuSi nanorod array by an oblique angle codeposition technique. Crystal Growth Des. **10**, 4954–4958 (2010)
93. He, Y., Brown, C., Lundgren, C.A., Zhao, Y.: The growth of CuSi composite nanorod arrays by oblique angle co-deposition, and their structural, electrical and optical properties. Nanotechnology **23**, 365703 (2012)
94. Seo, K., Bagkar, N., Kim, S.-i., In, J., Yoon, H., Jo, Y., Kim, B.: Diffusion-driven crystal structure transformation: synthesis of heusler alloy Fe3Si nanowires. Nano Lett. **10**, 3643–3647 (2010)
95. Jung, S.J., Lutz, T., Bell, A.P., McCarthy, E.K., Boland, J.J.: Free-standing, single-crystal Cu_3Si nanowires. Crystal Growth Des. **12**, 3076–3081 (2012)
96. Eberhardt, J., Kasper, E.: Ni/Ag metallization for SiGe HBTs using a Ni silicide contact. Semicond. Sci. Technol. **16**, L47 (2001)
97. Kim, J., Shin, D.H., Lee, E.S., Han, C.S., Park, Y.C.: Electrical characteristics of single and doubly connected Ni silicide nanowire grown by plasma-enhanced chemical vapor deposition. Appl. Phys. Lett. **90**, 253103–253103 (2007)
98. Paschen, S., Felder, E., Chernikov, M.A., Degiorgi, L., Schwer, H., Ott, H.R., Young, D.P., Sarrao, J.L., Fisk, Z.: Low-temperature transport, thermodynamic, and optical properties of FeSi. Phys. Rev. B **56**, 12916–12930 (1997)
99. Aeppli, G., DiTusa, J.F.: Undoped and doped FeSi or how to make a heavy fermion metal with three of the most common elements. Mater. Sci. Eng. B **63**, 119–124 (1999)
100. Hung, S.W., Yeh, P.H., Chu, L.W., Chen, C.D., Chou, L.J., Wu, Y.J., Chen, L.J.: Direct growth of β-$FeSi_2$ nanowires with infrared emission, ferromagnetism at room temperature and high magnetoresistance via a spontaneous chemical reaction method. J. Mater. Chem. **21**, 5704–5709 (2011)
101. Kang, S., Brewer, G., Sapkota, K.R., Pegg, I.L., Philip, J.: Electrical and magnetic properties of higher manganese silicide nanostructures. IEEE Trans. Nanotechnol. **11**, 437–440 (2012)
102. Kim, J.J., Shindo, D., Murakami, Y., Xia, W., Chou, L.J., Chueh, Y.L.: Direct observation of field emission in a single $TaSi_2$ nanowire. Nano Lett. **7**, 2243–2247 (2007)
103. Kim, J., Lee, E.S., Han, C.S., Kang, Y., Kim, D., Anderson, W.A.: Observation of Ni silicide formations and field emission properties of Ni silicide nanowires. Microelectron. Eng. **85**, 1709–1712 (2008)
104. Liang, S., Islam, R., Smith, D.J., Bennett, P.A., O'Brien, J.R., Taylor, B.: Magnetic iron silicide nanowires on Si(110). Appl. Phys. Lett. **88**, 113111–113113 (2006)
105. Kim, T., Bird, J.P.: Electrical signatures of ferromagnetism in epitaxial FeSi[sub 2] nanowires. Appl. Phys. Lett. **97**, 263111–263113 (2010)
106. Gottlieb, U., Sulpice, A., Lambert-Andron, B., Laborde, O.: Magnetic properties of single crystalline Mn_4Si_7. J. Alloy. Compd. **361**, 13–18 (2003)
107. Hou, T.C., Han, Y.H., Lo, S.C., Lee, C.T., Ouyang, H., Chen, L.J.: Room-temperature ferromagnetism in $CrSi_2$(core)/SiO_2(shell) semiconducting nanocables. Appl. Phys. Lett. **98**, 193104–193103 (2011)
108. DeGrave, J.P., Schmitt, A.L., Selinsky, R.S., Higgins, J.M., Keavney, D.J., Jin, S.: spin polarization measurement of homogeneously doped $Fe_{1-x}Co_xSi$ nanowires by andreev reflection spectroscopy. Nano Lett. **11**, 4431–4437 (2011)
109. Zhou, F., Szczech, J., Pettes, M.T., Moore, A.L., Jin, S., Shi, L.: Determination of transport properties in chromium disilicide nanowires via combined thermoelectric and structural characterizations. Nano Lett. **7**, 1649–1654 (2007)
110. Zou, C., Jing, G., Yu, D., Xue, Y., Duan, H.: Mechanical properties of $TiSi_2$ nanowires. Phys. Lett. A **373**, 2065–2070 (2009)
111. Weber, W.M., Geelhaar, L., Graham, A.P., Unger, E., Duesberg, G.S., Liebau, M., Pamler, W., Chèze, C., Riechert, H., Lugli, P., Kreupl, F.: Silicon-nanowire transistors with intruded nickel-silicide contacts. Nano Lett. **6**, 2660–2666 (2006)

112. Wang, W., Ari, A.M., Wong, W.K.: Ieee, On-chip interconnects and repeaters based on NiSi nanowires. Ieee, New York (2006)
113. Maex, K., De Keersmaecker, R.F., Ghosh, G., Delaey, L., Probst, V.: Degradation of doped Si regions contacted with transition-metal silicides due to metal-dopant compound formation. J. Appl. Phys. **66**, 5327–5334 (1989)
114. Hua, Q., Du, N., Zhang, H., Liu, Z., Wang, L., Yang, D.: Direct growth of Ni_2Si nanowire array electrodes for high-performance reversible lithium-ion batteries. J. Mater. Sci. Eng. **28**, 645 (2010)
115. Zhou, S., Wang, D.: Unique lithiation and delithiation processes of nanostructured metal silicides. ACS Nano **4**, 7014–7020 (2010)
116. Zhou, S., Simpson, Z.I., Yang, X., Wang, D.: Layered titanium disilicide stabilized by oxide coating for highly reversible lithium insertion and extraction. ACS Nano **6**, 8114–8119 (2012)
117. Kang, K., Song, K., Heo, H., Yoo, S., Kim, G.-S., Lee, G., Kang, Y.-M., Jo, M.-H.: Kinetics-driven high power Li-ion battery with a-Si/$NiSi_x$ core-shell nanowire anodes. Chem. Sci. **2**, 1090–1093 (2011)
118. Qi, Y., Du, N., Zhang, H., Fan, X., Yang, Y., Yang, D.: CoO/$NiSi_x$ core-shell nanowire arrays as lithium-ion anodes with high rate capabilities. Nanoscale **4**, 991–996 (2012)
119. Xie, J., Yang, X., Zhou, S., Wang, D.: Comparing one- and two-dimensional heteronanostructures as silicon-based lithium ion battery anode materials. ACS Nano **5**, 9225–9231 (2011)
120. Zhou, S., Yang, X., Lin, Y., Xie, J., Wang, D.: A nanonet-enabled Li ion battery cathode material with high power rate, high capacity, and long cycle lifetime. ACS Nano **6**, 919–924 (2011)
121. Lin, Y., Zhou, S., Liu, X., Sheehan, S., Wang, D.: TiO_2/$TiSi_2$ heterostructures for high-efficiency photoelectrochemical H_2O splitting. J. Am. Chem. Soc. **131**, 2772–2773 (2009)
122. Banerjee, S., Mohapatra, S.K., Misra, M.: Water photooxidation by $TiSi_2$-TiO_2 nanotubes. J. Phys. Chem. C **115**, 12643–12649 (2011)
123. Hung, S.W., Wang, T.T.J., Chu, L.W., Chen, L.J.: Orientation-dependent room-temperature ferromagnetism of FeSi nanowires and applications in nonvolatile memory devices. J. Phys. Chem. C **115**, 15592–15597 (2011)
124. Kim, D.J., Seol, J.K., Lee, M.R., Hyung, J.H., Kim, G.S., Ohgai, T., Lee, S.K.: Ferromagnetic nickel silicide nanowires for isolating primary $CD4^+$ T lymphocytes. Appl. Phys. Lett. **100**, 163703–163704 (2012)

Chapter 13
Integration of Strain Free III–V Quantum Dots on Silicon

Stefano Sanguinetti, Sergio Bietti and Giovanni Isella

Abstract The great interest in the implementation of GaAs quantum nanostructures (QNs) on silicon substrates is mainly due to the possibility of integrating specialized high efficiency optoelectronic and photonic devices on the existing complementary metal-oxide semiconductor technology developed on Si. This would allow the realization of specialized III–V devices such as nanoemitters and intersubband detectors directly embedded with a large number of existing Si devices. Of particular, technological interest is the possibility of carrying out the III–V device fabrication after the integrated circuit has been already realized, i.e., as a back-end process. In this case, the compatibility with the underlying integrated circuit is possible only imposing strict constraints on thermal budget for growth and processing of the epilayer.

13.1 Introduction

Gallium arsenide and silicon are the most widely used materials in device fabrication for optoelectronics and electronics. In particular, about 95 % of electronic devices are fabricated by silicon. Silicon, one of the most abundant element in the Earth's crust, it is robust, thus allowing the fabrication of large diameter wafers, and has a good thermal conductivity. It also gives the possibility to integrate silicon dioxide as insulator and has a high hole mobility. These properties made it possible the development of Complementary Metal-Oxide-Semiconductor (CMOS) technology which nowadays allows the integration on a single chip of billion components. This technology is

S. Sanguinetti (✉) · S. Bietti
Dipartimento di Scienza dei Materiali, Universita' degli Studi di, Milano-Bicocca,
via R. Cozzi 53, 20125 Milano, Italy
e-mail: stefano.sanguinetti@unimib.it

G. Isella
Politecnico di Milano, Polo di Como, Via Anzani 42, 22100 Como, Italy

H. Li et al. (eds.), *Silicon-based Nanomaterials*,
Springer Series in Materials Science 187, DOI: 10.1007/978-1-4614-8169-0_13,

© Springer Science+Business Media New York 2013

the base for the fabrication of microprocessors, microcontrollers, memories, image sensors, and data converters.

On the other hand, the presence of an indirect gap makes silicon less effective than direct gap material for optoelectronic applications such as fabrication of detector and light emitting devices. For this reason III/V compounds, like GaAs, are widely used in optoelectronics. Also for the fabrication of solar cells and for high mobility devices direct gap materials have better performances than silicon.

The possibility to integrate devices for applications in optoelectronics and photonics based on III–V semiconductors directly on Si substrates would allow the use of the highly refined silicon infrastructure, based on CMOS technology, to be maintained and would offer the option of integrating a few specialized III–V devices within a large number of Si devices.

13.2 Gallium Arsenide on Silicon Substrates

The first attempts to integrate GaAs on Si were performed in the early 1980s, growing a GaAs film directly on Si by MBE (e.g., [1, 2]), but only by the 1990s all the basic problems involved in GaAs epitaxy on Si were identified and reviewed [3]. Three main issues were identified: the growth of a polar semiconductor on a nonpolar substrate, the difference in lattice parameter and in the thermal expansion coefficient between GaAs and Si.

The first issue is related to the formation of anti-phase domains, due to the growth of a polar semiconductor on a nonpolar substrate. Because there are no preferential nucleation site for Ga and As on Si, if the growth of GaAs is started opening simultaneously both beam fluxes, different areas of the Si substrate are covered by As and Ga, thus resulting in the nucleation of GaAs with two different orientations of the sublattice, rotated by 90° from one another. Also the preparation of the substrate with an As or a Ga layer is not a sufficient condition to obtain an anti-phase domain free sample. It was theoretically predicted, and then confirmed by tunneling microscopy [4, 5], that on the clean (001) surface of silicon, monoatomic steps are presents. Si atoms on the surface form dimers in such a way that the dimers on one terrace are perpendicular to those of the neighboring one (panel (a) of Fig. 13.1). Passivating a Si surface with a monolayer of As leads to the formation of As dimers with perpendicular direction on neighboring steps as firstly proposed in 1987 [6]. The structure of the surface is maintained as the epitaxial growth proceeds. If neighboring Si terraces consist of As dimers perpendicularly directed to each other, then the GaAs islands that form on the neighboring terraces will be mutually rotated by 90° around the vertical axis. As the islands grow and coalesce, the contacting planes will results in atoms of the same species that give raise to crystal structure defects known as anti-phase defects. The Ga–Ga and As–As bonds break the periodicity of the lattice and act as centers of scattering for the carriers and centers of nonradiative recombination. The problem was solved by using Si substrates misoriented 4–6° from the (001) plane. Bringans et al. [7] discovered that under certain annealing conditions, two atomic

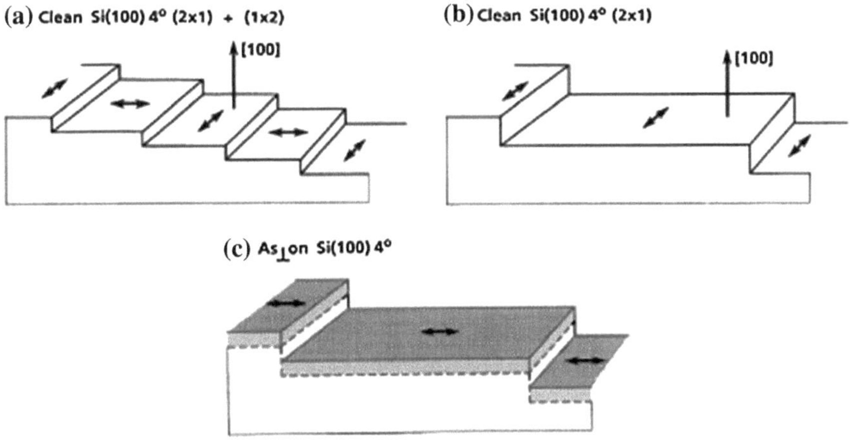

Fig. 13.1 Direction of dimers on Si(100) surface few degree miscut toward [011] with single steps (**a**), double steps (**b**), and with double steps and As passivation (**c**). From Ref. [7]

steps form on the silicon (001) vicinal surface and on these terraces Si–Si dimmers have an identical orientation. For this reason, on the As monolayer deposited before GaAs starts to grow, arsenic dimers also have an identical orientation and GaAs crystallites have the same orientation on either of these two Si surfaces.

The second issue is the formation of dislocations in the GaAs layer grown on Si due to the lattice mismatch between GaAs and Si. The lattice parameter of GaAs is 5.653 Å, for Si is 5.431 Å, so that the mismatch between the two materials is about 4 %. The mismatch causes strain on the lattice of deposited film and the energy associated with the strain is proportional to the thickness of the epilayer. If the thickness of the epilayer is small and below a critical value, the mismatch is accommodated by elastic deformation of the lattice, i.e., the GaAs lattice is compressed in the plane of growth and expanded in the direction perpendicular to the plane of growth.

Since the middle of the 1980s it is well-known that, in the initial stages of the deposition, GaAs on Si does not form a layer but gives raise to islands [8] and the growth follows the Volmer–Weber mode. In fact, crystalline material coherently adsorbed on a substrate which has a different lattice parameter tries to minimize the elastic energy accumulated creating misfit dislocations and/or with a deformation of the surface called 2D–3D transitions [9]. If the misfit is moderate, this deformation leads to alternating hills and valleys while for larger misfit the adsorbed film splits into clusters. The energetic advantage is that at the top and on the sides of the islands the adsorbate lattice constant is closer to that of the free adsorbate. The density and the size of the island is controlled through adatom diffusion length by the initial substrate temperature (see e.g., [3, 8]). This is important because the dislocations are most likely to nucleate in the high-stress regions at the edges of big islands near the island–substrate interface [10, 11]. The bigger is the island, the higher is the stress on the external edges, and the higher is the probability to nucleate a huge number of dislocations. To minimize the density of threading dislocation, GaAs is deposited on Si initially at low temperature to obtain small islands, then annealed and only after these steps a standard temperature GaAs growth (two-step growth) is performed. But also in this case, when islands coalesce the rough surface obtained becomes a source of defects.

When the epilayers become thick [12, 13], the associated strain energy is larger than the misfit dislocation energy, then dislocations are generated to relieve strain energy. Misfit dislocations, concentrated at the GaAs/Si interface, enable the release of strain energy while threading dislocations propagate to the surface and degrade the quality of the epilayer creating nonradiative recombination centers and eventually increasing the impurity diffusion along the dislocation line [3]. Dislocations are often described in terms of Burgers vectors. In a perfect crystal, it is always possible to choose a closed path defined by a series of displacements of the Bravais lattice. In a crystal with a defect, the same path fails to return to its starting point. The Bravais lattice vector that joins the starting point and the end point is defined as the Burgers vector of the dislocation. The Burgers vector is not dependent on the path chosen. The misfit accommodated by a dislocation is the projection of its Burgers vector onto the substrate plane. The commonly occurring dislocations in zincblende and diamond semiconductors can be classified into two different ways [14]: those whose Burgers vector are parallel to the growth plane (type I dislocations) and others (type II dislocations). Type I dislocations are pure edge type, type II dislocations have their Burgers vector at an angle of 45° to the plane of the substrate and 60° to the dislocation line. Type II dislocations can easily move up through the GaAs epilayer and reach the surface.

The last issue is the difference in the thermal expansion coefficients, $6.63 \times 10^{-6}\,K^{-1}$ for GaAs, and $2.3 \times 10^{-6}\,K^{-1}$ for Si. This difference promotes the nucleation of a huge amount of dislocations and of thermal cracks in the GaAs film when the substrate is cooled after the growth or when the sample is heated or cooled [15]. As already stated, the formation of anti-phase domains can be easily prevented by using Si substrates with surface a few degrees miscut from the singular plane (e.g., [16]). Bringans et al. [7] described the passivation with As of a Si(100) surface oriented at an angle of 4° from [100] direction toward [011]. Such a surface consists of (100) terraces arranged in a stairway fashion and the step height is usually one atomic layer, so that each successive terrace will have its Si–Si dimers direction rotated by 90°. Annealing such substrate at 1030 °C for a few minutes makes it possible to obtain a surface with predominant step height of two atomic layers. This is due to the lower energy of a double step compared to the sum of the energy of two single steps in this configuration. Depending on the substrate temperature, As valve opening time and As flux irradiated, it is possible to obtain a surface coverage mainly composed by As dimers parallel or perpendicular to the steps, but in both cases the GaAs grown will have the same orientation and will not give raise to anti-phase boundaries.

Lattice mismatch and difference in thermal expansion coefficient are more serious problems that can introduce threading dislocation density as high as $10^9 - 10^{10}\,cm^{-2}$. The two-step growth technique, the introduction of thermal cycling, or annealing treatment decrease the dislocation density but do not allow to reach a value useful for the fabrication of high optical quality III–V nanostructures.

A last technological problem must also be considered for the fabrication of III–V devices directly on Si-based CMOS technology, the compatibility with the low thermal budget required to integrate the nanostructures growth with CMOS technology. Two different strategies can be adopted. The back-end integration, i.e., when the CMOS circuitry is already realized, or the front-end integration, i.e., when the devices are patterned in the semiconductor, but the metal interconnect layer is not created yet. Keeping the structure at high temperature for long time deteriorates the interconnect layer and reduces the transistor performances. The adoption of thin Ge layer grown by LEPECVD as virtual substrate and of droplet epitaxy technique for the fabrication of GaAs quantum nanostructures will be here described to obtain high quality and high optical efficiency GaAs quantum nanostructures using a thermal budget suitable for the integration with CMOS technology.

13.3 Heteroepitaxy of GaAs on Si Through a Ge Layer

A different approach to solve the issues related to lattice mismatch and thermal expansion coefficients is the fabrication of a buffer layer between Si and GaAs. In the late 1980s, several studies were made with Ge buffer layers (e.g., [17]). Germanium has a lattice parameter of 5.658 Å, and a thermal expansion coefficient very close to the one of GaAs. Sieg et al. [18] deposited GaAs on a Ge substrate and by using an initial low temperature (350 °C) deposition, reduced the diffusion of Ge in GaAs, and thus the concentration of Ge in the epilayer to less than 2×10^{15} cm^{-3}. To maintain a high GaAs crystal quality at such a low growth temperature, deposition by migration enhanced epitaxy (MEE) was performed [19]. This technique consists in alternating the supply of Ga and As and allows to grow high quality GaAs films also at low temperature. Ting et al. [20] studied the formation of a single domain GaAs film on pure Ge substrates or on SiGe virtual substrates. In particular, they investigated the role of the substrate pre-growth annealing on the formation of a single domain film and the role of the temperature during the first As irradiation in the sublattice orientation of the GaAs film. The condition to obtain a single domain epilayer is an annealing at $T > 600\,°C$. This behavior is explained with the energy required for the formation of double steps on Ge. The subsequent substrate temperature set before As irradiation selects the orientation of As dimers on the surface (parallel or perpendicular to the steps) and then the orientation of the GaAs sublattice.

To better understand the possibility to integrate GaAs on Si through a Ge virtual substrate, several experiments were performed to test the quality of a GaAs epilayer directly grown on a Ge substrate. Brammertz et al. [21] studied the quality of GaAs films with different thicknesses by low-temperature photoluminescence spectroscopy. Several points should be remarked. First of all, also on Ge substrates an anti-phase domain free growth it is possible only on miscut substrates. The formation of anti-phase domains on nonmiscut substrates was confirmed by Nomarski microscope images and improvement of photoluminescence signal of four orders of magnitude from GaAs grown on miscut substrate. For anti-phase domain free samples, measurements showed a band-to-band structure at an energy of ~1.5 eV (similarly to GaAs grown on GaAs substrate) and a broad inner band gap structure at an energy of ~1.1 eV. For undoped GaAs films with thickness smaller than 200 nm, the band-to-band structure showed a broadening with values of full width half maximum (FWHM) up to 56 meV, which corresponds to high doped layer (estimated to $2 \times 10^{18}\,cm^{-3}$). For thick GaAs layers, the band-to-band structure is separated into two narrow peaks due to the split of light-hole and heavy-hole bands. This behavior is related to the strain introduced in the GaAs film by the mismatch between Ge and GaAs lattice parameter. The broad inner band gap structure is, in general, composed by two gaussian peaks centered around 1.04 and 1.17 eV and with FWHM of about 160 meV, respectively. Figure 13.2 shows these structures for three undoped GaAs films with thickness equal to 140, 300, and 600 nm with a fit for every structure that shows the individual Gaussian contributions to the structure.

The photoluminescence detected is explained with deep trapping states formed when the GaAs is heavily doped at levels in excess of $5 \times 10^{18}\,cm^{-3}$. At this high doping level the GaAs reaches a compensation regime, so that additional Ge atoms, instead of replacing Ga atoms in the GaAs lattice as donors, start to occupy the position of the As atoms as acceptors. This behavior was observed in GaAs heavily doped with Si [22, 23]. When the compensation regime is reached, the doping density stabilizes, because of the mutual elimination of the effects of Si or Ge atoms replacing Ga atoms (Si_{Ga} or Ge_{Ga}) and Si or Ge atoms occupying the As positions (Si_{As} or Ge_{As}) in the GaAs lattice. Even though the exact nature of the deep trapping states is not generally accepted, there is a prevailing consensus that the lower energy

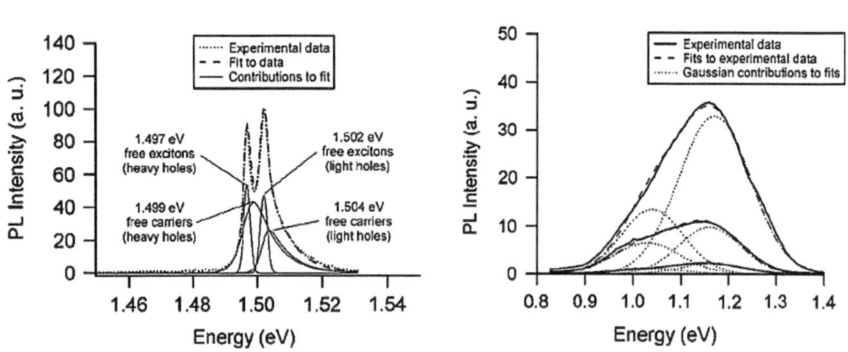

Fig. 13.2 *Left panel* 77 K PL spectrum of a 600 nm thick GaAs film on Ge (*dotted line*). A fit to the experimental data is shown as well (*dashed line*). The *solid lines* show the different components of the fit. *Right panel* 77 K spectra of three GaAs films with different thicknesses on Ge substrates (*solid lines*). The *top solid curve* is the spectrum of a 140 nm thick film, the *middle curve* corresponds to a 300 nm thick film, and the *bottom curve* corresponds to a 600 nm thick film. For every structure, a fit to the experimental data is shown (*dashed curves*) as well as the individual gaussian contributions to the fit with peak positions at 1.04 and 1.17 eV (*dotted curves*). From Ref. [21]

peak is related to the Si_{Ga}–Si_{As} complex, whereas the higher energy peak is related to the Si_{Ga}–V_{Ga} complex, where V_{Ga} is a Ga-vacancy site. The same mechanism is reported in [21] for Ge atoms to attribute the 1.04 and 1.17 eV peaks.

In the 1990s, Fitzgerald and et al. [24, 25] studied the growth of a graded GeSi buffer layer on Si substrate. Because Ge lattice parameter is similar to the one of GaAs, linearly increasing the content of Ge from 0 up to 100 % makes the lattice parameter of the buffer layer suitable for the deposition of a GaAs epilayer. The threading dislocation density on such kind of samples was reduced to $2 \times 10^6\ cm^{-2}$. The preparation of a low threading dislocation density film required a slow increase of Ge content (no more than 10 % per micron), leading to a large material and time consumption, wafer bowing and layer cracking because of several micrometer thick epilayers. Deposition at high substrate temperature reduced growth time but led to very high roughness on the final Ge surface (up to hundreds of nanometers [25]).

A different technique for growing graded buffer layers of GeSi on Si substrates is low-energy plasma-enhanced chemical vapor deposition (LEPECVD) of Si, Ge, and SiGe [26, 27]. The chemical vapor deposition is a chemical process used to produce thin films with high purity. The substrate is exposed to one or more gaseous precursors which react on the substrate surface to produce the desired epilayer. Plasma enhancement is an effective way to increase the growth rates for Si and Ge deposition, generating highly reactive neutral which reach the substrate surface greatly enhancing the adatom surface mobilities and the H-removal rate from the surface [28]. In a LEPECVD camera (Fig. 13.3), the discharge gas (Ar) is fed directly into the plasma source where a Ta filament is heated by a current of typically 120 A. Vacuum is provided by a turbo-molecular pump connected to the growth chamber by an orifice of ~1 cm in diameter. The plasma is sustained by applying a voltage of 20–30 V between the grounded chamber walls and the hot filament. The reactive gases, SiH_4, GeH_4, are fed into the chamber by means of a gas dispersal ring placed 5 cm below the substrate. This technique allows to reach a high growth rate (up to 10 nm/s) and low growth temperatures (about 550 °C). The density of threading dislocation in Ge/Si graded buffer was lowered to $1 \times 10^5\ cm^{-2}$ [29, 30] and the surface roughness reduced to few nanometers [31].

By means of LEPECVD, it is possible to deposit pure Ge directly on Si surface without any intermediate SiGe layer. This allows to reduce thickness of the film, not to exceed the critical thickness for the formation of thermal cracks [15]. Different groups have shown that growing a 2 μm Ge layer on Si, and performing a postgrowth annealing reduces threading dislocation density down to $2 \times 10^7\ cm^{-2}$ [32–34]. This is possible also using few annealing cycles at low temperature in order to meet the thermal budget requirements for back-end CMOS technology [35].

Colombo et al. [36, 37] studied the formation of thermal cracks in a GaAs layer deposited on such a Ge virtual substrate. The thickness of the GaAs epilayer (2.2 μm) was greater than critical thickness for GaAs/Ge epitaxy (0.24 μm), so that the sample was considered fully relaxed at the growth temperature. The low temperature photoluminescence measurements showed a split of the degenerate valence band and a shift toward lower energy of the energy gap due to biaxial strain. The origin of the strain is attributed to the difference in thermal expansion coefficient between

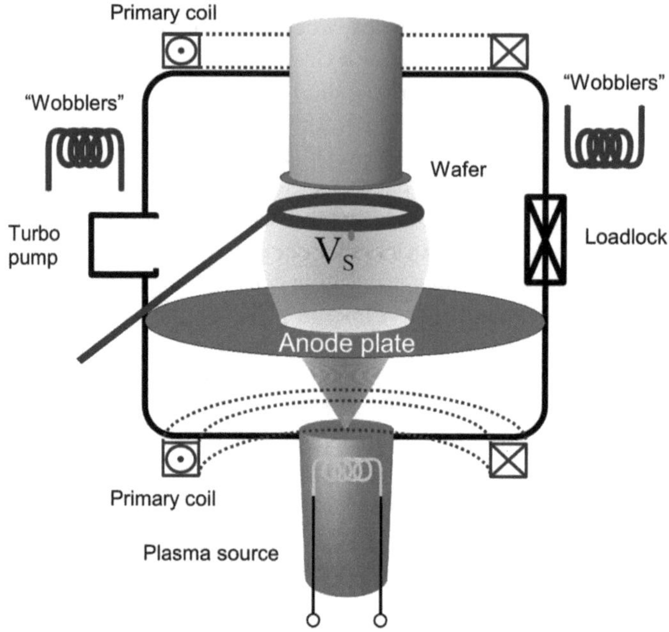

Fig. 13.3 Schematics for a Low-energy plasma-enhanced chemical vapour deposition camera

GaAs/Ge and Si. Also at room temperature, images of the GaAs epilayer obtained by optical microscopy showed a crack density of $4 \pm 1\,\text{mm}^{-1}$. After several thermal cycles the value increased to $12 \pm 1\,\text{mm}^{-1}$. Measurements of μ-photoluminescence showed that the position of the main GaAs peak is changing with the distance from the crack. Only near the intersection between two cracks, the structure showed a fully relaxed GaAs peak position.

13.4 Germanium-on-Silicon Substrates

It is now possible to analyze in deeper detail the fabrication of thin Ge layers on Si substrate by LEPECVD as a starting point for the integration of III–V on Si.

The threading dislocation density of Ge layers deposited directly on Si(100) at temperatures above 400 °C is of the order of $10^8\,\text{cm}^{-2}$ or more. This density is unacceptable for most applications. Efficient threading dislocation density reduction could be achieved also by cycling the sample temperature between 750 and 900 °C. The procedure needs to be repeated several times in order to be effective [35]. This leads to a correspondingly larger thermal budget which, indeed, could be a complication for the integration with CMOS technology and favors interdiffusion at the Ge/Si interface [38]. In [39], the characterization of thin Ge layers deposited by LEPECVD on Si substrate and submitted to cyclical annealing process is described.

As described in the previous sections, 6° misoriented (toward [110]) Si(100) substrates were used for the growth of Ge by LEPECVD. Different layers with thickness between 1 and 3 μm were deposited at a rate of 4.8 nm s^{-1} and substrate temperature of 500 °C. Cyclic thermal annealing was performed by increasing the temperature, at a rate of 1 °C per second, from 600 to 780 °C and then maintaining it constant for 4 min, before lowering it again. The procedure was repeated three times for a total annealing time of 30 min.

The surface of as-grown Ge layers deposited on Si misoriented substrates display a periodic undulation (Fig. 13.4, panel a) which has already been observed for Si homoepitaxy and SiGe heteroepitaxy on vicinal surface [40]. After the annealing procedure, a fragmented cross-hatch pattern develops on the deposited Ge layers with a root-mean-square (rms) roughness of about 0.8 nm. The cross-hatch lines are split perpendicularly to the off-cut direction [25] as commonly observed in SiGe deposition on misoriented Si(100). In thicker layers (Fig. 13.4 panel c), relatively long cross-hatch lines are visible while in thinner layers (Fig. 13.4 panel b), they are barely distinguishable and interrupted by pits and trenches. These defects are probably originated from the interaction between the strain field of threading dislocations and the Ge atoms diffusing on the surface during the annealing.

The threading dislocation density was measured by defect etching and Nomarski microscopy. During etching the temperature of the iodine-based solution [41] employed [CH_3COOH(67 ml) + HF(10 ml) + HNO_3(20 ml) + I(30 mg)] was stabilized with an ice water bath in order to improve the reproducibility. The etching depth was about 200 nm. Images collected with Nomarski microscope and AFM revealed a threading dislocation density of about 10^7 cm^{-2}, being slightly lower for the thicker layers. The reduction of threading dislocation density is about one order of magnitude if compared to the one observed in as-grown substrate, thus confirming the efficiency of the annealing process.

To characterize the crystalline quality, high resolution X-ray diffraction measurements were performed. Reciprocal space maps around the 004 reflection of the Ge

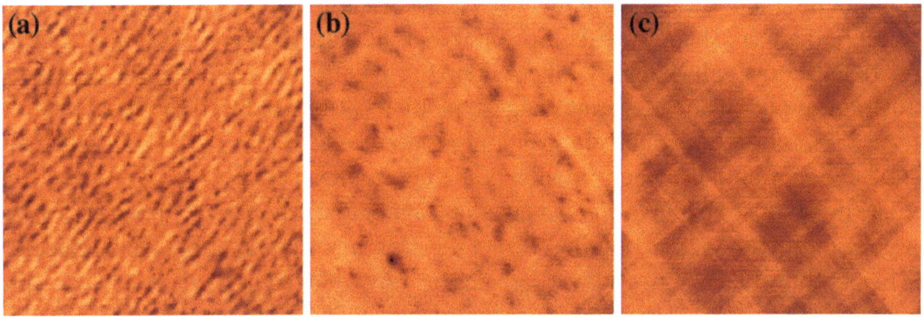

Fig. 13.4 20 μm × 20 μm AFM images of Ge on Si substrates. Before annealing (**a**), 1 μm annealed Ge layer (**b**) and 4.4 μm annealed Ge layer (**c**). From [39]

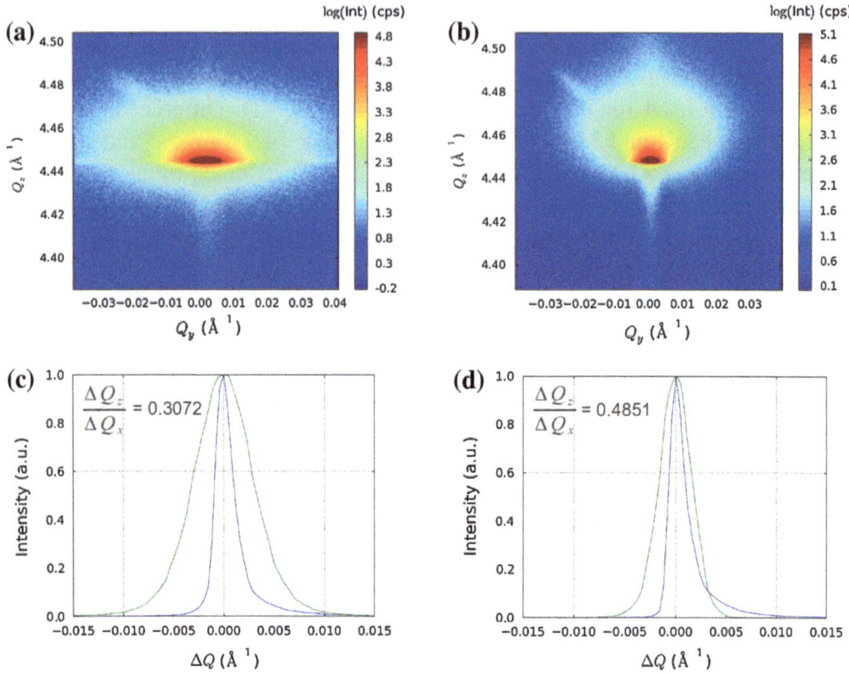

Fig. 13.5 In **a** and **b** reciprocal space maps (RSM) around the 004 reflection of the Ge layer are shown, respectively for an as-grown and annealed sample. In **c** and **d** we show the peak profiles, extracted from the RSM **a** and **b**, along Q_Z (*blue lines*) and Q_X (*green light*). From [39]

layer are reported in Fig. 13.5 panels a and b, for an as-grown sample and after the in-situ annealing process, respectively. The ellipsoidal shape of the peak (ΔQ_X is roughly three times larger than ΔQ_Z) is due to diffuse scattering. This scattering is correlated to the presence of randomly distributed 60°-type misfit dislocations (see Fig. 13.5, panel c) [42]. The asymmetry of the peak along Q_Z is attributed to interdiffusion of Ge at the interface between the epilayer and Si substrate. The effect of the cyclic thermal annealing on Ge layer is clearly visible comparing panel c and d of Fig. 13.5: the peak shape is still ellipsoidal, but the ratio between ΔQ_X and ΔQ_Z is now reduced to a value slightly bigger than 2. This effect is correlated to the partial transition of 60° dislocations into edge-type dislocations induced by the annealing process. Moreover, after annealing a more pronounced tail along Q_Z is present. This is due to a more intense interdiffusion at the heterointerface.

The thickness of the investigated layers is higher than the critical thickness for the onset of plastic relaxation, but the (004) $\omega - 2\theta$ scans from as-grown and annealed Ge layers show that as-grown layers are almost fully relaxed but for a residual tensile strain $\varepsilon \approx 7.5 \times 10^{-4}$. This strain is accumulated at the end of the growth during the reduction of substrate temperature and is due to the difference between the thermal expansion coefficient of Ge and Si [43]. As a consequence, annealed samples exhibit higher levels of tensile strain. The shift toward higher Bragg angles of the Ge peak indicates a residual tensile strain $\varepsilon \approx 1.8 \times 10^{-3}$.

The data presented demonstrate the possibility to deposit a thin Ge layer by LEPECVD with a low TD density and reduced roughness. As we will see in the proceeding, the Ge-on-Si is capable to accommodate the mismatch between GaAs and Si and can be used as a substrate for the growth of high quality AlGaAs/GaAs layers and of GaAs quantum nanostructures, still maintaining a low thermal budget compatible with the implementation of CMOS devices.

13.5 Deposition of a GaAs Buffer Layer

After the preparation of the substrate with the Ge layer and the subsequent annealing, the substrate is removed from the LEPECVD growth chamber, and loaded inside an MBE system. After a degas at 300 °C in buffer chamber, the substrate is loaded in the growth chamber. The substrate temperature is increased up to 650 °C to desorb the oxide layer formed during the transfer in atmosphere and to allow the transition from single to double steps on Ge surface.

The formation of double steps is verified by reflection high energy electron diffraction (RHEED). In the RHEED pattern reported in Fig. 13.6, if we define a as the distance between two atoms on the surface of the sample and W as the length of the step, the distances corresponding to $2\pi/a$ and $2\pi/W$ observed on RHEED pattern have a ratio $a/W = 8/41$. The step height d can be $a/2$ for double step or $a/4$ for single step, the angle α is about 6° from the wafer specifications. From the relation $d = W \sin\alpha$, the resulting angle we obtain is $\alpha \sim 5.6 \pm 0.7°$ in good agreement with the wafer specification.

In order to deposit GaAs, substrate temperature is then reduced to 580 °C, and the As flux opened to obtain a beam equivalent pressure of 5×10^{-6} torr to passivate the Ge surface with As dimers. After few seconds, the Ga shutter is opened and GaAs deposition start. Figure 13.7 shows the RHEED pattern along [110] and [1–10] directions of the Ge surface before passivation (left column), few seconds after starting GaAs deposition and after the growth of about 50 nm of GaAs (right column). The surface reconstruction at this point is clearly a (2 × 4), so that we can assert that the growth is almost anti-phase domain free. The confirmation of double step formation and the presence of a (2 × 4) surface reconstruction are of utter importance, because they allow to assess the quality of the GaAs epilayer in real-time during growth.

13.6 Formation of Quantum Dots by Droplet Epitaxy

After the growth of the GaAs buffer layer, it is possible to continue the growth exactly as on GaAs substrate, as reported in [45, 46] and [47].

Our interest focuses in particular on GaAs/AlGaAs quantum dots grown by droplet epitaxy [48, 49], a variant of molecular beam epitaxy. In droplet epitaxy, firstly a beam

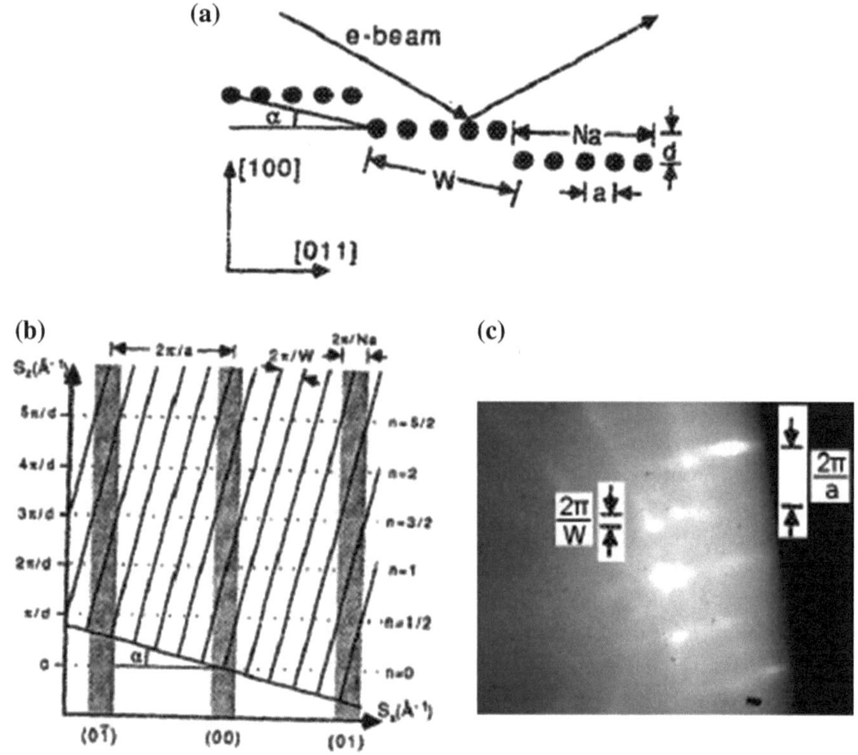

Fig. 13.6 Schematics of vicinal surface with relevant dimensions (**a**). Reconstruction of the reciprocal lattice of vicinal surface. The *shaded regions* represent a single terrace (**b**). RHEED pattern of Ge buffer surface 6° miscut (**c**). (**a**) and (**b**) from Ref. [44]

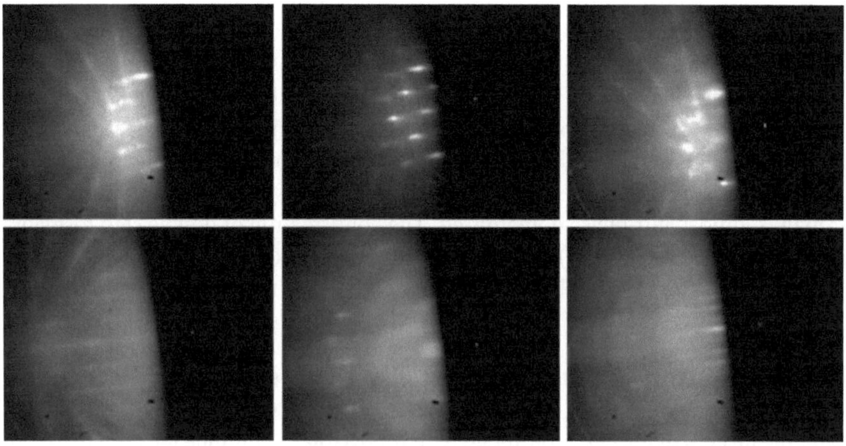

Fig. 13.7 RHEED pattern along [110] direction (*upper row*) and [1–10] direction (*lower row*) of the Ge virtual substrate surface (*left column*), after opening As and Ga cell shutters (*central column*) and after a few minutes of GaAs growth (*right column*)

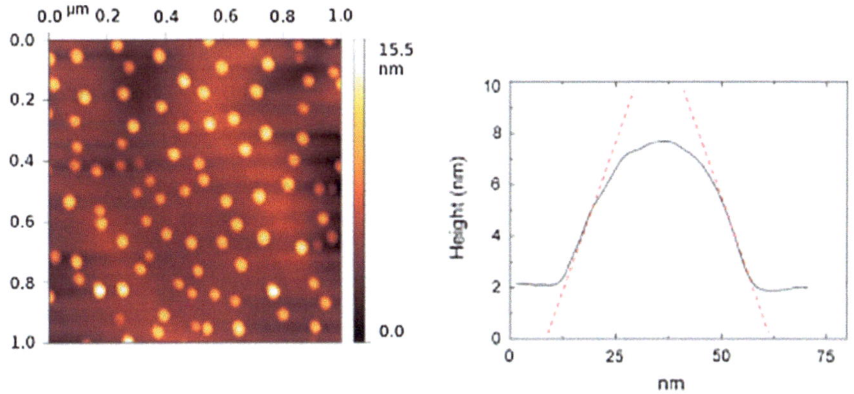

Fig. 13.8 AFM image of GaAs quantum dots on quantum dot sample (*left panel*). The profile along [110] direction is mostly compatible with {113} facets (*dashed red line* in *right panel*)

of Ga impinges on the sample surface in absence of arsenic, leading to the formation of gallium droplets. Changing the substrate temperature, it is possible to control the density of the droplets [50]. Then, an As flux is irradiated in order to crystallize the droplets forming GaAs nanocrystals. Changing the arsenization conditions (substrate temperature and beam equivalent flux of As), it is possible to control the diffusivity of Ga on the surface and to obtain different classes of quantum nanostructures [51, 52]. In particular, this technique is able to maintain a low thermal budget during the growth. Being essentially performed at temperatures below 350 °C, droplet epitaxy is perfectly suited for the strict thermal requirements of CMOS integration.

Reference [45] reports the fabrication of GaAs/AlGaAs quantum dots on a silicon substrate with a reduced thermal budget, as first step toward a growth procedure compatible with back-end integration of nanostructured GaAs devices on CMOS technology. For these experiments, a silicon (001) substrate 6° misoriented toward [110] was used. A fully relaxed 2 μm layer of Ge buffer layer was deposited by LEPECVD to accommodate the lattice mismatch between Si and GaAs. Just after deposition, the substrate was cyclically annealed six times between 600 and 780 °C to reduce the threading dislocation density down to 2×10^{7} cm^{-2}, respectively. The substrate was then loaded in an MBE machine. After the GaAs buffer, a 150 nm barrier layer of $Al_{0.3}Ga_{0.7}As$ is deposited to obtain quantum confinement in GaAs quantum dots. A standard droplet epitaxy growth was then perfomed, reducing substrate temperature down to 200 °C and the As pressure in chamber was reduced below 10^{-9} Torr, then 3.75 MLs of Ga were deposited to create Ga droplets on the surface. After another substrate temperature reduction to 150 °C, the As valve was opened to irradiate a 5×10^{-5} Torr flux on the substrate to crystallize the Ga droplets into quantum dots. The quantum dot formation was confirmed by RHEED pattern changes during the Ga deposition and arsenization process from halo to spotty pattern, in agreement with conventional droplet epitaxy experiments on GaAs substrates (e.g., [53]).

The surface of uncapped sample analyzed by AFM Fig. 13.8, shows a $1 \times 1\ \mu m^2$ area scan in which it is possible to observe quantum dots. GaAs nanostructures have a density of 9.1×10^9 dots/cm^2 with a mean base size of 44 nm and a mean height of 9 nm. The profile along the [1–10] direction is compatible with {113} family facets while along the [110] direction slope was not so clearly defined. The formation of clear surfaces along the [1–10] direction was favored by the geometry of the GaAs lattice, as reported in [54]. The As absorption sites on (111) surface are more stable than on (1–11). For this reason, As atoms on the (1–11) surface easily dissolves and the surface is less regular. It is also interesting to see that the size and density of GaAs quantum dots is not constant across the surface of the sample, being smaller and with higher density in the substrate valleys. Qualitatively, this behavior can be explained with the interaction of gallium droplets with surface steps. In fact, surface steps are preferential nucleation sites for gallium droplets, as shown by Mano et al. [55] on GaAs (110) substrates.

The density is about $1 \times 10^{10}\ cm^{-2}$, in good agreement with data reported for growths in a fully III–V environment [53]. The quantum dot ensemble shows a mean height of $\approx$5 nm and a base radius of $\approx$20 nm with a size dispersion for both values around 25 %.

A second sample was capped with 150 nm $Al_{0.3}Ga_{0.7}As$ and 10 nm GaAs layers grown by conventional MBE at 580 °C and then submitted to an annealing step at 700 °C in As atmosphere for 30 min.

PL spectra were measured on the capped sample at 14 K using a closed-cycle cold-finger cryostat. The PL was excited with a Nd:YAG laser ($E_{exc} = 2.33$ eV) with an excitation power density $P_{exc} = 50\ W/cm^2$. The spectra were measured by a grating monochromator operating with a Peltier-cooled CCD detector. The PL spectrum is shown in Fig. 13.2. The QD sample shows an intense emission peak at 1.78 eV, whose intensity is comparable with that of droplet epitaxy quantum dot samples grown in a fully III–V environment. The PL band lies in an energy range which is compatible with the emission from droplet epitaxy quantum dots [53, 56]. In order to attribute this line, electronic structure calculations were performed following the method outlined in Refs. [56, 57] within the effective mass using the material parameters reported in Ref. [58] for GaAs and $Al_{0.3}Ga_{0.7}As$. In these calculations, the potential for quantum confinement was derived by imposing a cylindrical symmetry to the actual shape of a randomly chosen quantum dot, measured by AFM. The predicted quantum dot ground state energy is $E_{GS} = 1.75$ eV, thus within the bandwidth of the line at 1.78 eV, which allows us to safely attribute such band to the ensemble emission from quantum dots.

To better understand the origin of the two shoulders around the main peak at 1.78 eV, a third sample was prepared as a blank following the same recipe of capped sample, but without the deposition of Ga droplets. The comparison between capped sample and its blank showed different common features, as reported in Fig. 13.9. PL measurements were took at the same temperature with similar condition of excitation power density. The GaAs related peaks around 1.5 eV showed the same shape and intensity. The $Al_{0.3}Ga_{0.7}$ As related peak around 1.9 eV was more intense in the blank sample and was split between bound exciton and donor–acceptor lines. The small

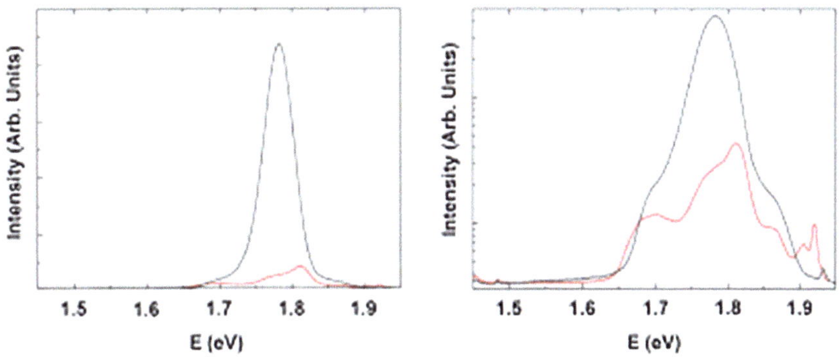

Fig. 13.9 PL measurements at low temperature for the QD sample (*black line*) and blank sample (*red line*) grown on a Ge virtual substrate. Linear (*left panel*) and logarithmic (*right panel*) are reported. The excitation power density is 10 W/cm^2

shift observed was attributed to small compositional difference in Al content. At 1.86 and 1.69 eV satellite peaks were presented in both spectra. The 1.78 eV band in the QD sample, related to GaAs quantum dots, is much stronger than any other band observed in the two samples and disappears in the blank sample. Anyway, the blank sample showed radiative recombination from states deep inside the $Al_{0.3}Ga_{0.7}As$ band gap between 1.81 and 1.76 eV. Probably, these bands are present also in the QD sample but overwhelmed by the signal from the GaAs quantum nanostructures.

As already reported, a strong contamination of Ge is expected, and this gives raise to radiative recombination from the impurities in the GaAs and AlGaAs layers. The comparison with scientific literature results [59, 60] reports an acceptor binding energy of about 25 and 80 meV for carbon and germanium in $Al_{0.3}Ga_{0.7}As$, respectively. For GaAs, epilayers grown on Ge substrate Brammertz et al. [21] attributed some deep level (~300 meV below the gap) to complex defects involving Ge impurities. The high density of Ge atoms in GaAs (density $> 10^{18}$ atoms/cm^2) creates the so-called "compensation regime" in which part of the Ge atoms, not only replace Ga atoms in the GaAs lattice as donors (Ge_{Ga}), but also start to occupy the position of the As atoms as acceptors (Ge_{As}). The behavior is similar to the one known for Si in GaAs and in the scientific literature does exist a prevailing consensus about the attribution of two strong photoluminescence peaks to the complex of two Ge atoms acting as donor and acceptor and to the Ge_{Ga}–V_{Ga} complex, where V_{Ga} is a Ga-vacancy site. Considering these data, it is possible to attribute the right shoulder in the spectra to C acceptors and the central peak in spectrum of the blank sample to Ge acceptors. The shoulder at about 1.7 eV can be attributed to deep states due to Ge complexes.

From the data presented emerge clearly the possibility to fabricate optically active GaAs quantum nanostructures on silicon substrate through a Ge layer. The signal from the dots is at least one order of magnitude stronger than the signals arising from other radiative recombinations. But it is also clear that the region between 1.7 and

1.9 eV is not a good choice for the emission of the nanostructures, due to the presence of several radiative recombination bands inside the $Al_{0.3}Ga_{0.7}As$ band gap.

13.6.1 Reducing Ge Contamination and Thermal Budget

In a second experiment described in [46], a different emission energy was selected in order to observe a clear emission from the GaAs nanostructures, and modified procedure for the growth of the barrier layer was employed in order to reduce the diffusion of Ge atoms and the thermal budget. The region between 1.55 and 1.7 eV is not affected by the radiative recombination due to Ge and C related impurities, and can be reached fabricating nanostructures with weaker quantum confinement. With droplet epitaxy, the fabrication of GaAs nanostructures fitted for this needing is quite simple task. The spectral emission of coupled ring disk [52] fabricated on GaAs substrate can be easily tuned to the selected region. The potential profile predicted for these nanostructures resembles the one of the dot-in-a-well nanostructures (i.e., each quantum dot is fabricated on a quantum well) and is expected to reduce the threshold when ring disks are embedded in laser diodes structures [61], and to improve the detection mechanism for infrared detectors [62]. Moreover, coupled ring disk grown on GaAs substrate have been reported to be efficient detectors in the infrared region [63]. It should be notice that coupled ring disk are fabricated keeping the substrate temperature at 350 °C, so that is also possible to grow the whole optically active region (AlGaAs barrier layers and quantum nanostructures) at the constant temperature of 350 °C, making it possible not only to fabricate GaAs nanostructures integrated on Si, but also to maintain a low thermal budget for the integration with CMOS technology. As matter of fact, this temperature is compatible with back-end CMOS technology, and it is also expected to slightly reduce the thermally activated diffusion of Ge in the optically active layer.

The new recipe is described in [46]. The Si wafer with 2 μm pure Ge layer was annealed up to 650 °C to desorb the oxide layer and to allow the formation of double steps on the Ge surface. The substrate temperature was then decreased to 580 °C and a GaAs buffer layer of 1 μm was deposited. The temperature was then reduced to 350 °C, observing the change in surface reconstruction to c(4 $\times$ 4). An $Al_{0.3}Ga_{0.7}$ As barrier layer of 80 nm was deposited by MEE and at the same substrate temperature of 350 °C, the As valve was closed to prepare a suitable environment for the Ga droplet formation on the AlGaAs surface. Droplet was formed by depositing 10 monolayers of Ga. The As flux of 8×10^{-6} torr was irradiated for 20 min keeping the same substrate temperature.

An uncapped sample was analyzed by AFM after this step as reported in Fig. 13.10. The resulting nanostructures were characterized by a regular, nanometers high, flat disk with a diameter of about 300 nm and a hole at the center of about 80 nm. The rim of the inner hole is protruded over the disk surface by some nanometers. The density of the nanostructures was 6×10^{8} cm^{-2}. The formation mechanism of the coupled ring disk (CRD) nanostructures can be explained in the details with the model

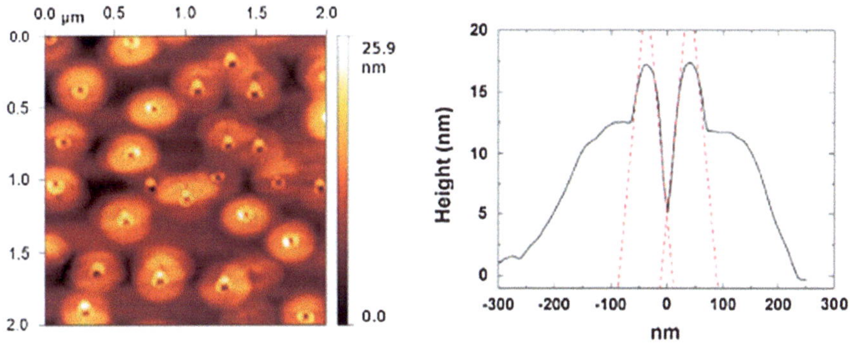

Fig. 13.10 AFM image of sample CRD (*left panel*). Profile of a single CRD along [1–10] direction (*right panel*). *Red-dashed lines* indicates {113} planes

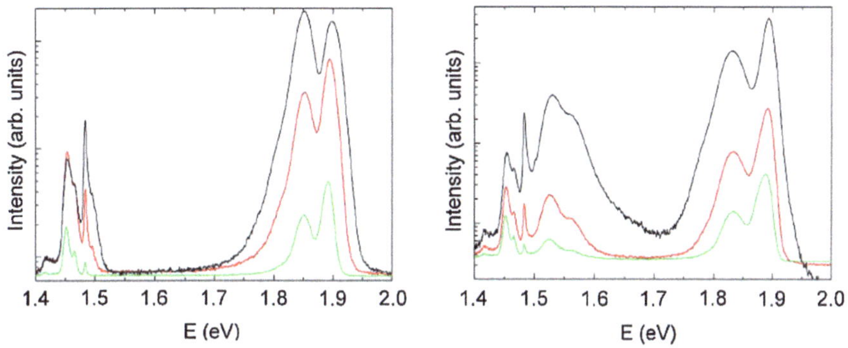

Fig. 13.11 PL measurements at low temperature for the blank sample (*left panel*) and for the CRD sample not annealed (*right panel*). The power density was 60 (*black line*), 6 (*red line*), and 0.6 W/cm^2 (*green line*)

proposed in [52]. A second sample was covered with a 80 nm $Al_{0.3}Ga_{0.7}As$ barrier layer deposited again by MEE, keeping the substrate temperature at 350 °C. The whole optically active region of the sample (barrier layers and GaAs nanostructures) was thus grown at a constant temperature of 350 °C. A last sample was prepared as a blank following the same recipe used for the CRD sample, but without the deposition of Ga droplets. The CRD sample was removed from the growth chamber, cutted and some parts underwent a Rapid Thermal Annealing (RTA) step. The RTA procedure consisted in a rapid heating (200 °C/min) in nitrogen atmosphere up to selected temperature of 500 °C (sample RTA500) or 600 °C (RTA600). After 4 min, the temperature was reduced to room temperature with the same rate. The RTA procedure has proved to be as effective as in situ postgrowth annealing [86], but, due to the reduced annealing time, it allows to maintain a lower thermal budget.

Photoluminescence spectra for samples were measured at 14 K with different excitation power densities (see Fig. 13.11). The two spectra showed sets of similar peaks around 1.9 and 1.5 eV, the former related to radiative transition in the AlGaAs barrier layer (1.89 and 1.83 eV), the latter to the GaAs buffer layer (1.50,1.48 and 1.45 eV). The photoluminescence spectrum of the CRD sample also showed an intense peak at 1.53 eV with a satellite peak at 1.56 eV. These radiative transitions were not observed in the blank sample and were thus attributed to recombination inside CRDs. A comparison between the photoluminescence spectra reported in Figs. 13.11 and 13.9 shows that the reduced temperature used for the growth of the optically active region decreased the photoluminescence signal related to Ge atom contamination inside the AlGaAs barrier. The two main peaks at 1.82–1.83 and 1.87–1.89 eV can be now attributed to neutral donor to neutral Ge acceptor and neutral donor to neutral carbon acceptor, respectively [59, 60]. The deep bands inside the AlGaAs gap disappeared, probably for the lower Ge diffusion. The introduction of different quantum nanostructures also allowed to select an emission region not affected by impurity-related recombinations. Photoluminescence spectra measured at 14 K with an excitation power density of 6 W/cm^2 of annealed samples, (RTA500 and RTA600) are reported in the left panel of Fig. 13.11. It is interesting to observe that the GaAs related peaks are unchanged by the annealing procedure.

The GaAs layer was grown at high temperature, so that the annealing procedure was not expected to introduce any important modification. The AlGaAs related region was subject to different changes. The peak at 1.89 eV was not modified by the 500 °C annealing step and decreased after the 600 °C annealing step.

The peak at 1.83 eV was decreased by the annealing step at 500 °C but greatly enhanced by the annealing step at 600 °C and at least two satellite peaks appeared on the low energy tail. Also the nanostructure emission band was modified by the annealing treatment: a blue-shift of the emission peak and an enhancement in the integrated intensity was observed (reported in detail in the right panel of Fig. 13.12). The behavior outlined can be explained as follows. The postgrowth thermal annealing step is known to improve the poor crystalline quality of the as-grown structure. In fact, the low temperature kept during the growth of the quantum nanostructures fabricated by droplet epitaxy and of the layers grown by MEE introduced a variety of lattice defects, including type III and type V vacancies, interstitials, and defect complexes. On the other hand, in addition to lattice recovery, the RTA treatment induced also two types of interdiffusion. The first was the diffusion of group III species, promoted by any of the lattice defects introduced during the growth and which changes the compositional profile of the nanostructures. The second was the diffusion of Ge atoms through threading dislocations. In fact, we expect that threading dislocations could act as preferential channels where Ge atoms diffuse, thermally activated, as reported for Si in [3].

Considering these effects, it is possible to explain the spectra reported in Fig. 13.12. The AlGaAs barrier quality was at the same time improved by the diffusion of group III atoms and contaminated by the diffusion of Ge atoms. The AlGaAs barrier quality was surely improved for both the RTA procedures, but the higher temperature used also introduced a diffusion of contaminants, resulting in the broad band around 1.8 eV. The CRD band was improved by the RTA treatment as expected. The integrated intensity enhancement and the blue-shift observed are phenomena well-known for GaAs

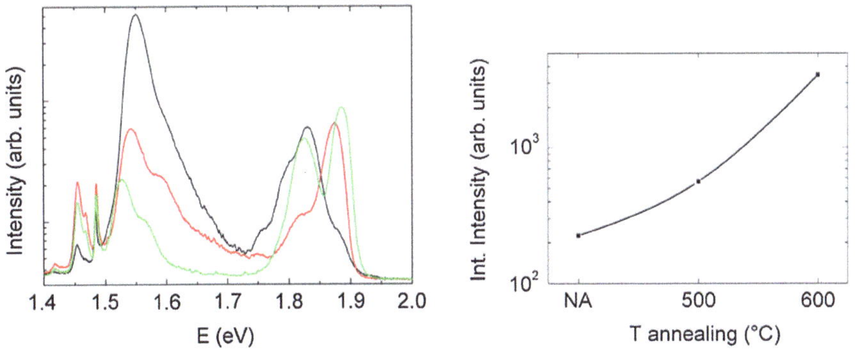

Fig. 13.12 PL measurements at low temperature for the samples CRD without annealing (*green line*), with RTA treatment at 500 °C (*red line*) and RTA treatment at 600 °C (*black line*). The excitation power density was 6 W/cm^2 (*left panel*). Integrated intensity for the CRD emission band versus the annealing temperature (*right panel*)

quantum nanostructures grown by droplet epitaxy (e.g., [64]). Following the method outlined the theoretical emission energy of CRDs (sample RTA600) was calculated in the effective mass approximation using as confinement potential the profile of the nanostructures obtained by AFM measurements. The theoretically calculated CRD ground electronic and hole states appeared to be confined in the ring structure, which is formed at the edge of the inner CRD hole. The calculated emission energy fits well the observed PL peak value ($E_{GS} = 1.56\,\text{eV}$). The CRD excited state was attributed to the quantum well-like state extended along the disk ($E_{EX} = 1.59\,\text{eV}$). The low confinement energy ($\sim$30 meV) is due to the relatively large, but still capable of quantum confinement, thickness of the disk. In the calculations, the interdiffusion of Al atoms at the CRD interface was taken into account.

The photoluminescence spectrum evolution with temperature of sample RTA600 is reported in the upper panel of Fig. 13.13. As predicted by the Varshni law, the energy gap of semiconductor materials is reduced by increasing temperature, so that also the CRD emission band red shifted, as expected, with the increasing temperature. As the temperature was raised, the ratio between the emission from the ground state and the emission from the excited state was decreased. The CRD emission band was still clearly visible at room temperature where the main contribution was due to the excited state emission. The ratio between the integrated intensity of ground state and excited state, reported in panel (c) of Fig. 13.13, showed an activation energy of $\sim$45 meV, which corresponds to the energy difference between ground and excited state emission. It is possible to attribute the relative increase of excited state emission to a change in equilibrium population of ground and excited states. As the temperature increased, the population ratio of the two states evolves according to the Fermi law. The predominance of disk emission with respect to the ring emission at room temperature was related to the different density of states (much higher in the disk case) attributable to the different dimensionality of the two CRD subsystems (0D

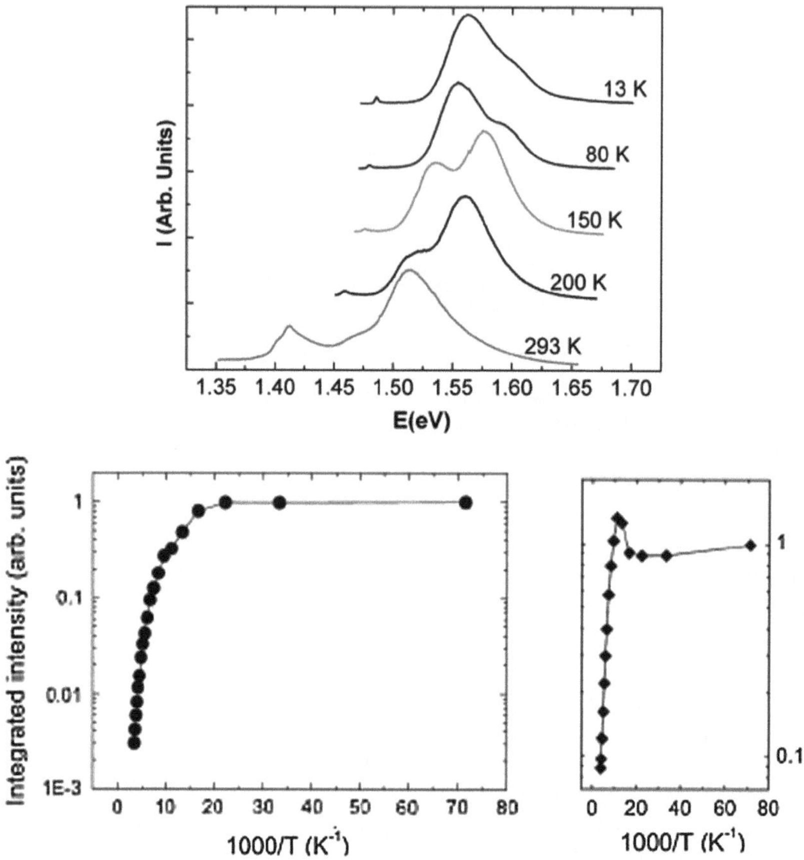

Fig. 13.13 (*Upper panel*) Normalized PL spectra of CRD band at different temperatures for the CRD sample after the RTA treatment at 600 °C. The excitation power density was 6 W/cm^2 . (*Lower left panel*) Arrhenius plot of the integrated intensity of the CRD band. (*Lower right panel*) Arrhenius plot of the intensity ratio between the ground state emission peak and the excited state. From Ref. [65]

for the ring and 2D for disk). The CRD band integrated intensity was reduced by a factor ~400 raising temperature from 14 to 293 K. The Arrhenius plot of photoluminescence integrated intensity, reported in lower left panel of Fig. 13.13, shows a clear temperature activated quenching, with a measured activation energy $E_{QUE} \sim$ 100 meV. The low quenching energy, much smaller than any energy barrier in the CRD system, and the relatively mild reduction in the signal between 14 K and room temperature, suggests attribution to a nonradiative recombination due to the defects directly accessible from CRD, or to the quenching active during the carrier diffusion and capture process [64]. It is possible to observe that some defectivity is expected at the interface between the internal ring of the CRDs structures and the MEE growth barrier layer, due to the unusual ring geometry, which prevents the correct diffusion of adatoms in the inner part of the ring.

Finally, to complete the characterization of the fabricated GaAs quantum nanostructures, the ratio η between the number of photogenerated carriers in the GaAs/AlGaAs active layer and the number of photons emitted by the CRDs was calculated. The value obtained for the RTA600 sample was about 3×10^{-3} at $T = 14$ K and $P_{exc} = 6W/cm^2$. This value calculated for GaAs CRDs grown by droplet epitaxy on Si substrate through a Ge virtual substrate well compares with $\sim 1 \times 10^{-2}$ obtained for a standard quantum dot sample (sample D680 in Ref. [66]) grown by droplet epitaxy on a GaAs substrate with a similar nanostructure density ($\rho = 1.2 \times 10^9 cm^{-2}$) confirming the high quality of the GaAs grown on Si substrate.

The value estimated for η dropped to about 8×10^{-6} at room temperature due to temperature activated nonradiative recombination channels. While at low temperature η was independent from the excitation power density, at room temperature a clear dependence $\eta \sim P_{exc}^{a}$ with $a \sim 2$ emerged. Such a behavior has been attributed to the saturation of nonradiative recombination channels in the barrier as reported in [67].

13.6.2 High Quality Quantum Dots on Si Substrate as Single Photn Emitters

Finally, the integration of high quality GaAs quantum nanostructures by droplet epitaxy on Si substrate as single photon emitters has been demonstrated [47].

The integration of quantum photonic nanodevices on Si-based circuitry is of revolutionary importance, as it would open the possibility to pursue integration between “classical” electronics and “quantum” photonics on a single technological Si-based platform. Quantum photonic effects have been first demonstrated with single atoms trapped in optical cavities [68]. Recently, the quantum dot approach allowed the semiconductor optoelectronic technology to be used for the generation, control, and detection of photons [69]. The implementation of semiconductor quantum light sources opened the possibility of integrating into single semiconductor chips a complete photonic integrated circuit (IC) for quantum information processing, including logic and memory elements, which are essential elements in quantum information processing.

In [47], it was demonstrated the fabrication of a III–V single photon emitter based on GaAs quantum dots, working at liquid nitrogen temperatures, on Si substrate obtained by means of IC compatible fabrication protocols. The fabricated single photon emitter consists of singly addressable GaAs/AlGaAs quantum dot deposited by droplet epitaxy on a GeonSi thin virtual substrate. Very bright and sharp individual lines of exciton and biexciton are resolved in the spectra and can be followed up to 150 K. The nature of quantum emitters of single photon pulses is demonstrated up to 80 K by Hanbury-Brown and Twiss (HBT) interferometric correlations. The CMOS compatible fabrication protocols as well as the high temperature stability of the single photon emission allow for the implementation of this protocol in present day silicon IC technology.

The structure of the device consists of a 2.5 μm thick Ge layer deposited by LEPECVD on a Si(001) substrates, 6° misoriented toward [110] providing a thin virtual substrate. Subsequent thermal annealing treatments lead to a reduced density of threading dislocations. Transmission electron microscopy (TEM) studies, performed in the two-beam diffraction contrast mode, indicate that the defects are mainly confined within few hundred nanometers from the Ge/Si interface inside the Ge layer (Fig. 13.14 panel c). Threading dislocations crossing the whole Ge layer were not seen by TEM suggesting that their density could even be somewhat smaller than 10^7 cm^{-2}. The dislocation segments visible in (Fig. 13.14 panel c) inside the Ge layer above the dislocation band at the Ge/Si interface are residues of the dislocation annihilation process. It is worth mentioning that the deposition and thermal treatment protocols used for the GeonSi substrate are fully compatible with the existing IC fabrication lines. A half micron thick GaAs buffer separates the active layer from the Ge/GaAs interface. TEM analysis do not evidence threading dislocations in the III–V volume. As expected, anti-phase domains are confined in the first few nm from the GaAs/Ge interface (Fig. 13.14 panel b).

The single photon emitters are realized embedding GaAs quantum dots into an $Al_{0.3}Ga_{0.7}As$ barrier. The dots are fabricated by droplet epitaxy. For this specific application, droplet epitaxy advantages are the extreme and independent tunability

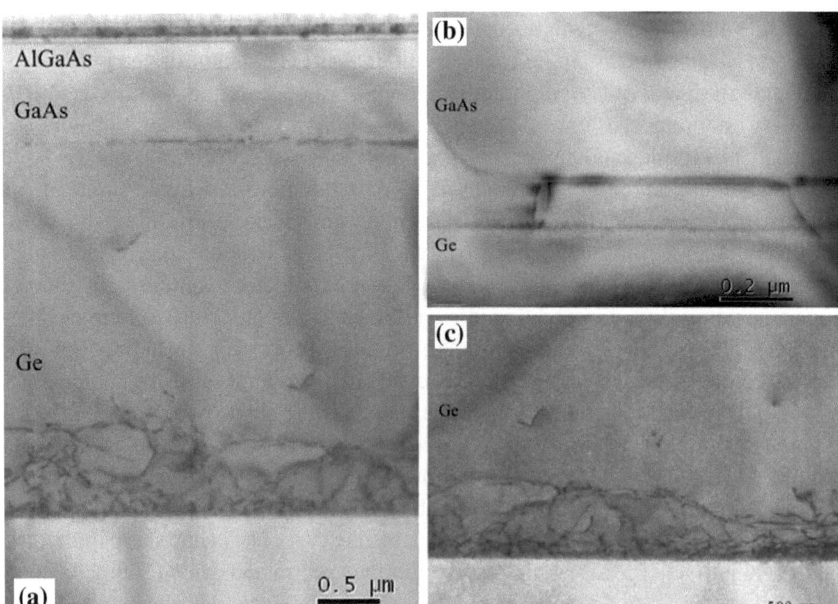

Fig. 13.14 **a** bright field TEM image of the whole structure. **b** magnification of Ge/GaAs interface in a different region. **c** magnification of Si/Ge interface

of the quantum dot density and size, which permits to obtain densities down to $10^8 cm^{-2}$, which, in turn,allows to take the PL signal from a single nanostructure, and tune the emissions between 1.55 and 1.85 eV.

Atomic force microscopy images of the uncapped quantum dots (Fig. 13.15) show a dot areal density 2.5×10^8 cm^{-2} and a dot shape with a mean height and radius of 8 and 40 nm respectively (20 % relative dispersion).

The demanding task of fabricating high efficiency quantum dot material at low thermal budget, required the optimization of the droplet epitaxy process as well of in-situ and ex-situ annealing processes compatible with present day CMOS technology. This was achieved in particular changing the grow procedure outlined in [45]. The substrate temperature was then increased to 180 °C during arsenization, but using the same beam equivalent flux of 5×10^{-5} torr. After As irradiation, a new step was introduced to improve the quality of GaAs quantum dots. The As beam equivalent pressure was reduced to 3×10^{-6} torr and the temperature increased to 400 °C. At the same temperature and As flux, the sample then underwent a 10 min before the growth of the AlGaAs barrier. A postgrowth, ex-situ rapid thermal annealing at 750 °C for four minutes completed the fabrication procedure. As already reported in the previous sections, annealing is required to remove the defects introduced in GaAs quantum dots and AlGaAs barrier growth at low temperature.

The PL measurements were performed placing the sample in a cold-finger helium flow cryostat. Continuous wave experiments were done by a frequency-doubled Nd-yttrium aluminum garnet laser $\lambda_{exc} = 532$ nm, while time-resolved measurements were performed exciting the sample with the second harmonic $\lambda_{exc} = 400$ nm of a

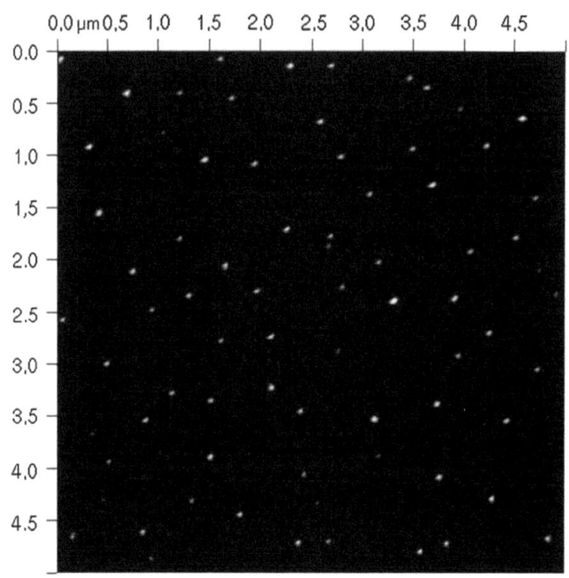

Fig. 13.15 5×5 μm AFM image of GaAs quantum dots

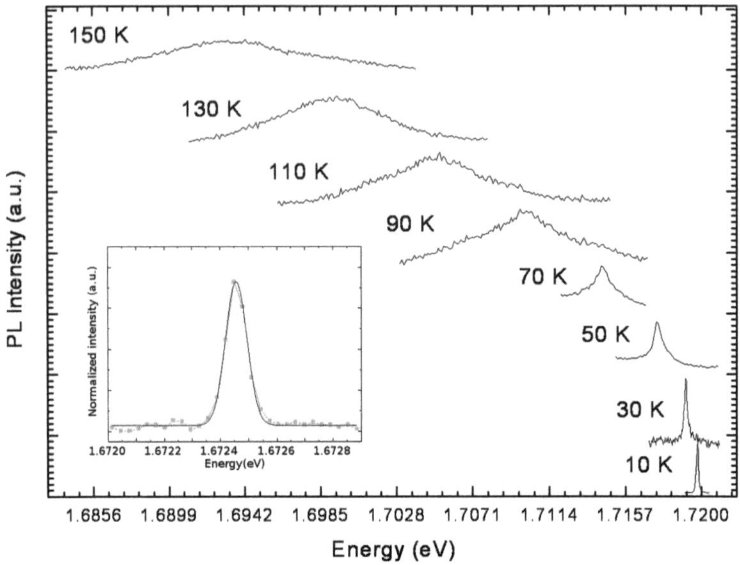

Fig. 13.16 PL spectra of a single quantum dot at different temperatures. Each spectrum is normalized to its maximum and vertically shifted for clarity. *Inset* magnification of the PL spectrum taken at 10 K. The FWHM of the line is 80 μeV

picosecond Ti-sapphire laser. Micro-PL experiments by means of a confocal setup (main objective 100x, $NA = 0.7$) leading to a lateral resolution of about 0.7 μm in the visible wavelength range we used for detecting the individual nanostructure. The small areal density allows collection of the PL from a few nanoemitters and it is enough for accessing the single quantum dot emission. The collected PL was fed either into a 30 cm focal length single monochromator (for time resolved and HBT measurements, with a spectral resolution of the order of 250 μeV) or into a 100 cm focal length double monochromator (for high resolution spectral resolved measurements, with a spectral resolution of the order of 40 μeV) and detected by a silicon-based charge coupled device camera. In time-resolved experiments, the PL was detected by an avalanche Si photodiode, providing a time resolution of 80 ps. HBT experiments were performed by splitting the signal out of the monochromator in two parts (by a fiber optical splitter) and sending them into two identical avalanche Si photodiodes.

The PL emission of individual quantum dots consists of sharp peaks, whose line shape at low T is well reproduced by a Gaussian profile. The broadening of the line at low T depends on the environment surrounding the nanostructure, due to spectral diffusion effects. Details of a typical PL spectra at 10 K are reported in the inset of Fig. 13.16, and the full width at half maximum linewidth (FWHM), as obtained from a Gaussian fit, is of the order of 80 μeV. The emission is as bright as the typical emission of GaAs quantum dots from a reference sample grown on a GaAs substrate. It was possible to follow the emission from a single quantum dot increasing the temperature, as reported in Fig. 13.16. The narrow line clearly visible in the low T is attributed to the so-called zero phonon line (ZPL) emission from the quantum dots. As the temperature increases, the quantum dot emission band redshifts and is strongly deformed. For all the quantum dots the PL band remains characterized, up to roughly 50 K, by a single ZPL, whose width increases with temperature. Around 70 K, the nonLorentzian sidebands start to appear and their relative intensity increases with temperature. Finally, above 100 K, the quantum dot emission is dominated by the sideband contribution. Also, the thermal quenching is not dramatic and it is possible to follow the emission from the single quantum dot up to $T = 150$ K, as reported in Fig. 13.16, denoting the high optical quality of the dots and indicating that the presence of threading dislocations in the substrate, do not significantly degrade the performance of quantum dot emission.

Increasing the excitation power density the quantum dot PL emission, at low T, turns into a multiplet. The spatial analysis of the emission PL bands at high excitation power density is reported in Fig. 13.17. It can be noted that the spatial localization of the emission lines is of the order of 1 μm which stems from our experimental resolution. The multiplet PL band clearly arises from the same spatial position, and therefore is originated from the multiexcitonic nature of the emission at high injection levels. This is confirmed by the power dependence analysis of the two main PL line X and XX of the multiplet band, attributed to the exciton and biexciton emissions. The PL intensity of the exciton line increases almost linearly at low excitation power, up to saturation and quenching at high power. Accordingly with the random population model, this behavior was fitted with Poissonian distributions $P(n) = \langle n \rangle \exp(-\langle n \rangle / n!)$ with $n = 1$ for the X and $n = 2$ for XX [70]. The value $\langle n \rangle$ is the statistical average over many realizations of the exciton population inside the quantum dots. Following the results of Ref. [70], we set $\langle n \rangle = \alpha P_{\text{exc}}^{\beta}$ (with P_{exc} in mW), where the phenomenological parameters α and β depend on the quantum dots carrier capture kinetics. The experimental data agree very well with the Poissonian

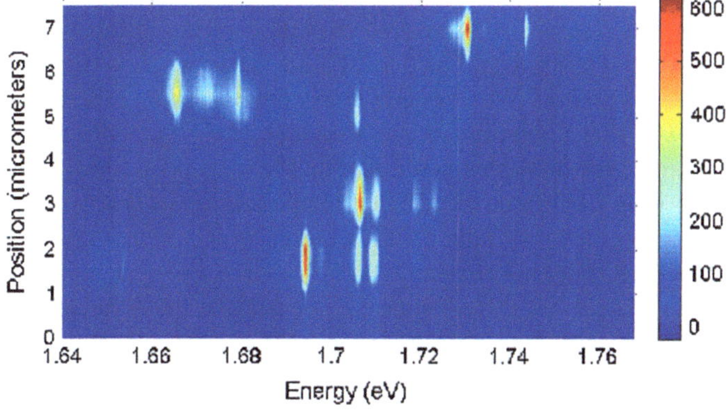

Fig. 13.17 Spatially and spectrally resolved PL map. *Brighter areas* indicate more intense emission

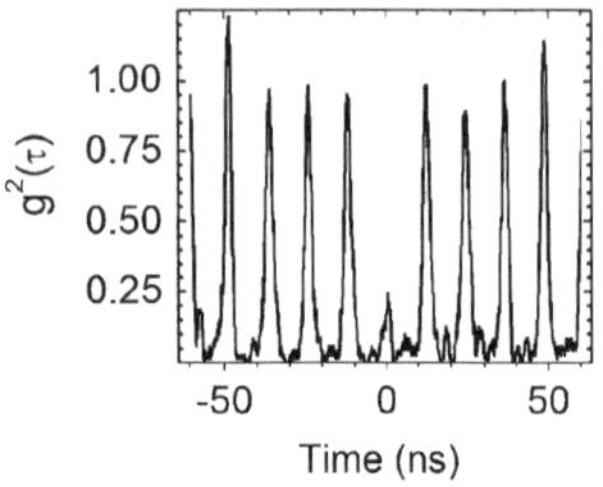

Fig. 13.18 Pulsed excitation measurement of $g^{(2)}(\tau)$ of a single quantum dot emission line at 80 K. The observed $g^{(2)}(0)$ values always remain below 0.27 in all the 10–80 K temperature range

distribution, for $\alpha = 8 \pm 1$ and $\beta = 0.9 \pm 0.01$, confirming the X and XX attribution of the PL lines. These multiexciton measurements, not only state the high quality of the dots, but also give important information on the recombination and capture dynamics. The emission intensity of the X line at saturation directly probes the radiative efficiency of the QDs when the QDs statistical population is $\langle n \rangle = 1$. At the same time, the corresponding value of excitation power $P_{\text{exc}}^{\text{sat}}$ provides an estimation of the capture efficiency. By comparing these quantities with the values obtained for GaAs quantum dots grown by droplet epitaxy on GaAs substrates, we conclude that the GaAs quantum dots grown on Si have a radiative efficiency quite similar to that of GaAs QDs grown on GaAs substrates. Furthermore, the value of $P_{\text{exc}}^{\text{sat}}$ is very much the same for the QDs on Si and on GaAs, denoting a similar QD capture efficiency, and therefore an effective high quality AlGaAs barrier surrounding the dots.

The time-resolved measurements of the X and XX lines of the single QDs give further confirmation of the high quality of the quantum dots. The lifetime of the X emission is of the order of 700 ps, comparable with the lifetime of quantum dots grown on GaAs substrates, denoting the lack of nonradiative channels in the QDs grown on Si. It is therefore demonstrated that possible threading dislocations do not affect the majority of dots. A further interesting feature is that increasing the excitation power density, the risetime of the X emission becomes longer and longer due to the state filling of the quantum dots.

For demonstrating the quantum nature of the light emitted from the quantum dots grown on Si substrate, we measured the autocorrelation function, $g^{(2)}(\tau) = \langle I(t) * I(t+\tau)\rangle / \langle I(t)\rangle^2$, as a function of time delay τ, in a HBT experiment, by using pulsed excitation. With respect to continuous wave experiments, pulsed excitation allows a direct determination of the value of $g^{(2)}(\tau)$ without the need of a fitting procedure taking into account the exciton lifetime and detector response [69]. Measurements were performed at different temperatures. In order to highlight the possibilities of integrating a single photon emitter on IC technology we report the data up to $T = 80$ K. For an ideal single-photon source, the peak at zero time delay is absent, $g^{(2)}(\tau = 0) = 0$; as the source cannot produce more than one photon per excitation period, clearly the two detectors cannot fire simultaneously. The single photon character is, however, demonstrated for $g^{(2)}(\tau = 0) < 0.5$ which can be reached only for quantum source of individual photons.

The recorded $g^{(2)}(\tau)$ for nonresonant pulsed optical excitation of the X emission of a single quantum dot in the 10–80 K temperature range shows a clear antibunching of the peak at zero delay, with the $g^{(2)}(\tau = 0)$ value always below 0.27. Figure 13.18 reports the $g^{(2)}(\tau)$ at 80 K. The low value of $g^{(2)}(\tau = 0)$ found for all the temperatures below 80 K is an evident and clear signature of an almost ideal single photon source.

13.7 Conclusions

In conclusion, it has been demonstrated the possibility to integrate high quality quantum GaAs nanostructures monolithically integrated on Si substrates through a thin Ge layer, using a CMOS compliant fabrication techniques and able to operate at microelectronics compatible temperatures. This achievement demonstrates the viability of an integrated quantum photonic-electronic approach for a new generation ICs on the silicon platform.

References

1. Fischer, R., Masselink, W.T., Klem, J., Henderson, T., McGlinn, T.C., Klein, M.V., Morkoc, H., Mazur, J.H., Washburn, J.: Growth and properties of GaAs/AlGaAs on nonpolar substrates using molecular beam epitaxy. J. Appl. Phys. **58**, 374 (1985)
2. Wang, W.I.: Molecular beam epitaxial growth and material properties of GaAs and AlGaAs on Si (100). Appl. Phys. Lett. **44**, 1149–1151 (1984)
3. Fang, S.F., Adomi, K., Iyer, S., Morkoc, H., Zabel, H., Choi, C., Otsuka, N.: Gallium arsenide and other compound semiconductors on silicon. J. Appl. Phys. **68**, R31 (1990)
4. Chadi, D.J.: Atomic and electronics-structures of reconstructed Si(100) surfaces. Phys. Rev. Lett. **43**, 43 (1979)
5. Hamers, R.J., Tromp, R.M., Demuth, J.E.: Scanning tunneling microscopy of Si(001). Phys. Rev. B **34**, 5343 (1986)
6. Chadi, D.J.: Stabilities of single-layer and bilayer steps on Si(001) surfaces. Phys. Rev. Lett. **59**, 1691 (1987)
7. Bringans, R.D., Biegelsen, D.K., Swartz, L.E.: Atomic-Step rearrangement on Si(100) by interaction with Arsenic and the implication for GaAs-on-Si epitaxy. Phys. Rev. B **44**, 3054 (1991)
8. Biegelsen, D.K., Ponce, F.A., Smith, A.J., Tramontana, J.C.: Initial stages of epitaxial growth of GaAs on (100) silicon. J. Appl. Phys. **61**, 1856 (1987)
9. Politi, P., Grenet, G., Marty, A., Ponchet, A., Villain J.: Instabilities in crystal growth by atomic or molecular beams. Phys. Rep. **328**, 271 (2000)
10. LeGoues, F.K., Reuter, M.C., Tersoff, J., Hammar, M., Tromp R.M.: Cyclic growth of strained-relaxed islands. Phys. Rev. Lett. **73**, 300 (1994)
11. Guha, S., Madhukar, A., Rajkumar, K.C.: Onset of incoherency and defect introduction in the initial stages of molecular beam epitaxy growth of highly strained In. Appl. Phys. Lett. **57**, 2110 (1990)
12. Van Der Merwe, J.H.: Crystal interfaces. Part II. Finite overgrowths. J. Appl. Phys. **34**, 123 (1963)
13. Matthews, J.W., Blakeslee, A.E.: Defects in epitaxial multilayers. Misfit dislocations. J. Crystal Growth **27**, 118 (1974)
14. Otsuka, N., Choi, C., Kolodziejski, L.A., Gunshor, R.L., Fischer, R., Peng, C.K., Morkoc, H., Nakamura, Y., Nagakura, S.: Study of heteroepitaxial interfaces by atomic resolution electron microscopy. J. Vacuum Sci. Technol. B **4**, 896 (1986)

15. Yang, V.K., Groenert, M., Leitz, C.W., Pitera, A.J., Currie, M.T., Fitzgerald, E.A.: Crack formation in GaAs heteroepitaxial films on Si and SiGe virtual substrates. J. Appl. Phys. **93**, 3859 (2003)
16. Fischer, R., Chand, N., Kopp, W., Morkoc, H., Erickson, P., Youngman R.: GaAs bipolar transistors grown on (100) Si substrates by molecular beam epitaxy. Appl. Phys. Lett. **47**, 397 (1985)
17. Baribeau, J.M., Jackman, T.E., Houghton, D.C., Maigné, P., Denhoff M.W.: Growth and characterization of $Si_{1-x}Ge_x$ and Ge epilayers on (100) Si. J. Appl. Phys. **63**, 5738 (1988)
18. Sieg, R.M., Ringel, S.A., Ting, S.M., Samavedam, S.B., Currie M., Langdo T., Fitzgerald, E.A.: Toward device-quality GaAs growth by molecular beam epitaxy on offcut $Ge/Si_{1-x}Ge_x/Si$ substrates. J. Vac. Sci. Technol. B **16**, 1471 (1998)
19. Horikoshi, Y., Kawashima, M.: Growth mechanism of GaAs during migration-enhanced epitaxy at low growth temperatures. Jpn. J. Appl. Phys. **28**, 200 (1989)
20. Ting, S.M., Fitzgerald, E.A.: Metal-organic chemical vapor deposition of single domain GaAs on $Ge/Ge_xSi_{1-x}/Si$ substrates. J. Appl. Phys. **87**, 2618 (2000)
21. Brammertz, G., Mols, Y., Degroote, S., Motsnyi, V., Leys, M., Borghs, G., Caymax M.: Low-temperature photoluminescence study of thin epitaxial GaAs films on Ge substrates. J. Appl. Phys. **99**, 93514 (2006)
22. Visser, E.P., Tang, X., Wieleman, R.W., Giling, L.J.: Deep-level photoluminescence studies on Si-doped, metalorganic chemical vapor deposition grown $Al_xGa_{1-x}As$. J. Appl. Phys. **69**, 3266 (1991)
23. Hudait, M.K., Modak, P., Hardikar, S., Krupanidhi, S.B.: Photoluminescence studies on Si-doped GaAs/Ge. J. Appl. Phys. **83**, 4454 (1998)
24. Fitzgerald, E.A., Xie, Y., Green, M.L., Brasen, D., Kortan, A.R., Michel, J., Mii, Y.J., Weir, B.E.: Totally relaxed Ge_xSi_{1-x} layers with low threading dislocation densities grown on Si substrates. Appl. Phys. Lett. **59**, 811 (1991)
25. Samavedam, S.B., Fitzgerald, E.A.: Novel dislocation structure and surface morphology effects in relaxed Ge/Si-Ge(graded)/Si structures. J. Appl. Phys. **81**, 3108 (1997)
26. Rosenblad, C., Deller, H.R., Dommann, A., Meyer, T., Schroeter, P., von Kaenel H.:. Silicon epitaxy by low-energy plasma enhanced chemical vapor deposition. J. Vac. Sci. Technol. A **16**, 2785 (1998)
27. Novikov, P.L., Le Donne, A., Cereda, S., Miglio, L., Pizzini, S., Binetti, S., Rondanini, M., Cavallotti, C., Chrastina, D., Moiseev, T., von Kaenel, H., Isella, G., Montalenti, F.: Crystallinity and microstructure in Si films grown by plasma-enhanced chemical vapor deposition: a simple atomic-scale model validated by experiments. Appl. Phys. Lett. **94**, 051904 (2009)
28. Cereda, S., Zipoli, F., Bernasconi, M., Miglio, L., Montalenti F.: Thermal-hydrogen promoted selective desorption and enhanced mobility of adsorbed radicals in silicon film growth. Phys. Rev. Lett. **100**, 046105 (2008)
29. Thomas, S.G., Bharatan, S., Jones, R.E., Thoma, R., Zirkle, T., Edwards, N.V., Liu, R., Wang, X.D., Xie, Q., Rosenblad, C., Ramm, J., Isella, G., von Kaenel H.: Structural characterization of thick, high-quality epitaxial Ge on Si substrates grown by low-energy plasma-enhanced chemical vapor deposition. J. Electron. Mater. **32**, 976 (2003)
30. Marchionna, S., Virtuani, A., Acciarri, M., Isella, G., von Kaenel, H.: Defect imaging of SiGe strain relaxed buffers grown by LEPECVD. Mater. Sci. Semicond. Process. **9**, 802 (2006)
31. Isella, G., Chrastina, D., Rossner, B., Hackbarth, T., Herzog, H.-J., Konig, U., von Kaenel, H.: Low-energy plasma-enhanced chemical vapor deposition for strained Si and Ge heterostructures and devices. Solid-State Electron. **48**, 1317 (2004)
32. Isella, G., Osmond, J., Kummer, M., Kaufmann, R., von Kaenel, H.: Hetero-junction photodiodes fabricated from Ge/Si (100) layers grown by low-energy plasma-enhanced CVD. Semicond. Sci. Technol. **22**, S26 (2007)
33. Osmond, J., Isella, D., Chrastina, D., Kaufmann, R., Acciarri, M., von Kaenel H.: Ultralow dark current Ge/Si(100) photodiodes with low thermal budget. Appl. Phys. Lett. **94**, 201106 (2009)

34. Luan, H.-C., Lim, D.R., Lee, K.K., Chen, K.M., Sandland, J.G., Wada, K., Kimerling, L.C.: High-quality Ge epilayers on Si with low threading-dislocation densities. Appl. Phys. Lett. **75**, 2909 (1999)
35. Colace, L., Masini, G., Cozza, S., Assanto, G., DeNotaristefani, F., Cencelli, V.: Near-infrared camera in polycrystalline germanium integrated on complementary-metal-oxide semiconductor electronics. Appl. Phys. Lett. **90**, 011103 (2007)
36. Colombo, D., Grilli, E., Guzzi, M., Sanguinetti, S., Fedorov, A., von Kaenel, H., Isella, G.: Study of thermal strain relaxation in GaAs grown on Ge/Si substrates. J. Lumin. **121**, 375 (2006)
37. Colombo, D., Grilli, E., Guzzi, M., Sanguinetti, S., Marchionna, S., Bonfanti, M., Fedorov, A., von Kaenel, H., Isella, G., Muller, E.: Analysis of strain relaxation by microcracks in epitaxial GaAs grown on Ge/Si substrates. J. Appl. Phys. **101**, 103519 (2007)
38. Halbwax, M., Bouchier, D., Yam, V., Debarre, D., Nguyen, L.H., Zheng, Y., Rosner, P., Benamara, M., Strunk, H.P., Clerc, C.: Kinetics of Ge growth at low temperature on Si(001) by ultrahigh vacuum chemical vapor deposition. J. Appl. Phys. **97**, 064907 (2005)
39. Bietti, S., Cecchi, S., Frigeri, C., Grilli, E., Fedorov, A., Vinattieri, A., Gurioli, M., Isella, G., Sanguinetti, S.: Fabrication of Ge-on-Si substrates for the integration of high-quality GaAs nanostructures on Si. ECS Trans. 50 (**9**), 783, (2012)
40. Berbezier, I., Ronda, A.: Self-assembling of Ge dots on nanopatterns: experimental investigation of their formation, evolution and control. Phys. Rev. B **75** 195407 (2007)
41. Sakai, A., Tatsumi, T., Aoyama, K.: Growth of strain-relaxed Ge films on Si(001) surfaces. Appl. Phys. Lett. **71**, 3510 (1997)
42. Kaganer, V.M., Köhler, R., Schmidbauer, M., Opitz, R.: X-ray diffraction peaks due to misfit dislocations in heteroepitaxial structures. Phys. Rev. B **55**, 1793 (1997)
43. Tsui, R.K., Gershenzon, M.: Plastic deformation and fracture resulting from stresses caused by differential thermal contraction in GaP/Si heterostructures. Appl. Phys. Lett. **37**, 218 (1980)
44. Chalmers, S.A., Gossard, A.C., Petroff, P.M., Gaines, J.M., Kroemer, H.: A reflection high-energy electron-diffraction study of (100) GaAs vicinal surfaces. J. Vac. Sci. Technol. B **7**, 1357 (1989)
45. Bietti, S., Sanguinetti, S., Somaschini, C., Koguchi, N., Isella, G., Chrastina, D., Fedorov, A.: Fabrication of GaAs quantum dots by droplet epitaxy on Si/Ge virtual substrate. IOP Conf. Series: Mater. Sci. Eng. **6**, 012009 (2009)
46. Bietti, S., Somaschini, C., Sanguinetti, S., Koguchi, N., Isella, G., Chrastina, D.: Fabrication of high efficiency III-V quantum nanostructures at low thermal budget on Si. Appl. Phys. Lett. **95**, 241102 (2009)
47. Cavigli, L., Bietti, S., Accanto, N., Minari, S., Abbarchi, M., Isella, G., Frigeri, C., Vinattieri, A., Gurioli, M., Sanguinetti, S.: High temperature single photon emitter monolithically integrated on silicon. Appl. Phys. Lett. **100**, 231112 (2012)
48. Koguchi, N., Takahashi, S., Chikyow, T.: New MBE growth method for InSb quantum well boxes. J. Cryst. Growth **111**, 688 (1991)
49. Koguchi, N., Ishige, K., Takahashi, S.: New selective molecular-beam epitaxy growth method for direct formation of GaAs quantum dots. J. Vac. Sci. Technol. B **11**, 787 (1993)
50. Mano, T., Koguchi, N.: Nanometer-scale GaAs ring structure grown by droplet epitaxy. J. Cryst. Growth **278**, 108 (2005)
51. Somaschini, C., Bietti, S., Koguchi, N., Sanguinetti, S.: Fabrication of multiple concentric nanoring structures. Nano Lett. **9**, 3419 (2009)
52. Somaschini, C., Bietti, S., Sanguinetti, S., Koguchi, N., Fedorov, A.: Self-assembled GaAs/AlGaAs coupled quantum ring-disk structures by droplet epitaxy. Nanotechnology **21**, 125601 (2010)
53. Watanabe, K., Koguchi, N., Gotoh, Y.: Fabrication of GaAs quantum dots by modified droplet epitaxy. Jpn. J. Appl. Phys. **39**, 79 (2000)
54. Koguchi, N., Ishige, K.: Growth of GaAs epitaxial microcrystal on an S-terminated GaAs substrate by successive irradiation of Ga and As molecular beams. Jpn. J. Appl. Phys. **32**, 2052 (1993)

55. Mano, T., Kuroda, T., Noda, T., Sakoda, K.: Ordering of GaAs quantum dots by droplet epitaxy. Phys. Status Solidi B-Basic Res. **246**, 729 (2009)
56. Sanguinetti, S., Watanabe, K., Kuroda, T., Minami, F., Gotoh, Y., Koguchi, N.: Effects of post-growth annealing on the optical properties of self-assembled GaAs/AlGaAs quantum dots. J. Cryst. Growth **242**, 321
57. Bastard, J.-Y., Marzin, G.: Calculation of the energy levels in InAs/GaAs quantum dots. Solid State Commun. **92**, 437 (1994)
58. Kuroda, T., Mano, T., Ochiai, T., Sanguinetti, S., Sakoda, K., Kido, G., Koguchi, N.: Optical transitions in quantum ring complexes. Phys. Rev. B **72**, 205301 (2005)
59. Pavesi, L., Guzzi, M.: Photoluminescence of Alx Ga1-xAs alloys. J. Appl. Phys. **75**, 4779 (1994)
60. Oelgart, G., Schwabe, R., Fieseler, H., Jacobs, B.: Photoluminescence of Ge-doped $Al_xGa_{1-x}As$. J. Appl. Phys. **3**, 943 (1988)
61. Lester, L., Stintz, A., Li, H., Newell, T., Pease, E., Fuchs, B., Malloy, K.: Optical characteristics of 1.24-μm InAs quantum-dot laser diodes. IEEE Photonics Technol. Lett. **11**, 931 (1999)
62. Raghavan, S., Rotella, P., Stintz, A., Fuchs, B., Krishna, S., Morath, C., Cardimona, D.A., Kennerly, S.W.: High-responsivity, normal-incidence long-wave infrared $InAs/In_{0.15}Ga_{0.85}As$ dots-in-a-well detector. Appl. Phys. Lett. **81**, 1369 (2002)
63. Wu, J., Li, Z., Shao, D., Manasreh, M.O., Kunets, V.P., Wang, Z.M., Salamo, G.J., Weaver, B.D.: Multicolor photodetector based on GaAs quantum rings grown by droplet epitaxy. Appl. Phys. Lett. **94**(17), 171102 (2009)
64. Sanguinetti, S., Mano, T., Gerosa, A., Somaschini, C., Bietti, S., Koguchi, N., Grilli, E., Guzzi, M., Gurioli, M., Abbarchi, M.: Rapid thermal annealing effects on self-assembled quantum dot and quantum ring structures. J. Appl. Phys. **104** 113519 (2008)
65. Bietti, S., Somaschini, C., Sarti, E., Koguchi, N., Sanguinetti, S., Isella, G., Chrastina, D., Fedorov, A.: Photoluminescence study of low thermal budget III-V nanostructures on silicon by droplet epitaxy. Nanoscale Res. Lett. **5**, 1650 (2010)
66. Mantovani, V., Sanguinetti, S., Guzzi, M., Grilli, E., Gurioli, M., Watanabe, K., Koguchi, N.: Low density GaAs/AlGaAs quantum dots grown by modified droplet epitaxy. J. Appl. Phys. **96**, 4416 (2004)
67. Sanguinetti, S., Colombo, D., Guzzi, M., Grilli, E., Gurioli, M., Seravalli, L., Frigeri, P., Franchi, S.: Carrier thermodynamics in $InAs/In_xGa_{1-x}$ As quantum dots. Phys. Rev. B **74**, 1 (2006)
68. Kimble, H.J.: The quantum internet. Nature **453**, 1023 (2008)
69. Shields, A.: Semiconductor quantum light sources. Nat. Photonics **1**, 215 (2007)
70. Abbarchi, M., Mastrandrea, C., Kuroda, T., Mano, T., Vinattieri, A., Sakoda, K., Gurioli, M.: Poissonian statistics of excitonic complexes in quantum dots. J. Appl. Phys. **106**, 053504 (2009)

Chapter 14
III–V Quantum-Dot Materials and Devices Monolithically Grown on Si Substrates

Huiyun Liu

Abstract The integration of III–V photonics materials and devices with Si microelectronics will enable the fabrication of complex optoelectronic circuits, which will permit the creation of the long-dreamed chip-to-chip and system-to-system optical communications. Direct epitaxial growth of semiconductor III–V compounds on Si substrates is one of the most promising candidates for the fabrication of photonics devices on the Si platform. III–V quantum dots (QDs) offer an attractive alternative to conventional quantum wells(QWs) for building III–V lasing devices on a Si platform due to their unique advantages. We developed the long-wavelength InAs/GaAs QD materials and devices monolithically grown on Si, Ge, and Ge-on-Si substrates by the use of Molecular Beam Epitaxy. Room-temperature(RT) lasing at a wavelength of around 1.3 μm has been achieved with threshold current densities of 64.3 A/cm^2 and lasing operation up to 83 °C for Si-based ridge-waveguide InAs/GaAs QD lasers with as-cleaved facets. The optical and electrical properties of InAs/GaAs QDs grown on Si substrates were further investigated to evaluate the potential for Si-based photodiodes. A peak responsivity of 5 mA/W was observed at 1.28 μm, while the dark current was two orders of magnitude lower than those reported for Ge-on-Si photodiodes. These studies ultimately form the basis for the monolithic integration of 1.3-μm InAs/GaAs QD lasers and detectors on the Si platform.

14.1 Introduction

Si microelectronics has been the engine of the modern information for almost 50 years. In our everlasting quest to process more data at faster speeds, while using the smallest components the $100 billion Si industry has successfully overcome many

H. Liu (✉)
Department of Electronic and Electrical Engineering, University College London, Torrington Place, London WC1E 7JE, UK
e-mail: huiyun.liu@ucl.ac.uk

H. Li et al. (eds.), *Silicon-based Nanomaterials*,
Springer Series in Materials Science 187, DOI: 10.1007/978-1-4614-8169-0_14,

© Springer Science+Business Media New York 2013

critical issues. The next critical challenge in the evolution of modern information system is the limitation of metal interconnects [1–4]. The continuous scaling of very-large-scale integration (VLSI) circuits and beyond pose significant problems for interconnects, especially for those responsible for long-distance communication on high-performance chips. Both power and delay of these interconnects will rise significantly in the near future, which are threatening the industry's future. The International Technology Roadmap for Semiconductor predicts that in spite of new materials like Cu and low-*k* dielectric, the performance improvement of advance VLSI is likely to begin to saturate unless a paradigm shift from present IC architecture is introduced.

The merger of photonics and electronics into a dual-function platform, the opto-electronic integrated circuit (OEIC) fabricated with existing Si infrastructure can overcome the interconnection issue while pushing forward Si microelectronics beyond the classical CMOS era [1, 4]. Incorporating photonic components into Si microelectronics has been the impetus behind the development of Si photonics for the last 30 years. Although Si-based light generation and modulation technologies have been studied extensively [5–7], a Si-based laser has been considered as the holy grail of Si photonics because (i) it is the most important active photonic device, (ii) the potential payoff is most significant, and (iii) it is one of the most difficult challenges to realize among all the Si photonic components [8]. Si and Ge have an indirect band structure, the significant challenge posed by nature. This means radiative recombination events do not occur frequently, and accordingly, radiative recombination processes for emitters are insignificant compared to nonradiative recombination [8].

Direct bandgap III–V compounds have robust photonic properties that can be tailored for III–V emitters operating at various wavelengths with high efficiency, large direct modulation bandwidth, and sufficient optical power output for many photonic applications. Integrating III–V emitters with Si microelectronics and photonics would thus provide the ideal solution for Si photonics [9]. The introduction of optical interconnects on a chip will speed up Si chip speeds very significantly to solve the impending Si CMOS interconnect challenge. However, there are huge implementation challenges for III–V compound materials on Si (III–V/Si) substrates [10, 11]. The main problem in realizing III–V devices on Si stems is the formation of three types of defects, i.e., anti-phase boundaries (APBs), threading dislocations (TD) and micro-cracks, as shown in Fig. 14.1.

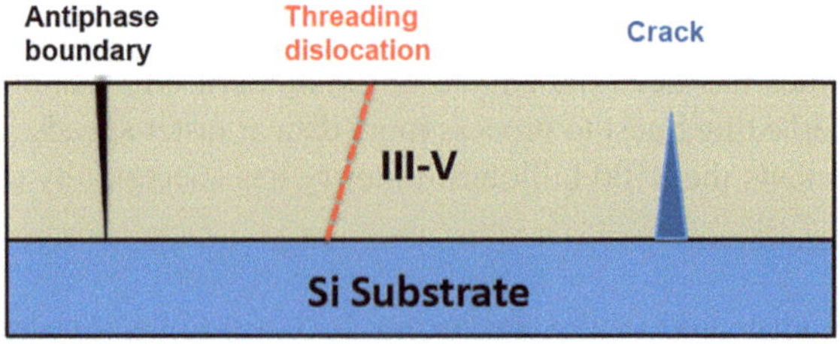

Fig. 14.1 Three types of defects generated during III–V/Si epitaxial growth

(1) Anti-phase boundaries: III–V compound materials are composed of two different atomic sublattices. Sublattice shift may nucleate during epitaxial growth of III–V on Si and Ge. Sheets of wrong nearest neighbor bonds, i.e., APBs or inversion domain boundaries (IDBs) can occur. IDBs are planar defects, which debilitate device performance by acting as nonradiative recombination centers.
(2) Micro-cracks: A mismatch in thermal expansion between III–V epilayers and Si substrate will result in tensile stress in III–V layer, and could lead to crack formation during cooling from growth temperature (> 500 °C) to room temperature. The use of thin compressive materials, such as InAlGaAs layers with In <6 % in the III–V buffer layers to compensate the thermal expansion difference and slowly cooling down the substrates after the growth has been used to suppress the formation of micro-cracks.
(3) Threading dislocations: The primary challenge of III–V/Si integration is the formation of high-density TDs due to the lattice mismatch between III–V compounds and Si. The lattice mismatch between GaAs (InP) and Si is 4.1 % (7.5 %). Typically, the TD density resulting from direct growth of GaAs or InP on Si is of the order of 10^{10}/cm^2, which is much higher than the residual TD density ($\sim 10^3$/cm^2) found in bulk GaAs and InP wafers. TDs are well known to act as nonradiative recombination centers and to promote dark line defects in the operation of semiconductor laser diodes, hence reducing optoelectronic conversion efficiency and device lifetime.
Despite the very great difficulties involved, great efforts have been made worldwide for more than 30 years because of the massive potential if successful. To date, although conventional III–V QW lasers have been demonstrated, their practicality is still a great challenge because of their very high threshold current densities, J_{ths} (such as 6.9 A/cm^2 in Ref. [10]).

14.2 The Advantages of Quantum Dots on the Direct Epitaxy of III–V on Si Substrates

III–V QDs are zero-dimensional crystal whose size is in nanometer (1 nm = 10^{-9} m) scale in three spatial directions. This small size of QDs leads to the confinement of electrons and holes in all three spatial directions giving unique property (a discrete energy spectrum) compared to other semiconductor structures, such as QWs. The QDs behave as artificial atoms with properties, which can be engineered by the crystal growth techniques, such as Molecular Beam Epitaxy. This unique QD property theoretically leads to revolutionary characteristics for QD lasers [12, 13]. These include temperature-independent operation, reduced drive current, reduced sensitivity to damage, and higher operating speeds, which are significantly superior to any conventional semiconductor QW lasers. In the last few years, QD lasers have been demonstrated with a significantly lower J_{th} relative to traditional semiconductor QW lasers and offers temperature insensitive operation (Characteristic Temperature, T_0

$\sim \infty$) up to 100 °C [12, 13]. This unique attribute of QD technology promotes us to develop III–V QD lasers on the Si platform. Furthermore, QD structures offer other special advantages over QWs, not directly related to their unique density of states, for the production of better III–V/Si photonics devices [14–16]. These include:

(1) Lower sensitivity to defects: Some of the TDs and APBs generated at the III–V epilayers grown on Si substrates will propagate through the III–V layers. Every TD propagating through QWs will become a nonradiative recombination center, leading to dramatically degraded performance of III–V QW devices. However, in the case of QD devices, one TD in the active region can only "kill" one or a few dots. It will not affect the majority of dots [16]. Furthermore, the electrons and holes within the active region are spatially localized in dots. This carrier localization could result in carriers not "seeing" the TDs. QD devices thus exhibit dramatically reduced sensitivity to defects.
(2) High resistance to degradation: The strain field of QDs gives greater mechanical strength to prevent the movement of dislocations, even in the presence of high-density nonradiative recombination centers. Consequently, in the case of III–V/Si devices with relative high density of TD, QD lasers should exhibit much longer lifetimes and higher operation temperature than QW devices [14, 16].
These advantageous properties of QDs led to the recent demonstrations of III–V QD lasers monolithically grown on Si substrates [17–20].

14.3 InAs/GaAs QD Lasers Grown on Si Substrates

The epitaxial materials were fabricated by solid-source Molecular Beam Epitaxy (MBE) system. InAs/GaAs QD samples were grown on phosphorus-doped (100)-oriented Si substrates with 4 ° offcut toward the [110] planes. The schematic layer structure of InAs/GaAs QDs is illustrated in Fig. 14.2. The oxide desorption was performed by holding the Si substrate at a temperature of 900 °C for 10 min. The Si substrate was then cooled down to the nucleation temperature of 30-nm GaAs

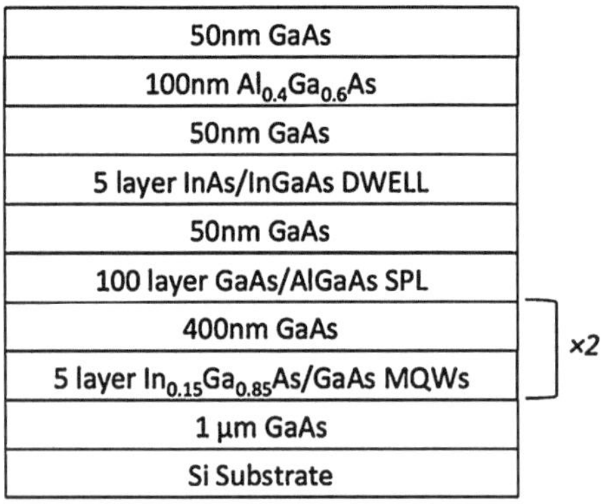

Fig. 14.2 Cross-sectional schematic of fabricated InAs/InGaAs DWELL structures grown on a Si substrate

layer with a low growth rate of 0.1 Monolayer (ML)/s, followed by 970-nm thick GaAs buffer layer grown at high temperature. Multilayer InGaAs/GaAs dislocation filter layers, consisting of two repeats of a five-period (10-nm $In_{0.15}$ $Ga_{0.85}As$/10-nm GaAs) superlattices (SPLs) and 400-nm GaAs, were deposited on the GaAs buffer layer. Finally, 1-μm SPL layers of alternating 2-nm GaAs/2-nm $Al_{0.4}Ga_{0.6}As$ layer completed the III–V buffer layers. A five-layer InAs/InGaAs dot-in-a-well (DWELL) structure was then grown at optimized conditions as on GaAs substrates, with each layer consisting of 3.0 MLs of InAs grown on 2 nm of $In_{0.15}Ga_{0.85}As$ and capped by 6 nm of $In_{0.15}Ga_{0.85}As$ at ~510 °C. 45-nm GaAs barriers separated the five DWELLs [21–25]. The InAs/GaAs DWELL structure was then embedded between 100-nm GaAs layer grown at 580 °C for GaAs, which was further confined by two 50-nm AlGaAs layers to prevent the photogenerated charge carriers from migrating to the substrate or surface at high temperatures.

The structural properties of InAs/GaAs QDs grown on Si substrates were studied by atomic force microscopy (AFM) and cross-sectional transmission electron microscope (TEM). AFM measurements were performed on uncapped samples in which the growth was halted after the formation of the first-layer InAs QDs. A typical AFM image for the sample grown on Si substrates is shown in Fig. 14.3a, from which a QD density of $\sim 4.3 \times 10^{10} cm^{-2}$ is obtained. Figure. 14.3b shows a typical cross-sectional TEM image of the InGaAs/GaAs QDs on a Si substrate, which indicates a dot height of ~6 nm and a base width of ~25 nm for this wafer. No dislocations are observed in a number of similar images.

The effects of GaAs nucleation temperature on the optical properties of InAs/GaAs QDs was studied with the initial 30-nm GaAs layer grown at 380 °C, 400 °C, and 420 °C, respectively. Photoluminescence (PL) measurements were performed under 532 nm excitation from a diode-pumped solid-state laser. The PL spectra were dispersed by a 0.25 m monochromator and detected by a TE-cooled Ge detector. Temperature-dependent PL measurements were taken from 10 to 300 K at a laser

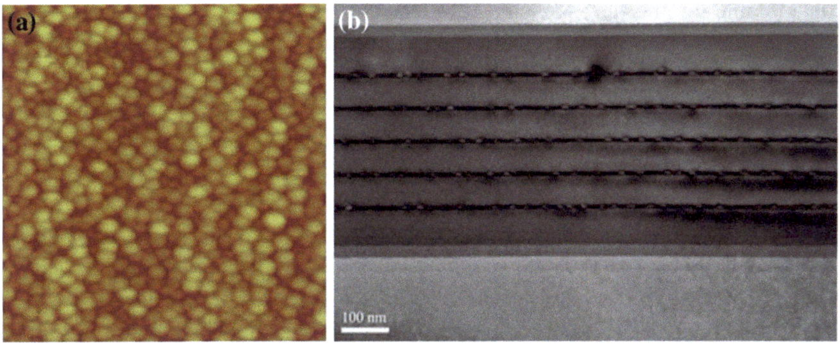

Fig. 14.3 **a** $1 \times 1\ \mu m^2$ AFM image of InAs/GaAs QDs grown on Si substrate with the GaAs nucleation temperature at 400 °C. **b** Dark field (200) TEM cross-sectional images of five-layer InAs/InGaAs DWELL structure grown on Si substrate with initial GaAs layer nucleated at 400 °C

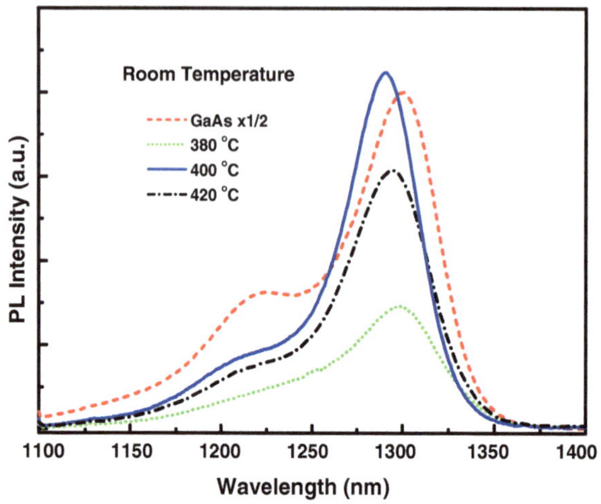

Fig. 14.4 Room-temperature PL spectra of 1.3-μm InAs/GaAs QDs directly grown on Si substrates with different growth temperatures for GaAs nucleation layer. The RT PL spectrum of GaAs-based InAs QDs grown with the same growth conditions is also shown as a reference

power of 30 mW. Figure 14.4 shows RT PL spectra of 1300-nm InAs/GaAs QDs grown on Si substrates with different GaAs nucleation temperatures. The GaAs-based InAs QD sample with the same growth conditions is also shown in Fig. 14.4 as a reference. It should be mentioned that the sample grown on GaAs substrate was fabricated under optimized conditions and represents high optical quality as the 1.3-μm InAs/GaAs QD laser diodes based on identical growth parameters give extremely low threshold current density, J_{th} and high output power at RT [13, 21]. The RT emission around 1.29 μm has been observed for all the samples shown in Fig. 14.4 with full width at half maximum (FWHM) of ~30 meV. These PL linewidths obtained in Fig. 14.4 are comparable to the values obtained for GaAs-based InAs QDs [23]. Of considerable significance is that the PL intensity of Si-based InAs/GaAs QD ground-state transition is strongly dependent on the nucleation temperature of initial GaAs layer on the Si substrate, with strongest PL intensity at 400 °C.

Further cross-sectional TEM measurements were performed to compare the structural properties of GaAs/Si interfaces with different GaAs nucleation temperatures, as shown in Fig. 14.5. The defects are generated at the GaAs/Si interface for all the samples shown in Fig. 14.5, due to the mismatch between GaAs and Si. While most defects are confined within 50 nm of GaAs/Si interface, some of them propagate into the III–V buffer layers. The density of defects propagating into GaAs buffer layer is highly dependent on the growth temperature of GaAs nucleation layer. The lowest defect density propagating into the GaAs buffer layer is obtained at 400 °C, which is also confirmed by the etch-pit density (EPD) measurements [26]. The defect densities of $1.03 \times 10^7/cm^2$, $6.03 \times 10^6/cm^2$, $8.17 \times 10^6/cm^2$ were obtained for the samples with the initial GaAs layer nucleated at 380, 400, and 420 °C, respectively.

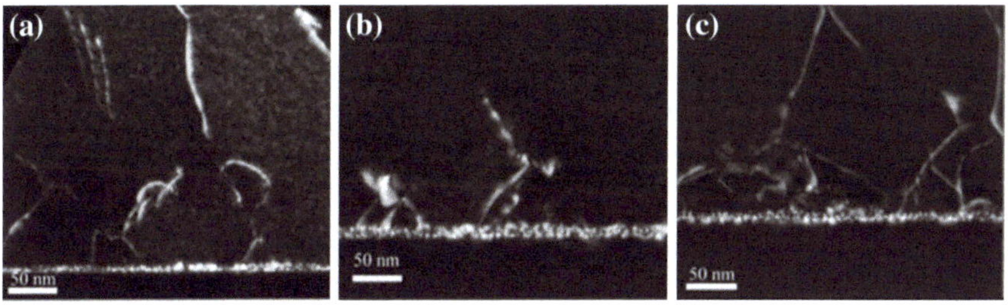

Fig. 14.5 Cross-sectional TEM images of GaAs/Si interface for the initial GaAs nucleation layer grown at **a** 380 °C, **b** 400 °C, and **c** 420 °C, respectively

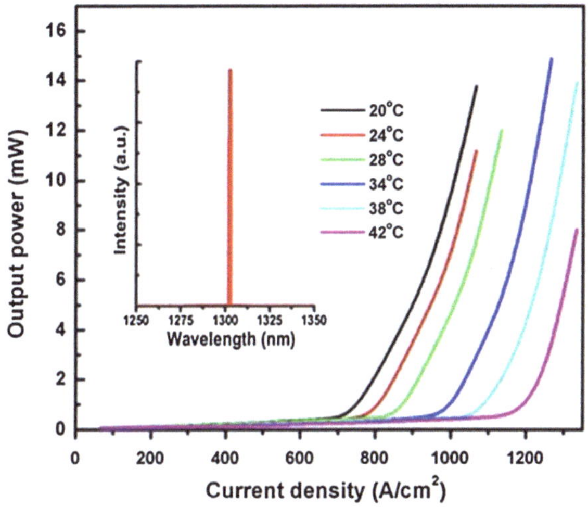

Fig. 14.6 Light output against current characteristic for InAs/GaAs quantum-dot laser on a Si substrate at various temperatures. The inset shows the laser optical spectrum above threshold at RT

With the use of the optimized GaAs nucleation temperature of 400 °C, a standard 1.3-μm DWELL laser structure was then grown on a Si substrate. The III–V buffer layer is similar to that used for studying the effects of the GaAs nucleation layer shown in Fig. 14.2 and consisted of the following layer sequence: GaAs buffer layer, InGaAs/GaAs dislocation filter layers, and GaAs/AlGaAs SPL layers. The five-layer InAs/InGaAs DWELL structure was grown at the center of an undoped 150-nm GaAs/AlGaAs waveguide layer with n-type lower and p-type upper cladding layers consisting of 1.5-μm $Al_{0.4}Ga_{0.6}As$ deposited at 610 °C. A 300-nm P^+-GaAs contact layer completed the growth. Broad-area 50-μm stripe lasers with cleaved facets were then formed by applying Ti/Au on the top GaAs contact layer and Cr/Au on the bottom of Si wafer. Devices of 3-mm cavity length were bar-tested, being directly probed without any mounting and bonding. Our device exhibits ground-state lasing at 1.302 μm by electrical carrier injection under pulsed operation with a pulse width of 0.1 μs and duty cycle of 0.01 %. The RT threshold current density, J_{th} is about 725 A/cm^2 at RT. QD laser operation is achieved for heatsink temperatures up to 42 °C, with a characteristic temperature, T_0, of ~44 K between 20 and 42 °C.

14.4 InAs/GaAs QD Lasers Grown on Ge Substrates

The main issue for the direct growth of GaAs on Si is the generation of high-density defects due to the lattice mismatch between GaAs and Si. Ge-on-Si (Ge/Si) technique can effectively bridge the lattice constant gap between GaAs and Si, because of the relatively small lattice mismatch (0.08 %), and closely matched thermal expansion between GaAs and Ge, and the completely miscibility between Ge and Si [27–29]. Furthermore, the fabrication of Ge/Si substrate is becoming a mature technology with low-density threading dislocations [28, 29]. It should also be noticed that, as the scaling of Si microelectronics devices approach the 22 nm node, Ge epilayers will replace Si as p-channel materials in CMOS device on Si substrates, because Ge has a much higher hole mobility than Si [30]. A promising method for incorporating III–V laser diodes on a Si platform is therefore to fabricate III–V materials and devices on Ge/Si substrates.

However, the growth of GaAs on Ge substrates is not well established so far because of the generation of APBs due to the nonpolar/polar interface between Ge/GaAs [11]. Therefore, the MBE growth of GaAs buffer layers on Ge substrates, off-cut 6 ° toward the [111] plane was first investigated. Oxide desorption was performed by holding the Ge substrate at a temperature of 400 °C. The substrate temperature was then increased to 650 °C and held at that temperature for 20 min. The substrate was then cooled to 380 °C for the growth of the III–V epitaxial layers. For the Ga prelayer, the base pressure was reduced to below 10^{-10} Torr in the MBE growth chamber before loading the Ge wafer into the growth chamber. To ensure total Ga coverage on the Ge substrate, 1.08 ML Ga was first deposited. Alternatively, the Ge surface was terminated with As by opening the valve of the As cracker for 1 min. After either As or Ga prelayers were deposited, 20 ML of GaAs were grown by migration enhanced epitaxy using alternating Ga and As_4 beams, and then the addition of III–V buffer layer at higher temperature. Initiation of GaAs with the typical procedure of using self-terminating As layer produces poor GaAs surface morphology due to APBs, see Fig. 14.7a, despite the large miscut of the (100) surface. However, initiation of GaAs growth with approximately 1 ML of Ga, i.e., Ga prelayer results in a very much smoother surface morphology for the GaAs layer, see Fig. 14.7b, which indicates the formation of a single-domain GaAs buffer layer on the Ge substrate [15]. Further TEM studies confirm the conclusion obtained from AFM images. The defects are only observed within 150 nm of the GaAs/Ge interface for the sample using the Ga prelayer approach, as shown in Fig. 14.7d, due to the relaxation of strain between the GaAs epilayer and the Ge substrate in the case of a thick GaAs layer. APBs are observed in the TEM image of the sample with an As prelayer, as shown in Fig. 14.7c. APBs are generated at the GaAs/Ge interface and

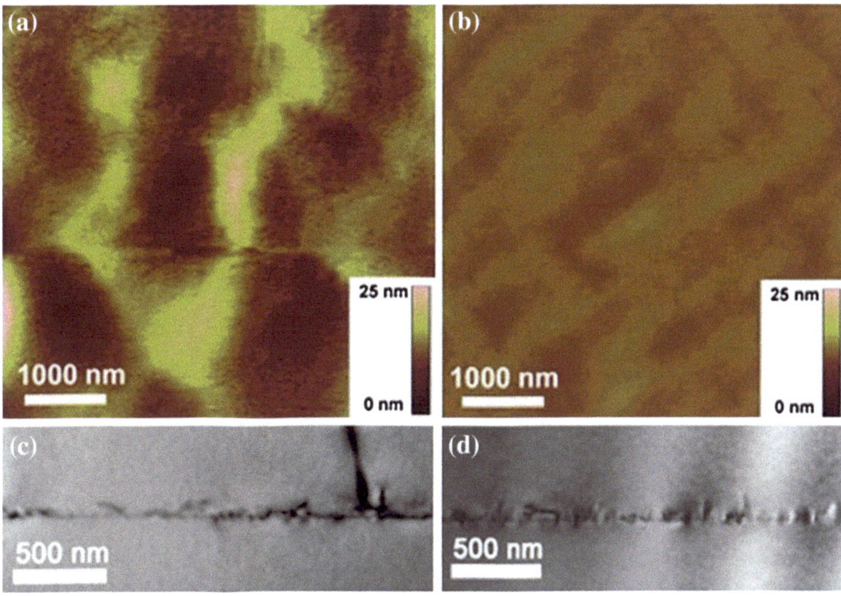

Fig. 14.7 $5 \times 5\ \mu m^2$ Atomic force microscopy images of the surface morphology for 1.2 μm GaAs on Ge substrates with As prelayer (**a**) and Ga prelayer (**b**) growth techniques. TEM images of the interface between the GaAs buffer layer and the Ge substrate with As prelayer (**c**) and Ga prelayer (**d**) growth techniques

propagate into the GaAs buffer layer, causing the surface roughness as observed in the AFM image in Fig. 14.7a.

The effects of nucleation growth of the initial GaAs layer on the structural and optical properties of InAs/GaAs QDs were studied at growth temperatures of 350, 380, and 410 °C, respectively. Cross-sectional TEM studies on the GaAs/Ge interface suggest that the defects generated at the interface are confined to within 100 nm of the GaAs/Ge interface and no defects are observed in InAs/GaAs QD regions for all three samples. Figure 14.8a shows an example TEM image for the sample with a GaAs nucleation temperature at 380 °C, and indicates that the density of defects propagating into the III–V active region is too low to be detected by cross-sectional TEM. 5-layer InAs/InGaAs DWELL structures as the one in Sect. 14.3 were fabricated on the single-domain GaAs buffer layer on Ge substrates. The InAs/InGaAs DWELL structure was then embedded between 100 nm GaAs layer grown at 580 °C which were further confined by two 50-nm AlGaAs layers grown at 610 °C. The growth temperatures were measured by infrared pyrometer, calibrated to thermocouple readings.

The structural properties of Ge-based InAs/GaAs QDs for all three samples with different GaAs nucleation temperatures are similar to those of GaAs-based QDs grown under similar growth conditions, such as the cross-sectional TEM image of five-layer DWELL structure for the sample with the GaAs nucleated at 410 °C, as

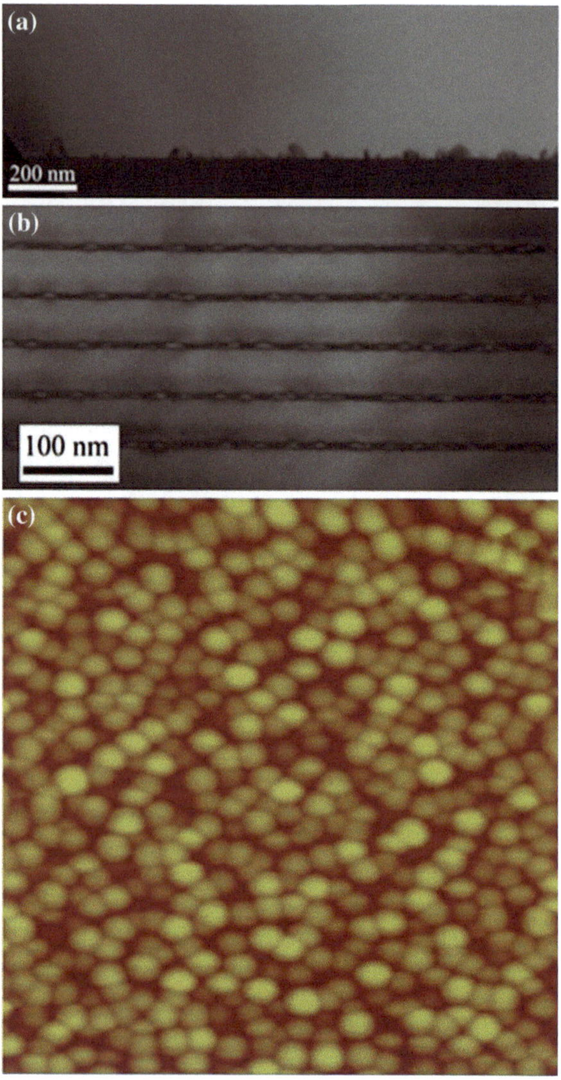

Fig. 14.8 Dark field (200) TEM cross-sectional images of (**a**) the interface between the GaAs buffer layer and the Ge substrate for the sample with initial GaAs layer nucleated at 380 °C, and (**b**) five-layer InAs/InGaAs DWELL sample on Ge substrate with the initial GaAs layer nucleated at 410 °C; (**c**) 1.0 μm × 1.0 μm AFM image of uncapped InAs QD sample with the initial GaAs layer nucleated at 380 °C

shown in Fig. 14.8b and the $1 \times 1\ \mu m^2$ AFM image of the uncapped InAs QDs grown on GaAs/Ge with the GaAs nucleation temperature of 380 °C, as shown in Fig. 14.8c, from which a QD density of about $4.3 \times 10^{10}\ cm^{-2}$ is obtained. The AFM result indicates that InAs QDs randomly distribute on the surface, as on GaAs substrates. These results suggest that the structural properties of Ge-based InAs/GaAs QDs were not affected by the GaAs nucleation temperature. This morphology of InAs QDs in Fig. 14.8c is significantly different from that of InAs/GaAs QDs grown on a Ge-on-insulator-on-Si substrate by metal organic chemical vapor deposition, in which APBs were observed and the InAs QDs were lined with a preferential orientation along the [1$\bar{1}$0] direction with bimodal size distribution [15].

The optical properties of InAs/GaAs QDs was measured by temperature-dependent PL. PL measurements were performed in a close-cycle He cryostat under 532 nm excitation from a diode-pumped solid-state laser. The PL spectra were dispersed by a 0.25 m monochromator and detected by a TE-cooled Ge detector. Temperature-dependent PL measurements were taken from 10K to 300K at a laser power of 5 mW. The comparison of the RT PL spectra of InAs/GaAs QDs grown on Ge substrates at the different GaAs nucleation temperatures and with a reference QD sample grown on GaAs substrate is shown in Fig. 14.9a. The InAs/GaAs QDs yield RT emission at $\sim$1.3 μm for all the samples shown in Fig. 14.9a. The PL linewidth of all the Ge-based InAs/GaAs QDs is about 30 meV and comparable to the values reported for 1.3-μm InAs QDs grown on GaAs substrates. The PL intensity of the ground-state transaction of InAs/GaAs QDs grown on the Ge substrate is strongly dependent on the growth temperature of the initial GaAs nucleation layer. The strongest PL intensity for InAs/GaAs QDs on a Ge substrate is obtained from the sample grown with a GaAs nucleation temperature of 380 °C, and is almost identical to that of reference sample grown on GaAs substrate.

The IPLIs over the temperature range 10–300 K for all three Ge-based InAs/GaAs QD samples were further investigated. Figure. 14.9b shows the Arrhenius plot of the IPLI signals of all three Ge-based samples. The slopes of the line at high temperatures yield thermal activation energies of about 258, 265, and 270 meV for the samples with the initial GaAs layer nucleated at 350, 380, and 410 °C, respectively. The thermal activation energies are similar and are close to the energy difference (such as $\sim$290 meV in Ref. [31]) between the QDs ground state and InGaAs QW, confirming that there are no significant band structure variations caused by varying the GaAs nucleation temperature on Ge substrates. It is well established that the reduction in IPLI is attributed to the thermal escape of carriers from QD ground states into the InGaAs QWs followed by nonradiative recombination in the barriers [31]. Therefore, the IPLI differences among three samples shown in Fig. 14.9b could be understood in term of the different density of defects generated at GaAs/Ge interface and propagating into the III–V active region. This is further confirmed by EPD measurements. The etchant used for the EPD delineation is a mixture of H_3PO_4, H_2O_2, and H_2O (in a 1:1:3 ratio) [26]. The defect densities of 1.7×10^7, 8.0×10^5, and 3.7×10^6 cm^{-2} were obtained for the samples with the initial GaAs layer nucleated at 350, 380, and 410 °C, respectively.

Standard five-layer 1.3-μm InAs/InGaAs DWELL lasers were fabricated on the GaAs/Ge substrate with the initial GaAs layer nucleated at the optimized temperature (380 °C). The device structure is shown in Fig. 14.10. The laser was grown on a p-doped Ge substrate and consisted of the following layer consequence. First, a 1.5-μm p-doped GaAs buffer layer and a 1.5-μm $Al_{0.4}Ga_{0.6}As$ p-doped cladding layer were deposited, followed by a 55-nm $Al_{0.2}Ga_{0.8}As$ guiding layer and a 70-nm GaAs barrier layer. Next five-layer InAs/InGaAs DWELLs with 45 nm GaAs spacer layer were grown with the same conditions as the samples shown in Fig. 14.8. Following that, a 70-nm GaAs barrier layer, a 55-nm $Al_{0.2}Ga_{0.8}As$ guiding layer,

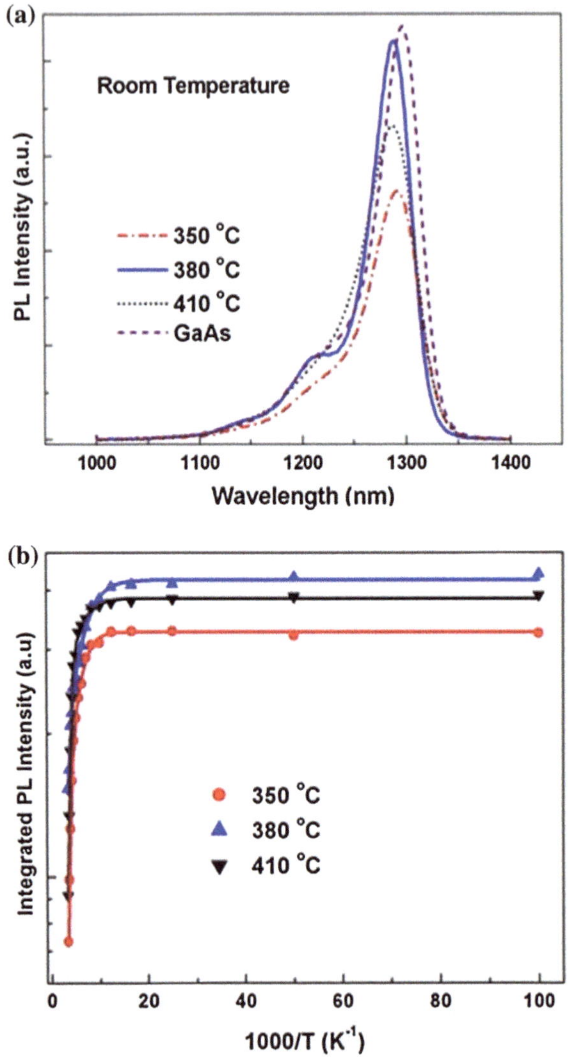

Fig. 14.9 **(a)** Room-temperature PL spectra for Ge-based InAs/GaAs quantum-dot samples with different growth temperatures for initial GaAs nucleation layers. The RT PL spectrum of InAs/GaAs QDs grown on a GaAs substrate is also shown as a reference. **(b)** The Arrhenius plots of the temperature dependence of integrated PL intensity (IPLI) for Ge-based InAs/GaAs quantum-dot samples with initial GaAs nucleated at 350, 380, and 410 °C, respectively

a 1.5-μm $Al_{0.4}Ga_{0.6}As$ n-doped cladding, and a 300-nm n^+-doped GaAs contact layer were grown. Broad-area laser devices were formed with a shallow ridge etch after deposition of an Al contact layer on the bottom of the wafer for contacting the Ge substrate. Wet chemical etching was terminated when the upper n-doped GaAs

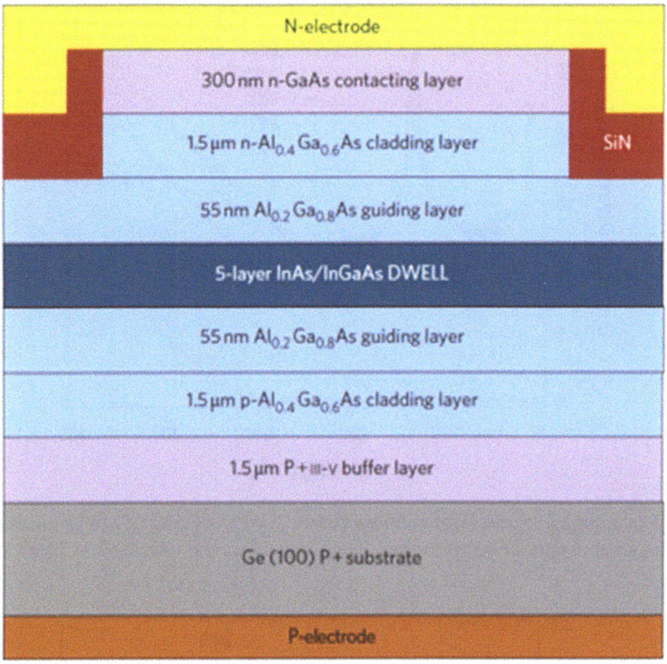

Fig. 14.10 Schematic showing the layer structure of an InAs/InGaAs dots-in-a-well laser diode on a Ge substrate

and n-doped AlGaAs layers above the QD active region were removed to a depth of 1.8 μm. A 500-nm-thick layer of SiNx was deposited on the sample surface and contact windows opened on the ridge top, followed by deposition of the In/Ge/Au contact layer. Broad-area devices with cavities of width 50 μm were fabricated with as-cleaved facets.

The laser diode with 5-mm cavity length was measured in continuous-wave (cw) condition. Figure. 14.11 shows light output against current characteristic for InAs/GaAs QD laser grown on a Ge substrate for operating temperatures between 20 and 60 °C. The inset in Fig. 14.11 shows the laser optical spectrum above threshold at RT, in which lasing at 1.305 μm is observed. The measured RT output power from one facet is close to 28 mW for an injection current of 500 mA, with no evidence of power saturation up to this current. The lasing threshold is 138 mA. J_{th} is 55.2 A/cm^2, which corresponds to about 11 A/cm^2 for each of the five QD laser layers, which is comparable to the best-reported values for GaAs-based QD laser diodes [32, 33]. The QD laser has a characteristic temperature T_0 of ~40 K between 20 and 60 °C.

The Ge-based InAs/GaAs QD laser didoes were further measured in the pulsed mode using a pulse width of 0.1 μs and duty cycle of 0.01 %. Devices with a cavity length of 2.5 mm were bar-tested and directly probed without any mounting and bonding. The inset of Fig. 14.12 shows the laser optical spectrum above threshold

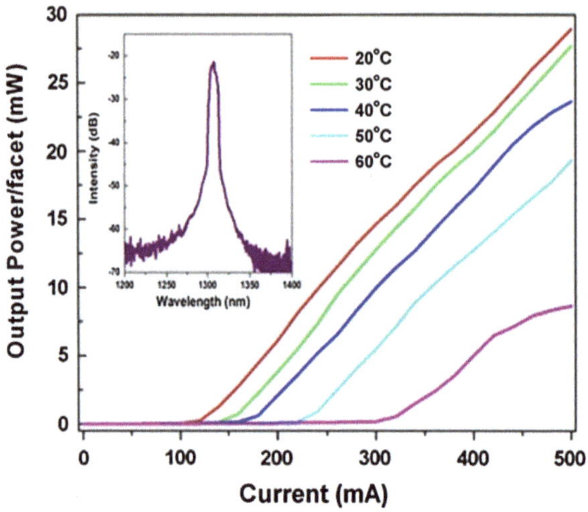

Fig. 14.11 Light output against current characteristic for InAs/GaAs QD laser on a Ge substrate under cw operation at various temperatures. The *inset* shows the laser optical above threshold at RT

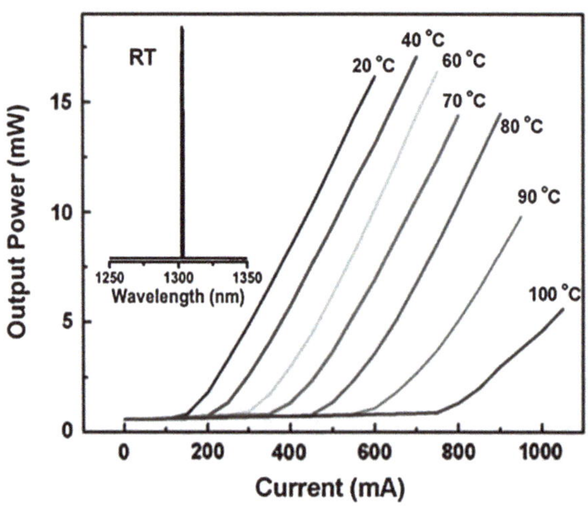

Fig. 14.12 Light output against current characteristic for InAs/GaAs quantum-dot laser grown on Ge substrate under pulsed condition for operating temperatures between 20 and 100 °C. The inset shows the laser optical spectrum above threshold at RT

at RT. The main part of Fig. 14.12 shows the output power against current at various temperatures. A RT J_{th} of ~106 A/cm^2 is obtained, with lasing up to 100 °C. The characteristic temperature, T_0 is about 47 K between 20 and 100 °C, which is comparable to the performance of 1.3-μm InAs/GaAs QD lasers grown on GaAs substrates. In general, 1.3-μm InAs/GaAs QD lasers show poor T_0 value in the range of 35–60 K above room temperature, mainly due to the hole excitation out of the lasing state [34]. The temperature stability of Ge-based QD lasers could be significantly enhanced by using p-type modulation doping of the QDs, which has been well established to dramatically increase the value of T_0, even to $T_0 \sim \infty$ above room temperature for GaAs-based InAs QD lasers [35, 36].

14.5 InAs/GaAs QD Lasers Grown on Ge-on-Si Substrates

After high quality QD lasers have been demonstrated on Ge substrates, the identical growth method was further explored on Ge-on-Si (Ge/Si) substrates. To form the Ge/Si virtual substrate, a 2-μm Ge layer was grown using chemical vapour deposition on phosphorus-doped (100)-oriented Si substrates with a 6° offcut toward the [111] planes. The dislocation density in the Ge epitaxial layer is about $5 \times 10^6/\text{cm}^2$. The device structure is shown in Fig. 14.13. Oxide desorption procedure for Ge/Si is similar to the one for Ge substrates described as in Sect. 14.4. After the oxide desorption, the Ge/Si substrate was then cooled to 380 °C for the growth of the III–V epitaxial layers [17]. The Ga prelayer growth technique was used for the nucleation of GaAs buffer layers. 20 MLs of GaAs were then grown by migration enhanced epitaxy using alternating Ga and As_4 beams, and then the addition of III–V buffer layer at higher temperature. QD laser structures containing five InAs/InGaAs DWELL structures were then grown at optimized conditions as on GaAs substrates, with each layer consisting of 3.0 MLs of InAs grown on 2 nm of $In_{0.15}Ga_{0.85}As$ and capped by 6 nm of $In_{0.15}Ga_{0.85}As$. 45 nm GaAs barriers separated the five DWELLs

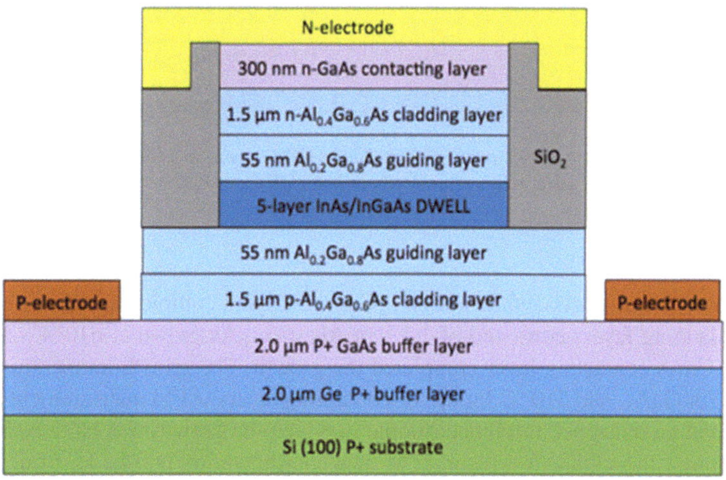

Fig. 14.13 The schematic showing the layer structure of an InAs/GaAs QD laser diode on a Ge-on-Si substrate

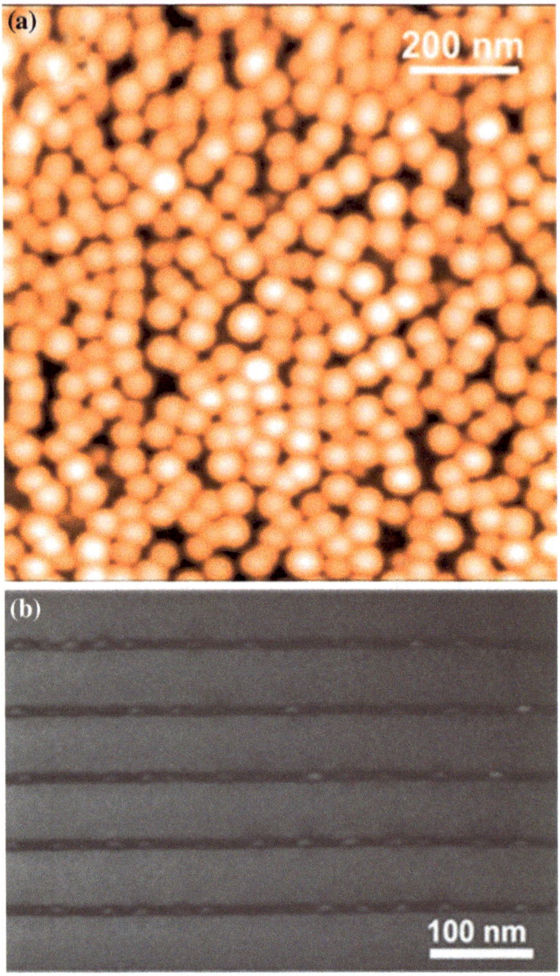

Fig. 14.14 **a** AFM image ($1 \times 1\ \mu m^2$) of InAs/GaAs QDs grown on a Ge/Si substrate. **b** Cross-sectional TEM image of laser active region, where InAs QDs are separated by 8-nm InGaAs and 45-nm GaAs spacer layers

with outer 70 nm GaAs and 55 nm $Al_{0.2}Ga_{0.8}As$ layers completing the waveguide core. Cladding layers consisted of 1.5 μm $Al_{0.4}Ga_{0.6}As$ grown at 610 °C. A 300-nm n^+-GaAs contacting layer completed the growth. The growth temperature was 580 °C for GaAs, and 510 °C for the In-containing layers. AFM measurements were performed on uncapped reference sample in which the growth was halted after the formation of InAs QDs. A typical AFM image is shown in Fig. 14.14a, from which a QD density of about $3.5 \times 10^{10}/cm^2$ is obtained. The laser active region consists of five InAs/GaAs DWELLs separated by 45-nm GaAs spacer layers, as shown in the TEM image of Fig. 14.14b.

Ridge-waveguide edge-emitting laser devices were fabricated using standard photolithography, wet etching, and metallization techniques. After form 20 μm ridge stripe and exposing the p-AlGaAs contacting layer by conventional wet etch, Ni/Ge/Au and Ti/Pt/Au contact layers were deposited on the n-GaAs and the exposed p-AlGaAs contacting layer, respectively. No highly reflective coatings were used for the device mirror facets. Devices of 3.0-mm and 3.5-mm length were bar-tested, being directly probed without any bonding. Laser device characteristics were measured under both cw and pulsed conditions. For the pulsed measurements, a pulse width of 1 μs and duty cycle of 0.2 % was used.

The lower RT threshold current densities are observed for 3.5-mm-long devices. The inset of Fig. 14.15a shows a series of RT spontaneous and lasing spectra for a 3.5-mm-long QD laser on Ge/Si substrate operating below and above threshold under pulsed condition. Spontaneous emission can be observed at a peak wavelength of approximately 1.29 μm with FWHM of 36 meV at a current of 30 mA. Lasing emission with peak wavelength of 1.28 μm can be observed at a current of 50 mA. The

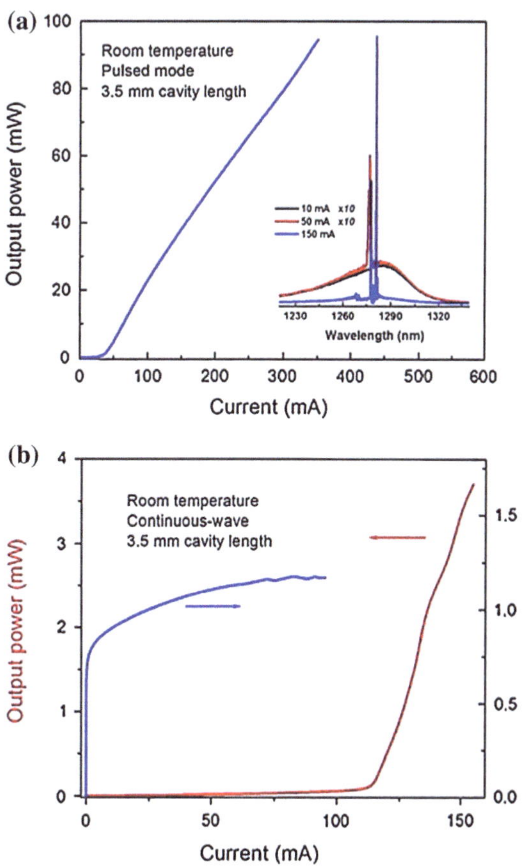

Fig. 14.15 Room-temperature emission spectra, light output power, and electrical characteristics of 3.5-mm-long laser diode. **a** Light output power versus current for 3.5-mm-long QD laser diodes grown on a Ge/Si substrate under pulsed mode. The *inset* shows the emission spectra of InAs/GaAs QD laser for different drive currents below and above threshold. **b** Light output power against current and voltage against current for 3.5-mm-long QD laser diodes grown on a Ge/Si substrate under cw conditions

multimode lasing spectrum appears with increasing the injection current to 150 mA. The main part of Fig. 14.15a shows the RT light output power against current (*L-I*) characteristics of 3.5-mm-long device under pulsed conditions. The lasing threshold current is 45 mA at RT, and the maximum output power is close to 93 mW for an injection current of 350 mA (external quantum efficiency $\eta_{ex} = 37.5\,\%$), with no evidence of power saturation up to this current. The threshold current density, J_{th} under pulsed mode is as low as 64.3 A/cm^2, which corresponds to about 12.9 A/cm^2 for each of the five QD laser layers. This very low J_{th} is comparable to the best-reported values for GaAs-based InAs QD laser diodes with as-cleaved facets [32–34].

The RT light output power/current characteristic of a 3.5-mm-long device under cw operation is shown in Fig. 14.15b. The cw lasing threshold is 114 mA, which corresponds to a RT cw J_{th} of 163 A/cm^2. A maximum cw output power of ~3.7 mW (for both facets) is obtained for a current 150 mA. The voltage/current (V-I) characteristics of 3.5-mm-long device is also shown in Fig. 14.15b, from which it can be seen that the turn-on voltage is ~0.8 V with series resistance of ~5.0 Ω. The device exhibits higher resistive characteristics compared to GaAs-based devices. This could be understand in terms of higher defect density of $\sim 5 \times 10^6$/cm^{2}in the Ge buffer layer on Si substrate used in this work, compared to the typical value of defect density (about 5×10^3/cm^2) for GaAs substrate. The device performance could be further improved through using a low-defect-density Ge buffer layer on Si substrates and/or dislocation filter layer in III–V buffer layer [19].

While the lower threshold current densities are obtained for 3.5-mm-long device, 3.0-mm-long laser diodes operate at higher temperature. Figure. 14.16a shows the output power versus current for 3-mm-long Si-based InAs/GaAs QD laser in pulsed mode at various substrate temperatures. This Si-based InAs/GaAs QD laser has a 84 °C maximum lasing temperature with a characteristic temperature, T_0, of ~37 K between 21 and 84 °C. The temperature-dependent output power versus current under cw operation is shown in Fig. 14.16b. The cw output powers were measured under maximum drive current of 350 mA, which is limited by our experimental setup. Continuous-wave laser operation is achieved up to 30 °C with a T_0 of ~23 K between 12 and 30 °C. A next step toward increasing cw operation temperature for Si-based QD lasers are expected by reducing the defect density in III–V active region and further exploiting p-type modulation doping of the QDs into these laser structures.

14.6 InAs/GaAs QD Detectors Grown on Si Substrates

Although Ge/Si infrared detectors have been demonstrated for telecom applications, their usefulness is seriously limited by relative high dark current. For example, dark current densities of 5 A/cm^2 for a Ge diode operating at 1.3 μm [37] and 2 mA/cm^2 for a Ge diode operating at 1.5 μm [38], were reported at 1 V reverse bias, while a typical commercial InGaAs photodiode has a dark current density of 0.1 μA/cm^2 at 5 V reverse bias. Due to QDs' zero-dimensional density of states, lower dark current

Fig. 14.16 Temperature-dependent light output power characteristics of 3.0-mm-long laser diode. **a** Light output against current for Si-based InAs/GaAs quantum-dot laser diode at various substrate temperatures under pulsed mode. **b** Light output against current for Si-based InAs/GaAs quantum-dot laser diode at various heatsink temperatures under cw conditions

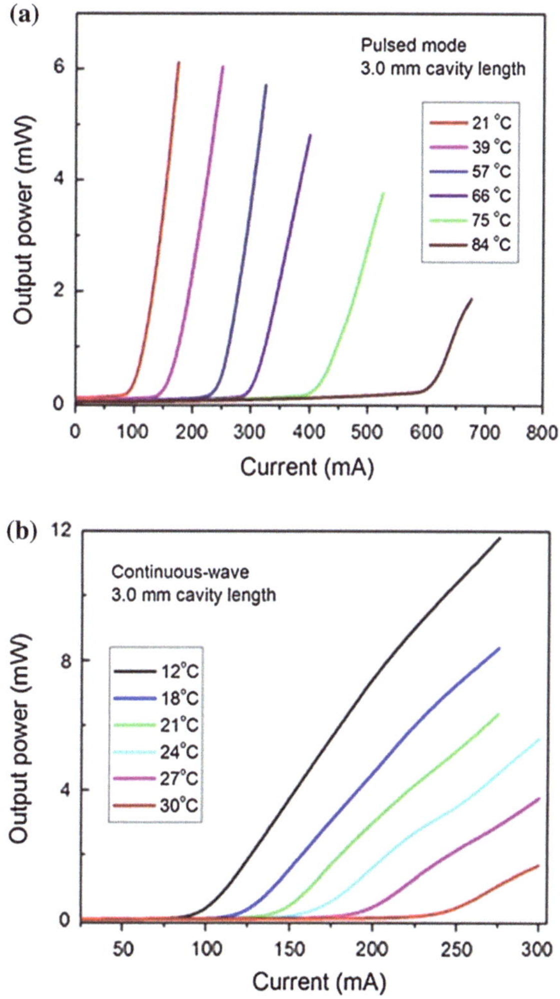

is expected for QD detector. Consequently, Si-based InAs/GaAs QD detectors are developed by exploiting the techniques used in producing RT lasing from InAs/GaAs QDs grown on Si substrates.

The structure investigated for photodetector was initially designed for an LED so was not optimized for detection purposes. InAs/GaAs QD were grown under the same conditions as the laser structures in Sect. 14.3. A schematic diagram summarizing the wafer grown is shown in Fig. 14.17. It comprises a GaAs buffer layer followed by 50 periods of an AlGaAs/GaAs SPL. Five periods of InAs/InGaAs DWELL structures by a GaAs spacer layer are sandwiched in a p-i-n structure. Circular mesa diodes with radii of 200, 100, 50, and 25 μm were fabricated from the wafer via wet chemical etching and metal depositions of p-contact (Au-Zn-Au) and n-contact (InGe-Au).

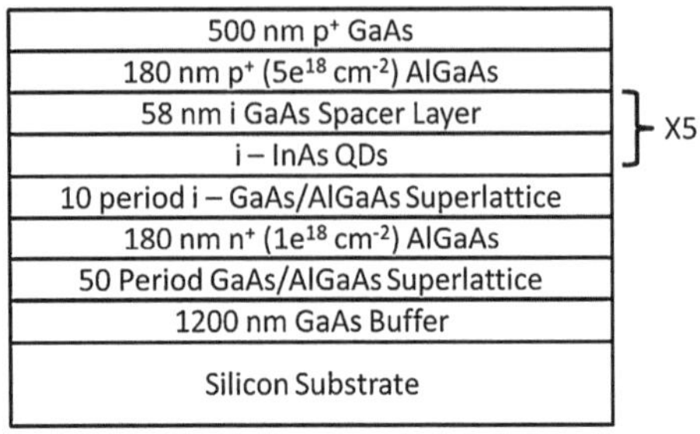

Fig. 14.17 Schematic diagram of wafer structures used for fabricating Si-based InAs/GaAs detectors

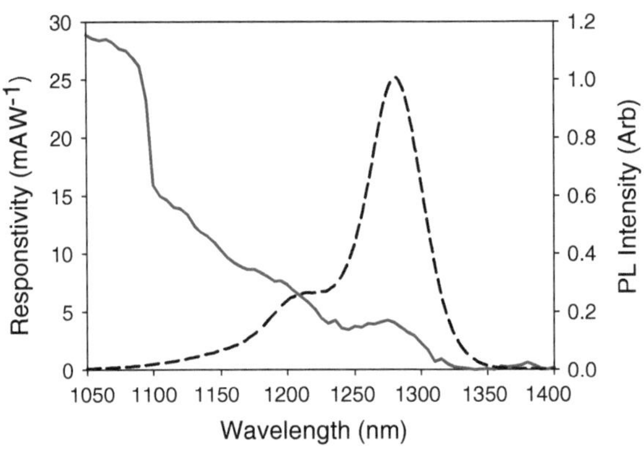

Fig. 14.18 Room temperature responsivity of a 200 μm radius device (*left axis, solid line*) and PL spectra (*right axis, dashed line*) of the as-grown sample

RT responsivity of the device and RT PL spectra from as-grown material were accessed and are shown in Fig. 14.18. For the responsivity measurements, white light from a tungsten lamp was passed through a monochromator so that light at a chosen wavelength can be focused onto the mesa diode to enable measurements of photocurrent as a function of wavelength. The responsivity for a given wavelength was given by the ratio of the photocurrent, measured using a phase sensitive detection method, to the optical power on the diode, measured using an optical power meter. For the PL spectra, a separate piece of sample was excited with a 532 nm cw laser. The emission from the sample was then collected and focused onto the input slit of a monochromator before being dispersed in wavelength and measured by a liquid nitrogen-cooled Ge detector.

The main PL peak and a small peak (5 mA/W) at the wavelength of 1,280 nm are obtained in Fig. 14.18, which suggest that the ground state transition of InAs QDs occurs at 1280 nm. There is a secondary peak in the PL at a wavelength of 1,210 nm, which could be originated from the smaller dots for a bi-modal dot distribution in the sample. In the responsivity data, there is a sharp decrease in responsivity at

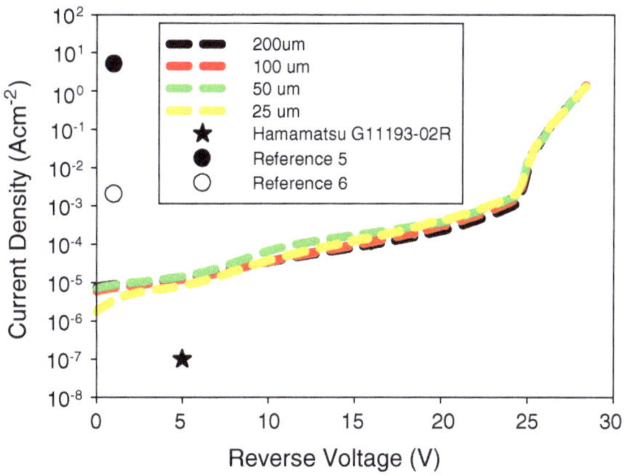

Fig. 14.19 Dark current densities for 200, 100, 50, and 25 μm radius devices. The data is compared with Ge on Si and commercial InGaAs photodiodes

1,100 nm, which is attributed to the wetting layer and InGaAs QWs due to the lack of corresponding PL feature. The responsivity due to ground state transition of InAs QDs in Fig. 14.18 is less than that for Ge grown on Si at comparable wavelengths [37], because there are only five layers of dots in our wafer, resulting in incomplete light absorption. In addition to the obvious need to increase the number of dot layers, incorporation of QD in a resonant cavity structure can increase the quantum efficiency. For instance, an external quantum efficiency of 90 % at ~1 μm wavelength was reported for a resonant cavity enhanced photodiode with 23 layers of In(Ga)As QDs [39].

The reverse bias dark current densities of diodes with different radii are compared in Fig. 14.19. The reverse dark current densities are in good agreement indicating that dark currents are dominated by bulk leakage mechanism(s). At low voltages, our diodes exhibited a dark current density ~ two orders of magnitude higher than that of commercially available InGaAs photodiodes which respond to 1.3 μm light. However, our dark currents are two orders of magnitude lower than some Ge on Si devices.

14.7 Conclusions

The growth of InAs/GaAs QD materials and devices on Si, Ge, and Ge/Si substrates has been investigated by using Molecular Beam Epitaxy for Si-based laser didoes and photodetectors. For InAs/GaAs QD device directly grown on Si substrates, RT electrically pumped lasing at a wavelength of 1.3 μm has been achieved with RT J_{th} of 725 A/cm^2 and lasing operation up to 42 °C under pulsed mode. By using the Ga prelayer technique for the growth of GaAs on Ge substrates, the APBs within III–V layers grown on Ge are significantly suppressed. By further exploiting this technique into QD laser didoes, the very low RT cw J_{th} of 11A/cm^2 for per QD layer has been achieved with lasing operation up to 100 °C for five-layer InAs/GaAs QD laser diodes monolithically grown on Ge substrates. The performance of Ge-based InAs/GaAs QD lasers is comparable to that of best GaAs-based devices reported in the literature. The first RT cw operation of InAs/GaAs QD laser diodes was demonstrated by transferring the growth approaches on Ge substrates to Ge/Si substrates. RT lasing around 1.28 μm has been achieved with cw J_{th} of 163 A/cm^2 and lasing operation up to 83 °C for ridge-waveguide InAs/GaAs lasers grown on Ge/Si substrates with as cleaved facets. The potential of InAs/GaAs QD materials grown on Si substrates for Si-based photodiodes was further evaluated. The responsivity measurement showed a peak related to the QD ground-state transition at 1.28 μm of 5mA/W for a five-QD-layer device. The dark current of Si-based InAs/GaAs QD detectors were two order of magnitude lower than those reported for Ge on Si photodiodes. These results clearly demonstrate that the direct, monolithic growth of InAs/GaAs QD materials and devices on the Si platform is a strong contender for the realisation of an optical source integrated on an electronics platform.

Acknowledgments The author wishes to thank Professor Alwyn Seeds (University College London) for valuable discussion and support. The author also thanks Mr. Andrew Lee, Dr. Qi Jiang, Dr. Ting Wang, Dr. James Wilson, Dr. Kris Groom, Dr. Ian Sandall, Dr. Chin-Hing Tan, Dr. Jo Shien Ng, and Professor Richard Hogg for collaboration and help. This study is supported by The Royal Society, UK Defence Science and Technology Laboratory (Dstl), UK Engineering and Physics Science Research Council (EPSRC).

References

1. Jalali, B., Fathpour, S.: Silicon photonics. J. Lightwave Technol. **24**, 4600–4615 (2006)
2. Young, I.A., Mohammed, E., Liao, J.T.S., Kern, A.M., Palermo, S., Block, B.A., Reshotko, M.R., Chang, P.L.D.: Optical I/O technology for tera-scale computing. IEEE J. Solid-State Circuits **45**, 235–248 (2010)
3. Miller, D.: Device requirements for optical interconnects to silicon chips. Proc. IEEE **97**, 1166–1185 (2009)
4. Savage, N.: Linking with light high-speed optical interconnects. IEEE Spectr. **39**, 32–36 (2002)
5. Reed, G.T., Mashanovich, G., Gardes, F.Y., Thomson, D.J.: Silicon optical modulators. Nat. Photon **4**, 518–526 (2010)
6. Michel, J., Liu, J., Kimerling, L.C.: High-performance Ge-on-Si photodetector. Nat. Photon **4**, 527–534 (2010)
7. Leuthold, J., Koos, C., Freude, W.: Nonlinear silicon photonics. Nat. Photon **4**, 535–544 (2010)
8. Liang, D., Bowers, J.E.: Recent progress in lasers on silicon. Nat. Photon **4**, 511–517 (2010)
9. Fan, J., Poate, J.M. (eds.): Heteroepitaxy on Silicon. Materials Research Society, Pittsburgh (1986)
10. Fischer, R., Kopp, W., Morkoc, H., Pion, M., Specht, A., Burkhart, G., Appelman, H., McGougan, D., Rice, R.: Low threshold laser operation at room temperature in GaAs/(Al, Ga)As structures grown directly on (100)Si. Appl. Phys. Lett. **48**, 1360–1361 (1986)
11. Fischer, R., Masselink, W.T., Klem, J., Henderson, T., McGlinn, T.C., Klein, M.V., Morkoc, H., Mazur, J.H., Washburn, J.: Growth and properties of GaAs/AlGaAs on nonpolar substrates using molecular beam epitaxy. J. Appl. Phys. **58**, 374–381 (1985)

12. Sugawara, M., Usami, M.: Quantum dot devices:handling the heat. Nat. Photon **3**, 30–31 (2009)
13. Liu, H., Sellers, I.R., Badcock, T.J., Mowbray, D.J., Skolnick, M.S., Groom, K.M., Gutierrez, M., Hopkinson, M., Ng, J.S., David, J., Beanland, R.: Improved performance of 1.3 μm multilayer InAs quantum-dot lasers usinga high-growth-temperature GaAs spacer layer. Appl. Phys. Lett. **85**, 704–706 (2004)
14. Yang, J., Bhattacharya, P., Mi, Z.: High-performance $In_{0.5}Ga_{0.5}As$/GaAs quantum-dot lasers on silicon with multiple-layer quantum-dot dislocation filters. IEEE Trans. Electron Devices **54**, 2849–2855 (2007)
15. Bordel, D., Guimard, D., Rajesh, M., NishiokA, M., Augendre, E., Clavelier, L., Arkawa, Y.: Growth of InAs/GaAs quantum dots on germanium-on-insulator-on-silicon (GeOI) substrate with high optical quality at room temperature in the 1.3 μm band. Appl. Phys. Lett. **96**, 043101 (2010)
16. Beanland, R., Sanchez, A.M., Childs, D., Groom, K.M., Liu, H., Mowbray, D.J., Hopkinson, M.: Structural analysis of life tested 1.3 μm quantum dot lasers. J. Appl. Phys. **103**, 014913 (2008)
17. Liu, H., Wang, T., Jiang, Q., Hogg, R., Tutu, T., Pozzi, F., Seeds, A.: Long-wavelength InAs/GaAs quantum dot laser diode monolithically grown on Ge substrate. Nat. Photon. **5**, 416–419 (2011)
18. Wang, T., Liu, H., Lee, A., Pozzi, F., Seeds, A.: 1.3-μm InAs/GaAs quantum-dot lasers monolithically grown on Si substrates. Opt. Express **19**, 11381–11386 (2011)
19. Mi, Z., Yang, J., Bhattacharya, P., Huffaker, D.L.: Self-organised quantum dots as dislocation filters: the case of GaAs-based lasers on silicon. Electron. Lett. **42**, 121–123 (2006)
20. Lee, A., Jiang, Q., Tang, M., Seeds, A., Liu, H.: Continuous-wave InAs/GaAs quantum-dot laser diodes monolithically grown on Si substrate with low threshold current densities. Opt. Express **20**, 22181–22187 (2012)
21. Liu, H., Childs, D.T., Badcock, T.J., Groom, K.M., Sellers, I.R., Hopkinson, M., Hogg, R.A., Robbins, D.J., Mowbray, D.J., Skolnick, M.S.: High-performance three-layer 1.3-μm InAs/GaAs quantum-dot lasers with very low continuous-wave room-temperature threshold currents. IEEE Photon. Technol. Lett. **17**, 1139–1141 (2005)
22. Liu, H., Xu, B., Chen, Y.H., Ding, D., Wang, Z.G.: Effects of seed layer on the realization of larger self-assembled coherent InAs/GaAs quantum dots. J. Appl. Phys. **88**, 5433–5436 (2000)
23. Liu, H., Hopkinson, M., Harrison, C., Steer, M., Frith, R., Sellers, I.R., Mowbray, D.J., Skolnick, M.S.: Optimizing the growth of 1.3 μm InAs/InGaAs dots-in-a-well structure. J. Appl. Phys. **93**, 2931–2936 (2003)
24. Jin, C., Badcock, T., Liu, H., Groom, K., Royce, R., Mowbray, D.J., Hopkinson, M.: Observation and modelling of a room-temperature negative characteristic temperature 1.3-μm p-type modulation-doped quantum-dot laser. IEEE J. Quant. Electron. **42**, 1259–1265 (2006)
25. Badcock, T., Royce, R., Mowbray, D., Skolnick, M., Liu, H., Hopkinson, M., Groom, K., Jiang, Q.: Low threshold current density and negative characteristic temperature 1.3 μm InAs self-assembled quantum dot lasers. Appl. Phys. Lett. **90**, 111102 (2007)
26. Clawson, A.R.: Guide to references on III-V semiconductor chemical etching. Mat. Sci. Eng.: R: Rep. **31**, 1–438 (2001)
27. Wang, T., Lee, A., Tutu, F., Seeds, A., Liu, H., Jiang, Q., Groom, K., Hogg, R.: The effect of growth temperature of GaAs nucleation layer on InAs/GaAs quantum dots monolithically grown on Ge substrates. Appl. Phys. Lett. **100**, 052113 (2012)
28. Akatsu, T., Deguel, C., Sanchez, L., Allibert, F., Rouchon, D., Signamarcheix, T., Richatarch, C., Boussagol, A., Loup, V., Mazen, F., Hartmann, J., Campidelli, Y., Clavelier, L., Letertre, F., Kernevez, N., Mazure, C.: Germanium-on-insulator (GeOI) substrates - A novel engineered substrate for future high performance devices. Mat. Sci. Semicond. Proc. **9**, 444–448 (2006)
29. Currie, M., Samavedam, S., Langdo, T., Leitz, C., Fitzgerald, E.: Controlling threading dislocation densities in Ge on Si using graded SiGe layers and chemical-mechanical polishing. Appl. Phys. Lett. **72**, 1718–1720 (1998)

30. Brammertz, G., Caymax, M., Meuris, M., Heyns, M., Mols, Y., Degroote, S., Leys, M.: GaAs on Ge for CMOS. Thin Solid Films **517**, 148–151 (2008)
31. Chen, R., Liu, H., Sun, H.D.: Electronic energy levels and carrier dynamics in InAs/InGaAs dots-in-a-well structure investigated by optical spectroscopy. J. Appl. Phys **107**, 013513 (2010)
32. Sellers, I.R., Liu, H., Groom, K.M., Childs, D.T., Robbins, D.J., Badcock, T.J., Hopkinson, M., Mowbray, D.J., Skolnick, M.S.: 1.3 μm InAs/GaAs multilayer quantum-dot laser with extremely low room-temperature threshold current density. Electron. Lett. **40**, 1412–1413 (2004)
33. Deppe, D.G., Shavritranuruk, K., Ozgur, G., Chen, H., Freisem, S.: Quantum dot laser diode with low threshold and low internal loss. Electron. Lett. **45**, 54–55 (2009)
34. Ustinov, V.M., Zhukov, A.E.: GaAs-based long-wavelength lasers. Semicond. Sci. Technol. **15**, R41–R54 (2000)
35. Jin, C.Y., Liu, H.Y., Groom, K.M., Jiang, Q., Hopkinson, M., Badcock, T.J., Royce, R., Mowbray, D.J.: Effects of photon and thermal coupling mechanisms on the characteristics of self-assembled InAs/GaAs quantum dot lasers. Phys. Rev. B **76**, 085315 (2007)
36. Liu, H., Badcock, T.J., Jin, C.Y., Nabavi, E., Groom, K.M., Hopkinson, M., Mowbray, D.J.: Reduced temperature sensitivity of lasing wavelength in near-1.3 μm InAs/GaAs quantum-dot laser with stepped composition strain-reducing layer. Electron. Lett. **43**, 670–672 (2007)
37. Jutzi, M., Berroth, M., Wohl, G., Oehme, M., Kasper, E.: Zero biased Ge-on-Si photodetector on a thin buffer with a bandwidth of 3.2GHz at 1300nm. Mater. Sci. Semicond. Process. **8**, 423–427 (2005)
38. Colace, L., Ferrara, P., Assanto, G., Fulgoni, D., Nash, L.: Low dark-current germanium-on-silicon nearinfrared detectors. IEEE Photon. Technol. Lett. **19**, 1813–1815 (2007)
39. Sun, X.M., Zhang, H., Zhu, H., Xu, P., Li, G.R., Liu, J., Zheng, H.Z.: High responsivity resonant-cavity enhanced InGaAs/GaAs quantum-dot photodetector for wavelength of $\sim$1 μm at room temperature. Electron. Lett. **45**, 329–330 (2009)

Chapter 15
Cubic GaN on Nanopatterned 3C-SiC/Si (001) Substrates

Ricarda Maria Kemper, Donat Josef As and Jörg K. N. Lindner

Abstract In this chapter we demonstrate the growth and characterization of nonpolar relaxed cubic GaN by plasma-assisted molecular beam epitaxy on prepatterned 3C-SiC/Si (001) substrates. Nanopatterning of 3C-SiC/Si (001) was achieved by two different fabrication techniques: nanosphere lithography (NSL) to generate large-area pattern, and conventional electron beam lithography (EBL) for tailoring particular surface morphologies. Both methods were followed by a lift-off and a reactive ion etching (RIE) process. We analyze the influence of the substrate on the GaN growth and show that it is possible to grow single phase and defect-reduced cubic GaN crystals on 3C-SiC nanostructures. Furthermore cubic GaN/AlN multi-quantum wells were grown on 3C-SiC nanostructures, which is a further step toward nanoscaled device applications.

15.1 Introduction

Wide band-gap materials, especially group III-nitrides like GaN, AlN and InN, have a great potential for optoelectronic and electronic device applications like laser diodes and transistors [1–3]. Up to date there is a huge interest in nitride technologies to fabricate dislocation free epilayers for high-quality GaN-based devices.

Devices show best performance if fabricated on lattice-matched substrates since lattice-mismatch leads to the generation of misfit dislocations at the interface which may cause scattering of charge carriers, limited carrier mobility, nonradia-

R. M. Kemper (✉) · D. J. As · J. K. N. Lindner
Department of Physics, University of Paderborn, Warburger Str. 100, 33098 Paderborn, Germany
e-mail: rkemper@mail.uni-paderborn.de

D. J. As
e-mail: d.as@uni-paderborn.de

J. K. N. Lindner
e-mail: lindner@physik.uni-paderborn.de

H. Li et al. (eds.), *Silicon-based Nanomaterials*,
Springer Series in Materials Science 187, DOI: 10.1007/978-1-4614-8169-0_15,

© Springer Science+Business Media New York 2013

tive recombination in optical devices and device heating. However, the choice of suitable substrate materials is limited and thus additional techniques have to be developed to overcome persisting mismatch issues.

One method for the elimination of misfit dislocations is the reduction of the growth area [4]. Defect reduction in mismatched systems was also demonstrated by the theory of nanoheteroepitaxy (NHE), which describes the selective growth on a nanopatterned (10–100 nm) substrate [5]. In contrast to planar heteroepitaxy, NHE includes a partitioning of strain between the epilayer and the substrate and a distribution of the mismatch strain in the epilayer in three dimensions. NHE shows the possibility of growing dislocation free epilayers on nonlattice-matched nanopatterned substrates.

So far the most frequently used material among all group-III nitrides is the hexagonal polytype of GaN. In hexagonal GaN because of the crystal symmetry inherent spontaneous and piezoelectric polarization fields are present along the c-axis, which influence device characteristics [6, 7]. Therefore, nonpolar and semi-polar systems have attracted growing interest in the past few years [8]. One method to produce real nonpolar materials is the growth of cubic GaN (c-GaN). Although the hexagonal polytype is the thermodynamically favorable phase, it has been shown that there is a narrow window of experimental conditions to achieve the heteroepitaxial growth of c-GaN [9]. As there is a shortage in low defect density bulk GaN substrates [10] on which one might grow lattice matched GaN epilayers, the most adequate substrate is cubic silicon carbide 3C-SiC (001) exhibiting a lattice misfit of 3.5 %. The dislocation density [11] of c-GaN deposited on 3C-SiC (001) by plasma-assisted molecular beam epitaxy (MBE) is $\sim (10^9 - 10^{10})\text{cm}^{-2}$, which is in the order of magnitude of what has been achieved in hexagonal GaN epitaxy a few years ago. First, field effect transistors (FET) [3] and infrared detectors [12] based on c-GaN have been successfully realized. However, a reduction of the dislocation density would allow the fabrication of high performance devices.

In this chapter, we will describe current attempts (and obstacles) to improve c-GaN growth on 3C-SiC by using nanopatterned surfaces. For this purpose we first briefly introduce the theory of nanoheteroepitaxy before we describe experiments and results. We will demonstrate the growth of nonpolar c-GaN by molecular beam epitaxy (MBE) on nanopatterned 3C-SiC/Si (001) substrates with an extremely small content of the hexagonal phase. Nanopatterning of 3C-SiC/Si (001) is achieved either with self-organized colloidal masks or using electron beam lithography. Both processes are followed by a lift-off and a reactive ion etching process. We will then highlight the influence of the 3C-SiC substrate on the selective-area-grown c-GaN epilayer [13]. It is shown that c-GaN grown on 3C-SiC/Si (001) by MBE forms two types of structural domains which are induced by the substrate. They can be distinguished by their surface facets which are oriented either parallel or perpendicular to the [110] direction. Several independent measurement techniques like electron backscatter diffraction and microphotoluminescence demonstrate the formation of hexagonal inclusions preferentially in one type of domains. Finally, the realization of a complex cubic AlN/GaN multiquantum well layer structure on 3C-SiC nano–structures is demonstrated.

The studies presented here aim to give more insight into the basic growth mechanism of c-GaN on 3C-SiC/Si. They thus may help to improve the structural quality of future electrical and optoelectronical devices based on c-GaN grown on 3C-SiC (001). And they give the opportunity toward the fabrication of nanoscaled devices.

15.2 Theory of Nanoheteroepitaxy

In this section we shall address the possibility to avoid extended defects in heteroepitaxial layers by the use of substrates with tailored surface morphologies. In general there are three basic types of defects in epitaxial c-GaN thin films grown on 3C-SiC (001). The first type is misfit dislocations due to relaxation of a pseudomorphically strained thin layer on the substrate. The second type is dislocations originating in the substrate, which extend into the epilayer. The third type is stacking faults (SFs) on {111} planes. A change of the stacking sequence leads to hexagonal inclusions (h-GaN), which locally change the crystal symmetry and result in polar faces. As we will see later in Chap. 6, there are also other defects in the substrate which may adversely affect the epi-growth of c-GaN.

A common method for the elimination of misfit dislocations is the reduction of the growth area [4]. Even the reduction of the growth area in the μm-range reduces the nucleation sites of dislocations and filters substrate defects. Furthermore the probability of defect termination at local free surfaces is increased.

Additional defect reduction mechanisms become available, if the growth area is further reduced, as proposed in the theory of nanoheteroepitaxy (NHE) introduced by Hersee et al. [5]. This theory deals with the growth on nanostructures with a size of 10–100 nm. NHE offers the possibility of growing dislocation free epilayers on nonlattice-matched nanopatterned substrates if the growth areas are sufficiently small. In this case, the mismatch-induced stress leads to a three-dimensional strain partitioned between both, the growing film and the substrate. In consequence, the strain within the growing film can be kept below a threshold at which the strain relaxation by the formation of extended defects becomes energetically favorable.

In order to illustrate the difference between conventional epitaxy and nanoheteroepitaxy, Fig. 15.1 displays a schematic drawing of a planar epilayer and a selective-area-grown epilayer on a patterned substrate. In conventional planar epitaxy, the lattice mismatch at the interface leads to lateral strain in the growing film and in consequence to transversal lattice distortions (in growth direction). In contrast, on a two-dimensionally patterned surface both the substrate and the growing film are not subjected to lateral confinements and therefore a three-dimensional partitioning of the strain in the epilayer and in the substrate and consequently a reduced strain in the epilayer is expected. Calculations of the strain energy by Sun and coworkers [14] have shown that the 3.2 % lattice mismatch between hexagonal GaN and hexagonal SiC can be accommodated without defect generation when the substrate nanostructures have a size of about 40 nm. Our studies presented here are based on

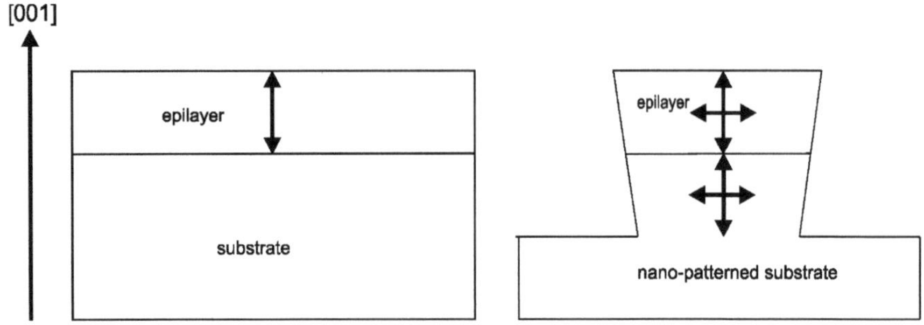

Fig. 15.1 Schematic drawing of a planar epilayer and a selective-area-grown epilayer on a patterned substrate

this theory but use slightly larger surface features. This allows us to observe clearly the structural changes occurring when the three-dimensional growth surfaces instead of planar ones are used.

15.3 Nanopatterning of 3C-SiC

In what follows we shall explain the different basic steps toward the heteroepitaxy of c-GaN by MBE on prepatterned 3C-SiC/Si including the nanopatterning of the substrate. Two techniques are used to obtain nanopatterns on the 3C-SiC/Si (001) substrate: nanosphere lithography (NSL) and electron beam lithography (EBL). Both methods are followed by a lift-off process and reactive ion etching (RIE) process.

15.3.1 Nanopatterning by Nanosphere Lithography

NSL is a cost-effective method to produce regular nanoscale surface patterns of scalable size and with a variety of motifs [15]. NSL uses a self-organized monolayer of nanobeads as a shadow mask in order to create nanopatterns on a surface. The advantage of this technique is that it can easily produce nanopatterns on areas several centimeters large. Therefore, thin film growth studies on NSL-patterned surfaces can subsequently employ all of the standard characterization techniques with typical sampling areas of few millimeters diameter.

3C-SiC (001) thin films grown by chemical vapour deposition on Si (001) [16] were used in our experiments. Due to the hydrophobic character of SiC [17], the substrates were coated with ~16 nm of SiO_2 by plasma-enhanced chemical vapor deposition followed by a cleaning process ($H_2O : NH_3 : H_2O_2 = 3 : 1 : 1$). The surface was subsequently covered with a mask of 600 nm diameter polystyrene (PS) spheres forming a self-organized hexagonally closed packed monolayer. This was achieved by the controlled drying [18] of a droplet of an aqueous suspension of colloidal PS spheres. In the next step, the spheres were shrunken in oxygen plasma as outlined in [19]. On top of the colloidal mask 60 nm of Ni was thermally evaporated and the PS spheres were removed. The resulting Ni-mask-coated 3C-SiC/Si (001) was vertically etched by 300 nm using reactive ion etching (10 % O_2, 6, 1 % SF_6). After removing the metal by $FeCl_3$ and DI-water (1:3), the samples were cleaned with a buffered oxide etching solution (NH_4F:H_2O:HF = 4:6:1). Altogether, this procedure allows the creation of surface patterns with a lateral feature size of some ten to few hundred nanometers. Figure 15.2 shows a side view scanning electron microscopy (SEM) image of a cleaved nanopatterned 3C-SiC (001) surface. The hole-diameter is about 400 nm and the depth is about 300 nm. Since the cleaving edge cuts through some holes it is obvious that the etching process leads to a high surface roughness at the bottom of the holes.

15.3.2 Nanopatterning by Electron Beam Lithography

EBL is one of the most common techniques to fabricate nanostructure masks. Although for time reasons one would typically fabricate nanostructures only in small areas of few hundred μm^2, the high flexibility in motive shapes and the excellent precision at which motives can be reproduced make EBL attractive for patterning substrates for selected-area growth studies. Hiller and coworkers [20] have shown how to produce 3C-SiC nanostructures which are aligned parallel to the [110] directions of the substrate. Recently, the same group achieved side wall steepness of up to 83° in 3C-SiC (001) [20].

The used etching masks were formed by EBL and a following lift-off process with a Ni metal mask. The Ni metal mask was produced by thermal evaporation. Due to the high chemical stability of the SiC dry etching techniques are the methods of

Fig. 15.2 Side view SEM image of a 3C-SiC (001) surface nanopatterned by NSL

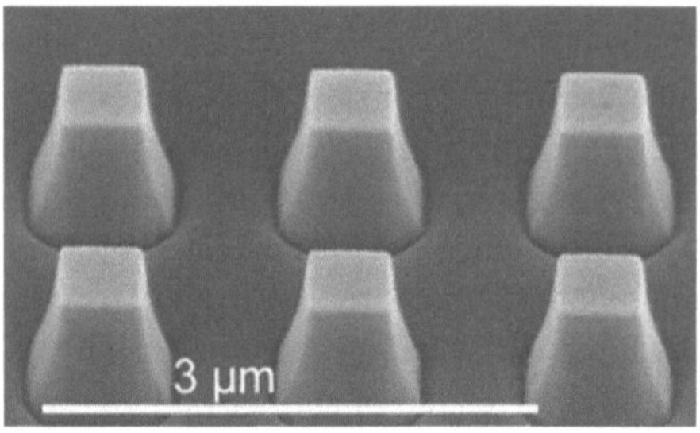

Fig. 15.3 *Side view* SEM image of the 3C-SiC (001) surface with *post-shaped* nanostructures [21]

choice to transfer the mask pattern into the three dimensions. Therefore an electron cyclotron resonance (ECR) plasma system was modified by a special gas inlet unit suitable for injecting SF_6 into the high-density downstream Ar/O_2 ECR plasma. An etching rate of about 100 nm/min was achieved. Figure 15.3 displays a side view SEM image of the 3C-SiC (001) surface with post-shaped nanostructures. Arrays of 5×5 nanoposts were repeated several times on the substrate surface with etched down spaces in between. The posts have top edges with a length of about 500 nm and a height of about 700 nm.

15.4 Molecular Beam Epitaxy of Cubic GaN

In this section we shall explain the growth of c-GaN on 3C-SiC (001) substrates by MBE. First we will discuss the growth process and the growth parameter of planar c-GaN films. Afterwards the MBE of selective-area-grown c-GaN on prepatterned 3C-SiC/Si (001) is introduced.

15.4.1 Molecular Beam Epitaxy of Planar Cubic GaN Films

Cubic group III-nitride samples were grown on 200 µm thick, free standing 3C-SiC (001) substrates by molecular beam epitaxy (MBE) [22, 23]. An Oxford Applied Research HD25 radio frequency plasma source was used to provide activated nitrogen atoms and gallium was evaporated from a Knudsen cell. Prior to growth, the 3C-SiC substrates were chemically etched by organic solvents and a buffered oxide etch (BOE) and were annealed for 10 h at 500 °C. Cubic GaN layers were deposited at 720 °C directly on 3C-SiC substrates. The adsorption and desorption of metal (Ga) layers on the c-GaN surface was investigated using the intensity of a reflected high energy electron beam (RHEED) as a probe. The structural and morphological properties of both the 3C-SiC substrates and the group III-nitride epilayers were measured by high resolution X-ray diffraction (HRXRD) and atomic force microscopy (AFM).

Reciprocal space mapping (RSM) has been performed to determine the content of hexagonal inclusions and strain in the epilayers.

As an important step to improve the GaN surface morphology in a systematic way, it is essential to understand the surface structure and the underlying growth process on an atomic scale. In particular, the kinetic processes of adsorption and desorption on the surface are considered as key parameters that govern the surface morphology, incorporation kinetics and consecutively the overall material quality. In MBE of GaN, two-dimensional surfaces are commonly achieved under Ga-rich conditions, with theoretical [24] and experimental [25, 26] evidence suggesting that the growth front is stabilized by a metallic Ga adlayer. The optimum conditions for the epitaxial growth of c-GaN are mainly determined by two parameters, the surface stoichiometry and the substrate temperature [22]. Both parameters are interrelated; therefore an in situ control of substrate temperature and surface stoichiometry is highly desirable. The study of the surface reconstruction by RHEED was one of the key issues in understanding the c-III nitride growth [22, 27, 28]. First principle calculations by Neugebauer et al. [29] show that all energetically favored surface modifications of the nonpolar (001) c-GaN surface are Ga-stabilized and therefore optimum growth conditions are expected under slightly Ga-rich conditions.

By recording the RHEED intensity transient of the (0,0) reflexion of the (2×2) reconstruction the Ga-coverage in the range between 0 and 1 monolayer can be estimated with an accuracy of 0.1 monolayer [9]. Optimum growth conditions of c-III nitrides were found, when a one monolayer Ga coverage is formed at the growing surface [9]. This is in contrast to what has been observed with h-GaN, where the optimum growth conditions with regard to surface morphology are related to the formation of a Ga bilayer (c-plane, [30, 31]) or a trilayer (m-plane, [26]), respectively. Under one monolayer Ga coverage the root mean square (RMS) roughness of the c-GaN surface measured by a $5 \times 5\,\mu m^2$ AFM-scan is decreasing to a minimum value of 2.5 nm. The smoother the surface of the epilayer the narrower is also the rocking curve line width (ω-scan) of the cubic GaN and the higher is the structural quality of the cubic epilayer.

In Fig. 15.4 the rocking curve line width (ω-scan) of all our cubic GaN epilayers (full triangles) grown on 3C-SiC is plotted versus c-GaN layer thickness. Two clear effects can be seen from this plot. First a reduction of the full width at half maximum (FWHM) with increasing epilayer thickness is observed. This linewidth dependence is consistent with the defect annihilation process observed in c-GaN grown on GaAs (001) substrates [32]. Since in zinc-blende structure the stacking faults (SFs) lie on the (111) planes, an annihilation mechanism is possible, when two SFs, lying, for example on the (111) and on the (-1-11) planes intersect and annihilate simultaneously with the creation of a sessile dislocation aligned along [110] directions. For the case of 3C-SiC, where the lattice mismatch is only $-3.5\,\%$ to c-GaN, the full line shows the theoretical calculated FWHM as a function of layer thickness using the dislocation glide model by Ayers [33]. This model implies that the dislocation density N_{disl} is inversely proportional to the layer thickness d and that the FWHM is proportional to $d^{-1/2}$. Comparing the full curve with the experimental data (full triangles) a reduction of the FWHM by a factor 1.5 is still possible. For the sake

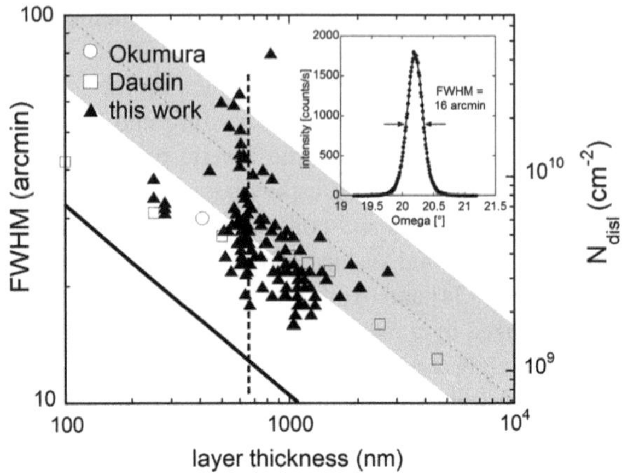

Fig. 15.4 Rocking curve linewidth of cubic GaN epilayers grown on 3C-SiC substrates versus thickness of the cubic GaN epilayers

of completeness the shaded area in Fig. 15.4 depicts the range of FWHM values as measured for cubic GaN grown on GaAs substrates. By comparing our data with data cited in literature, the dependence of the FWHM on film thickness has to be taken into account. Up to now, only two other groups reported data on 3–5 μm thick 3C-SiC/Si (001) pseudosubstrates grown by chemical vapor deposition (open circle Ref. [34] and open squares Ref. [35]). As can be seen in Fig. 15.4 we are clearly able to remain under the best cited values, indicating the improved structural quality of our c-GaN epilayers due to the availability of free standing, bulk like 3C-SiC (001) substrates. These experiments were performed on freestanding 3C-SiC (001) substrates, however, similar results were also obtained on 10 μm thick 3C-SiC (001) grown on Si (001) substrates (not included in Fig. 15.4).

As mentioned above, the second parameter which strongly influences the quality of the cubic epilayers is the roughness of the 3C-SiC substrate. In Fig. 15.4 the rocking curve linewidth of 600 nm thick c-GaN epilayers varies from 60 to 20 arcmin for 3C-SiC substrates with different surface roughness (indicated by the vertical dashed line). The RMS-roughness of the corresponding 3C-SiC substrates as measured by AFM on $5 \times 5\,\mu m^2$ large areas varied between 11 and 0.7 nm, respectively. The smoother the surface of the substrates the narrower is the FWHM and the higher is the structural quality of the cubic epilayer. This observation is in agreement with measurements of Ref. [34] and indicates that a further improvement of the surface preparation (atomic flat surfaces) may allow to achieve the theoretical limit indicated by the full line.

However, using the method of Gay et al. [11], which correlates the X-ray FWHM with the dislocation density, one sees from Fig. 15.4, that even with the best surface roughness a residual dislocation density in the order of $N_{disl} \sim 10^9/cm^2$ remains for about 1 μm thick c-GaN epilayers. For many device applications such a high density of dislocations severely reduces the device performances and it is urgently demanded to avoid them. As discussed above, nanoheteroepitaxy offer this possibility and avoids the defect generation when the substrate nanostructures have a size of about 40 nm.

15.4.2 Molecular Beam Epitaxy of Selective-Area-Grown Cubic GaN

The optimized growth conditions of cubic GaN described above are the basis for the growth of cubic GaN on prepatterned 3C-SiC/Si (001) substrates by MBE. Prepatterning of the substrate was achieved, i.e., by NSL (see Sect. 15.3), which offers a large area of a patterned substrate surface. Due to the large-patterned area it is guaranteed that the electron beam for RHEED touches the patterned surface and monitors information about the selective-area-growth of cubic GaN. Therefore, we discuss MBE growth only for substrates patterned by NSL.

On the patterned 3C-SiC/Si (001) substrate displayed in Fig. 15.2 a 440 nm c-GaN (001) thin film was grown [36]. For comparison a reference sample with a c-GaN epilayer of the same thickness was grown on an unstructured 3C-SiC substrate. For both patterned and unpatterned substrates the growth temperature was 720 °C and an one monolayer Ga coverage at the surface was used to grow under optimum conditions [9]. Figure 15.5a shows the RHEED pattern ([110] azimuth) of the planar cubic GaN layer of the reference sample. The long streaks indicate a two-dimensional surface. Figure 15.5b illustrates the [110] azimuth RHEED pattern of the surface of a 440 nm thin GaN layer grown on nanostructured 3C-SiC/Si (001). The RHEED pattern in Fig. 15.5b is similar to the one of the planar c-GaN layer clearly indicating the growth of c-GaN also on the prepatterned 3C-SiC. Obviously, the three-dimensional structure of the surface does not disturb the growth conditions in a way that would lead to the formation of hexagonal GaN. Figure 15.6a shows a side-view SEM image of the

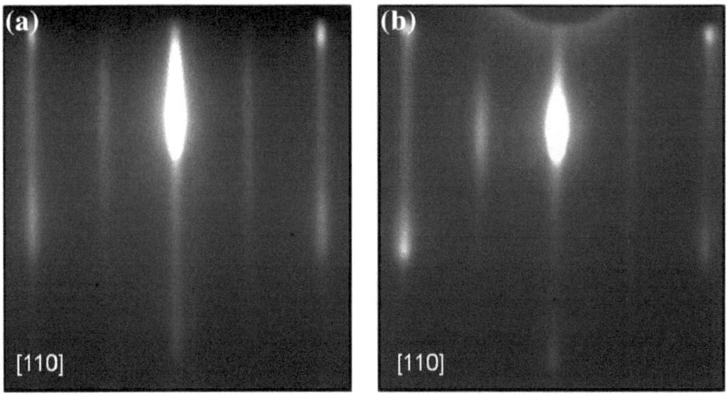

Fig. 15.5 **a** RHEED pattern of a 440 nm thin cubic planar GaN on 3C-SiC/Si (001), **b** RHEED pattern of a 440 nm thin cubic GaN layer grown on nanostructured 3C-SiC/Si (001) (both images in [110] azimuth) [36]

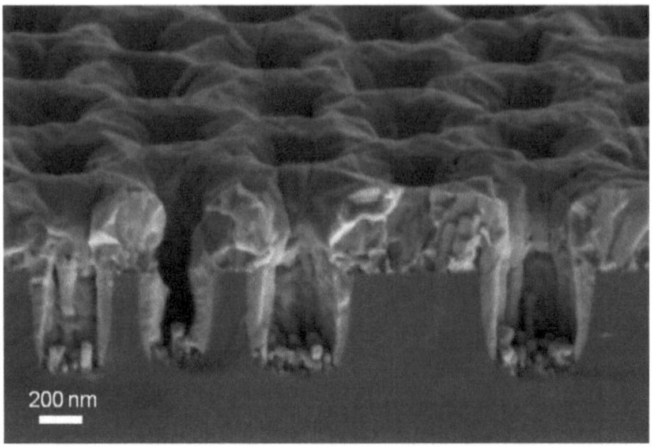

Fig. 15.6 *Side view* SEM image of a 440 nm cubic GaN layer grown on 3C-SiC/Si (001) nanopatterned by NSL

440 nm c-GaN on 3C-SiC/Si (001) nanopatterned by NSL. It is obvious from this image that there is GaN growth both on the SiC top surface and inside the SiC holes. The holes have not been closed by coalescence of the GaN top layer at this GaN layer thickness. The growth rate inside the holes is lower than on the (001) top surface due to geometrical effects. The large surface roughness at the bottom of holes observed before leads to a particularly grainy structure of the GaN film here.

15.5 Characterization of Selective-Area-Grown Cubic GaN

In Sect. 15.5, we shall discuss the characterization of the selective-area MBE grown c-GaN. HRXRD and cross-sectional transmission electron microscopy (XTEM) were used to characterize the structural properties. The optical properties were studied by microphotoluminescence ($\mu - \text{PL}$) measurements.

15.5.1 Cubic GaN on 3C-SiC/Si (001) Substrate Prepatterned by Nanosphere Lithography

HRXRD is a powerful tool to characterize the strain distribution and defect densities in heteroepitaxially grown films. Figure 15.7 shows an asymmetrical reciprocal space map around the (113) reflection of the selectively grown 440 nm thin c-GaN film. Position 1 marks the bulk value of 3C-SiC [37]. Position 2 presents the literature value [38] of planar c-GaN and point 3 shows the experimental data of selective-area grown c-GaN, respectively. The selective-area-grown c-GaN is relaxed to the 3C-SiC with a minimal residual tensile strain. The reason for the residual tensile strain can

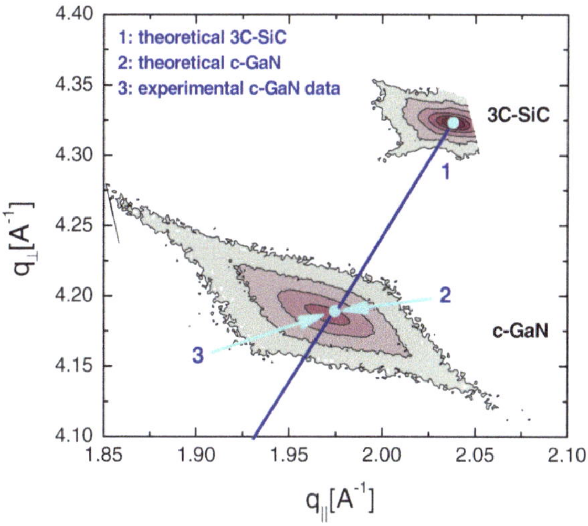

Fig. 15.7 Reciprocal space map around the (113) reflection of selective-area-grown c-GaN [36]

be found in the different thermal expansion coefficients of cubic GaN and 3C-SiC [39]. The same measurements were done for the planar reference layer and show that there is equal strain in both cases. The lattice parameter in growth direction was determined to 4.503 ± 0.001 Å. Additionally, a negligible amount of hexagonal inclusions in the selective-area-grown c-GaN is measured by symmetrical reciprocal space maps around the (002) reflection (not shown). The FWHM of the rocking curve of the selective-area-grown c-GaN is about 29 arcmin. and smaller than the FWHM of the rocking curve of the planar c-GaN reference layer with 36 arcmin. Using the method of Gay et al. [11] the defect density of selective-area-grown c-GaN is determined to be $4.5 \times 10^9/\text{cm}^2$ and $7.4 \times 10^9/\text{cm}^2$ for a planar layer. It should be noted that all HRXRD data measured on selective-area grown films are values averaging over the GaN top layer and the GaN at the sidewalls and bottoms of the SiC nanoholes.

In order to investigate the micro- and nano-structural features of the MBE grown GaN nanostructures, cross-sectional TEM samples were prepared by mechanical grinding and polishing followed by an ion beam thinning step. The samples were subsequently examined in a Philips CM200ST TEM operated at 200 kV.

Figure 15.8 shows a TEM bright-field micrograph of the nanostructured sample taken along the [110] zone axis under a two-beam condition. The cross section cuts through a row of ~300 nm deep and roughly 400 nm wide holes in the SiC substrate overgrown with about 400 nm GaN. Selective area diffraction (SAD) confirms that the GaN is cubic and is epitaxially oriented in full registry with the 3C-SiC/Si (001) substrate. The c-GaN film on the top SiC surface exhibits lateral facets which are inclined by ~10° with respect to the growth direction and additional facets on the top which are inclined by ~20° with respect to the sample template surface.

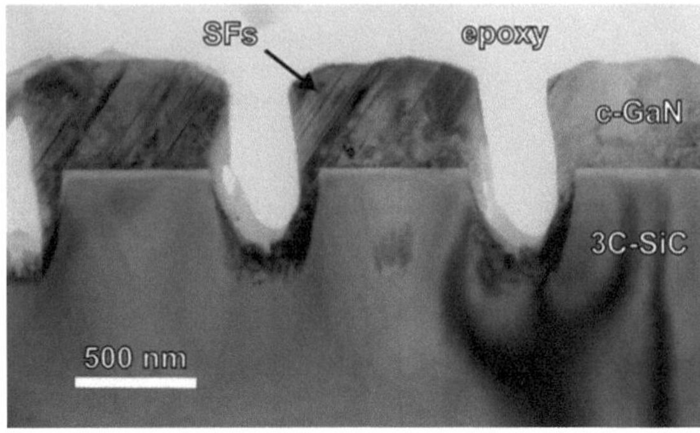

Fig. 15.8 Cross-sectional TEM image of a nanostructured cubic GaN film taken along the [110] zone axis under two-beam condition. Stacking faults along the $[\bar{1}1\bar{1}]$ direction can be observed in the cubic GaN and are marked with 'SFs'. [36]

Sanorpim et al. [40] have observed similar faceting of cubic GaN in their studies of epitaxial lateral overgrowth of (001)-oriented stripe-patterned GaAs with cubic GaN by MOVPE. The mechanism of facet formation is not yet fully understood and will be a topic of further studies.

The GaN film shown in Fig. 15.8 contains stacking faults ('SFs') mainly on the $(\bar{1}1\bar{1})$ planes. Due to the [110] specimen orientation these stacking faults are seen edge on and therefore appear as bright and dark straight lines. In [110] SAD patterns like the one in Fig. 15.9a taken from the middle island of Fig. 15.8 they lead to streaks along the $[\bar{1}1\bar{1}]$ direction. Additional streaks can be found along the $[\bar{1}11]$ direction but they are significantly weaker indicating a much lower stacking fault density on these planes. This is consistent with observations in weak-beam dark-field images (not shown) which also support an asymmetric stacking fault density. The same finding can be made in HRXRD symmetrical reciprocal space maps around the (002) reflection (not shown).

In addition to the reflections originating from cubic GaN and the stacking faults therein, weaker spots along the $[\bar{1}1\bar{1}]$ direction can be found. All these additional reflections can be explained by the formation of a small fraction of hexagonal inclusions, which are predominantly formed with their c-axis along the $[\bar{1}1\bar{1}]$ direction. Figure 15.9b shows the expected positions of reflections in a SAD pattern originating from cubic GaN (filled circles) and from hexagonal inclusions (crosses) as simulated using the JEMS code [41], assuming an epitaxial relationship of [0002]h-GaN || $[\bar{1}1\bar{1}]$c-GaN and [110]c-GaN || $[11\bar{2}0]$h-GaN. The absence of similar diffraction spots along the $[\bar{1}11]$ direction indicates a negligible amount of hexagonal inclusions along this direction, in perfect agreement to a much smaller density of stacking faults on the $(\bar{1}11)$ plane. This indicates that high stacking fault densities and hexagonal inclusions are correlated. While the contrasts in TEM bright-field images are dominated by stacking faults, it should be noted that there is also a limited number of dislocations close to the GaN/SiC interface, the density of which decreases over the first hundred nanometers.

The optical properties of the patterned cubic GaN layer were studied by room temperature $\mu-$PL measurements. Luminescence was excited by the 325 nm line of a HeCd laser with a power of 400 μW on the sample surface (beam diameter 1–2 μm).

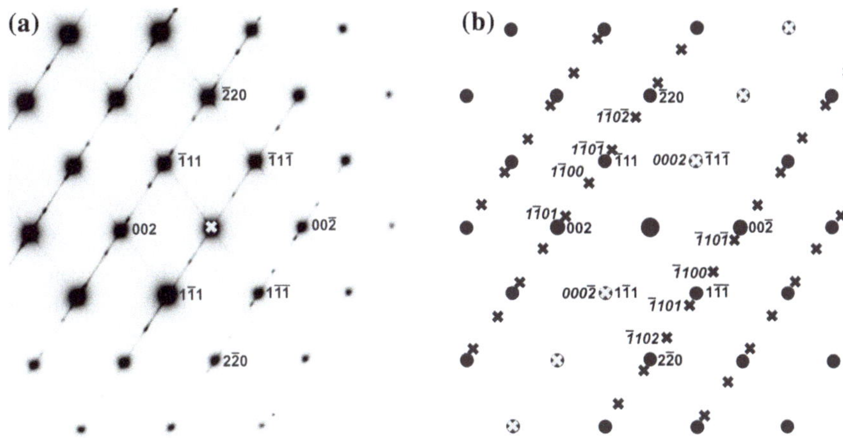

Fig. 15.9 **a** SAD pattern taken from the middle cubic GaN "island" in Fig. 15.8 along the [110] cubic GaN zone axis with the main reflections indexed. Streaks along [$\underline{1}1\underline{1}$] and [$\underline{1}11$] indicate the formation of stacking faults in these directions. **b** Simulated SAD pattern of cubic GaN along the [110] direction (*filled circles*) together with the simulation of h-GaN viewed along the [11$\underline{2}$0] direction (*crossed*). The good match of the experimental pattern (**a**) with the simulation (**b**) indicates the presence of a small fraction of hexagonal inclusions along the [$\underline{1}1\underline{1}$] direction within the cubic GaN "islands" [36]

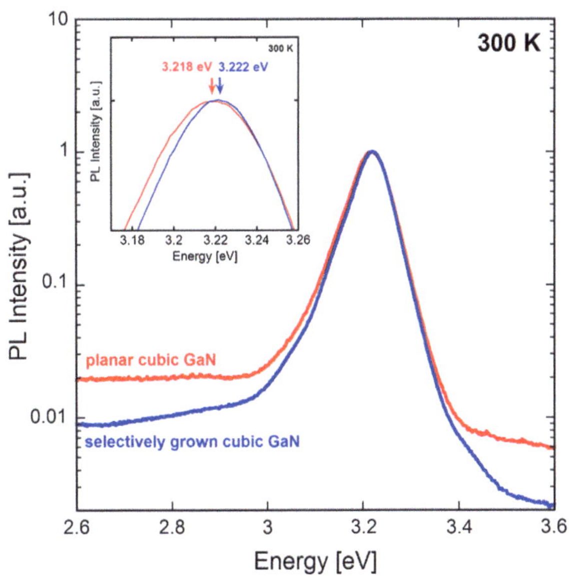

Fig. 15.10 Room temperature μ − PL spectra of a planar cubic GaN epilayer (*red*) and selective-area-grown c-GaN film (*blue*) [36]

In Fig. 15.10 the spectra of the c-GaN planar reference layer sample (red curve) and the selective-area-grown cubic GaN sample (blue curve) are depicted. The spectra show that there are no luminescence bands of the hexagonal phase indicating again that amount of hexagonal inclusions found by XTEM in our nanopatterned GaN is negligible.

The peak of the band-edge luminescence of the reference layer is at 3.218 eV. This value agrees well with the literature value of 3.21 eV for cubic GaN grown on GaAs [42]. There is a blue shift of 4 meV in the band-edge transition (3.222 eV) of the selective-area-grown cubic GaN sample. The shift to a higher bandgap energy indicates that the tensile strain in the selective-area-grown cubic GaN that was shown by HRXRD is lower than in the planar cubic GaN layer.

15.5.2 Cubic GaN on 3C-SiC/Si (001) Substrate Prepatterned by Electron Beam Lithography

Figure 15.11 shows a SEM top view image of post-shaped 3C-SiC (001) nanostructures (see Fig. 15.3) overgrown with 400 nm c-GaN [21]. The edge length of the post structures is increased after GaN overgrowth from about 500 nm to about 1 μm. Furthermore there are darker areas in between neighboring posts, indicating deepenings in the GaN layer. They arise probably due to shadowing in particular when the sample holder rotation is stopped during MBE growth for RHEED inspection of the growth surface.

To have a closer look to the GaN on the post structures, cross-sectional TEM specimens were prepared using a focused ion beam (FIB) and analyzed using either a JEOL 2000 FX or a FEI Technai F20 microscope at 200 kV. Figure 15.12a displays

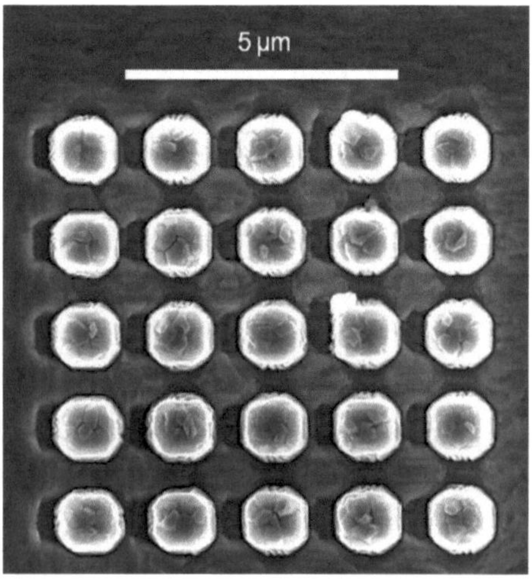

Fig. 15.11 *Top view* SEM image of a nanostructure array overgrown with 400 nm c-GaN [21]

a bright-field TEM micrograph of an individual 3C-SiC (001) post overgrown with 400 nm c-GaN taken along the [110] zone axis. The GaN surface is covered with a Pt protection film (dark in Fig. 15.12a) necessary in the TEM preparation process. The image shows that the GaN grows on the top of the 3C-SiC post and on its sidewalls. Very generally speaking, we observed in this and all similar samples that GaN growth on the sidewalls is characterized by a huge irregularity of the stacking sequence in one of the ⟨111⟩ directions, while the stacking fault density in the c-GaN on top of the SiC posts is comparatively small. The GaN growth on the sidewalls of the 3C-SiC posts is dominated by {111} planar defects. One reason for this could be the sidewall surface roughness resulting from the etching process during substrate patterning. Another reason is probably the deviation from an ideal 90° slope of the flanks. The GaN layer thickness on the sidewalls is smaller than on top of the post. Hence there is a lower GaN growth rate on the sidewalls than on top of the 3C-SiC post.

Since the TEM lamella in Fig. 15.12a was cut through the sidewall of the post and not through the center as sketched in Fig. 15.12b, the GaN is visible on three sidewalls of the post. This leads to the strongly varying contrasts within the upper

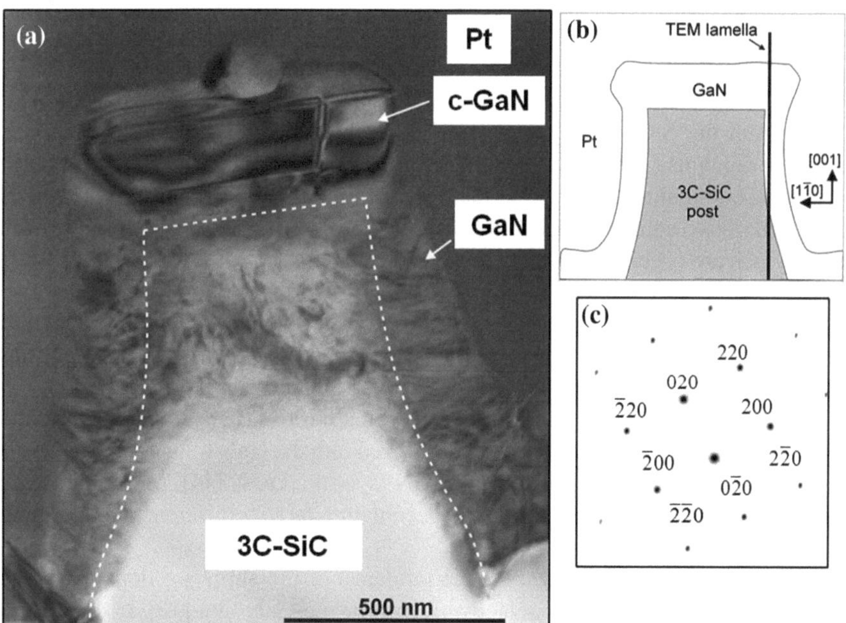

Fig. 15.12 **a** Cross-sectional TEM image of an individual 3C-SiC post overgrown with 400 nm cubic GaN taken along the [110] zone axis. **b** A schematic cross-sectional view through the center of a typically overgrown SiC post parallel to the [1$\bar{1}$0] direction. The TEM lamella cuts through the sidewall of the post perpendicular to the [1$\bar{1}$0] direction. **c** SAD pattern (contrast inverted) taken from the middle cubic GaN island in (**a**) along the [001] cubic GaN zone axis with the main reflections indexed [21]

part of the SiC post, since here the almost defect free crystal lattice of the SiC post is superimposed with the defect rich GaN growing on the flanks of posts. The GaN on the top of the post grows in individual crystals. It should be emphasized that the part of the GaN film displayed in Fig. 15.12a is located at the edge of the GaN top layer. The homogeneous contrasts within this part of the GaN film indicates a very low defect density of the GaN growing on top of the SiC post. There is only one dislocation like defect running vertically in the GaN top bit. The corresponding SAD pattern taken from the central GaN area on top of the post in Fig. 15.12a is shown in Fig. 15.12c. The diffraction pattern can be completely assigned to the lattice of cubic GaN with the [010] zone axis-oriented parallel to the electron beam. The SAD pattern contains no further reflections indicating the presence of a crystal with a pure cubic phase. These are the first steps toward defect-reduced cubic GaN grown via MBE on a dimensionally limited 3C-SiC (001) growth area. A small reservation is, however, that we also found a small misaligned GaN crystal centered on top of the epitaxial GaN film, not contributing to the SAD pattern shown.

15.6 Influences of the 3C-SiC/Si Substrate on the Selective-Area-Grown Cubic GaN

In addition to the defects discussed above, there is yet another class of defects originating from the SiC substrates. 3C-SiC/Si (001) substrates exhibit two types of antiphase domains (APDs) sharing equal parts of the growth surface. The presence of antiphase domains is due to the fact that the zinc blende structure of 3C-SiC is composed of two equivalent fcc lattices (shifted by 1/4, 1/4, 1/4 with respect to each other) which are occupied by either one or the other atomic species. Thus both (001) or (111) growth surfaces can be either Si or C terminated, corresponding to the two different APDs. During the nucleation of 3C-SiC on an elementary Si substrate APDs are assumed to be formed at terrasses on the Si surface. The C-face {111} plane has a lower surface energy and thus has a higher growth rate than the Si-face. [43] Our studies [44] show that GaN nucleation on 3C-SiC/Si (001) is governed by the APDs of the substrate and that the APDs are transferred into the GaN films. Thus, the presence of the APDs is independent of the GaN layer thickness [44]. More important, however, is that the two types of domains contain a largely different stacking fault density. [21]

In the following, we will discuss the influence of the antiphase domains in the 3C-SiC/Si substrate on a thick selective-area-grown c-GaN film [13]. In contrast to the former sections we chose here a large layer thickness of about 1.4 μm to achieve a pronounced lateral growth and to promote the formation of surface facets.

Figure 15.13a shows a top view SEM image of the surface of a 1.4 μm thick c-GaN layer grown on 3C-SiC/Si (001) nanopatterned using the NSL technique (see Sect. 15.3). Even though the film thickness is as large as 1.4 μm the GaN film is still discontinuous and exhibits a close-packed array of residual holes. In the SEM

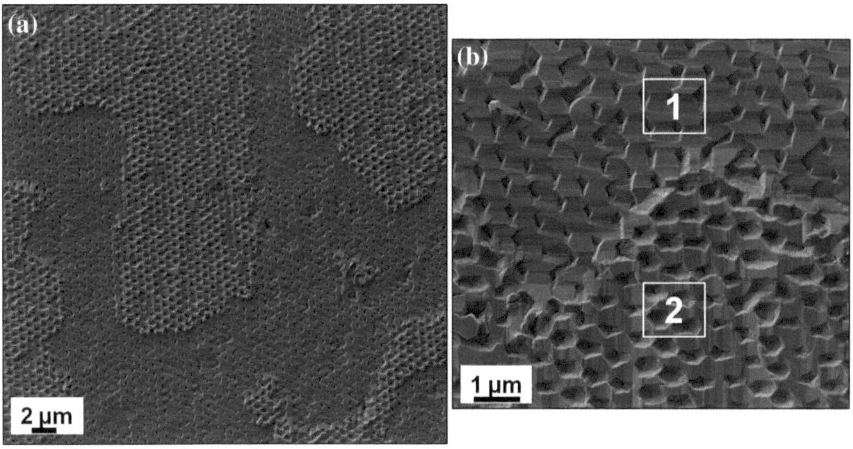

Fig. 15.13 **a** Top view SEM image of the surface of a 1.4 µm thick cubic GaN layer grown on a NSL nano-patterned 3C-SiC/Si (001) substrate. The surface consists of domains, which have an arbitrary shape with lateral dimensions varying from a few micrometers up to about 100 µm. **b** Enlarged section of (**a**) showing the different size and orientation of residual holes in the two domains (1 and 2)

domains with an arbitrary shape with lateral dimensions varying from a few microns up to about 100 µm can be seen. Different surface morphologies are present in the two types of domains, as is clearly visible in the detail SEM image of Fig. 15.13b. Note that the residual hole openings in domain 1 are smaller than in domain 2 and that their main axis is rotated by 90° with respect to each other in the two domains. NSL patterned surfaces may contain distortions of the periodicity (e.g., grain boundary-like mask defects) but a close inspection of the domains reveals that irregularities in the periodicity of the nanopattern on the substrate do not influence the surface morphology of the GaN grown within the same domain. This is an indication that the different morphologies originate from the domain structure of the substrate. The reason for these different GaN growth types is assumed to be differently terminated surfaces of the SiC substrate, leading to different polarities of the GaN film and thus a different growth behavior.

Figure 15.14 shows an AFM image of two adjacent domains on the surface of GaN grown on the prepatterned substrate. In domain 1 GaN grows with atomically smooth facets, indicating that in this region the Frank-Van-der-Merwe growth mechanism dominates. Due to the lateral growth the shape of the nanostructures changes from circular holes before GaN growth into triangle-shaped openings afterwards. The diameter of the holes is reduced from 500 nm before to 100–200 nm after GaN growth. In contrast to the layer-by-layer growth mechanism in domain 1, cubic GaN seems to grow in form of islands in domain 2 (Fig. 15.14). The shape of the holes after GaN growth has changed into squares with diameters from 200 to 350 nm.

The structural quality, averaged over all domains of the patterned cubic GaN film, is investigated by HRXRD measurements. Figure 15.15 shows a symmetrical reciprocal space map around the c-GaN (002) reflection. The FWHM of the (002)

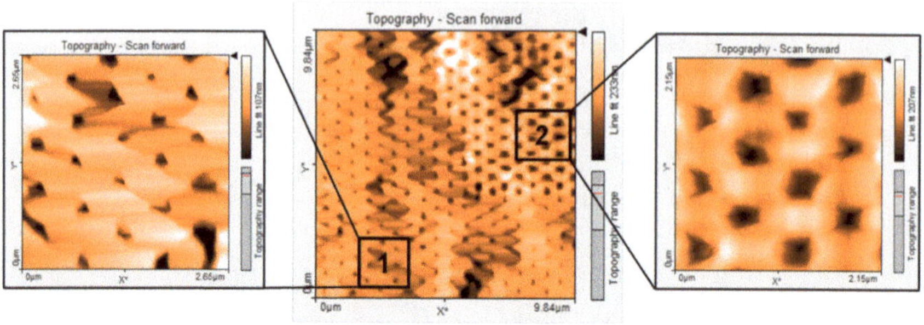

Fig. 15.14 AFM image of the surface of the selective-area-grown c-GaN with two adjacent domains. Zoom-in sections on the *left* and on the *right* show a step-flow-like flat surface in domain 1 and a more three-dimensional surface morphology in domain 2 [13]

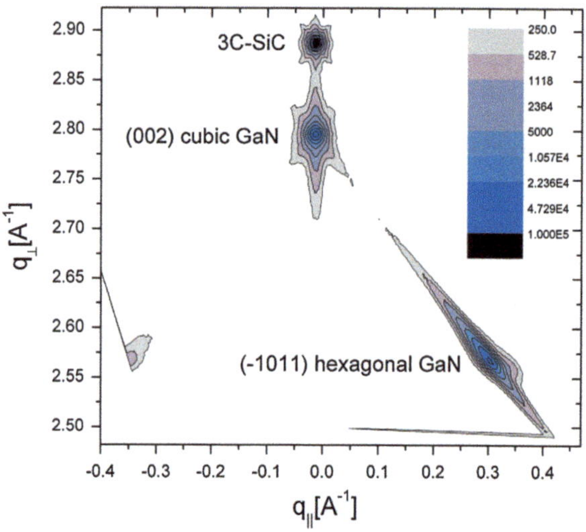

Fig. 15.15 A symmetrical reciprocal space map around the (002) reflection of a selective-area-grown 1.4 µm thick GaN film on NSL prepatterned 3C-SiC/Si (001) [13]

rocking curve of the selective-area-grown cubic GaN is about 25 arcmin (compared to 29 arcmin in a 440 nm thin film). In addition to the (002) c-GaN reflection, there is a strong (-1011) reflection of hexagonal GaN visible in Fig. 15.15. From this, an amount of 78 % hexagonal inclusions in the selective-area-grown cubic GaN is estimated. This value is very high compared to the cubic GaN reference layer, where only 17 % hexagonal inclusions and a FWHM of the (002) rocking curve of 21 arcmin were observed at the same film thickness. Since the quality of the cubic phase of the selective-area-grown GaN and the reference layer are identical and in view of the enhanced formation of stacking faults on the side walls of SiC nanostructures (see Sect. 15.5) and since the hexagonal volume fraction in thin GaN films is negligible (see Sect. 15.5) it is concluded that substrate patterning enhances the formation of stacking faults and hexagonal inclusions mainly at the sidewalls, leading to substantial fractions of hexagonal GaN upon the growth of thick layers.

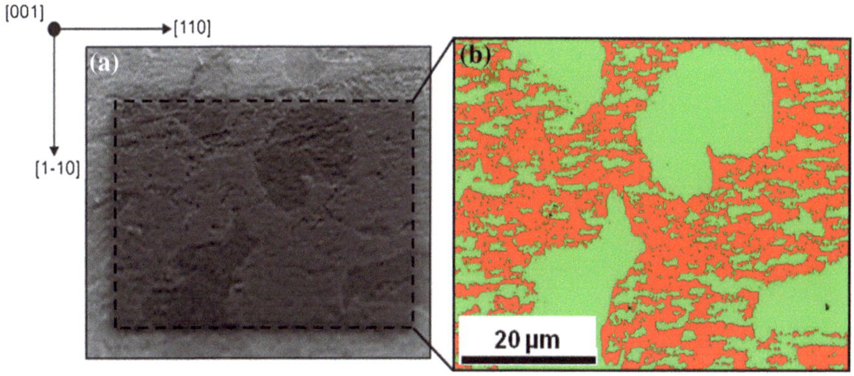

Fig. 15.16 **a** SEM image of the EBSD-measurement area on the surface of the 1.4 μm thick selective-area-grown c-GaN film, which is tilted to about 70° in the measurement setup. **b** Phase map showing regions of predominantly cubic (*red*) and hexagonal (*green*) GaN [13]

Electron backscatter diffraction (EBSD) measurements were used for phase identification of the selective-area-grown GaN domains. For this method the primary electron beam (20 keV) from a tungsten cathode of a XL40 ESEM is inelastically and elastically scattered by the crystal planes of the sample. This yields Kikuchi diffraction patterns, in which each pair of Kikuchi lines represents a crystal plane [45]. By this, areas of predominantly cubic and hexagonal phase of GaN can be distinguished and mapped. Figure 15.16a shows a top view SEM image of the EBSD-measurement area on the surface of the selective-area-grown c-GaN sample, which is tilted to about 70° in the microscope. The domains described above are clearly visible in this image. The diffraction pattern of the cubic and the hexagonal phase of GaN exemplify measured raw data and lead to the phase map in Fig. 15.16b. The crystal plane identification was achieved by a Hough-transformation followed by a least-square fit to both the cubic and the hexagonal phase, yielding a phase identification based on the smallest error. The data illustrated in the phase map were detected with a step size of 0.6 μm and an angle resolution of 1–2°. The cubic phase is displayed in red color and the hexagonal phase of GaN in green, where a 50 % threshold is used to assign either phase to a measurement position. That is, even though it is assumed that both phases are present in the domains, a sharp discrimination of the two pure phases is made in this diagram, and it obviously strongly correlates with the SEM image in Fig. 15.16a. It is clearly visible that the fraction of hexagonal phase between close-by domains varies. The domains with mainly cubic phase (red) contain hexagonal inclusions (green). Additionally there are domains where the hexagonal phase (green) dominates. The correlation between SEM images and AFM measurements demonstrates that in domains with smaller lateral growth the cubic phase is dominant. Furthermore, the domains with a smooth layer-by-layer growth can be identified to contain a high fraction of the hexagonal phase.

The optical properties of the domains of the selective-area-grown GaN layer were studied by room temperature μ – PL measurements. The spot diameter of this method is smaller than the domain size. In the inset of Fig. 15.17 a top view SEM image of the

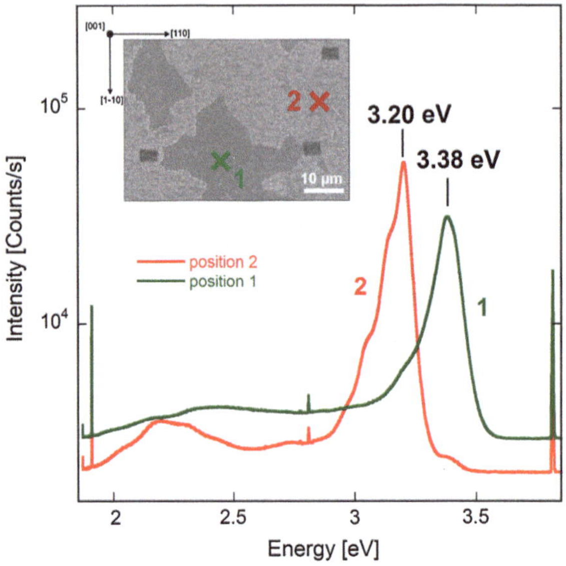

Fig. 15.17 Micro-PL spectra of adjacent domains (position 1 *green curve* and position 2 *red curve*) of the selective-area-grown cubic GaN at room temperature. *Inset*: *Top view* SEM image of the μ – PL measurement area on the sample surface. Each domain is identified with markers (*dark rectangles*) positioned by a SEM [13]

μ – PL measurement area on the sample surface is displayed. The different domains show a different secondary electron yield and thus can be clearly distinguished. In order to find appropriate positions in the optical measurements, contamination markers (dark rectangles in the inset of Fig. 15.17) were deposited with a SEM. Position 1 (green) is allocated to a domain which according to EBSD results is mainly hexagonal GaN, while position 2 (red) marks a domain consisting mainly of cubic GaN.

The μ – PL luminescence was excited by the 325 nm line of a HeCd laser with a power of 5 mW (beam diameter 1 μm). The spectra of two domains (position 1 green curve and position 2 red curve) of the selective-area-grown GaN are depicted in Fig. 15.17. The band-edge luminescence of position 1 (green curve) is at 3.38 eV and is assigned to defect luminescence related to the hexagonal phase. It should be noted that a pure hexagonal phase is expected to show a luminescence peak at 3.42 eV, which would be clearly separated from the peak observed here. By comparison, the μ – PL peak of domain 2 has more intensity and occurs at 3.20 eV. This value agrees well with the band-edge transition at 3.21 eV for cubic GaN grown on GaAs [42]. Furthermore there is an emission shoulder with low intensity at 3.38 eV, which is an indication for defects and hexagonal inclusions in the selective-area-grown cubic GaN. A further luminescence peak in the red curve at 2.23 eV is related to defect luminescence. These spectra demonstrate clearly that in addition to the structural differences shown by EBSD the optical properties between two adjacent domains are quite different.

15.7 Cubic GaN/AlN Multi-Quantumwells on 3C-SiC Nanostructures

Due to the large band discontinuity between AlN, GaN and InN, these materials are qualified for novel nitride devices like quantum well infrared photodetectors (QWIP) and quantum cascade lasers (QCL). In the cubic system Zainal and coworkers [46] have shown a resonant tunneling through cubic GaN/AlGaN double barriers on GaAs substrate. Furthermore, Mietze et al. [47] demonstrated resonant tunnel diodes with reproducible I–V characteristics and recoverable negative differential resistance (NDR) based on Al(Ga)N/GaN double barrier structures on 3C-SiC (001).

The first basic steps toward nanoscale devices without any etching process are the growth of cubic AlN/GaN multiquantum wells (MQWs) on prepatterned 3C-SiC/Si (001) substrates by means of MBE. Growth conditions are based on the experimental data in Sect. 15.4 and ref. [47]. The substrate patterning has been realized by electron beam lithography and a reactive ion etching process using the same parameters as described in Sect. 15.3. The sample structure consists of 10 periods of 2 nm c-AlN barriers with a 4 nm c-GaN layer in between, which were grown on 3C-SiC post-shaped nanostructures. The 3C-SiC posts have a length of about 550 nm and a hight of about 700 nm. Figure 15.18 displays cross-sectional TEM images of an individual 3C-SiC post overgrown with 10 periods of c-AlN/GaN MQWs taken along the [110] zone axis. Due to the difference in extinction lengths the c-GaN layers (dark) and the c-AlN barriers (bright) can be clearly separated. In general the sequential arrangement of the MQWs shows an undulated form, resulting from the high density of stacking faults. In regions without any {111} planar defects a straight sequential arrangement of c-AlN and c-GaN is observed. However, bunches of {111} stacking faults occur at which the growth rate is reduced leading to the undulating shape. The stacking

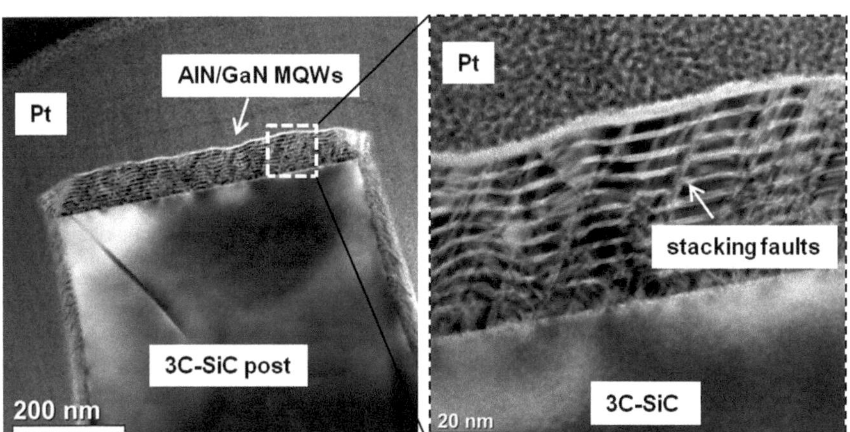

Fig. 15.18 Cross-sectional TEM images of an individual 3C-SiC post overgrown with 10 periods of c-AlN/GaN MQWs taken along the [110] zone axis

faults originate from the 3C-SiC/AlN interface and propagate towards the surface, accompanied by an increasing amplitude of the multilayer undulation.

These are first experimental results of cubic AlN/GaN MQWs grown on 3C-SiC nanostructures. We have shown that it is possible to grow cubic AlN/GaN MQWs on prepatterned substrates by MBE. It is obvious, however, that similar to pure c-GaN layers on prepatterned SiC substrates, the biggest challenge will be the elimination of the high stacking fault density. A possible approach is the reduction of the growth area to less than $100 \times 100\,\text{nm}^2$ towards nanoheteroepitaxy to reduce the density of crystalline defects and to avoid the undulated shape of the MQWs.

15.8 Conclusions

We conclude that the heteroepitaxy of nonpolar c-GaN on prepatterned substrates is a promising method for developing the structural quality of cubic nitrides. Our studies have shown that it is possible to grow phase-pure c-GaN on 3C-SiC nanostructures under Ga-rich growth conditions via MBE. Nanopatterning of the 3C-SiC/Si (001) substrate surface was realized by NSL and EBL followed by an RIE etching process. With both techniques different types of 3C-SiC nanostructures have been fabricated.

The selective-area-grown GaN, which was deposited on a surface patterned by NSL, crystallizes in the zincblende structure. HRXRD measurements exhibit that the selective-area-grown cubic GaN as well as planar epilayers are relaxed on 3C-SiC/Si (001) with a minimal residual tensile strain. The selective-area-grown cubic GaN displays lateral facets which are evidenced by cross-sectional TEM measurements. The GaN "islands" show stacking faults and a small fraction of hexagonal inclusions. Micro-PL measurements at room temperature demonstrate that the selective-area-grown cubic GaN is less tensile strained than a planar reference layer.

Not only stacking faults but also antiphase domains of the substrate influence the structural quality of the selective-area-grown c-GaN. The c-GaN grows in two different domains. The nanopatterning of the substrate does not influence the shape of these domains. HRXRD measurements show that at large layer thicknesses the hexagonal fraction in selective-area-grown cubic GaN is strongly enhanced. Several independent measurement techniques reveal that there is a largely differing amount of hexagonal inclusions and defects in the two sets of domains. Depending on the fraction of cubic or hexagonal phase the Volmer-Weber or the Frank-Van-der-Merwe growth mechanism dominates and the lateral growth rate varies accordingly.

Another focus lies on cubic GaN crystals on top of post shaped 3C-SiC nanostructures, fabricated by EBL and a RIE etching process. We have demonstrated that they grow nearly defect-free with excellent structural properties. Here the nanostructures are aligned parallel and perpendicular to the [110] directions of the substrate. On the sidewalls of these posts, however, the GaN layer contains a high density of {111} planar defects.

Moreover, a complex cubic AlN/GaN MQW layer structure on prepatterned 3C-SiC substrate has been successfully realized. However, the sequential arrangement of the MQWs shows an undulated form due to the high stacking fault density.

All in all the reduction of the stacking fault density must be the main issue of further studies. In order to draw a conclusion about the possibility of a defect reduction in GaN and the AlN/GaN MQW films on nanostructured surfaces according to the theory of nanoheteroepitaxy the size of the nanostructures still has to be scaled down further.

Acknowledgments The authors would like to thank L. Hiller, Th. Stauden and J. Pezoldt (TU Ilmenau) for patterning the substrates with electron beam lithography and reactive ion etching. The authors also wish to thank Th. Niendorf, K. Duschik and H.-J. Maier (University of Paderborn) for EBSD and some of the TEM measurements. We thank M. Ruth and C. Meier (University of Paderborn) for the micro-photoluminescence measurements. Furthermore we thank the team of the Ernst Ruska-Centre for Microscopy and Spectroscopy with Electrons (ER-C) at Forschungszentrum Jülich, in particular D. Meertens, M. Luysberg and K. Tillmann for access to and comprehensive support at the FIB and TEM facilities of ER-C. Part of the work at Paderborn was financially supported by German Science Foundation (As(107/4-1)).

References

1. Nakamura, S., Mukai, I., Senok, M.: Candelaclass highbrightness InGaN/AlGaN double-heterostructure bluelight emitting diodes. Appl. Phys. Lett. **64**, 1687 (1994)
2. Rajan, S., Waltereit, P., Poblenz, C., Heikman, S.J., Green, D.S., Speck, J.S., Mishra, U.K.: Power Performance of AlGaN-GaN HEMTs Grown on SiC by Plasma-Assisted MBE. IEEE Electron Device Lett. **25**, 247 (2004)
3. Tschumak, E., Granzer, R., Lindner, J.K.N., Schweiz, F., Lischka, K., Nagasawa, H., Abe, M., As, D.J.: Nonpolar cubic AlGaN/GaN heterojunction field-effect transistor on Ar+implanted 3C-SiC (001). Appl. Phys. Lett. **96**, 253501 (2010)
4. Fitzgerald, E.A., Watson, G.P., Proano, R.E., Ast, D.G.: Nucleation mechanisms and the elimination of misfit dislocations at mismatched interfaces by reduction in growth area. J. Appl. Phys. **65**, 2220 (1989)
5. Zubia, D., Hersee, S.D.: The Application of nanostructuring and substratecompliance to the heteroepitaxy of mismatched semiconductor materials. J. Appl. Phys. **85**, 6492 (1999)
6. Ambacher, O., Majewski, J., Miskys, C., Link, A., Hermann, M., Eickhoff, M., Stutzmann, M., Bernardini, F., Fiorentini, V., Tilak, V., Schaff, B., Eastman, L.F.: Pyroelectric properties of Al(In)GaN/GaN hetero- and quantum well structures. J. Phys. Condens. Matter **14**, 3399–3434 (2002)
7. Bernardini, F., Fiorentini, V., Vanderbilt, D.: Spontaneous polarization and piezoelectric constants of III-V nitrides. Phys. Rev. B **56**, R10 024 (1997)
8. Waltereit, P., Brandt, O., Trampert, A., Grahn, H.T., Menninger, J., Reiche, M., Ploog, K.H.: Nitride semiconductors free of electrostatic fields for efficient white light-emitting diodes. Nature **406**, 865 (2000)
9. Schörmann, J., Potthast, S., As, D.J., Lischka, K.: In situ growth regime characterization of cubic GaN using reflection high energy electron diffraction. Appl. Phys. Lett. **90**, 041918 (2009)
10. Novikov, S.V., Stanton, N.M., Campion, R.P., Foxon, C.T., Kent, A.J.: Free-standing zinc-blende (cubic) GaN layers and substrates. J. Crystal Growth **310**, 3964 (2008)
11. Gay, P., Hirsch, P.B., Kelly, A.: The estimation of dislocation densities in metals from x-ray data. Acta Metallurgica **1**, 315 (1953)

12. DeCuir Jr, E.A., Manasreh, M.O., Tschumak, E., Schörmann, J., As, D.J., Lischka, K.: Cubic GaN/AlN multiple quantum well photodetector. Appl. Phys. Lett. **92**, 201910 (2008)
13. Kemper, R.M., Häberlen, M., Schupp, T., Weinl, M., Bürger, M., Ruth, M., Meier, C., Niendorf, T., Maier, H.J., Lischka, K., As, D.J., Lindner, J.K.N.: Formation of defects in cubic GaN grown on nano-patterned 3C-SiC (001). Phys. Stat. Sol. (c) **9**(3–4), 1028 (2012)
14. Sun, X.Y., Bommena, R., Burckel, D., Frauenglass, A., Fairchild, M.N., Brueck, S.R.J., Garett, G.A., Wraback, M., Hersee, S.D.: Defect reduction mechanisms in the nanoheteroepitaxy of GaN on SiC. J. of Appl. Phys. **95**, 1450 (2004)
15. Haynes, C.L., Van Duyne, R.P.: A versatile nanofabrication tool for studies of size-dependent nanoparticle optics. J. Phys. Chem. B **105**(24), 5599–5611 (2001)
16. Chassagne, T., Leycuras, A., Balloud, C., Arcade, P., Peyre, H., Juillaguet, S.: Investigation of 2 inch SiC layers grown in a resistively-heated LP-CVD reactor with horizontal hot-walls. Mater. Sci. Forum **457–460**, 273–276 (2004)
17. Cicero, G., Catellani, A., Galli, G.: Interaction of Water Molecules with SiC(001) Surfaces. J. Phys. Chem. B **108**, 16518 (2004)
18. Lindner, J.K.N., Seider, C., Fischer, F., Weinl, M., Stritzker, B.: Regular surface patterns by local swelling induced by He implantation into silicon through nanosphere lithography masks. Nucl. Instr. Meth. B **267**, 1394 (2009)
19. Gogel, D., Weinl, M., Lindner, J.K.N., Stritzker, B.: Plasma modification of nanosphere lithography masks made of polystyrene beads. J. Optoelectron. Adv. Mater. **12**, 740 (2010)
20. Hiller, L., Stauden, T., Kemper, R.M., Lindner, J.K.N., As, D.J., Pezoldt, J.: ECR-etching of submicron and nanometer sized 3C-SiC(100) mesa structures. Mater. Sci. Forum **717–720**, 901 (2012)
21. Kemper, R.M., Hiller, L., Stauden, T., Pezoldt, J., Duschik, K., Niendorf, T., Maier, H.J., Meertens, D., Tillmann, K., As, D.J., Lindner, J.K.N.: Growth of cubic GaN on 3C-SiC/Si (001) nanostructures. J. Cryst. Growth (2012). doi:10.1016/j.jcrysgro.2012.10.011
22. As, D. J.: Growth and characterization of MBE-grown cubic GaN, $In_x Ga_{1-x}N$, and $Al_y Ga_{1-y}N$. In: Manasreh, M.O. (ed.) Optoelectronic Properties of Semiconductors and Superlattices, Vol. 19, Chap. 9, pp. 323–450. Taylor and Francis, New York (2003)
23. As, D.J., Potthast, S., Schörmann, J., Li, S.F., Lischka, K., Nagasawa, H., Abe, M.: Molecular Beam Epitaxy of Cubic Group III-Nitrides on free-standing 3C-SiC substrates. Mater. Sci. Forum **527**, 1489 (2006)
24. Northrup, J.E., Neugebauer, J., Feenstra, R.M., Smith, A.R.: Structure of GaN (0001): The laterally contracted Ga bilayer model. Phys. Rev. B **61**, 9932 (2000)
25. Koblmüller, G., Brown, J., Averbeck, R., Riechert, H., Pongratz, P., Speck, J.S.: Continuous evolution of Ga adlayer coverages during plasma-assisted molecular-beam epitaxy of (0001) GaN. Appl. Phys. Lett. **86**, 041908 (2005)
26. Brandt, O., Sun, Y.J., Däweritz, L., Ploog, K.H.: Ga adsorption and desorption kinetics on M-plane GaN. Phys. Rev. B **69**, 165326 (2004)
27. Schikora, D., Hankeln, M., As, D.J., Lischka, K., Litz, T., Waag, A., Buhrow, T., Henneberger, F.: Epitaxial growth and optical transitions of cubic GaN films. Phys. Rev. B **54**, 8381 (1996)
28. Feuillet, G., Hamaguchi, H., Ohta, K., Hacke, P., Okumura, H., Yoshida, S.: Arsenic mediated reconstructions on cubic (001) GaN. Appl. Phys. Lett. **70**, 1025 (1997)
29. Neugebauer, J., Zywietz, Z., Scheffler, M., Northrup, J.E., Van der Walle, C.G.: Clean and As-Covered Zinc-Blende GaN (001) Surfaces: Novel Surface Structures and Surfactant Behavior. Phys. Rev. Lett. **80**, 3097 (1998)
30. Mula, G., Adelmann, C., Moehl, S., Oullier, J., Daudin, B.: Surfactant effect of gallium during molecular-beam epitaxy of GaN on AlN (0001). Phys. Rev. B **64**, 195406 (2001)
31. Adelmann, C., Brault, J., Jalabert, D., Gentile, P., Mariette, H., Mula, G., Daudin, B.: Dynamically stable gallium surface coverages during plasma-assisted molecular-beam epitaxy of (0001) GaN. J. Appl. Phys. **91**, 9638 (2002)
32. Nagayama, A., Sawada, H., Takuma, E., Katayama, R., Onabe, K., Ichinose, H., Shiraki, Y.: Structural study on stacking faults in GaN/GaAs (001) heterostructures. Inst. Phys. Conf. Ser. **170**, 749 (2002)

33. Ayers, J.E.: New model for the thickness and mismatch dependencies of threading dislocation densities in mismatched heteroepitaxial layers. J. Appl. Phys. **78**, 3724 (1995)
34. Okumura, H., Ohta, K., Feuillet, G., Balakrishnan, K., Chichibu, S., Hamaguchi, H., Hacke, P., Yoshida, S.: Growth and characterization of cubic GaN. J. Cryst. Growth **178**, 113 (1997)
35. Daudin, B., Feuillet, G., Hübner, J., Samson, Y., Widmann, F., Philippe, A., Bru-Chevallier, C., Guillot, G., Bustarret, E., Bentoumi, G., Deneuville, A.: How to grow cubic GaN with low hexagonal phase content on (001) SiC by molecular beam epitaxy. J. Appl. Phys. **84**, 2295 (1998)
36. Kemper, R.M., Weinl, M., Kemper, R.M., Weinl, M., Mietze, C., Häberlen, M., Schupp, T., Tschumak, E., Lindner, J.K.N., Lischka, K., As, D.J.: Growth of cubic GaN on nano-patterned 3C-SiC/Si (001) substrates. J. Cryst. Growth **323**, 84 (2011)
37. Taylor, A., Jones, R.M.: The crystal structure and the thermal expansion of cubic and hexagonal silicon carbide, Silicon Carbide-A High Temperature Semiconductor, edited by J.R.O Connor, J. Smiltens, Oxford, Symposium Publications Division, Pergamon Press, 1960, Section III, Chap.1, p.147 (1960)
38. Strite, S., Juan, J., Li, Z., Salvador, A., Chen, H., Smith, D.J., Choyke, W.J., Morkoc, H., Vac, J.: An investigation of the properties of cubic GaN grown on GaAs by plasma-assisted molecular-beam epitaxy. J. Vac. Sci. Technol. **B9**(4), 1924 (1991)
39. Wu, J., Yaguchi, H., Zhang, B.P., Segawa, Y., Onabe, K., Shiraki, Y.: Optical properties of cubic GaN grown on 3C-SiC (100) substrates by metalorganic vapor phase epitaxy. Phys. Stat. Sol. (a) **180**, 403 (2000)
40. Sanorpim, S., Takuma, E., Ichinose, H., Katayama, R., Onabe, K.: Structural transition control of laterally overgrown c-GaN and h-GaN on stripe-patterned GaAs (001) substrates by MOVPE. Phys. Stat. Sol. (b) **244**(6), 1769 (2007)
41. Stadelmann, P.A.: EMS - A software package for electron diffraction analysis and HREM image simulation in materials science. Ultramicroscopy **21**(2), 131 (1987)
42. As, D.J., Schmilgus, F., Wang, C., Schöttker, B., Schikora, D., Lischka, K.: The near band edge photoluminescence of cubic GaN epilayers. Appl. Phys. Lett. **70**, 1311 (1997)
43. Nagasawa, H., Abe, M., Yagi, K., Kawahara, T., Hatta, N.: Fabrication of high performance 3C-SiC vertical MOSFETs by reducing planar defects. Phys. Stat. Sol. (b) **245**(7), 1272–1280 (2008)
44. Kemper, R.M., Schupp, T., Häberlen, M., Niendorf, T., Maier, H.-J., Dempewolf, A., Bertram, F., Christen, J., Kirste, R., Hoffmann, A., Lindner, J., As, D.J.: Anti-phase domains in cubic GaN. J. Appl. Phys. **110**, 123512 (2011)
45. Reimer, L.: Scanning Electron Microscopy, 2nd edn, pp. 368–374. Springer, New York (1998)
46. Zainal, N., Novikov, S.V., Mellor, C.J., Foxon, C.T., Kent, A.J.: Current-voltage characteristics of zinc-blende (cubic) Al0.3Ga0.7N/GaN double barrier resonant tunneling diodes. Appl. Phys. Lett. **97**, 112102 (2010)
47. Mietze, C., Lischka, K., As, D.J.: Current-voltage characteristics of cubic Al(Ga)N/GaN double barrier structures on 3C-SiC. Phys. Stat. Sol. (a) **209**(3), 439 (2012)

Index

A
ab–initio, 141
Absorption cross section, 47
Acoustic speaker, 67, 69, 80, 85
Amorphous silicon shells, 45
Anode, 2–5, 7, 10–13, 15, 18–20, 26–31, 33, 35, 37, 38, 40, 41, 292, 311–314, 319
Anodic aluminum oxide membrane, 162

C
Carbon coating, 27, 39
Cell seperation, 318, 319
Chemical vapor deposition, 19, 27, 70, 120, 162, 179, 180, 186, 187, 191, 192, 194, 208, 267, 366
Chemical vapor transport, 278
Coaxial nanowire arrays, 58
Coaxial nanowires, 45, 52, 273, 302
Conductive paths, 121, 130, 131, 133, 135
Crystalline silicon cores, 68

D
Defects, 32, 69, 89, 90, 94, 98, 99, 102–104, 106, 107, 109, 110, 113, 114, 132–135, 243, 244, 249, 250, 299, 317, 328, 335, 341, 344, 346, 348, 349, 358–360, 362, 364, 365, 367, 383, 395–397, 400–402
DFT, 140–142, 146
Diffraction order, 49, 51
Diffusion, 2, 26, 31, 46, 53, 63, 68, 74, 131, 163, 165, 268–270, 272, 277, 281, 284, 288, 294, 329–331, 342, 344, 347, 350
Doping, 5, 18, 74, 89, 244, 253, 256, 277, 332, 371, 374
Droplet epitaxy, 331, 337, 342, 345, 347, 349

E
EBSD, 399, 400
Electrical, 3, 5, 14, 20, 36, 37, 56, 79, 80, 121–124, 130–132, 135, 140, 180, 245, 266, 295, 297–299, 301, 303, 309, 311, 315, 319, 357, 363, 373, 383
Electrochemical etching, 4–7, 10, 14, 17
Electroluminescence, 7, 120, 121
Electron collection, 311
Endotaxial, 291
Epitaxial, 163, 271, 283, 285, 286, 288–300, 319, 328, 357–360, 364, 371, 383, 387, 392, 396

F
Fabry–Pérot resonances, 60
Ferromagnetic, 266, 276, 301–304, 306, 318
Field effect transistor, 309, 382

G
Geometrical cross section, 47
Graphene, 32–34, 140, 141, 146, 157, 244–248
Grating, 30, 50, 51, 58, 93
Group–IV elements, 145
Growth mechanism, 171, 176, 190, 267, 269, 272, 277, 283, 291, 293–295, 319, 383, 397, 402

H
High–efficiency, 67, 69, 71, 74, 75, 78–80
III–V on Si, 334, 359
III–V/Si integration, 151, 359

H. Li et al. (eds.), *Silicon-based Nanomaterials*,
Springer Series in Materials Science 187, DOI: 10.1007/978-1-4614-8169-0,

© Springer Science+Business Media New York 2013

L

Leaky–mode resonances, 45, 53
Light absorption, 45–47, 49, 50, 52–54, 57–64, 68, 75, 79, 86, 377
Light absorption efficiency, 46, 47, 52, 86
Light emitting capacitor, 121, 123, 132, 135
Light interference, 93, 95
Light trapping, 46, 52, 57, 58, 67–69, 71, 73, 75, 80, 83
Lithium ion battery, 2–4, 7, 10, 11, 15, 18, 20, 25, 311, 312
Lorenz–Mie, 45, 46, 52

M

Magnéli–phase Ti_nO_{2n-1} Nanobelts, 38–42
Maxwell's equations, 47, 50
Metallic, 143, 176, 177, 180, 187, 191, 208, 243, 244, 247–250, 252, 254, 255, 266, 292, 295–297, 299, 306, 313, 314, 387
Methyltrichlorosilane, 179, 186, 187, 193, 194, 208
Molecular beam epitaxy, 357, 359, 360, 377, 381, 382, 386

N

Nanoelectronic, 56, 57, 64, 161
Nanoheteroepitaxy, 382, 383, 402, 403
Nanomaterials, 27, 46, 114, 140, 162, 163, 168, 185, 186, 190
Nanostructures, 1, 30, 35, 50, 69, 71, 75, 90, 131, 139, 140, 151, 156, 161–163, 165, 168, 171, 177, 180, 181–185, 189, 266–268, 271, 281, 283, 289–291, 295, 299, 309, 311, 313, 315, 319, 327, 331, 337, 339–342, 344, 345, 347, 353, 381–383, 385, 386, 389, 394, 397, 398, 401–403
Nanowire, 28, 30–34, 40, 41, 57, 58, 68–70, 76–80, 82, 140, 151, 162, 168, 171–177, 179–195, 197, 198, 201, 204, 206, 208, 244, 267–279, 281–307, 310–319, 306, 307, 309, 309–314, 317, 318
Negative electrode, 28
NEGF
NEMS, 67, 69
Nitrides, 180, 185, 382, 387, 402
Nucleation and growth, 180, 185, 188

O

Ohmic contact, 266, 310, 319
Omnidirectional light absorption, 63
One dimensional nanostructures, 180, 184, 185
Open–circuit voltage, 37, 57
Optical, 4, 6, 45, 49, 50, 52, 58, 67–69, 76–82, 84–86, 89, 90, 92, 94–96, 99, 105–107, 113, 114, 119, 121, 124, 126, 139–141, 143–147, 149, 152–155, 157, 167, 183, 219–221, 225, 232–238, 248, 266, 307, 308, 310, 319, 331, 334, 347, 350–352, 357, 358, 361–363, 365, 367, 369, 370, 376, 378, 382, 390, 392, 399, 400

P

Paramagnetic, 301, 302
Photocurrent density, 54, 55, 60, 62, 316
Photocurrent enhancement factor, 55
Photoluminescence, 4, 89, 90, 92–94, 96, 105, 112, 114, 134, 186, 190–192, 331–333, 341, 344, 345, 361, 376
Photoluminescence spectroscopy, 90, 92
Photovoltaic cells, 45
Planar coaxial nanowire arrays, 57
Planar defects, 359, 395, 401, 402
Porous silicon, 2–15, 17–20, 89, 108, 120, 190
Precursors, 29, 165, 167, 182, 188, 190, 203, 267, 268, 272, 274, 276, 279, 333

Q

Quantum dots, 89, 90, 92, 93, 100, 110, 113, 114, 162, 163, 168, 180, 337, 339, 340, 347–352, 357, 359, 375, 377

R

RCWA, 50, 52, 58
Refractive index, 59, 70, 122, 124–126
Relative permittivity, 50
Remote actuation

S

Selective epitaxy, 382
Semiconductor, 53, 69, 76, 89, 113, 140, 150, 182, 244, 247, 253, 276, 297, 299, 306, 307, 327, 328, 331, 347, 357–359
Semiconductor detector, 327, 342, 361, 367, 374, 376
Semiconductor laser, 357–360
Short–circuit current density, 54, 57, 60
SiCNT, 243, 249–254, 256–258, 260
Silicidation, 281, 283–285, 287, 296, 309–311, 318, 319

Silicon carbide, 90, 140, 161, 179, 185, 187, 189–192, 206, 208, 244, 248, 382
Silicon carbide, plasma, 162, 163, 172, 177, 188, 215
Silicon nanoparticles, 27, 90, 109, 111, 120–122, 124, 129, 130, 135
Silicon nanowires, 25, 31, 281
Silicon photonics, 358
Silicon rich oxide, 120
Si nanowire array, 667, 77, 287, 318
Single coaxial nanowires, 52
Solar cells, 7, 63, 67–69, 74, 75, 328
Solid state reaction, 288
Space–harmonic fields, 51, 52
Structural supporter, 311, 315, 316, 319
Subwavelength, 45, 46, 50, 52, 53, 60, 62, 53, 60, 62
Synchrotron radiation, 93, 94, 105, 113

T
Taguchi method, 179, 193–195, 198
TEM, 32, 34, 36, 38, 39, 105, 169–175, 217, 218, 267, 271–275, 283–286, 291–294, 306, 307, 313, 348, 361–366, 372, 391, 392, 394, 395, 401, 402
TE–polarization, 50, 62
Thin films, 58, 60, 89, 93, 94, 193, 215, 217, 274, 292, 296, 297, 333, 383, 384
TM–polarization, 50, 62
Transition metal silicide, 265

V
Vapor–liquid–solid, 28, 68, 184, 267
Vapor–solid, 184, 267, 271, 277
Vapor solid mechanism, 206, 267, 277

MIX
Papier aus verantwortungsvollen Quellen
Paper from responsible sources
FSC® C105338

If you have any concerns about our products,
you can contact us on
ProductSafety@springernature.com

In case Publisher is established outside the EU,
the EU authorized representative is:
Springer Nature Customer Service Center GmbH
Europaplatz 3, 69115 Heidelberg, Germany

Printed by Libri Plureos GmbH
in Hamburg, Germany